Man and River Systems

Developments in Hydrobiology 146

Series editor
H. J. Dumont

Man and River Systems

The Functioning of River Systems at the Basin Scale

Edited by

J. Garnier and J.-M. Mouchel

Reprinted from Hydrobiologia, volume 410 (1999)

Springer-Science+Business Media, B.V.

Library of Congress Cataloging-in-Publication Data

A C.I.P. Catalogue record for this book is available from the Library of Congress.

ISBN 978-90-481-5393-0 ISBN 978-94-017-2163-9 (eBook)
DOI 10.1007/978-94-017-2163-9

Printed on acid-free paper

Originally published by Kluwer Academic Publishers in 1999
Softcover reprint of the hardcover 1st edition 1999

Hydrobiologia **410:** v–vii, 2000.
J. Garnier & J.-M. Mouchel (eds), Man and River Systems.

Contents

Introduction

A basin scale framework for the study of Human pressure on river system functioning
by Josette Garnier & Jean-Marie Mouchel .... ix–xii
List of reviewers .... xiii–xvii

Part one: Biological communities

Plankton dynamics in the river Rhine during downstream transport as influenced by biotic interactions and hydrological conditions
by Th. Ietswaart, L. Breebaart, B. van Zanten & R. Bijkerk 1–10
Aquatic bryophyte assemblages along a gradient of regulation in the river Rhine
by A. Vanderpoorten & J.-P. Klein 11–16
A macrophyte communities sequence as an indicator of eutrophication and acidification levels in weakly mineralised streams in north-eastern France
by Gabrielle Thiébaut & Serge Muller 17–24
Use of oligochaete communities for assessment of ecotoxicological risk in fine sediment of rivers and canals of the Artois-Picardie water basin (France)
by J. Prygiel, A. Rosso-Darmet, M. Lafont, C. Lesniak, A. Durbec & B. Ouddane .. 25–37
Contribution of the macrobenthic compartment to the oxygen budget of a large regulated river: the Mosel
by Vincent Bachmann & Philippe Usseglio-Polatera 39–46
Contribution to the ecological quality of Aliakmon river (Macedonia, Greece): a multivariate approach
by M. Lazaridou-Dimitriadou, V. Artemiadou, G. Yfantis, S. Mourelatos & Y. Mylopoulos 47–58
Multiple human impacts by the City of Paris on fish communities in the Seine river basin, France
by Philippe Boët, Jérôme Belliard, Romuald Berrebi-dit-Thomas & Evelyne Tales . 59–68
Rehabilitation efforts for Atlantic salmon in the Meuse basin area: Synthesis 1983–1998
by C. Prignon, J.C. Micha, G. Rimbaud & J.C. Philippart 69–77
Biophysical interactions and the structure and dynamics of riverine ecosystems: the importance of biotic feedbacks
by Robert J. Naiman, Scott R. Elliott, James M. Helfield & Thomas C. O'Keefe ... 79–86

Part two: Eutrophication, nutrient dynamics and organic pollution

Biogeochemical nutrient cycles and nutrient management strategies
by Daniel J. Conley 87–96
Application of the 'snapshot' methodology to a basin-wide analysis of phosphorus and nitrogen at stable low flow
by M. Salvia, J.F. Iffly, P. Vander Borght, M. Sary & L. Hoffmann 97–102
Nutrient runoff and transfer from land and rivers to the Gulf of Riga
by Per Stålnacke, Nils Vagstad, Timo Tamminen, Paul Wassmann, Viesturs Jansons & Enn Loigu 103–110
Retention of nutrients in river systems: dependence on specific runoff and hydraulic load
by Horst Behrendt & Dieter Opitz 111–122
Modelling nutrient fluxes from source to river load: a macroscopic analysis applied to the Rhine and Elbe basins
by Marcel de Wit 123–130
Integrated modelling of hydrological processes and nutrient dynamics at the river basin scale
by Valentina Krysanova & Alfred Becker 131–138
Nitrogen transfers through the Seine drainage network: a budget based on the application of the 'Riverstrahler' model
by G. Billen & J. Garnier 139–150
Understanding the oxygen budget and related ecological processes in the river Mosel: the RIVERSTRAHLER approach
by Josette Garnier, Gilles Billen & Laurent Palfner 151–166
Model-based analysis of oxygen budget and biological processes in the regulated rivers Moselle and Saar: modelling the influence of benthic filter feeders on phytoplankton
by Andreas Schöl, Volker Kirchesch, Tanja Bergfeld & Dieter Müller 167–176
Optimal design and real time control of the integrated urban run-off system
by P. Harremoës & W. Rauch 177–184
Quantification of nitrate removal by a flooded alluvial zone in the Ill floodplain (Eastern France)
by José Miguel Sánchez Pérez, Michèle Trémolières, Naima Takatert, Philippe Ackerer, Arianne Eichhorn & Gérard Maire 185–193
The impact of hydrological fluctuations on shallow groundwater hydrochemistry under two alluvial meadows
by L. Krebs, J. Corbonnois & S. Muller 195–206
Scenario analysis of nutrient management at the river basin scale
by B. Kronvang, L. M. Svendsen, J. P. Jensen & J. Dørge 207–212
Methodology for multifunctional assessment of riverine wetlands in the Seine river basin
by Eliane Fustec, Philippe Boët, Aïcha Amezal & Nils Fauchon 213–221

Part three: Erosion and sediment transport

Linking land use, erosion and sediment yields in river basins
by D. E. Walling 223–240

High frequency variations of water flux and sediment discharge during the Little Ice Age (1586–1725 AD) in the Rhône Delta (Mediterranean France). Relationship to the catchment basin
by G. Arnaud-Fassetta & M. Provansal 241–250

Estimation of sediment yield and soil loss using suspended sediment load and ^{137}Cs measurements on agricultural land, Brie Plateau, France
by S. Sogon, M.-J. Penven, P. Bonte & T. Muxart 251–261

Assessment of soil erosion in a small agricultural basin of the St. Lawrence River watershed
by L. Mabit, C. Bernard, M. R. Laverdière & S. Wicherek 263–268

Challenges in relating land use to sediment yield in the Upper Yangtze
by D. L. Higgitt & X. X. Lu 269–277

Bank erosion in a Danish lowland stream system
by A. Laubel, L. M. Svendsen, B. Kronvang & S. E. Larsen 279–285

Flood transport of dissolved and suspended matter in the El Abid river basin (Morocco)
by O. Cherifi & M. Loudiki 287–294

Spatial and temporal variability of Total Suspended Solids in the Seine basin
by M. Meybeck, Z. Idlafkih, N. Fauchon & V. Andreassian 295–306

Part four: Fate of micropollutants

Point and diffuse source analysis of heavy metals in the Elbe drainage area: Comparing heavy metal emissions with transported river loads
by R.J. Vink, H. Behrendt & W. Salomons 307–314

Annual emissions of pollutants from mine stone applications in drainage basins of Dutch rivers
by R. S. E. W. Leuven, P. H. Nienhuis, J. M. A. Kesseleer & W. A. Zwart 315–323

Role of the dams on the River Lot on two types of pollution: point-sources (cadmium) and non-point sources (phosphorus)
by Alain Dauta, Yvon Lapaquellerie & Noële Maillet 325–329

A cadmium budget for the Lot-Garonne fluvial system (France)
by Gérard Blanc, Yvon Lapaquellerie, Noèle Maillet & Pierre Anschutz 331–341

Contamination by Polycyclic Aromatic Hydrocarbons (PAHs) in French and European rivers
by M. B. Fernandes, M.-A. Sicre, I. Broyelle, A. Lorre & D. Pont 343–348

Variability of herbicides (triazines, phenylureas) and tentative mass balance as a function of stream order, in the river Marne basin (France) *Triazine and phenylurea flux and stream order*
by M. Chevreuil, M. Garmouma & N. Fauchon 349–355

Hydrobiologia 410: ix–xii, 1999.
J. Garnier & J.-M. Mouchel (eds), Man and River Systems.

Introduction

A basin scale framework for the study of Human pressure on river system functioning

The ecology of aquatic systems has developed most of its basic concepts and methods by studying small, homogenous and well delimited natural systems. Lakes, in this respect, are ideal objects of ecological study, and this is why the functioning of these systems has long been better understood than that of river systems. Moreover, the problem of lake eutrophication was a matter of concern as early as the 1950s, and prompted numerous studies (e.g. Vollenweider, 1968). Since then, there have been many cases of successful lake restoration. Understanding and obviating river alteration by human pressures represent a much bigger challenge. In some cases, such as organic pollution by point discharges of waste water, the alteration is local and concerns a limited stretch of river. As early as in 1925, with the work of Streeter and Phelps, the basic knowledge was acquired to understand and manage this kind of alteration. In most cases, however, the dimension of human impact on river systems is much greater. Hydraulic management (for navigation or flood and drought regulation), diffuse pollution by changing agricultural practices, landscape modifications, all alter the functioning of the whole drainage network in a way which is much more difficult to investigate scientifically. The need for basin-scale management of water resources is now recognized by most authorities. But, sometimes, the scientific basis is still insufficient to allow coherent actions to be decided at the basin and landscape scale, in order to preserve or restore biodiversity and water quality.

This volume reflects the discussions along these lines that were held during the Symposium ("Man and River Systems") in Champs-sur-Marne (Paris) in March 1998, as the joint initiative of three large interdisciplinary Research Projects implemented on large river basins in Europe, namely the PIREN-Seine programme (CNRS, Seine Basin, Meybeck et al. - eds-, 1998), BINOCULARS (CE-DGXII, Seine, Loire, Scheldt, Rhine, Aliakmon, Exe, Severn basins), the ICPMS project (ICPMS & CE-DGXI, Mosel basin). The purpose of all these projects is to better describe the functioning of large river systems, by considering biological, chemical and hydrological processes at the scale of the watersheds, and to evaluate the influence of humans on these processes. The knowledge gained in the course of these projects and of other broad-scale investigations has now been collected and discussed in 32 papers, grouped into four main topics.

1. Biological communities

The River Continuum concept (Vannote et al., 1980) offered the first framework for the study of the structure and function of a biological community along the whole gradient of environmental constraints within the drainage network of large river systems. This concept led to a break with traditional approaches concentrated to small stretches of rivers. Besides the longitudinal dimension that underlies the RCC, the concepts of ecotone and fluvial corridors have also inspired a number of studies of the role played by the riparian zones and alluvial plains, including wetlands and connected hydraulic annexes, essential for maintaining aquatic life within whole river systems.

The structure and dynamics of the biological communities have traditionally been seen as a response by hierarchical physical and chemical controls, and are therefore considered as indicators of ecosystem perturbations. Macrophyte sequences are thus used as indicator of trophic states by Thiébaut & Muller (this volume); Bryophyte responses in a gradient of regulation is examined by Vanderpoorten & Klein (this volume); Oligochaetes distribution is proposed as an indicator for assessing ecotoxicological risks (Prygiel et al., this volume). Nevertheless, biotic feedback, as argued by Naiman et al. (this volume), opens new perspectives for a deeper understanding of the ecological functioning of rivers.

Planktonic communities are known to show immediate response to modifications of the environment. As shown by Ietswaart et al. (this volume), interannual discharge variability may greatly modify the level of the zoo-phytoplankton biomass, as well as the situation of their maximums in the hydrographic network; biomass oscillations due to water residence time and growth rate of the species, can be observed at a large spatial scale (Garnier et al., 1995). Fish community behaviour represents a long-term integration of several types of constraints. At historical time-scales, navigation in large European rivers has led to great changes in the original fish community composition (Boët et al., this volume; Petts et al., 1989; Philippart, 1987). Water quality alteration, flow regulation by reservoirs, channelization are all factors that may affect fish community, sometimes leading to dramatic situations. Several rehabilitation projects aimed at restoring noble species such as atlantic salmon and sea trout (Prignon et al., this volume), have symbolically important objectives. Although the macrobenthic compartment has been studied from the point of view of water quality indexes (Prygiel et al., this volume; Lazaridou-Dimitriadou, this volume) or of its influence on planktonic communities (Ietswaart et al, this volume; Bachman et al, this volume; Garnier et al., this volume; Schöl et al., this volume), more studies are still needed to fully understand the role of the benthic compartment in the fish diet as well as in sediment bioturbation and nutrient recycling.

2. Eutrophication, nutrient dynamics and organic pollution

Regarding eutrophication in river systems, it is now recognised that the phytoplankton biomass at any point in the drainage network is the result of its development throughout the whole upstream network. Phosphorus is often the limiting factor of algal growth in most watersheds because of excess of nitrogen, even in headwaters in basins with intensive agriculture. Silica may become limiting for diatoms in the drainage network in eutrophication cases (Garnier et al., 1995; Billen & Garnier, 1997; Conley, this volume). Eutrophicated reservoirs in the river basin, that trap silica through diatom sedimentation, have been shown to lead to silica limitation (Conley et al., 1993; Humborg et al., 1997; Conley, this volume). Beside nutrient limitation, water residence time and light intensity are other major constraints on phytoplankton development. Clearly, the study of eutrophication in a river requires to link our knowledge at the microscopic scale with constraints at the macroscopic scale of the drainage network. As reported by Conley (this volume), larger-scale studies of coastal eutrophication and nutrient limitation have focused on the variability of nutrient the load in the drainage network (Stålnacke et al., this volume). To assess nutrient delivery to the coastal zone, it is imperative not only to properly quantify the source of nutrients but also to investigate transformation and retention in the drainage basin and hydraulic annexes. Several lines of investigation were presented during the symposium to arrive at an understanding of the ecological functioning of river systems: typologies (Fustec et al., this volume), material budgets (Salvia et al. this volume, etc.) and mathematical modelling. Whatever the types of the models (statistical, Behrendt & Opitz, this volume; calibrated and spatially distributed, Krysanova & Becker, this volume, De Wit, this volume; deterministic and stream-order based, Billen & Garnier, this volume; Garnier et al., this volume; deterministic on a defined axis, Schöl et al., this volume; combined models, Kronvang et al., this volume), they all concern a large spatial scale or can be applied to a whole drainage basin, or even extended to a multi-basin approach. In addition to providing a framework for a quantitative description of the ecological functioning, models may be powerful tools to realistically explore the effect of restoration measures being considered for the watershed (effects of phosphorus, or nitrogen abatement in wastewater treatment plants on eutrophication or nutrient delivery to the coastal zone, etc.).

Although the lateral (flood plain) and vertical (groundwater) scales of the river system have been discussed in relation to the ecological functioning of rivers, because of the complexity of interactions between surface, subsurface and underground water, modelling studies of nutrient transformations are scarce or focus on a limited sector of the basin (Sánchez Pérez et al., this volume). Whereas nutrient retention/elimination is an essential function in fluvial corridors, we have to go further into process studies to distinguish the rate of uptake by the riparian vegetation (see Krebs et al., this volume) from bacterial transformation or chemical immobilisation...

Urban areas are important sub-basins within river basins. They generally concentrate most of the inhabitants and human activities and are major sources of many pollutants such as phosphorus, metals, hydrocarbons as well as organic matter and ammonia. The traditional problem of pollution by point discharge of waste water, as addressed by Streeter & Phelps (1925), is now changing into a more diffuse urban runoff problem. In their paper, Harremoës & Rauch (this volume) describe problems encountered when all aspects of urban-runoff systems are considered together (including three main components: sewer systems, waste water treatment plants and receptor aquatic ecosystems). Due to the development of on-line data acquisition, it is suggested that integrated simple and adaptive models are best suited to real-time control of urban runoff impacts. At the river-basin scale, the sphere of influence of one urbanised area overlaps that of neighbouring cities within the same basin and is probably of varying size whatever type of pollution is considered (organic matter discharges, remanent pollutants, pathogen bacteria, etc.). Such cross-influences are still insufficiently taken into account.

3. Erosion and sediment transport

The transfer of particulate matter inside whole river systems was discussed by D.E. Walling (Walling, this volume), who emphasised the need for sediment budgets in catchments rather than simple output estimates. Rare surveys of total suspended matter (TSS) over 30 years shows multiple trends related to land use in the river basins of the world. In the absence of such records, the time frame of transfer processes in the last 100 years can be studied with the help of environmental radionucleides such as Cesium-137 (^{137}Cs) and lead-210 (^{210}Pb) Both techniques have been used in oceanography and limnology for 30 years and are now applied to flood plain deposits. The delayed response of the river basins to environmental changes, where particulate matter is concerned is discussed in many of the other studies in this field (Arnaud-Fassetta & Provansal, this volume). The variability of TSS has been shown to be enormous (up to 250 000 mg l^{-1} during flash hydrological events in a dry environment, Cherifi & Loudiki, this volume). Although such events are also noted in wet environments, they are rarely studied. In other types of climate and topography, mechanical erosion and TSS levels have been shown to be smaller by orders of magnitude. It is interesting to note that stream bank erosion may be responsible for 60% of the annual sediment transport (Laubel et al., this volume). The ^{137}Cs method was also used on slow erosion conditions to provide detailed spatialization of erosion processes at small geographical scales (Sogon et al., this volume; Mabit et al., this volume). Such studies can help to identify erosion risks and establish priorities for corrective action.

Whole basin studies of TSS transport are still uncommon. From the study on the Seine Basin, Meybeck et al. (this volume) demonstrate that although TSS levels are relatively low, significant differences can be attributed to sub-basin lithology and urbanisation and to the stream orders. Compared to an earlier survey in 1862, TSS are much lower today, probably because of navigation management. The records of the long-term evolution over a hundred years have proved to be very informative when combined with alluvial sand analysis, and show that changes can occur over periods of 40–50 years (Arnaud-Fassetta & Provansal, this volume). Land erosion in China is probably the one according most attention through comparison between spatial patterns of sediment yields and satellite imagery of land cover (Higgitt & Lu, this volume). As opposed to nutrients that are mostly dissolved and for which the kinetics are relatively easy to formulate mathematically, no methods for modelling suspended solids were presented.

4. Fate of micropollutants

The impact of human on riverborne pollutants becomes manifest in two ways: i) an increase of natural levels and/or fluxes of trace metals in water-suspended particles, sediments, biota, ii) the occurrence of xenobiotic substances in the aquatic system. Both issues are illustrated in this volume of European case studies. A case study of the long-term impact of a zinc mine, in operation since 1840 but strongly reduced in 1986, was fully documented in two papers (Dauta et al, this volume; Blanc et al., this volume); as a result of a succession of 57 small dams, the extensive contamination by cadmium of the river sediments, will remain for years after restoration measures have been completed. Many similar examples are likely to be found in major mining regions where a low dilution of contaminated particles by natural erosion is combined with a high sediment storage capacity. In large river systems with multiple sources of trace metals, a source apportionment must be realized (Vink et al., this volume): in the

Elbe catchment, major pathways of As, Cu, Hg, Pb and Zn from point sources, atmospheric fallout, groundwater and soil erosion were determined.

A new source of pollution is created by the use of alternative materials such as mine stone, steel and phosphorus slags, which are being promoted in some countries to replace sand and gravel from alluvial deposits. This recycled material is used in dams, dikes and river bank protection. It is interesting to note that a modelling approach was used to calculate the inputs to the aquatic environment of a number of heavy metals, polyaromatic hydrocarbons (PAH) and persistent organic pollutants of industrial and domestic origins (Leuven et al., this volume).

PAH have been shown to be associated with particles so that their fluxes and concentrations are increasing by one order of magnitude during flood events (Fernandes et al., this volume). Regarding herbicides, an attempt was made to budget the spread on small catchments, revealing that riverine export is presently very limited (Chevreuil et al., this volume); therefore the major preoccupation is rather with short-term peaks than with the yearly export.

Acknowledgments

We are grateful to all participants of the symposium for their active contributions, and to the authors for their precious help in producing this book. We especially acknowledge the referees for the huge amount of work they have done.

We hope that this volume will invite many scientists to continue the discussions opened in several fields of river ecology.

JOSETTE GARNIER
UMR Sisyphe 7619 CNRS/UPMC
4 Place Jussieu
75005 Paris
France

JEAN-MARIE MOUCHEL
CEREVE, Ecole Nationale des Ponts et Chaussées
6 & 8 av. Blaise Pascal, Cité Descartes
77455 Mame-la-Vallée
France

References *(in addition to those presented in this volume)*

Billen, G. & Garnier J., 1997. The Phison River Plume: coastal eutrophication in response to changes in land use and water management in the watershed. Aquat. Microb. Ecol. 13: 3–17.

Conley, D.J., C.L. Schelske & E.F. Stoermer, 1993. Modification of silica biogeochemistry with eutrophication in aquatic systems. Mar. Ecol. Prog. Ser. 81: 121–128.

Garnier, J., G. Billen & M. Coste, 1995. Seasonal succession of diatoms and Chlorophyceae in the drainage network of the river Seine: Observations and modelling. Limnol. Oceanogr. 40: 750–765.

Humborg, C., V. Ittekot, A. Cosiascu & B.V. Bodungen, 1997. Effect of Danube River dam Al on Black sea biogeochemistry and ecosystem structure. Nature, 386: 385–388.

Meybeck, M., E. Fustec & G. de Marsily (eds), 1998. La Seine en son bassin. Fonctionnement écologique d'un système fluvial anthropisé. Elsevier, Paris, 749 pp.

Petts, G.E., H. Möller & A.-L. Roux (eds), 1989. Historical change of alluvial rivers: Western Europe. John Wiley & Sons, Chichester, 355 pp.

Philippart, J.-C., 1987. Histoire de l'extinction et problématique de la restauration des Salmonidés migrateurs dans la Meuse. In M. Thibaut & R. Billard (eds), La restauration des rivières à saumons, NRA, Paris: 125–137.

Vannote, R.L., G.W. Minshall, K.W. Cummins, J.R. Sedell & C.E. Cushing, 1980. The river continuum concept. Can. J. Aquat. Sci. 37: 130–137.

Vollenweider, R.A., 1968. Scientific fundamentals of the eutrophication of lakes and flowing waters with particular references to nitrogen and phosphorus as factors of eutrophication. O.C.D.E. Paris, Technical Report, DA 5/SCI/68.27, 250 pp.

Streeter, W.H. & E.B. Phelps, 1925. A study of the pollution and natural purification of the Ohio River, Public Health Bull. 146, U.S. Public Health Service, Washington, DC.

Hydrobiologia **410:** xiii–xvii, 1999.
J. Garnier & J.-M. Mouchel (eds), Man and River Systems.

List of reviewers

Admiraal W., Aquatic Ecotoxicology, Univ. Amsterdam, Kruislaan 320, 1098 SM Amsterdam, The Netherlands

Alvarez Cobelas M., Centro de Ciensias Medioambientales (CSIC), Serrano 115 dpdo., E-28006 Madrid, Spain

Amoros C., URA CNRS 367, Ecologie des Eaux Douces, Université Cl. Bernard-Lyon 1, 43 Bd. du 11 Novembre 1918, 69622, Villeurbanne cedex, France

Andreassian V., CEMAGREF, Division Hydrologie, Parc de Tourvoie, BP 121, Antony cedex, France

Armitage P.D., Freshwater Biological Association River Laboratory, East Stoke, Wareham, Dorset BH20 6BB, UK

Behrendt H., Institute of Freshwater and Fish Ecology, Department of Limnology of Rivers and Shallow Lakes, P.O. Box 35, 12587 Berlin, Germany

Benoit M., INRA Station de Recherche SAD, Domaine du Joly, BP 29, 88501, Mirecourt cedex, France

Berman T., Kinneret Limnological Laboratory, P.O. Box 345, Tiberias 14102, Israel

Bernard C., Complexe Scientifique B-1-305, 2700, Einstein, Ste-Foy, Quebec G1P 3W8, Canada

Billen G., UMPC, UMR Sisyphe 7619, Boite 123, Tour 26 étage 4, 4 place Jussieu, 75005 Paris, France

Boët P.H., CEMAGREF – Division QHAN, Parc de Tourvoie, BP 121, Antony cedex, France

Bonte Ph., Centre des Faibles Radioactivités, Laboratoire mixte CNRS-CEA, Domaine du CNRS, 91198 Gif sur Yvette Cedex, France

Boudreau P.R., Department of Fisheries and Oceans, Biological Sciences Branch, Bedford Institute of Oceanography, P.O. Box 1006, Dartmouth, N.S. B2Y 4A2, Canada

Boyd R.A., 1996, US Geological Survey, Water Resources Division, P.O. Box 1230, Iowa city, IA 52245, USA

Brown A., Dept. of Geog., Univ. of Exeter, Exeter EX4 4RJ, UK

Brunstein D., CNRS, Laboratoire de Géographie Physique, 1 place Aristide Briant, 92195 Meudon cedex, France

Capblancq J., CESAC, Université Paul Sabatier 118, Route de Narbonne, 31062 Toulouse cedex, France

Caraco N., Inst. Ecosystem Studies, Box AB, Millbrook NY, 12545, USA

Caruso B.S., Otago Reg. Council, 70 Stafford St., Dunedin, New Zealand

Chappell A., Department of Geography, University College London, 26 Bedford Way, London, WC1H OAP, UK

Chestérikoff A., UMR Sisyphe 7619, Boite 123, Tour 26 étage 4, 4 place Jussieu, 75005 Paris, France

Chevreuil M., UMR Sisyphe 7619, Boite 123, Tour 26 étage 4, 4 place Jussieu, 75005 Paris, France

Chiffoleau J.-F., Rue de l'Ile d'Yeu, B.P. 21105, 44311 Nantes Cedex 03, France

Chiou C.T., US Geological Survey, Box 25046, MS 408, Denver Federal Center, Denver, CO 80225, USA

Cloern J.E., US Geological Survey, MS496, 345 Middlefield Road, Menlo Park, CA 94025, USA

Collins A., Dept. of Geog., Univ. of Exeter, Exeter EX4 4RJ, UK

Conley D., National Environmental Research Institute, Frederiksborgveg 399, P.O. Box 358, 4000 Roskilde, Denmark

Cooper C., Dept. of Oceanography, Texas A&M University, Eller Building, College Station, Texas 77843, USA

Cooper L.W., Oak Ridge Nat. Lab., Div. Environm. Sci., P.O. Box 2008, MS 6038, Oak Ridge, TN 37831, USA

Correll D.L., Smithsonian Environmental Research Center, 647 Contees Wharf Road, P.O. Box 28, Edgewater, MD 21037-0028, USA

Cossa D., IFREMER-Nantes, Laboratoire Chimie des Contaminants et Modélisation, BP 1105, 44311 Nantes Cedex 03, France

Coste M., CEMAGREF, 50 rue du Verdun, Cazinet, BP3, F-33610 Cestas, France

Crouzet Ph., IFEN, 61 bd Alexandre Martin, 45058 Orléans cedex, France

Dauta A., CESAC, Université Paul Sabatier 118, Route de Narbonne, 31062 Toulouse cedex, France

de Marsilly G., UMR Sisyphe 7619, Boite 123, Tour 26 étage 4, 4 place Jussieu, 75005 Paris, France

Descamps, H., Centre d'Ecologie des Systemes Aquatiques Continentaux, CNRS 29, rue Jeanne Marvig, B.P. 4349 31055 Toulouse Cedex 4, France

Descy J.-P., Unité d'Ecologie des Eaux Douces, Facultés Universitaires Notre-Dame de La Paix, 5000 Namur, Belgium

Dodds W.K., Kansas State Univ., Div. Biol., Ackert Hall, Manhattan, KS 66506, USA

Dorioz J.M., Institut de Limnologie, B.P. 11F, 74203 Thonon, France

Downing J.A., Animal Ecology, Iowa State University, Ames, IA 50011-322, USA

Duarte C.M., Centro de Estudios Avanzados de Blanes, Cami de Sra. Barbara s/n, 17300 Blanes, Gerona, Spain

Dufay O., Fondation Universitaire Luxembourgeoise, 185, av. de Longwy, 67000 Arlon, Belgium

Dumont P., CEMAGREF, Unité Ressources Aquatiques Continentales, 50 Avenue de Verdun, 33612 Cestas Cedex, France

Dunne Th., Dept of Geological Sciences, University of Washington, Seattle, WA 98915, USA

Ellis J.B., Natural Environmental Research Council, Polaris House, Northstar Avenue, Swindon Wills, SN21 EU, UK

Fejes J., IVL, P.O. Box 210 60, SE 100 31 Stockholm, Sweden

Foy R.H., 1995. Agricultural & Environmental Science Division, Newforge Lane, Belfast BT9 5PX, Northern Ireland

Fustec E., UPMC, UMR Sisyphe 7619, Boite 123, Tour 26 étage 4, 4 place Jussieu, 75005 Paris, France

Gerdeaux D., INRA, Institut de Limnologie, Avenue de Corzent, B.P. 11F, 74203 Thonon, France

Golterman H., Mas dou Bochi, 13200 Arles, France

Gotz R., Umweltbehörde, Umweltuntersuchungen, Marckmannstr 129B, D-20539 Hamburg, Germany

Grant G.E., USDA, Forest Service, Pacific Northwest Research Station, 3200 Jefferson Way, Corvallis, OR 97331, USA

Gunnison D., Ecosystem Research and Simulation Division/EL, USAE Waterways Experiment Station, P.O. Box 631, Vicksburg, MS 39180, USA

Haury J., Ecole Nationale Supérieure Agronomique de Rennes, 65 rue de St. Brieuc, 35042 Rennes cedex, France

He Q., Dept. of Geog., Univ. of Exeter, Exeter EX4 4RJ, UK

Higgitt D., Dept. of Geog., Univ. of Durham DH1 3LE, UK

House W.A., Inst. Freshwater Ecol., River Lab., Wareham BH20 6BB, Dorset, UK

Hubert P., EMP, 35, rue St. Honoré, 77305, Fontainebleau, France

Hupfer M., Institut für Gewässerökologie und Binnenfischerei, Müggelseedamm 310, 12561 Berlin, Germany

Jaworski N., US-EPA Laboratory, 27 Tarzwell Drive, Narragansett, RI-02882-1197, USA

Jordan T., Smithsonian Environmental Research Center, P.O. Box 28, Edgwater, MD 21037, USA

Karamanev D.G., Univ. Western Ontario, Dept. Chem. & Biochem. Engn., London, ON N6A 5B9, Canada

Karr J.R., Univ. Washington, Box 357980, Seattle, WA 98195, USA

Kelly M.G., Bowburn Consultancy, 23 Clarence St., Durham DH6 5BB, UK

Khalanski M., EDF, DER Département Environnement, 6 quai Watier, 78401 Chatou cedex, France

Köhler J., Institute of Freshwater Ecology and Fisheries, Dept. Limnology of Lowland Rivers and Shallow Lakes, Müggelseedamm 260, 12562 Berlin, Germany

Kolpin D.W., 1996. US Geological Survey, 400 S. Clinton St., Iowa City, IA 52244, USA

Kothyari U.C., Department of Civil Engineering, University of Roorkee 247 667, UP, India

Kuhnle R.A., National Sedimentation Laboratory, USDA/ARS, 598 McRoy Dr., P.O. Box 1157, Oxford, MS 388665, UK

Lafont M., CEMAGREF, Division Biologie des Ecosystèmes Aquatiques, 3 bis quai Chauveau, CP 220, F-69336 Lyon cedex 09, France

Lal R., 2021 Coffrey Road, Columbus, OH 43210-1085, USA

Larsen P., Institute of Hydraulic and Agricultural Engineering, University of Karlsruhe, Kaiserstrasse 12, D-76128 Karlsruhe, Germany

Li H.W., US Fish and Wildlife Service, Oregon State University, OR 97331, USA

Loague K., Dept. of Geological and Environmental Sciences, School of Earth Sciences, Stanford University, California 94305-2115, USA

Longert J.F., Institute of Freshwater Ecology, River Laboratory, East Stoke, Wareham, Dorset BH20 6BB, UK

Lorenz C.M., Free Univ. Amsterdam, Inst. Environm. Studies, de Boelelaan 1115, 1081 HV Amsterdam, The Netherlands

McLelland S.J., Dept. of Geog., Univ. of Exeter, Exeter EX4 4RJ, UK

Meade R., US Geological Survey, Branch of Regional Research, Denver, CO 80225, USA

Mertes L., Department of Geography, University of California, Santa Barbara, CA 93106-4060, USA

Meybeck M., UMPC, UMR Sisyphe 7619, Boite 123, Tour 26 étage 4, 4 place Jussieu, 75005 Paris, France

Millman J.D., Woods Hole Oceanographic Institute, Woods Hole, MA 02243, USA

Miquel J., EDF Production Transport, Service Ressources en Eau, 37, rue Diderot, B.P. 41, 38040 Grenoble cedex, France

Mitosava H., Geogaphic modeling and systems laboratory, Department of Geography, University of Illinois at Urbana-Champaign, 220 Davenport Hall, Urbana, IL 61801, USA

Mouchel J.-M., CEREVE, ENPC, 6-8 av. Blaise Pascal, Cité Descartes, F-77455 Mame La Vallée cedex 2, France

Müller D., Bundesanstalt für Gewässerkunde, Department of Microbiology and Water Quality, Postfach 309, 56003 Koblenz, Germany

Muxart T., CNRS, Laboratoire de Géographie Physique, 1 place Aristide Briant, 92195 Meudon cedex, France

Naiman R., School of Fisheries, University of Washington, Box 357980, Seattle, WA 98195, USA

Neal C., Institute of Hydrology, Crowmarsh Gifford, Wallingford, OX10 8BB, UK

Nicholas A., Dept. of Geog., Univ. of Exeter, Exeter EX4 4RJ, UK

Nixon S., School of Oceanography, University of Rhode Island, Narragansett, RI 02882-1197, USA

Owens P., Dept. of Geog., Univ. of Exeter, Exeter EX4 4RJ, UK

Pelton D.K., USA, Cold Reg. Res. & Engn. Lab., 72 Lyme Rd., Hanover, NH 03755, USA

Petts G., Department of Geography, University of Technology, Loughborough, Leicestershire, LE11 3TU, UK

Geoffrey P., Department of Geography, University of Technology, Loughborough, Leicestershire, LE11 3TU, UK

Phillips J., Dept. of Geog., Univ. of Exter, Exeter EX4 4RJ, UK

Pinay G., Université de Rennes I, Campus de Beaulieu, Av. Général Leclerc, bte 14, F-35042 Rennes cedex, France

Pont D., Univ. Claude Bernard-Lyon I, Lab. Ecologie des Eaux douces, 43 Bd. du 11 Novembre 1918, 69622 Villerbanne, France

Poulin M., CIG EMP, 35, rue St. Honoré 77305, Fontainebleau, France

Quine T., Dept. of Geog., Univ. of Exeter, Exeter EX4 4RJ, UK

Reid I., Loughborough Univ. Technol., Dept. Geog., Loughborough LE11 3TU, Leics., UK

Rewitt M., Urban Pollution Research Centre, Middlesex University, Bounds Green Road, London N11 2NQ, UK

Reynolds C.S., NERC, Institute of Freshwater Ecology, Windermere Laboratory, GB-LA22 0LP, Ambleside, UK

Rodriguez M.A., Univ. Quebec, Dept. Biol. & Sci. Sante., CP 3300, Rimouski, PQ G5L 3A1, Canada

Russell M., Dept. of Geog., Univ. of Exeter, Exeter EX4 4RJ, UK

Sabater S., Departamento de Ecologia, Facultade de Biologia, Universidad de Barcelona, Avda. Diagonal, 645, E08028, Barcelona, Spain

Sánchez Pérèz J.M., Ecologie Vegetale CEREG-URA95 CNRS Université Louis Pasteur, 28 rue Goethe F-67083 Strasbourg Cedex, France

Schmidt A., Environmental Protection Authority, Southern Danube Region, P.O. Box 353, H-6501 Baja, Hungary

Schöl A., Bundesanstalt für Gewässerkunde, Department of Microbiology and Water Quality. Postfach 309, 56003 Koblenz, Germany

School of Fisheries, University of Washington, Box 357980, Seattle, WA 98195, USA

Schuttelaar M., Delft University of Technology, Faculty of Civil Engineering, P.O. Box 5048, 2600 GA Delft, The Netherlands

Servais P., GMMA, Université Libre de Bruxelles, Campus de la Plaine-CP 221, B-1050 Bruxelles, Belgium

Simon M., Limnological Institute, University of Konstanz, P.O. Box 5560 X 913, D-78434 Konstanz, Germany

Sogon S., Centre des Faibles Radioactivités, Laboratoire mixte CNRS-CEA, Domaine du CNRS, 91198 Gif sur Yvette Cedex, France

Soulié A., CRITT Verseau, 859 rue JF Breton, 34093 Montpellier cedex 5, France

Stålnacke P., Jordforsk-Centre for Soil and Environmental Research N-1432 Aas, Norway

Testard P., UPMC, UMR Sisyphe 7619, Boite 123, Tour 26 étage 4, 4 place Jussieu, 75005 Paris, France

Thévenot D., CEREVE, Laboratoire CEREVE, Université Paris XII-Val de Marne, Faculté des Sciences et Techniques, 61 avenue du General de Gaulle, 94010 Creteil Cedex, France

Thingstadt F., Department of Microbiology, University of Bergen, Jahnebakken, N-5020, Bergen, Norway

Thomas A., Institut de Biogéochimie Marine, 1 rue Maurice Arnoux, 91120 Montrouge, France

Tonderski A., Dept. Water & Environmental Studies, Linköping University, SE-581 83 Linköping, Sweden

Trémolières M., Ecologie Végétale CEREG-URA95 CNRS Université Louis Pasteur, 28 rue Goethe F-67083 Strasbourg Cedex, France

Tronczynski J., IFREMER, rue de l'Ile d'Yeu, BP 21105, 44311, Nantes Cedex 03, France

Tucker G.E., Ralph Parsons Laboratory, Dept. of Civil and Environmental Engineering, Massachussetts Institute of Technology, Cambridge, MA 02139, USA

Usseglio-Polatera P., Centre Recherches Ecologique, UPRES-EBSE, Univ. de Metz, 1, rue des Récollets, BP 4116, 57040 Metz cedex 01, France

Van den Berg G.A., Univ. Utrecht, Fac. Earth Sci., Dept. Gechem., P.O. Box 80-021, 3508 TA, Utrecht, The Netherlands

Vander Borght P., Fondation Luxembourgeoise, 185 Av. de Longwy, 6700 Arlon, Belgium

Wade P.M., International Center of Landscape Ecology, Department of Geography, University of Technology, Loughborough, Leicestershire LE11 3TU, UK

Wallbrink P.J., CSIRO Land and Water, P.O. Box 1666, Canberra 2601, A.C.T. Australia

Walling D., Dept. of Geog., Univ. of Exeter, Exeter EX4 4RJ, UK

Webb B., Dept. of Geog., Exeter, Exeter EX4 4RJ, UK

Webster J.R., Department of Biology, Virginia Politechnique Institute and State University, Blacksburg, VA 24061, USA

Whitehead P.G., Univ. Reading, Dept. Geog., Aquat. Environm. Res. Ctr., Reading RG6 6AB, Berks, UK

Wichert G.A., Ontario Ministry Nat. Resources, RR4, Picton, ON K0K 2T0, Canada

Wolanski E., Australian Inst. Marine Sci., PMB 3, Townsville MC, Townsville, QLD 4810, Australia

Zalewski M., Lodz University, Department of Applied Ecology, ul. Banacha, 6, PL-Lodz, Poland

Zhang Y., Dept. of Geog., Univ. of Exeter, Exeter EX4 4RJ, UK

Zobrist J., EAWAG, CH 8600 Dübendorf, Switzerland

Hydrobiologia **410:** 1–10, 1999.
J. Garnier & J.-M. Mouchel (eds), Man and River Systems.

Plankton dynamics in the river Rhine during downstream transport as influenced by biotic interactions and hydrological conditions

Th. Ietswaart[1], L. Breebaart[1], B. van Zanten[1] & R. Bijkerk[2]
[1]*National Institute of Public Health and the Environment, P.O. Box 1, 3720 BA Bilthoven, The Netherlands*
[2]*Koeman & Bijkerk BV, P.O. Box 14, 9750 AA Haren, the Netherlands*

Key words: river Rhine, phytoplankton, growth rate, zoobenthos

Abstract

Two water parcel studies were performed in the river Rhine from Maxau (Germany) to Maassluis (the Netherlands) in the spring and summer of 1995. Chlorophyll-*a* levels were low both in spring and in summer, probably due to the high discharge in 1995 as compared to other years. Nutrients were never depleted. Zooplankton densities were very low. Diatoms and green algae had positive net apparent growth rates of 0.38 d^{-1} and 1.01 d^{-1}, respectively, as calculated from cell counts. Grazing experiments estimated the zooplankton grazing rate in the order of 0.05 d^{-1} or 8% of the chlorophyll-based growth rate. As in previous years, the plankton densities decreased in the lower part of the Rhine. The loss rate could not be explained by zooplankton grazing, respiration and sedimentation. From earlier published data the potential impact of benthic filter feeders was calculated. This exercise suggests that filtration by zoobenthos can explain the observed decrease in plankton density in the Rhine.

Introduction

It is now well established that large rivers contain a native population of phytoplankton (Reynolds, 1988; de Ruyter van Steveninck et al., 1990; Rojo, 1994). The conditions to which phytoplankton is exposed in rivers, like turbulence (Reynolds, 1994), short residence times (Søballe & Kimmel, 1987), the connectivity with backwaters (Heiler et al., 1995; Reynolds, 1995) and regulation (Wehr & Thorp, 1997) have been studied in some detail. However, the nature and magnitude of carbon flows between primary producers and consumers remains a relevant issue. Some papers have shown a considerable influence of grazing by zooplankton (Descy et al., 1987; Gosselain et al., 1994a,b; Kobayashi et al., 1996), whereas others observed a more modest influence (Köhler, 1993, 1997; Lonsdale et al., 1996). Several studies suggest a significant impact from benthic filter feeders (Köhler, 1997; Basu & Pick, 1997).

Most studies of plankton dynamics in rivers are carried out by sampling at regular intervals at one or more locations. The measurements give an indication of the processes upstream of the sampling location. An alternative approach is to repeatedly sample a water parcel as it travels downstream. With this approach the relationship between observed changes in plankton densities and actual growth and loss processes is much more straightforward. Probably due to the considerable logistic effort involved in a water parcel study, there are only a few reports of such studies in rivers (de Ruyter van Steveninck et al., 1992; Köhler, 1997).

This report involves a series of water parcel studies performed in the river Rhine in 1995. The river Rhine is being regularly monitored in the framework of the Rhine Action Plan. In the present study, two water parcel experiments were conducted. The aim of the study was to describe the effects of river hydrodynamics and chemistry on phytoplankton and zooplankton, as well as their interactions, in an as quantitative way as possible. Special attention was paid to loss processes in the lower part of the river. In some large rivers phytoplankton densities decrease in the lower part, although nutrients are not depleted (De Ruyter van Steveninck et al., 1992; Garnier et al., 1995; Basu & Pick, 1997). An attempt was made to explain this decrease. The relative influence of the various possible loss factors will be discussed, as well of the influence of the hydrological conditions, notably discharge. In conjunction with these surveys, experiments were conducted to assess the loss of phytoplankton by zooplankton grazing.

Material and methods

Study area and sampling

Samples were taken at 12 locations, nine in the river itself and three in major tributaries. The locations and their distance from lake Constance are presented in Figure 1.

The spring experiment started on April 19th 1995 at 6 am, the summer experiment at September 9th, also at 6 am. The other sampling times were determined using the Alarm-model, a hydraulic transport model developed for the prediction of the transport of accidental toxicant spills (Van Mazijk et al., 1991). The model was fed by actual discharge data (see Table 1), and has been shown to predict measured travel times with a precision of 1–4% (Van Mazijk et al., 1991).

Samples were taken with a bucket, either from a bridge (Maxau, Neckar, Düsseldorf) or from a ferry crossing the river (Lobith, Werkendam, Krimpen, Maassluis, Kampen). At the river mouth station Maassluis sampling always took place during outgoing tide. Bucket samples were taken across the width of the river and pooled to obtain a mixed sample of approx. 60 l. This sample was taken by car in a cooled insulated container to the laboratory for subsampling and analysis. The time between sampling and analysis was at most 8 h for the farthest locations (Maxau, Neckar) and for the others less than 6 h. Upon arrival in the laboratory, temperature, pH and oxygen were measured in the pooled sample, and compared with the *in situ* values. No great differences were found.

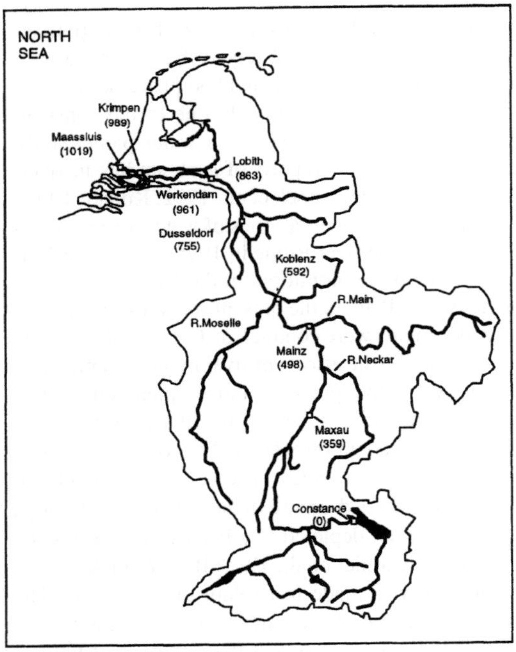

Figure 1. Catchment of the Rhine river, with sampling stations. Between parentheses the distance from Lake Constance (km).

Measurements

Temperature, conductivity, pH and light extinction were measured *in situ* at each sampling site. The light extinction was measured using a LiCor LI-1000 datalogger equipped with two UWQ underwater light sensors. The sensors were mounted on a frame at a fixed depth interval. The extinction coefficient was calculated with Lambert-Beer's law from at least 20 simultaneous readings.

Sub-sampling for all chemical and biological variables was done after returning to the laboratory. Chlorophyll-*a* was determined by filtration of water through glass fibre filters, extraction with hot 80% ethanol and spectrophotometric measurement of the chlorophyll-*a* concentration, following the Netherlands Standard Method (NEN No. 6520). Phytoplankton samples were preserved with Lugol (0.5%) and stored at room temperature until analysis. Phytoplankton was identified down to the lowest possible taxonomic level. Cells were counted and measured to obtain abundance and biovolume estimates. Zooplankton was harvested by filtering 7 – 20 l over a 36 μm mesh screen. This procedure has been found to yield sufficient numbers to the concentrated samples were preserved with 6% formaldehyde. Zooplankton was identified and counted according to Van Dijk & van Zanten (1995).

Nutrient samples were filtered (0.2 μm) and frozen. Nitrate and ammonium were measured using an ion chromatograph. Dissolved silicate and soluble reactive phosphate (SRP) were measured with ICP-AES.

Bacteria were counted in samples fixed with formaldehyde (1% w/v) with an epifluorescence microscope after staining with the fluorochrome Hoechst 22358. Bacterial activity was determined with the [^{3}H]-thymidine incorporation method according to Tubbing & Admiraal (1991).

Table 1. Sampling locations, dates, distances from Lake Constance and calculated discharge at the time of sampling. Discharges in the Dutch part of the Rhine are strongly influenced by regulation and are therefore not calculated

Location	River km	April			September		
		Sampling date	Transport time (h)	Discharge (m^3 s^{-1})	Sampling date	Transport time (h)	Discharge (m^3 s^{-1})
Maxau	359	19 April	0	1430	12 Sep.	0	1496
Mainz	498	20 April	32	2202	13 Sep.	33	1707
Koblenz	592	21 April	50	3730	14 Sep.	52	1909
Düsseldorf	755	22 April	77	3419	15 Sep.	81	1870
Lobith	863	23 April	101	3019	16 Sep.	108	1871
Werkendam	961	24 April	131		18 Sep.	148	
Maassluis	1019	25 April	148		19 Sep.	180	
Kampen	995	25 April	145		18 Sep.	151	
Krimpen	989	24 April	132		19 Sep.	169	
Neckar	428	19 April	3		12 Sep.	3	
Main	498	20 April	33		13 Sep.	33	
Moselle	592	21 April	52		14 Sep.	51	

The light-saturated primary production was measured by ^{14}C bicarbonate incorporation following de Ruyter van Steveninck et al. (1992). TOC and DOC were determined with a Shimadzu HTCO carbon analyzer. The bicarbonate content of the river water was determined titrimetrically.

Additional experiments

On five occasions during 1995, grazing experiments were performed with Rhine river water with the aim to determine the loss rate of phytoplankton through zooplankton grazing. Two of the experiments coincided with the water parcel studies.

Water was collected at Düsseldorf as described above. The samples were split. One half was filtered over a submerged 50 μ mesh screen to remove larger zooplankton. The other half was left untreated. The treated and untreated samples were incubated in duplicate in 7 l vessels equipped with a gentle stirring system (Techne Biological Stirrer, type MCL-101L) at the *in situ* temperature during sampling. The cultures were illuminated with circular fluorescent tubes (Philips TL EM 40W/33RS), in a 14/10 h light/dark cycle. Irradiance at the center of the empty bottles was adjusted to 40 μE m^{-2} s^{-1}, which approximates the average light intensity in the lower Rhine. During four to eight days, daily samples of chlorophyll-*a*, nutrients and zooplankton were taken. Analyses were as described above. The effect of reduced grazing was estimated by comparing the rate of increase in chlorophyll-*a* levels in the treated and untreated cultures. A test for the equality of slopes (Sokal & Rohlf, 1981) was used to detect statistically significant differences.

Results

The discharge of the river Rhine in 1995 was well above average. The yearly average discharge at Lobith was 2794 m^3 s^{-1}, which is 30% above the long term average of 2220 m^3 s^{-1}. This resulted in rather low travel times of 148 h in April and 169 h in September for the river stretch from Maxau to Maassluis.

In Figure 2, temperature, suspended matter and the light extinction coefficient are shown as a function of travel time. Suspended matter (Figure 2 C,D) in the April experiment increased until Düsseldorf, and decreased thereafter. The Main, one of the three major tributaries, had much higher levels of suspended matter than the Rhine. The two other tributaries carried suspended matter levels comparable to the Rhine. The extinction coefficient (Figure 2G,H) increased during downstream transport. The coefficient was highest during the April experiment, as expected from the higher levels of suspended matter and chlorophyll-*a*. The variations in conductivity at the downstream location of Maassluis indicate a varying marine influence at the two times of sampling.

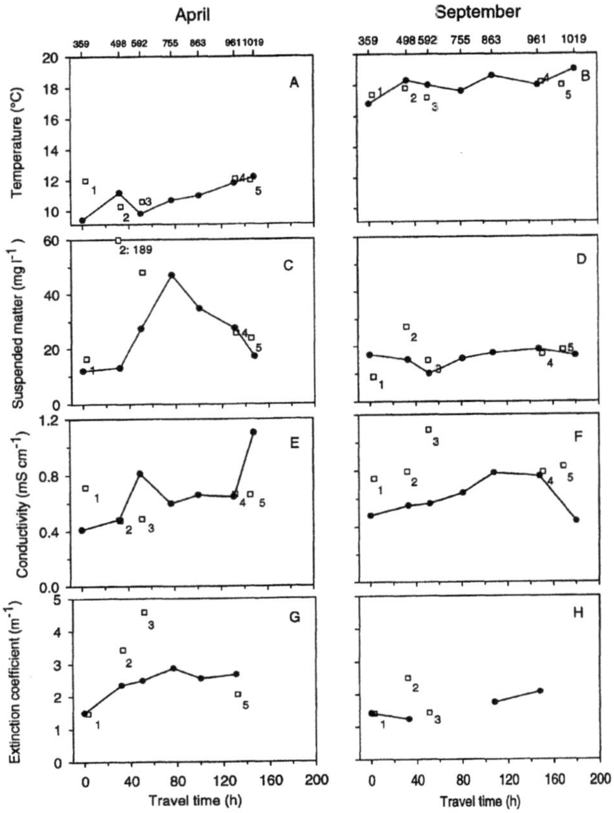

Figure 2. Temperature (A,B), suspended matter (C,D), conductivity (E,F) and light extinction coefficient (G,H) as a function of travel time. Open numbered squares denote levels in tributaries and branches: (1) Neckar; (2) Main; (3) Moselle; (4) Krimpen (r. Lek); (5) Kampen (r. IJssel) The top axis shows the river kilometres from Lake Constance.

Chlorophyll-*a* levels (Figure 3A,B) increased during the April experiment from Maxau until a maximum of 22 $\mu g\ l^{-1}$ was reached in Düsseldorf. Thereafter they decreased again. This decrease is observed regularly in the Rhine (de Ruyter van Steveninck, 1990; Admiraal et al., 1993). In the summer period, chlorophyll-*a* levels were much lower.

The decrease in the downstream part was much less pronounced than in spring. This was also the case for bacterial abundance.

The densities of rotifers (Figure 3E, F) and crustaceans (Figure 3G,H) were low during both experiments. The maximum value in the spring experiment in the river proper was 66 animals per l in Lobith. This is low even for rivers, which generally have rather low zooplankton densities. Only the last station Maassluis, which is under marine influence, had higher rotifer densities in spring.

Nutrient levels in the Rhine during the experiments are summarised in Table 2. Levels of phosphate and nitrate generally increased during downstream transport. Levels of silicate were more or less constant. Phosphate levels in the tributaries were much higher than in the Rhine itself.

Photosynthesis increased slightly over the course of the river in both experiments. The production per unit chlorophyll-*a* (Figure 4 A,B, triangles) was rather constant during the April experiment and had an optimum in the mid section of the river stretch during the September experiment. Levels of thymidine incorporation were constant during both experiments, despite the large differences in bacterial density (see Figure 3).

Figure 5 shows the abundances of the four main algal groups in the Rhine (diatoms, green algae, cyanobacteria and cryptophyceae) as a function of travel

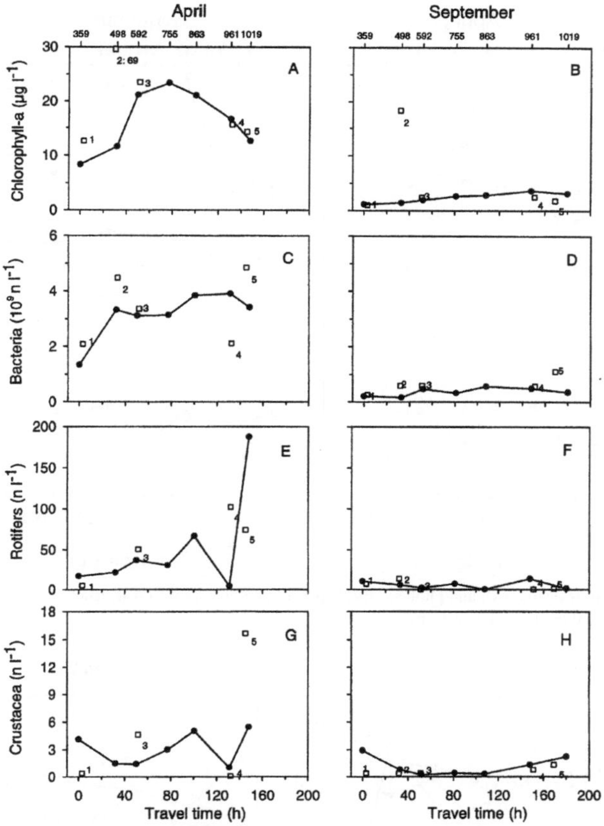

Figure 3. Levels of chlorophyll-*a*, (A,B), bacteria (C,D), rotifers (E,F) and crustaceans (G,H) as a function of travel time. Numbered symbols as in Figure 2.

Table 2. Concentrations of nutrients in the river Rhine during the water parcel experiments

	April				September			
	SRP	NO_3	NH_4	SI	SRP	NO_3	NH_4	SI
	µg P/l	mg N/l	mg N/l	mg/l	µg P/l	mg N/l	mg N/l	mg Si/l
Maxau	15.5	0.80	0.08	1.1	24.8	0.80	0.04	1
Mainz	27.9	1.36	0.13	1.0	37.2	1.01	0.06	1.1
Koblenz	43.4	1.78	0.08	n.a.	52.7	1.50	0.04	n.a.
Düsseldorf	62	2.59	0.10	1.9	77.5	1.67	0.03	1.4
Lobith	58.9	1.36	0.17	2.0	74.4	1.43	0.07	n.a.
Werkendam	55.8	3.70	0.10	1.7	62.0	1.34	0.06	1.4
Krimpen	52.7	1.37	0.08	1.7	89.9	1.67	0.08	1.5
Maassluis	74.4	1.48	0.11	1.5	6.2	1.48	0.10	1.3
Neckar	58.9	1.29	0.14	2.0	189.1	4.34	0.08	3.3
Main	80.6	1.78	0.08	n.a.	201.5	2.74	0.07	1.5
Moselle	102.3	2.77	0.13	1.8	303.8	2.18	0.04	1.9

time. Diatoms were the dominant group, both on a per-cell and a biovolume basis. In April, cell numbers increased until Düsseldorf and decreased thereafter. In the summer, no decrease was observed in the downstream part. During downstream transport, the relative proportion of green algae increased both in spring and in summer.

The phytoplankton abundance data can be used to calculate net apparent growth rates of algal groups and indeed individual algal species. Table 3 shows the cell number-based net apparent growth rates of the four main algal groups in the Rhine. The calculation was based upon the assumption of exponential growth. No correction was made for the influence of tributaries.

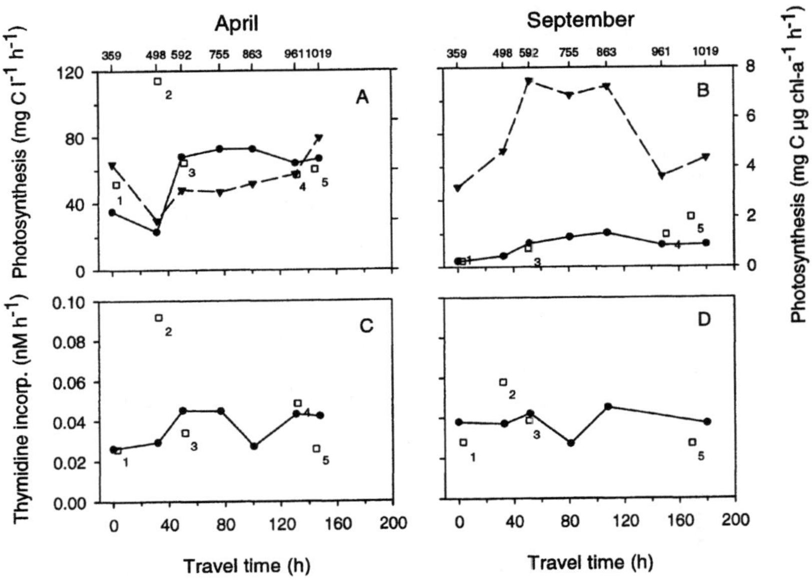

Figure 4. Photosynthesis (A,B, circles, squares), and photosynthesis per unit chlorophyll-*a* (A,B, triangles) and thymidine incorporation (C,D) as a function of transport time. Numbered squares as in Figure 2.

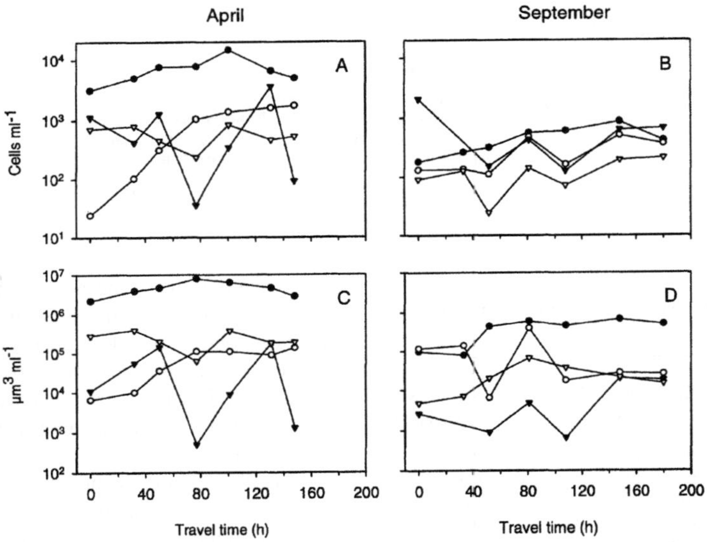

Figure 5. Cell numbers (A,B) and biovolumes (C,D) of the four main algal groups as a function of travel time. Closed circles: diatoms; open circles: green algae; closed triangles: cyanobacteria; open triangles: cryptophytes.

Additional grazing experiments

During the two water parcel studies and at four other occasions throughout the growth season of 1995, grazing experiments were performed. Filtration of the Rhine water through 50 μm mesh reduced densities of rotifers and crustacea by a factor of 5–10, without having large effects on the chlorophyll-*a* levels.

The difference in chlorophyll-based apparent growth rate between the untreated and filtered cultures can be attributed to the reduction of grazing by zooplankton. The experiments suggest a grazing rate in the order of 8% of primary production per day (Table 4). It should be noted that the chlorophyll-*a* levels in the incubations always reached values of about 200 μg l^{-1}. The corresponding samples in the river always showed a decreasing trend.

Table 3. Net apparent growth rates (d^{-1}, Maxau to Düsseldorf) of algal groups in the Rhine during the water parcel surveys. Statistically significant rates are marked with an asterisk

	April	September
Diatoms	0.35*	0.29*
Chlorophyceae	1.02*	0.16
Cryptophyceae	−0.08	−0.02
Cyanophyceae	−0.49	−0.54

Discussion

The discharge of the river Rhine in 1995 was well above average. The yearly average discharge at Lobith was 2794 m^3 s^{-1}, which is 30% above the long term average of 2220 m^3 s^{-1}. This resulted in rather low travel times of 148 h in April and 169 h in September. If we compare the levels of chlorophyll-*a* and zooplankton from the 1995 experiments with the ones performed in 1990 (de Ruyter van Steveninck et al., 1992), the influence of the high discharges in 1995 is clear. The high discharges did not only lead to shorter travel times, but also to lower water temperatures, especially in spring. As a result, maximum chlorophyll-*a* levels in 1995 were more than seven times lower than those during the May water parcel study in 1990. The 1995 spring experiment was held two weeks earlier than the one in 1990. This could have placed the experiment earlier relative to the spring peak, thus accounting for the lower chlorophyll levels. However, monitoring of the chlorophyll-*a* levels showed a spring peak (May 20th, 1995) which reached only 65 μg per l at Lobith (ICPR, 1997), as opposed to 140 μg chlorophyll-*a* per l in 1990 (Admiraal et al., 1994).

The difference between zooplankton densities in 1995 and 1990 were even greater than with chlorophyll-*a*. In 1990, rotifer densities reached a maximum value of more than 1500 animals per l, whereas the maximum density in 1995 was a meagre 66. Again, this could be the consequence of the early sampling period, but levels were low throughout 1995 (ICPR, 1997), as compared to earlier years (van Dijk & van Zanten, 1995).

Of the major nutrients, silicate is at times depleted to limiting levels in the Rhine (de Ruyter van Steveninck et al., 1992; Admiraal et al., 1994). During the 1995 surveys, nutrients were present in high concentrations throughout the river stretch. The absence of a deterioration of the physiological state of the algae is confirmed by the high chlorophyll-*a* specific photosynthetic rate, which did not decrease during travel downstream (Figure 4A,B).

During the spring experiment, diatoms had an apparent net growth rate of 0.38 per day, which was equal to the overall chlorophyll-*a* based rate (Table 4). Green algae had a much higher net apparent growth rate of about 1 per day. This last value is high considering the low temperatures and high turbidity of the river water. Assuming that our estimate of travel times was reasonably correct, such a high apparent growth rate may be explained by growth of phytoplankton in retentive zones alongside the river channel (e.g. Reynolds & Glaister, 1993) or input by tributaries. It remains unclear, however, why only green algae showed such a high net apparent growth rate.

During the summer experiment, diatoms were still the most abundant group, but green algae and cyanobacteria were relatively abundant as well (Figure 5). This is also found in other rivers, especially regulated ones (Descy et al., 1987; Marker & Collett, 1997). The chlorophyll-*a* levels during the surveys increased until Düsseldorf, and decreased again thereafter. The pattern was most pronounced during the

Table 4. Apparent chlorophyll-*a* based growth rates (μ) in untreated and filtered incubations of Rhine water. Experiments B and D coincided with respectively the April and September water parcel studies. Grazing rates are defined as difference between treated and filtered growth rates and are given as absolute rates and percentages of the untreated growth rate. Differences in bold were statistically significant

Exper.	μ untr. d^{-1}	μ filt. d^{-1}	grazing rate d^{-1} (%)
A	0.48	0.51	0.03 (6.2)
B	0.57	0.61	**0.04** (7.8)
C	0.96	0.96	0.00 (−0.25)
D	0.23	0.27	**0.04** (18.1)
E	0.92	1.02	**0.1** (11.0)
F	0.77	0.81	**0.04** (4.2)

Table 5. Loss rates between Lobith and Maassluis during the spring experiments of 1995 (this study) an 1990 (de Ruyter van Steveninck et al., 1992)

	April 1995	May 1990
Phytoplankton disappearance rate (d^{-1})	0.40	0.62
Respiration (d^{-1})	0.05	0.09
Zooplankton grazing (d^{-1})	0.04	0.18
Sedimentation (d^{-1})	0.03	0.03
Remaining (d^{-1})	0.28	0.32

spring survey, but it was present in September as well. This phenomenon has been observed regularly in the Rhine (de Ruyter van Steveninck et al., 1990, 1992; Admiraal et al., 1994) and also in other rivers of varying size (Garnier et al., 1995; Basu & Pick, 1997). Admiraal et al. (1993) conclude that silicate limitation of diatoms may explain the decline. However, the present study shows a decline in phytoplankton without signs of nutrient limitation. As nutrients are generally not depleted, the responsible loss processes must be found elsewhere.

The most obvious one, zooplankton grazing, had only a small impact in the grazing experiments that were performed (Table 4). The growth of phytoplankton in the experimental bottles as compared to the corresponding levels in the river suggest that the cause of phytoplankton loss must be sought outside the water column. The chlorophyll-*a* levels in the incubation vessels rose steadily whereas the levels in the water parcel in the river, from which the water was taken, decreased. This may at least in part be attributed to 'bottle effects', such as more favourable light and turbulence conditions. Although the illumination of the bottles was similar to the average depth-integrated light level in the Rhine, the algae in the river live in a highly fluctuating light field which may have impeded growth in the river itself as opposed to the incubations. Köhler (1997) has used an elegant technique to simulate this light climate, using dialysis bags filled with a river phytoplankton assemblage, being moved up and down in the river Severn. Even in this more realistic light climate, the caged algae increased in density whereas the densities in the river dropped. Köhler ascribes this to sedimentation and grazing by benthic filter feeders.

The results from the grazing experiments in the present study point in the same direction. Grazing loss rates during the experiments were around 0.05 per day, which is about 10% of the algal growth rate. In Table 5, the decrease in phytoplankton biomass and the estimated magnitude of the various loss processes are summarised. Zooplankton grazing, sedimentation, respiration and benthic activity are taken into account. As a comparison the values found by de Ruyter van Steveninck et al. (1992) in the same river stretch are also shown. Sedimentation and respiration rates were calculated as in De Ruyter van Steveninck et al. (1992). The 1995 estimate of grazing was taken from the grazing experiments. The lower respiration and grazing rates can be explained by the lower temperatures and zooplankton densities respectively. In both years a considerable fraction of the phytoplankton loss rate remains unexplained.

As mentioned above, several authors name zoobenthic filter feeders as a significant loss factor (Effler et al., 1996; Köhler, 1997; Basu & Pick, 1997). However, quantitative estimates of their impact in rivers are scarce. It is difficult to reliably estimate densities and filtration rates of benthic filter feeders on the river bed and thus to estimate the impact of these organisms on the water column above. There is one report of zoobenthos densities for the lower river Rhine (Bij de Vaate & Greijdanus-Klaas, 1993). We used the densities in this report to calculate a potential impact of benthic filter feeders in the lower Rhine.

The most important benthic species are the bivalves *Dreissena polymorpha*, *Corbicula* cf. *fluminea*, *Corbicula* cf. *fluminalis* and the amphipod *Corophium curvispinum*. Bij de Vaate and Greijdanus-Klaas (1993) cite densities for these organisms at the river bottom and at the banks, which are reinforced with stones. These values were used to calculate a total filtration rate per m of a transect across the river. A mean depth of 5 m and a mean transect length of 250 m was assumed, which means that a transect of a meter wide has 11 m^2 of bank area and 240 m^2 of bottom area. The specific grazing rate of *D. polymorpha* was estimated at 100 ml ind^{-1} h^{-1}, which is around the median value commonly observed. (cf. Reeders et al., 1989; Horgan & Mills, 1997). For *C. fluminea* and *C. fluminalis*, a value of 50 ml ind^{-1} h^{-1}(Boltovskoy et al., 1995). For *Corophium curvispinum* a value of 5 ml ind^{-1} h^{-1} was taken (Rajagopal et al., 1995). The zoobenthos densities of two locations in the free-flowing part of the Rhine were averaged and recalculated to represent the number of animals in a meter of cross-section. Multiplying these densities with the cited specific filtration rates results in an estimate of the total community filtration rate.

Table 6. Average densities (locations Hagestein and Vuren) and calculated filtration rates of zoobenthos in the lower Rhine. See text for explanation

	Spec. filtration rate ($l\ ind^{-1}\ h^{-1}$)	Density (10^6 ind m^{-1})	Community filtration rate ($m^3\ m^{-1}\ h^{-1}$)
Dreisssena polymorpha	0.1	0.38	3.17
Corbicula cf. *fluminalis*	0.05	0.02	0.87
Corbicula cf. *fluminea*	0.05	0.31	15.56
Corophium curvispinum	0.005	1.79	8.95
TOTAL			63.55

This exercise is summarized in Table 6 and yields an estimated filtration rate of 63 m^3 per m cross-section per h. Taking into account an average current velocity of 0.5 m s^{-1} and an average discharge of 2200 $m^3\ s^{-1}$, the calculated filtration rate represents a specific filtration rate of 0.38 d^{-1}. This figure is quite similar to the unexplained part of the phytoplankton loss rate (Table 5). The results support recent studies in other rivers. Köhler (1997) in the river Severn, who found that benthic filter feeders and sedimentation were the most important loss factors. In the river Spree Unioid mussels, living in the soft-bottom substrate, lead to grazing rates in the order of the total daily discharge (Welker & Walz, 1998). In the Meuse, however, zooplankton grazing was found to be much more important than benthic grazing (Gosselain et al., 1998)

Naturally both the densities and filtration rates of benthic organisms are variable in space and time. Over its course of more than 1000 km, both the substrate of the river bed and the zoobenthic assemblage vary greatly. The above exercise should be taken as an indication, not an absolute value. However, it illustrates the importance of the benthic compartment of the ecosystem of the Rhine. Investigations into the distribution of benthic filter feeders in the Rhine and into their activity will be a most necessary addition to our knowledge of the functioning of the Rhine ecosystem and other similar large rivers.

The present study shows again the importance of hydraulic conditions for plankton in eutrophic rivers. In the absence of nutrient limitation, discharge and linked to this variables like residence time, extinction coefficient and temperature, have a profound influence on plankton dynamics. However, it is evident that biotic interactions do play an important role as well in the Rhine ecosystem. Research is needed into those interactions, including better estimates of *in situ* primary production and consumption by the various functional groups in the river ecosystem.

Acknowledgements

The authors wish to thank Wim Willemsen and Erik Meuffels for assistance with the sampling and Lowie van Liere for critical comments. The contribution of TI was partly funded by the EU BINOCULARS project.

References

Admiraal, W., L. Breebaart, G. M. J. Tubbing, B. van Zanten, E. D. De Ruyter van Steveninck, E. D. & R. Bijkerk, 1994. Seasonal variation in composition and production of planktonic communities in the lower River Rhine. Freshwat. Biol. 32: 519–531.

Admiraal, W., S. D. Mylius, E. D. de Ruyter van Steveninck, & G. M. J. Tubbing, 1993. A model of phytoplankton production in the lower River Rhine verified by observed changes in silicate concentration.

Basu B. K. & F. R. Pick, 1997. Phytoplankton and zooplankton development in an lowland, temperate river. J. Plankton Res. 19: 237–253

Bij de Vaate, A. & M. Greijdanus-Klaas, 1993. Monitoring macroinvertebrates in the River Rhine. Results of a study executed in the Dutch part in 1990. Report no. 52–1993 of the project 'Ecological rehabilitation of the rivers Rhine and Meuse'. Institute for Inland Water Management and Waste Water Treatment, Lelystad, The Netherlands.

Boltovskoy D., J. Izaguirre, & N. Correa, 1995. Feeding selectivity of *Corbicula fluminea* (Bivalvia) on natural phytoplankton. Hydrobiologia 312: 171–182.

De Ruyter van Steveninck, E. D., W. Admiraal & B. van Zanten, 1990. Phases in the development of riverine plankton: examples from the Rhine and Meuse Hydrobiol. Bull. 24: 47–55.

De Ruyter van Steveninck, E. D., W. Admiraal, L. Breebaart, G. M. J. Tubbing & B. van Zanten, 1992. Plankton in the River Rhine: structural and functional changes observed during downstream transport. J. Plankton Res. 14: 1351–1368.

Descy, J.-P., P. Servais, J. S. Smitz, G. Billen & E. Everbeck, 1987. Phytoplankton biomass and production in the River Meuse, Belgium. Wat. Res. 21: 1557–1566.

Effler, S. W., C. M. Brooks, K. Whitehead, B. Wagner, S. M. Doerr, M. Perkins, C. A. Siegfried, L. Walrath & R. P. Canale, 1996. Impact of zebra mussel invasion on river water quality. Wat. Envir. Res. 68: 205–214.

Garnier, J., G. Billen & M. Coste, 1995. Seasonal succession of diatoms and Chlorophyceae in the drainage network of the Seine River. Limnol. Oceanogr. 40: 750–765.

Gosselain, V., J. -P. Descy & E. Everbecq, 1994a. The phytoplankton community of the River Meuse, Belgium: seasonal dynamics, year 1992. and the possible incidence of zooplankton grazing. Hydrobiologia 289: 179–191.

Gosselain V., C. Joaquim-Justo, L. Viroux, M. Mena, A. Metens, J.-P. Descy & J.-P. Thome, 1994b. Laboratory and in situ grazing rates of large river rotifers and their contribution to community grazing rates. Arch. Hydrobiol. Suppl. 113: Large Rivers 10, 1/4, 353–361.

Gosselain, V., L. Viroux & J. –P. Descy, 1998. Can a community of small-bodied grazers control phytoplankton in rivers? Freshwat. Biol. 39: 9–24.

Heiler, G., Th. Hein & F. Schiemer, 1995. Hydrological connectivity and flood pulses as the central aspects of a river-floodplain system. Reg. Riv. 11: 251–361.

Horgan, M. J., & E. L. Mills, 1997. Clearance rates and filtering activity of zebra mussel (*Dreissena polymorpha*): implications for freshwater lakes. Can. J. Fish. aquat. Sci. 54: 249–255.

ICPR., 1997. Plankton im Rhein 1995. International Commission for the Protection of the Rhine, Koblenz, Germany.

Köhler, J., 1993. Growth, production and losses of phytoplankton in the lowland river Spree. I. Population dynamics. J. Plankton Res. 15: 335–349.

Köhler, J., 1995. Growth, production and losses of phytoplankton in the lowland river Spree: carbon balance. Freshwat. Biol. 34: 501–512.

Köhler, J., 1997. Measurement of in situ growth rates of phytoplankton under conditions of simulated turbulence. J. Plankton Res. 19: 849–862.

Kobayashi, T., P. Gibbs, P. I. Dixon & R. J. Shiel, 1996. Grazing by a river zooplankton community: importance of microzooplankton. Mar. Freshwat. Res. 47: 1025–1036.

Lonsdale, D. J., E. M. Cosper & M. Doall, 1996. Effects of zooplankton grazing on phytoplankton size-structure and biomass in the lower Hudson River estuary. Estuaries 19: 874–889.

Marker, A. F. H. & G. D. Collett, 1997. Spatial and temporal characteristics of algae in the river Great Ouse. I. Phytoplankton. Reg. Riv. 13: 219–233.

Rajagopal, S., B. G. P. Paffen & G. Van der Velde, 1995. Particle removal from the water column by epilithic dominant filter feeders in the River Rhine and Meuse, The Netherlands. Report, Catholic University Nijmegen, The Netherlands.

Reeders, H H., A. Bij de Vaate & F. J. Slim, 1989. The filtration rate of *Dreissena polymorpha*, Bivalvia, in three Dutch lakes with reference to biological water quality management. Freshwat. Biol. 22: 133–141.

Reynolds, C. S., 1988. Potamoplankton: paradigms, paradoxes and prognoses. In E. F. Round (ed.), Algae and the Aquatic Environment. Biopress, Bristol.

Reynolds, C. S., 1994. The long, the short and the stalled: on the attributes of phytoplankton selected by physical mixing in lakes and rivers. Hydrobiologia 289: 9–21.

Reynolds, C. S., 1995. River plankton: the paradigm regained. In D. M. Harper & A. J. Ferguson (eds), The ecological basis for river management, Wiley, Chichester: 161–174.

Reynolds, C. S. & M. S. Glaister, 1993. Spatial and temporal changes in phytoplankton abundance in the Upper and Middle Reach of the River Severn. Arch. Hydrobiol. Suppl. Large Rivers 9: 1–22.

Rojo, C., M. A. Cobelas & M. Arauzo, 1994. An elementary, structural analysis of river phytoplankton. Hydrobiologia 289: 43–55.

Sokal, R. R. & F. J. Rohlf, 1981. Biometry W.H. Freeman, San Fransisco: 859 pp..

Søballe, D. M. & B. L. Kimmel, 1989. A large-scale comparison of factors influencing phytoplankton abundance in rivers, lakes and impoundments. Ecology 68: 1943–1954.

Tubbing, G. M. J. & W. Admiraal, 1991. Sensitivity of bacterioplankton in the Rhine river to various toxicants measured by thymidine incorporation and activity of exoenzymes. Envir. Toxicol. Chem. 10: 1161–1172.

Van Dijk, G. M. & B. van Zanten, 1995. Seasonal changes in zooplankton abundance in the lower Rhine during 1987–1991. Hydrobiologia 304: 29–38.

Van Mazijk, A., P. Verwoerdt, J. Van Mierlo, M. Bremicker & H. Wiesner, 1991. Rheinalarmmodell Version 2.0, Kalibrirung und Verifikation. Bericht nr. II-2 der IKSR/KHR Expertengruppe, Lelystad, The Netherlands.

Wehr, J. D. & J. H. Thorp, 1997. Effects of navigation dams, tributaries and littoral zones on phytoplankton communities in the Ohio River. Can. J. Fish. aquat. Sci. 54: 378–395.

Welker, M. & N. Walz, 1998. Can mussels control the plankton in rivers? A planktological approach applying Lagrangian sampling strategy. Limnol. Oceanogr. 43: 753–762.

Yang, J. R., B. K. Basu, P. B. Hamilton & F. R. Pick, 1997. The development of a true riverine phytoplankton assemblage along a lake-fed lowland river. Arch. Hydrobiol. 140: 243–260.

Hydrobiologia **410:** 11–16, 2000.
J. Garnier & J.-M. Mouchel (eds), Man and River Systems.

Aquatic bryophyte assemblages along a gradient of regulation in the river Rhine

A. Vanderpoorten[1] & J.-P. Klein[2]
[1]*Faculté Universitaire des Sciences Agronomiques de Gembloux, Zoologie Générale & Appliquée, 2, Passage des Déportés, B-5030 Gembloux, Belgium*
[2]*Université Louis Pasteur, Centre d'Analyses et de Recherches, Département Hydrologie & Environnement, 76, route du Rhin, BP 24, F-67400 Illkirch-Graffenstaden, France*

Key words: Rhine, river regulation, water quality, aquatic bryophyte zonation, refugia

Abstract

A study of the zonation of the aquatic bryophyte assemblages from the Anterior Rhine to the Middle Rhine is presented. The typology obtained differs from the hydrogeomorphological classification because the correction of the river has led to the homogenization of habitats. Since the straightening of the river in the course of the 19th century, the embankments have created more stable habitats on which mountain species formerly characteristic of the rapids of the High Rhine appeared. During the 20th century, regulation, the construction of hydroelectric powerstations even the canalization of the river in Alsace, have caused a great change in water level. On the Upper Rhine, the regularisation of the water level allowed the appearance and the spread of strict hydrophytes such as *Octodiceras fontanum* which is now common between Rastatt and Bonn. Conversely, species characteristic of variable water levels decreased or disappeared. The industrial development along the river has caused water quality to severely decrease downstream of the industrial areas, hence the decrease or the disappearance of their primary flora in favour of pollution-tolerant assemblages. The former side channels of the Upper Rhine in France that have been disconnected from the main river and fed by groundwater since canalization, possess a bryophyte flora very similar to that of good quality waters in the upstream part of the High Rhine or in the Alpine Rhine. This flora is the last regional relic of a flora that has disappeared from the main channel due to water pollution. These refugia should consequently be preserved from the flooding of alluvial forests, part of the flood management program.

Introduction

A natural frontier between Switzerland, Liechtenstein, Austria, Germany, France and The Netherlands, the Rhine joins the North Sea after a 1326 km long course beginning at 2350 m in the Swiss Alps. Its basin covers 190 000 km^2 where 50 million people live. Being navigable along 800 km from Schaffhausen, it is the second most important European river but also one of the most regulated: straightened from the city of Chur, it was used for producing hydroelectricity between Konstanz and Lauterbourg, canalized in France and embanked to the North Sea. A large industrial estate developed in its alluvial floodplain, causing severe water pollution in the 1970s. The aim of this paper is to describe the consequences of the human disturbances on the aquatic bryophyte assemblages in order to answer the following questions: what are the main ecological factors responsible for the gradation of the bryophyte vegetation, to what extent have disturbances caused by human activities modified this zonation, and what possible ecological solutions are there for conserving the most interesting floristic assemblages?

Methods and study area

The Rhine flow was surveyed on ca. 850 km from Lake Tuma in the Swiss Alps to Bonn in Germany (Figure 1). In this study area, the Rhine is divided into five hydrogeomorphological sectors. Many of these rapids formerly characteristic for the High Rhine have disappeared since the construction of hydroelectric powerstations that caused the water level to increase

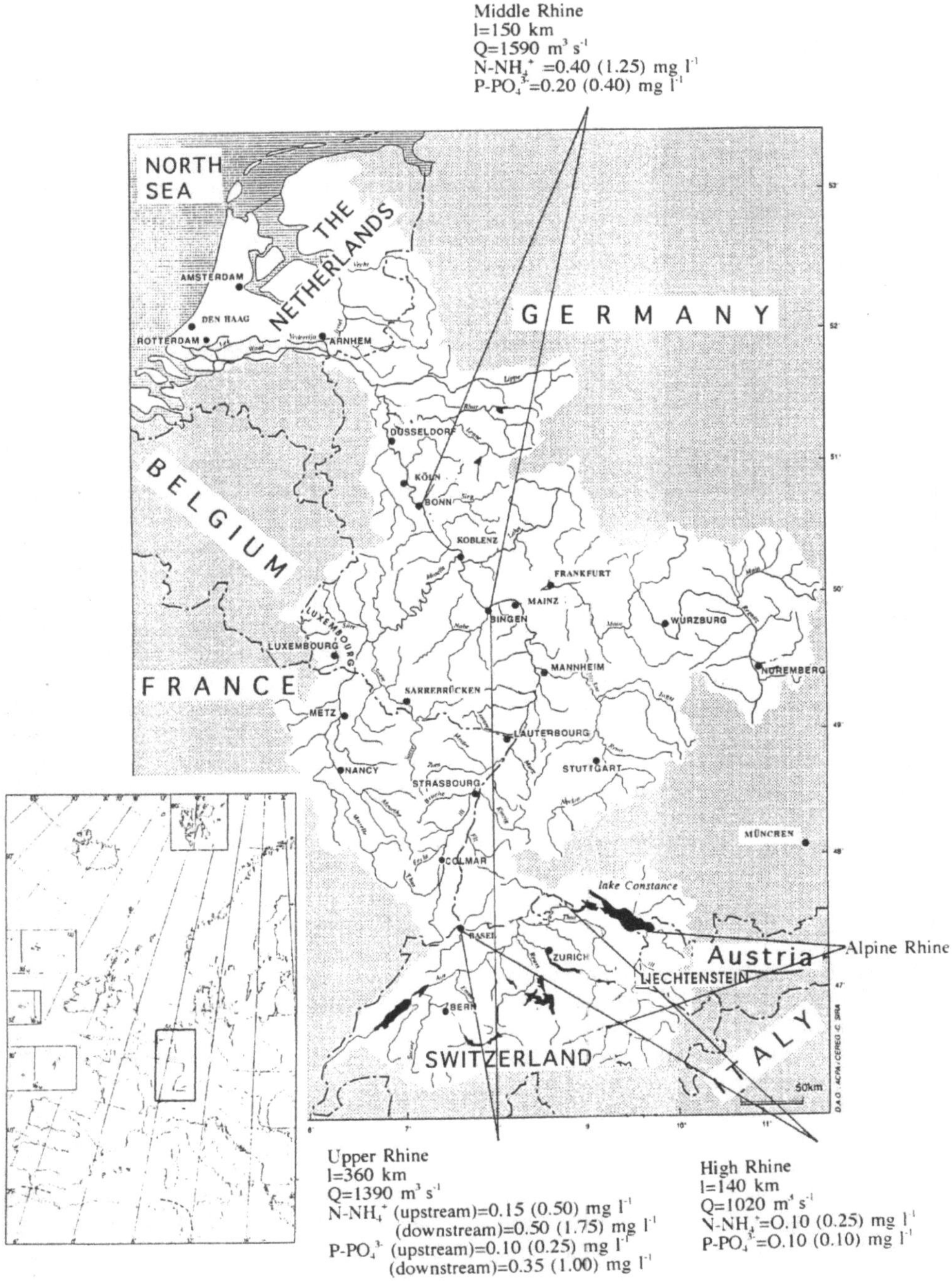

Figure 1. The Rhine basin. The concentrations in N-NH_4^+ and P-PO_4^{3-} are the means (and maxima) in 1987. Q = average annual discharge (from 1931 to 1980); l = length of the sector (after Tittizer & Krebs, 1996).

upstream the barrages. Until the 19th century, the Upper Rhine was flowing in an alluvial floodplain a few km broad and was characterized by its numerous side-channels and its strands of mobile pebbly alluvium annually disturbed by the floods. In order to provide protection from flooding, the correction of the Rhine began in the 19th century. The continual moving of the gravel and sandbanks, even after the rectification measures, and the low water depth in period of low discharges, were seen as an unacceptable disadvantage for navigation. The Upper Rhine regulation was therefore started at the beginning of the 20th century. A low water channel with a width of only 80 m was maintened by the construction of groins in the river bed (Dister et al., 1990). This caused the increase of the vertical erosive power of the river. Between 1870 and 1950, the water level was lowered 5–7 m in the areas downstream from Basel, and rapids progressively emerged at the level of the village of Istein. Consequently, the Upper Rhine was canalized in the course of the 20th century between France and Germany to allow the navigation between Basel and Strasbourg, so that most of the former side-channels were disconnected from the main river. Being almost exclusively fed by groundwater since the canalization, their discharges have become very constant. Some of them have a tendency to silt-up.

In the main channel, water quality is eutrophic downstream from Lake Constance and severely decreases from the area of Mannheim downstream from Karlsruhe. Severe pollution levels were reached in the decades 1970–1980. However, water quality has tended to improve (Malle, 1996; Tittizer & Krebs, 1996). In the former side channels disconnected from the main river since the canalization, water quality is oligotrophic (with annual mean concentrations of ammonium nitrogen and phosphates of ca. 15 $\mu g\ l^{-1}$) because of an almost exclusively phreatic feed.

Within this study area, the aquatic bryophytes (nomenclature of Corley et al., 1981; Corley & Crundwell, 1991 for the mosses; Grolle, 1983 for the liverworts) were recorded according to stratified sampling within the five investigated hydrogeomorphological sectors. Presence–absence of the bryophytes were recorded in 95 10–30 m long samples. The typology of the records according to their floristic composition was obtained by hierarchical aggregative numerical classification. The initial 0/1 data matrix was transformed into a Jaccard's distances matrix. The records were clustered from this distances matrix by using Ward's minimum variance criterion. The species fidelity to the clusters of records was measured by their relative conditional frequency in each cluster. This frequency is equal to the sum of the occurrences of the species i in the cluster of records j divided by the total number of records within the cluster j.

Results

Five clusters of records were generated by the classification (Figure 2). The relative conditional frequencies of their characteristic species are given in Table 1. The first cluster (I) is characterized by pollution-tolerant mosses of eutrophic status such as *Octodiceras fontanum* and *Amblystegium riparium.* It includes samples from the Upper Rhine downstream (including five records of Ahrens, 1995) and the Middle Rhine. The second cluster (II) is characterized by rheophilous mosses such as *Cinclidotus danubicus* and *Rhynchostegium riparioides.* It includes samples from the downstream part of the High Rhine and of the upstream part of the Upper Rhine. The third cluster (III) is characterized by mosses of still waters of mesotrophic status such as *Cratoneuron filicinum.* It includes samples from Lake Constance (including five records of Ahrens, 1992), the upstream part of

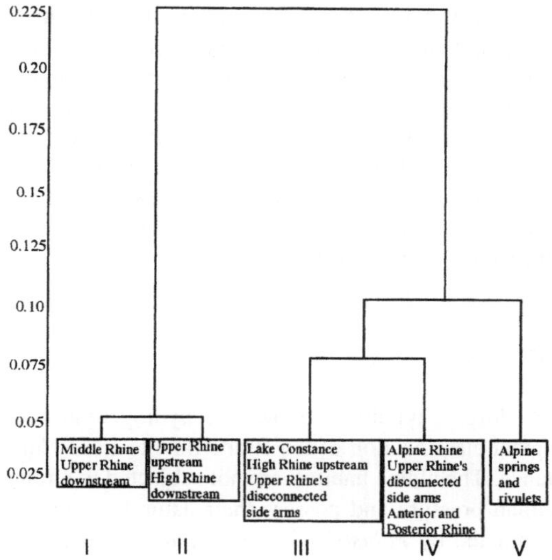

Figure 2. Summary cluster diagram based on 95 samples for the classification of the study sites according to their aquatic bryophyte assemblages using Ward's minimum variance criterion. Dissimilarity, indicated on the vertical axis, is the between clusters semi-partial R-squared. Roman numeral indicated cluster number.

Table 1. Conditional relative frequencies of the characteristic aquatic and subaquatic bryophytes of the floristic clusters

Cluster	I	II	III	IV	V
Octodiceras fontanum	0.64	-	-	-	-
Didymodon nicholsonii	0.48	-	-	-	-
Leskea polycarpa	0.61	0.28	-	-	-
Cinclidotus riparius	1.00	0.85	-	0.15	-
Amblystegium riparium	0.84	0.86	0.58	-	-
Cinclidotus danubicus	0.71	0.93	0.16	-	-
Amblystegium fluviatile	0.64	1.00	0.16	-	-
Fissidens crassipes	0.48	0.64	0.83	0.05	-
Rhynchostegium riparioides	0.09	0.78	0.41	0.05	-
Fontinalis antipyretica	0.09	0.71	0.35	0.25	-
Hygrohypnum luridum	0.06	0.14	0.25	0.75	-
Cratoneuron filicinum	0.09	0.42	0.83	0.80	-
Palustriella commutata	-	-	-	0.45	0.64
Philonotis fontana	-	-	-	0.45	0.64
Blindia acuta	-	-	-	0.10	0.57
Hygrohypnum smithii	-	-	-	0.05	0.21
Jungermannia exsertifolia	-	-	-	0.05	0.28
Philonotis seriata	-	-	-	0.05	0.78
Calliergon sarmentosum	-	-	-	-	0.21
Drepanocladus exannulatus	-	-	-	-	0.28
Dicranella palustris	-	-	-	-	0.71

Roman numeral relate to clusters as classified in Figure 2.

the High Rhine and disconnected Alsatian side arms. The fourth cluster (IV) is characterized by rheophilous mountain mosses such as *Hygrohypnum luridum* and *Schistidium alpicola*. It includes samples from the Alpine Rhine, the Anterior and Posterior Rhine and disconnected Alsatian side arms. The fifth cluster (V) is characterized by typical alpine bryophytes such as *Philonotis seriata* and *Hygrohypnum smithii.* It includes samples from alpine springs and rivulets at altitude higher than 2000 m.

Discussion

The floristic typology follows the hydrogeomorphological classification of the Rhine flow to Lake Constance. The alpine habitats are not subjected to strong human pressure and possess their natural vegetation dominated by *Palustriella commutata.* In the downstream parts of the Anterior Rhine and Posterior Rhine and in the upstream Alpine Rhine, the alpine flora naturally disappears in favour of a rheophilous mountain flora dominated by *Hygrohypnum luridum* and *Schistidium alpicola.* Dowstream from the city of Chur, the Alpine Rhine has been regulated but the embankments have substitute habitats for a flora characteristic of the upstream sectors. In Lake Constance, the rheophilous mountain flora disappears in favour of species characteristic of still waters such as *Cratoneuron filicinum.* It is downstream of Lake Constance that the floristic typology differs from the hydrogeomorphological classification, because the regulation of the river has led to the homogenization of the habitat conditions. At the beginning of the 19th century, the High Rhine was characterized by a succession of rapids and waterfalls (Lauterborn, 1916). These were natural places for rheophilous mosses presenting structural adaptations to counteract the strength of the flow: strong rhizoïds (*Rhynchostegium riparioides*), keeled leaves (*Fontinalis*) and multicellular leaf margin (*Cinclidotus*) (Vitt & Glime, 1984). Downstream of Basel, the Upper Rhine lies in a much broader alluvial floodplain. The bed of the river was consequently composed of pebbly strands, whose texture decreases with the upstream-downstream decreasing slope of the river (Lauterborn, 1910). Although substrate movement under high flow provides new places and allows less competitive species to establish (Vitt et al., 1986; Muotka & Virtanen, 1995), these sandy or pebbly strands, being disturbed every year by the floods, were obviously unfavourable for the aquatic bryoflora. In the Middle Rhine, in the area of St Goar, only four aquatic bryophyte species were found: *Cinclidotus fontinaloides* and *Amblystegium riparium* that were common, *Hygrohypnum luridum,* and *Fissidens crassipes* in one locality (Frahm, 1997). Since the straightening of the river, the embankments have created more stable habitats on which mountain species formerly characteristic of the rapids of the High Rhine such as *Cinclidotus danubicus* and *C. riparius* appeared. Others, such as *Fissidens crassipes,* have spread massively. Consequently, the straightening has enhanced the diversity of aquatic bryophytes since the last century in the Middle and Inferior Rhine (Frahm & Abst, 1993; Frahm, 1997), where ca. 15 aquatic bryophyte species are now commonly found. During the 20th century, the regulation, the construction of hydroelectric powerstation and even the canalization of the Rhine in Alsace, caused a great change in the water level. On the High Rhine, the increased water level upstream of the barrages caused the disapperance of almost all the primary habitats. This brought about a decrease in niche diversity, reduced the area of water level variations and, as already shown in regulated rivers (Glime & Vitt, 1987; Englund et al., 1997), a decrease in aquatic bryophyte diversity. Some species characteristic of the area, such as *Fissidens grandifrons* and *F. rufulus* (Philippi, 1961), were on the verge of extinction. Others, such as *Cinclidotus danubicus,* were able to adapt and to colonize substitute habitats-such as the weirs. In the Upper Rhine, the water level variations were greatly reduced after the regulation and the canalization of the river, so that strict hydrophytes such as *Octodiceras fontanum* appeared and have spread massively. Conversely, species characteristic of variable water levels, e.g. *Hyophila involuta,* were on the verge of extinction (Philippi, 1978).

The regulation of the Rhine has allowed the substantial development of an industrial estate that has caused water quality to markedly decrease. Downstream from the area of Mannheim, the water trophic level severely increases, causing the change of the aquatic bryophyte flora with species of eutrophic pollution-tolerant status like *Amblystegium riparium,* taking the place of *Rhynchostegium riparioides* in polluted waters (Holland & Harding, 1984). *Octodiceras fontanum* also becomes one of the dominant species from Mainz, hence the decrease of *Fissidens crassipes* which occupies the same niche in the Upper Rhine upstream from Mannheim. However, due to the severe water pollution levels reached in the 1970s in the Upper Rhine, many of the good water quality species also disappeared from the main channel between Basel and Mannheim. For example, *Fissidens grandifrons,* which was known in many localities between Basel and Strasbourg at the beginning of the century (Philippi, 1968), is on the verge of extinction in the area. This good quality water flora found refugia in the former side channels that have been disconnected from the main river and fed by the groundwater since the canalization. This flora consequently shows affinities with that of the upstream sectors with good quality waters, in the upstream High Rhine or in the Alpine Rhine, and is the last regional relic of a flora that has disappeared from the main channel due to water pollution.

Certain disconnected side-channels have a tendency to silt up. The reconnection of some of these disconnected side-arms to the main river is, therefore, being undertaken (Service de la Navigation de Strasbourg, 1995). Some of these disconnected side-arms are also located in alluvial forests that will become areas of lateral retention of the floods called 'polders'. The Rhine waters show a clearly eutrophic status and occasional pollution peaks by xenobiotic factors can occur. Mixing of these surface waters with the groundwater-fed streams could consequently lead to an increase of the pollution level of these brooks. Moreover, the floodpulse caused by the sudden input of surface water in the disconnected brooks would probably wash away most of the hydrophytes. In fact, the hydraulic works of the polders would allow input discharges of surface waters up to 140 m^3 s^{-1} (Burgéap, 1994) whereas the annual mean discharge of the phreatic brooks is currently ca. 350 l s^{-1} (Sánchez-Pérez & Maire, 1995). If algae such as *Chara* exhibited their ability to recolonize quickly the groundwater-fed stretches of waters in Polder Altenheim after the floods (Landesanstalt für Umweltschutz Baden-Württemberg, 1994), it is very likely that species with very slow recolonization strategies like the aquatic bryophytes would disappear under these conditions. It would consequently be highly desirable to protect the remaining oligotrophic brooks with their original hydrophyte assemblages from flooding. The brooks showing a tendency to silt up can be desilted by ecological engineering. It was demonstrated that the replacing of the concrete pipes under the paths by wooden bridges and the mechanical desilting by pumping out the mud enables the runoff to quicken again, hence the decrease of the duckweed carpets and the return of the red algae *Hildenbrandia rivularis* on the riffles (Klein et al., 1994). The choice of the sites for lateral retention of the floods should be focussed on eutrophic and silted-up areas, where sudden increasing discharges can re-create fluvial dynamics allowing the re-appearance of certain aquatic biocenoses, as shown in Polder Altenheim (Siepe, 1994). In the phreatic-fed streams areas that have already been designated as areas for flood retention such as the forest of Erstein, the hydraulic works should authorize inputs of surface water only in the silted-up brooks in order to preserve as far as possible the flowing oligotrophic streams.

Conclusion

The regulation of the Rhine has led to the longitudinal spread of aquatic bryophytes formerly restricted to the High Rhine into the downstream areas. Conversely, the decrease in water quality in the main channel in the decades 1960–1980 caused a lateral shift of the aquatic flora that found refugia in lateral disconnected systems. These refugia should be preserved from disturbance in order to conserve their very rich aquatic flora, which is the last witness of the natural aquatic vegetation of the area.

Ackowledgements

The authors would like especially to thank the Navigation Office of Strasbourg for providing hydrological data, A. Sotiaux and T. Arts for checking some determinations of bryophytes, R. Palm for his assistance with the statistical procedure, J.-P. Frahm for his friendly greeting in Bonn and Mrs Moreman for correcting the English of this paper.

References

Ahrens, M., 1992. Die Moosvegetation des nördlichen Bodenseegebietes. Dissertationes Botanicae 190: 1–681.

Ahrens, M., 1995. Das Laubmoos *Barbula nicholsonii* Culm. am nördlichen Oberrhein und am Neckar (Südwestdeutschland). Carolinea 53: 229–241.

Burgeap, 1994. Expertise écologique du projet de polder d'Erstein (67).

Corley, M. F. V. & A. C. Crundwell, 1991. Additions and amendments to the mosses of Europe and the Azores. J. Bryol. 16: 337–356.

Corley, M. F. V., A. C. Crundwell, R. Düll, M. O. Hill & A. J. E. Smith, 1981. Mosses of Europe and the Azores; an annotated list of species, with synonyms from the recent literature. J. Bryol. 11: 609–689.

Dister, E., D. Gomer, P. Obrdlik, P. Peterman & E. Schneider, 1990. Water management and ecological perspectives of the Upper Rhine's floodplains. Regulated Rivers: Research & Management 5: 1–15.

Englund, G., B.-J. Jonsson & B. Malmqvist, 1997. Effects of flow regulation on bryophytes in North Swedish rivers. Biol. Cons. 79: 79–86.

Frahm, J.-P. & U. W. Abst, 1993. Veränderungen in der Wassermoosflora des Niederrheins 1972–1992. Limnologica 23: 123–130.

Frahm, J.-P., 1997. Zur Ausbreitung von Wassermoosen am Rhein (Deutschland) und an seinen Nebenflüssen seit dem letzten Jahrhundert. Limnologica 27: 251–262.

Glime, J. M. & D. H. Vitt 1987. A comparison of bryophyte species diversity and niche structure of montane streams and stream banks. Can. J. Bot. 65: 1824–1837.

Grolle, R., 1983. Hepatics of Europe and the Azores: an annotated list of species, with synonyms from the recent literature. J. Bryol. 12: 403–459.

Holland, D. G. & J. P. C. Harding 1984. Mersey. In B. A. Whitton (ed.), Ecology of European Rivers. Blackwell Scientific Publications, Oxford: 113–144.

Klein, J.-P., G. Maire, F. Exinger, G. Lutz, J. M. Sánchez-Pérez, M. Trémolières & P. Junod, 1994. The restoration of former channels in the Rhine alluvial forest: the example of the Offendorf nature Reserve (Alsace, France). Wat. Sci. Technol. 29: 301–305.

Landesanstalt für Umweltschutz Baden-Württemberg, 1994. Flutungen der Polder Altenheim. Materialen zum Intergrierten Rheinprogramm.

Lauterborn, R., 1910. Die Vegetation des Oberrheins. Verh. Naturhist.-Med. Ver. 10: 450–502.

Lauterborn, R., 1916. Die geographische une biologische Gliederung des Rheinstromes. I. Teil. Sitzungberichte der Heidelberger Akademie der Wissenschaften, Math.- naturwiss. Klasse Abt. B.

Malle, K.-G., 1996. Cleaning up the river Rhine. Scientific American 274: 55–59.

Muotka, T. & R. Virtanen, 1995. Stream as habitat templet for bryophytes: distribution along gradients in disturbance and substratum heterogeneity. Freshwat. Biol. 33: 141–149.

Philippi, G., 1961. Die Wassermoosflora am Hochrhein zwischen Rekingen und Waldshut. Veröff. Landesst. Natursch. Landschaftpf. Bad.-Württ. 27–28: 168–177.

Philippi, G., 1968. Zur Verbreitung einiger hygrophytischer und hydrophiler Moose im Rheingebiet zwischen Bodensee und Mainz. Beitr. naturk. Forsch. Südw.-Dtl. 27: 61–81.

Philippi, G., 1978. Veränderungen der Wasser- und Uferflora im badischen Oberrheingebiet. Beih. Veröff. Naturschutz Landschaftpfl. Bad. Württ. 11: 99–134.

Sánchez-Pérez, J. M. & G. Maire, 1995. Etude du fonctionnement hydrologique des réserves naturelles d'Erstein et d'Offendorf. Conservatoire des Sites Alsaciens-Université Louis Pasteur-CNRS-CEREG.

Service de la Navigation de Strasbourg, 1995. Restauration des écosystèmes rhénans. Proposition de programme pour la restauration d'anciens bras du Rhin.

Siepe, A., 1994. Regeneration of floodplain biotopes on the High Rhine – The 'Polder Altenheim' case. Wat. Sci. Technol. 29: 281–288.

Tittizer, T. & F. Krebs (eds), 1996. Ökosystemforschung: Der Rhein und seine Auen- eine Bilanz. Springer Verlag, Berlin-Heidelberg.

Vitt, D. H. & J. Glime, 1984. The structural adaptations of aquatic Musci. Lindbergia 10: 95–110.

Vitt, D. H., J. M. Glime & C. LaFarge-England, 1986. Bryophyte vegetation and habitat gradient of montane streams in western Canada. Hikobia 9: 367–385.

Hydrobiologia **410:** 17–24, 1999.
J. Garnier & J.-M. Mouchel (eds), Man and River Systems.

A macrophyte communities sequence as an indicator of eutrophication and acidification levels in weakly mineralised streams in north-eastern France

Gabrielle Thiébaut & Serge Muller
Université de Metz, UPRES 'Ecotoxicité, Biodiversité et Santé Environmentale', Laboratoire de Phytoécologie, Ile du Saulcy, 57045 Metz Cedex, France

Key words: eutrophication, acidification, macrophyte communities, weakly mineralized streams, France

Abstract

The Northern Vosges streams are subjected to acidification upstream and to eutrophication downstream. A bio-indicator scale of the degree of eutrophication and of the level of acidification, based on four aquatic macrophyte communities, was established in these weakly mineralized streams. Aquatic plant communities therefore begin as Group A, develop into Group B and become Group C or D downstream. The A community was characterized by bryophytes (*Scapania undulata, Sphagnum* sp.) and *Potamogeton polygonifolius* in oligotrophic, acidified to weakly acidified, and poorly buffered streams. The B community, defined by *P. polygonifolius* and the appearance of *Ranunculus peltatus* and *Callitriche* species (*C. platycarpa, C. hamulata*) was found in oligotrophic waters with a higher buffer capacity than the A community. The C community was determined by the disappearance of *P. polygonifolius* and the appearance of *Elodea* species *(E. canadensis, E. nuttallii)* and rare species in mesotrophic and neutral streams. The D community, with very high nutrient loading, was characterized by the abundance of *Callitriche obtusangula*, by the presence of *Amblystegium riparium, Fissidens crassipes* and the development of filamentous algae. This macrophyte sequence corresponded to an upstream to downstream zonation, which was characterized by an increase in buffer capacity and in nutrient levels and a decrease in protons and aluminium load.

Introduction

Freshwaters are highly sensitive to any input of nutrients or acidifying substances. Increasing human activities, particularly urbanisation, agricultural and industrial activities, lead to an increase in nutrients. Large nutrient inputs have accelerated eutrophication. In many lakes and rivers world-wide, the input of large quantities of phosphorus from agricultural run-off and sewage, and also of nitrogen (mainly as runoff from agricultural land), produces ideal conditions for high algal growth (Lee, 1973) and macrophyte decline (Phillips et al., 1978). Thus, the main channels of the Rhine and the Meuse rivers are characterized by a dense plankton which develops rapidly in the nutrient rich river waters (Admiraal et al., 1993). In the Alsacian plain, aquatic macrophyte communities can be used as bioindicators of eutrophication in calcareous streams (Carbiener et al., 1990; Trémolières et al., 1994; Robach et al., 1996).

In the Netherlands, acidification is always accompanied by nitrogen enrichment (Arts et al., 1990). Research on the effects of acid deposition and acidification on aquatic biota has been ongoing in Europe and North America for the last 20 years, and many comprehensive studies have been published (Grahn, 1977; Roberts et al., 1985; Farmer, 1990). In the Vosges mountains, investigations of acidification started in 1985 with the aim of characterizing the degree of water acidification (Massabuau et al., 1987; Probst et al., 1995; Guérold et al., 1995). These studies have shown a drastic decrease in fish and aquatic animal biodiversity in acidified headwater streams.

Even though research has been conducted in several areas and concerned many different organisms, it has not been exhaustive and there are gaps in our understanding of the response of organisms to acidic or eutrophic conditions. These gaps have often occurred because funding has focused on chemical mechanisms and modelling the response of systems rather than on making resource inventories or resolving uncertainties in biological responses to acidification and eutrophication.

In the Northern Vosges, Muller (1990) has established a preliminary sequence of increasing eutrophication based on macrophyte communities in weakly mineralized waters. Later, Thiébaut & Muller (1996) described more precisely this eutrophication sequence by using multivariate analyses and by increasing the network. A comparison was done with the Alsacian plain sequence (Robach et al., 1996). In parallel, an other preliminary study has consisted in assessing the acidification degree and in researching the impact of acidification on macrophyte communities (Thiébaut et al., 1996). In this present study, the first objective was to research the relationships between aquatic plants and water quality in a widespread assessment. The purpose was to define an aquatic macrophyte communities sequence as an indicator of the degree of eutrophication and of acidification in the weakly mineralized streams of the Northern Vosges streams (N-E, France) by using a multivariate approach and by spatial and temporal comparisons between streams located in the same catchment.

Study site

The study area is situated in north-eastern France. The landscape pattern of the Northern Vosges consists of sandstone mountains of 200–580 m in altitude topped by rocky conglomerate, surrounded by steep cliffs. The regional climate is subcontinental. Winters are cold, with more than 100 days of frost and an annual temperature of 8.6 °C. Summers are relatively hot. The mean rain fall is 900 mm. The Northern Vosges represent an acid-sensitive area in France, receiving high levels of acid deposition (Dambrine et al., 1994). The streams are subject to eutrophication (Muller, 1990; Thiébaut & Muller, 1996; Thiébaut & Muller, 1998) and to acidification (Thiébaut et al., 1996).

The 23 chosen streams are located in the Moder catchment. They drain a forested area covered by acidic soils ranging from brown acidic soils to podzolic soils. This hydrological network flows in alluvial deposits characterized by low calcium carbonate concentration.

Forty two sites were selected in order to represent a wide range of physico-chemical parameters in sections of streams that were more or less homogenous in terms of their morphometric parameters (length, depth, width, slope, flow velocity...).

Material and methods

Two methods were used, a biological one consisting of a periodic survey of the macrophyte communities and a physico-chemical monitoring of the water.

Floristic data

At each sampling site, a river section of 50 m was chosen. A semi-quantitative inventory using cover percentage of abundance-dominance (Braun-Blanquet, 1964) was carried out.

The botanical survey was conducted during the growing season in 1993, 1994 and 1995. The nomenclature has followed the literature for the bryophytes (Smith, 1992) and for vascular plants (Lambinon et al., 1992). To avoid identification errors, *Sphagnum* and filamentous algae were not designated to species.

Physico-chemical data

Water samples were collected every three months from January 1993 to December 1995. Seven physico-chemical variables were analyzed: pH, conductivity, nutrients (PO_4^{3-}, NH_4^+ by spectrophotometry), NO_3^- (ion chromatography), ANC 'Acid Neutralising Capacity' (Gran's titration) and total aluminium (ICP after acidification with HNO_3).

Statistical procedures

The floristic data were analyzed by a Factorial Correspondence Analysis (FCA). Species and sites were distributed according to their position along the two principal axes.

An agglomerative hierarchical cluster analysis (Ward's method, Pearson squared distance measure) was used to identify physico-chemical groups and to arrange them in a hierarchical system. The results of this cluster analysis were combined with an ordination by a Principal Component Analysis (PCA). PCA was used to derive an ordination of the 7 selected physico-chemical variables. The data matrix X represents the mean value of physico-chemical parameters on 42

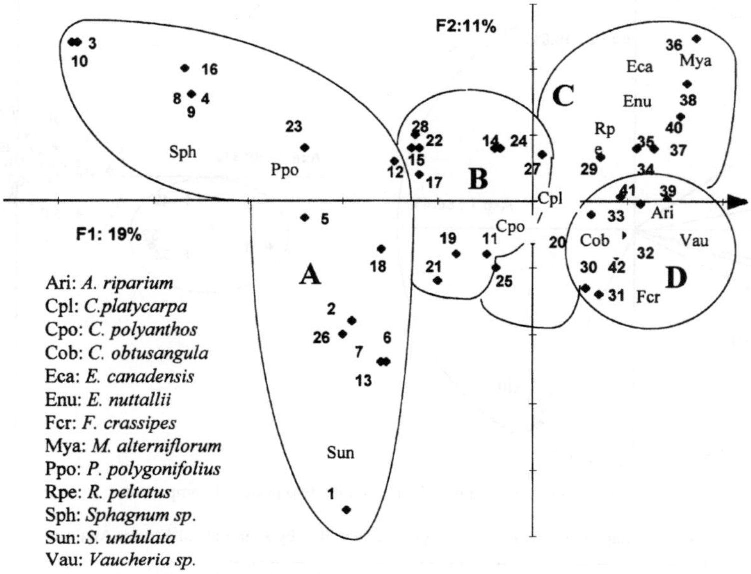

Figure 1. FCA ordination diagram of the Northern Vosges streams data. Species points on the edge of the diagram are rare species, whereas species at the centre may either be unimodal with optima at the centre or bimodal.

sites. The individual principal axes provide separate ordination of the sampling sites.

The assessment of the relationship between the macrophyte species and the physico-chemical variables was undertaken by calculating correlation coefficients (Student tests).

Results

Floristic composition

The FCA, carried out from the data matrix species × sites, showed an increasing gradient of acidification and a decreasing gradient of eutrophication. The first axis, having an eigenvalue of 0.19, was characterized by species typical of acidified and oligotrophic waters such as *Sphagnum* sp. and *Potamogeton polygonifolius* on the left, and species typical of neutral and eutrophic waters such as *Callitriche obtusangula*, *Amblystegium riparium* or *Fissidens crassipes* and *Vaucheria* sp. on the right.

The composition of each aquatic macrophyte unit confirmed the floristic communities, previously defined by Muller (1990) and Thiébaut & Muller (1996). Four plant communities, named A–D, were determined (Figure 1).

Community A was characterized by a vascular plant *Potamogeton polygonifolius,* often associated with *Glyceria fluitans* and *Scapania undulata* in spring sites. This floristic group was very heterogenous. Bryophyte species such as *Chiloscyphus polyanthos, Sphagnum* sp., *Dicranella palustris* and the algae *Batrachospermum* sp. had a narrow distribution in this group, whereas *Scapania undulata* was a wide distributed species.

Community B was defined by the appearance of new vascular plants such as *Callitriche* species (*C. platycarpa, C. stagnalis* and *C. hamulata*) and *Ranunculus peltatus* in upstream sites. Bryophyte species such as *Scapania undulata, Chiloscyphus polyanthos* were also found, when the habitat conditions (presence of rocks, no turbidity...) allowed their establishment.

Community C was mainly determined by the disappearance of *Potamogeton polygonifolius* and by the appearance of *Elodea* species (*Elodea canadensis, E. nuttallii*). Rare and protected species in this area, *Oenanthe fluviatilis*, *Potamogeton alpinus* and *Myriophyllum alterniflorum,* were found in this group. *Ranunculus peltatus* and *Callitriche* species were well developed in these downstream sites.

Community D was characterized by the appearance of *Callitriche obtusangula.* Filamentous algae

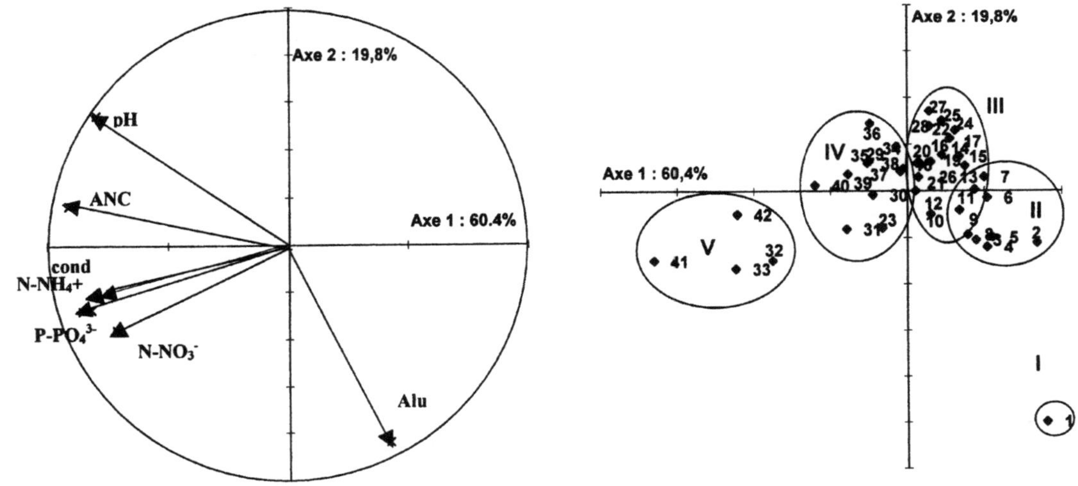

Figure 2. PCA- ordination diagram on the two principal components.

Table 1. Characteristics of chemical groups identified by Hierarchical Cluster Analysis

	Group I	Group II	Group III	Group IV	Group V
Number of sites	1	9	17	11	4
pH	4.4	5.7	6.5	7.0	7.0
Conductivity μS/cm	55	52	53	72	94
ANC μeq/l	−41	69	175	350	509
Al μg/l	510	165	63	79	56
$P\text{-}PO_4^{3-}$ μg/l	16	13	34	63	179
$N\text{-}NH_4^+$ μg/l	37	35	54	95	218

such as *Vaucheria* sp., bryophyte species such as *Fissidens crassipes* and *Amblystegium riparium* were often associated with *Callitriche* species, with *Ranunculus peltatus* and with *Elodea nuttallii*. Aquatic weeds such as *Vaucheria* sp. or *Callitriche obtusangula* were abundant in these downstream sites.

Physico-chemical composition of streams

The PCA was carried out on the physico-chemical data × sampling sites (Figure 2). The first axis of the PCA, defined principally by pH ($r = -0.80$), ANC ($r = -0.92$), conductivity ($r = -0.88$), ammonia-nitrogen ($r = -0.82$), phosphorus ($r = -0.82$) and to a lesser extent nitrate ($r = -0.69$), explains 60.4% of the physico-chemical variability. The second axis of the PCA is correlated with aluminium ($r = -0.83$) and explains 19.8% of the physico-chemical variance.

Five groups were identified by Hierarchical Cluster Analysis (Table 1). The group I represents an acidified, un-buffered, oligotrophic stream with high aluminium levels, whereas the group II was less acidified. In the group III, waters were more buffered than in the first two groups. The nutrient levels slightly increased whereas the aluminium load decreased in the group III. The first three weakly mineralized groups were characterized by an oligo-mesotrophic level, whereas the degree of acidification decreased. The group IV and V represent neutral sites, well buffered and more mineralized than other groups. The main difference between the group IV and the group V was the high nutrient levels in the last group.

This cluster method separated upstream sites (group I) and downstream sites (group V). The streams were subjected upstream to acidification (low ANC, high aluminium load, acidic pH) in the groups I and

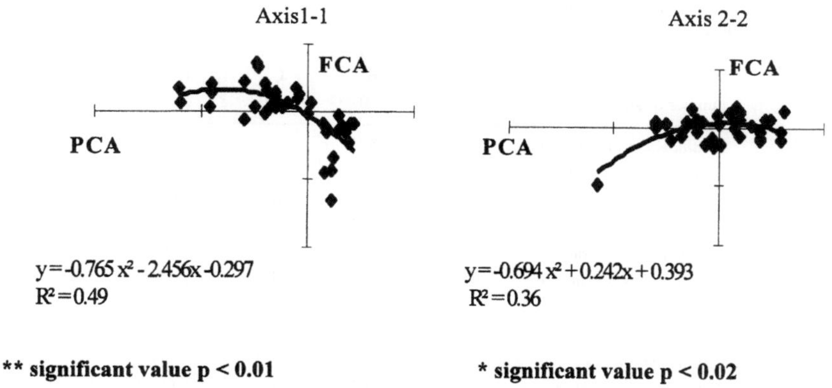

Figure 3. Relationships between PCA and FCA axes.

II and downstream to eutrophication (high nutrient levels) in groups IV and V. The group III sites were generally not sensitive to acidification or to eutrophication. They were buffered, weakly acidified and had moderate nutrient concentrations.

Relationships between macrophyte species and physico-chemical composition of the water

The FCA confirmed the ecological gradient suggested by the PCA. Aquatic macrophytes were distributed along an upstream to downstream zonation, which was characterized by an increase in eutrophication and a decrease in acidification of the water in the streams.

Relationships between the chemical data and the flora were tested (Figure 3). A significant negative correlation between axis 1 of the PCA and of the FCA (Student test $R^2 = -0.49$, $p < 0.01$) and between axis 2 of the PCA and of the FCA (Student test $R^2 = -0.36$, $p < 0.02$) was established.

Discussion

The A plant community corresponded to chemical group I, II and to a lesser extent to group III and IV. This heterogeneity reflected the variability of the physico-chemical conditions in which *Potamogeton polygonifolius* can be developed and the variability of the morphometric conditions (Table 2). Chemical group I (site 1), which is strongly acidified (pH = 4.4) and has a high aluminium load ([Al]= 510 μg/l), was mapped as A for vegetation composition (no plants apart from *Scapania undulata*). *Scapania undulata,* the only species present in the most acidified upstream sites (sites 1, 2, 5, 6, 7) has an exceptionally high tolerance to acidity and toxic metal levels in North America (Stephenson et al., 1995), in Europe (Tremp & Kohler, 1995) and Japan (Satake et al., 1989). The natural acidified sites near peat bogs (group II: sites 3, 4, 8, 9) were colonized by *Potamogeton polygonifolius* and to a lesser extent by *Sphagnum* sp. The upstream sites belonging to group III (10, 13, 16, 17, 18) and to group IV (site 23) were mapped as A because of the presence of *Potamogeton polygonifolius* and the absence of *Callitriche* species and *Ranunculus peltatus.* These non acidified upstream sites could be considered as the reference A aquatic plant community. The relative rarity of bryophyte species could be explained by the presence of sand in the stream bed in the upper catchment area and the lack of rocks (Stephenson et al., 1995).

The B plant community corresponded generally to group III. Bryophytes were found, when the habitat conditions allowed their establishment. Site 11, classified in group II and not in group III, was mapped as B because of the presence of *Ranunculus peltatus* and *Callitriche* species. Located downstream of a peat bog, it was a site sensitive to acidification, in which *Ranunculus peltatus* was present but not abundant. Tremp & Kohler (1991) have experimentally shown that *Ranunculus peltatus* is sensitive to acid pH. The absence of *Callitriche* species (*C. platycarpa, C. stagnalis, C. hamulata)* and *Ranunculus peltatus* from spring sites could be explained by a proton sensitivity or by a too low bicarbonate concentration (low ANC). ANC can be a limiting factor in a poorly buffered water for aquatic vegetation (Roelofs et al., 1984; Arts et al., 1990).

Table 2. Synoptic table of communities A to D based on frequency of species and physical-chemical characteristic

Plant community	A	B	C	D
number of sites	15	10	10	7
Potamogeton polygonifolius	IV(2)	V(1)	.	.
Batrachospermum sp.	I (+)	I(+)	.	.
Scapania undulata	III (+)	IV (+)	II (+)	I (+)
Glyceria fluitans	IV(2)	V(1)	V(+)	V(+)
Sparganium emersum	I(+)	III(1)	IV(1)	III(1)
Chiloscyphus polyanthos	I(+)	III (+)	II +)	III (+)
Rhynchostegium riparioides		I (+)	II(+)	III (+)
Fontinalis antipyretica		II (+)	III (1)	III (1)
Callitriche hamulata	.	III(1)	V(1)	V(2)
Callitriche platycarpa	.	III(2)	III(2)	II(1)
Callitriche stagnalis	.	IV(+)	IV(+)	III(+)
Lemna minor	.	II(+)	III (+)	IV(+)
Ranunculus peltatus	.	II(1)	II(2)	II(3)
Berula erecta	.	II(1)	I(1)	I(2)
Potamogeton crispus	.	.	I(1)	.
Myriophyllum alterniflorum	.	.	II(+)	.
Potamogeton alpinus	.	.	I(+)	.
Potamogeton variifolius	.	.	I(+)	.
Elodea canadensis	.	.	III(3)	I(1)
Oenanthe fluviatilis	.	.	I(2)	I(+)
Potamogeton berchtoldii	.	.	I(2)	I(+)
Elodea nuttallii	.	.	I(2)	III(3)
Vaucheria sp.	.	.	I(+)	I (1)
Amblystegium riparium	.	.	I(+)	IV(+)
Callitriche obtusangula	.	.	.	V(3)
Fissidens crassipes	.	.	.	IV (+)
pH	5.8 (0,5)	6.4 (0.2)	6.9 (0.2)	7.0 (0.2)
conductivity (μS/cm)	56 (6)	50 (6)	69 (10)	85 (16)
ANC (μeq/l)	82 (68)	150 (44)	323(90)	412 (139)
[N-NH_4^+] μg/l	43 (13)	49(14)	86 (35)	170 (109)
[P-PO_4^{3-}] μg/l	20 (26)	22 (11)	53 (19)	139 (46)
[N-NO_3^-] mg/l	0.4 (0.2)	0.3 (0.1)	0.4 (0.1)	0.8 (0.3)
Total aluminium (μg/l)	154 (109)	66 (33)	49 (12)	95 (35)

(V: 80–100%; IV: 60–80%; III: 40–60%; II: 20–40%; I: 0–20%). In brackets, the mean coefficient of abundance -dominance and standard deviation of chemical variables.

The C plant community was usually associated with group IV. The site 20 classified in group III, was mapped as C (absence of *Potamogeton polygonifolius*). The large development of *Callitriche* species could be attributed to the high phosphorus level ([P-PO_4^{3-}] = 100 μg/l). Phosphorus is the main limiting factor for aquatic plants (Lee, 1973). This increase in the phosphorus level was due to the domestic sewage of a village. This upstream site was weakly mineralized (conductivity = 64 μS/cm). This weak mineralisation explains the classification in chemical group III rather than in group IV. Rare and threatened species were essentially found in this C community, in which there are no limiting chemical factors (high phosphorus and bicarbonate).

The D plant community corresponded to group V. Sites 30 and 31, mapped as D, were classified in group IV although they were as rich in nutrients as the sites of group V. The regrouping of these sites in group IV can be explained in the same way as above. They contained high aluminium and nutrient loads. The high aluminium load in disturbed sites (sites 30, 31) was responsible for the increase in the mean aluminium value of the D community (Table 2). The high level of nutrients can be explained by the installation of a fish farm and domestic effluents upstream of these sites. The role of fish farms and domestic sewage in the enrichment of water nutrient content is clearly established in the literature (Trémolières et al., 1994). The bryophytes *Fissidens crassipes* and *Amblystegium riparium,* found in nutrient rich waters, express the deterioration of the water quality (Haury & Muller, 1991). Species such as *Callitriche obtusangula* and *Amblystegium riparium* are found in eutrophic waters in weakly mineralized waters, whereas they appear in mesotrophic waters in mineralized waters (Carbiener et al., 1990; Robach et al., 1996). Two hypotheses could be proposed to explain the difference of ecological niche for *Amblystegium riparium* and *Callitriche obtusangula* in these two types of environment: a more efficient absorption of phosphorus by plants in alkaline waters, or the requirement of a higher phosphorus concentration in calcium poor waters (Robach et al., 1996).

The diversity and the abundance of aquatic macrophytes rose along an increasing gradient of nutrient and mineral loading and a decreasing sequence of proton and aluminium load. A scale based on the succession of macrophyte communities, A–D, along an upstream to downstream zonation, reflects the degree of increasing eutrophication and of decreasing acidification in streams. The upstream A and B plant communities were developed in acidified waters, whereas the downstream C and D plant communities were found in non acidified waters. The main difference between the A and B communities was the degree of acidification (acidified for A, weakly acidified for B), whereas between the C and D communities it was the nutrient level (mesotrophic for C, eutrophic for D).

For 65% of the sites, the classification according to the type of plant community corresponded to the chemical composition of the water. The relationships between the chemical data and the flora were significant but weak (Figure 3). The few floristic sites misclassified in their chemical group corresponded to highly disturbed sites. Acidification and eutrophication were responsible for a disruption in the functioning of the ecosystem. Chemical analyses only represent one point in time and space. Aquatic macrophyte surveys, on the other hand, reflect the inherent temporal variability. Macrophytes also have the advantage that they exist throughout the season and respond to disturbances. The spatial integration of eutrophication and acidification processes at the basin scale as well as the links between these two types of human disturbances is reflected by the composition and the structure of aquatic macrophyte communities. One example was given by the analysis of the sites 30 and 31. These upstream weakly acidified and weakly buffered sites with high aluminium load contained high nutrient levels (sites 30, 31: [$P\text{-}PO_4^{3-}$]: 130 μg/l; [$N\text{-}NH_4^+$]: 115 μg/l). They were subjected to acidification (relatively high aluminium load) and to eutrophication. The impacts of these perturbations were superimposed. This situation lead to a decrease in the aquatic vascular plant richness: the absence of oligo-acidophilous species such as *Potamogeton polygonifolius* and of meso-neutrophilous species such as *Elodea* sp., *Myriophyllum alterniflorum, Ranunculus peltatus.* Tolerant pollutant species such as *Callitriche obtusangula, C. platycarpa* and *Amblystegium riparium* were abundant in these eutrophicated sites. The composition of the aquatic communities became impoverished. However, the biomass was important.

Conclusions

The Northern Vosges streams are subjected to acidification upstream and to eutrophication downstream. A scale of bio-indication of eutrophication and acidification by aquatic macrophyte communities was established. Upstream, bryophytes (*Scapania undulata,* or *Sphagnum* sp.) and *Potamogeton polygonifolius* were found in oligotrophic, acidified streams, whereas downstream, *Callitriche obtusangula,* filamentous algae such as *Vaucheria* sp. and bryophyte species such as *Amblystegium riparium* and *Fissidens crassipes* were found in the most eutrophic sites.

Uncertainties remain, particularly in our understanding of the responses of aquatic plants to eutrophication and acidification. The biological impact of interactions between acidification and eutrophication on aquatic plants should also be researched. The reconstruction of changes in communities should identify the initial floristic communities present in these weakly mineralized streams. A comparison with others weakly mineralized areas such as Limousin or such as Brittany should be done to discuss the relative effects of acidification and eutrophication. Only by examining a complex problem, such as acidification and eutrophication, using many different approaches and perspectives can we develop a comprehensive view of the responses of aquatic macrophyte communities to stress.

References

Admiraal., W., G. van der Velde, H. Smit & W. G. Cazemier, 1993. The rivers Rhine and Meuse in The Netherlands: present state and signs of ecological recovery. Hydrobiologia 265: 97–128.

Arts, G. H. P., J. G. M. Roelofs & M. J. H. De Lyon, 1990. Differential tolerance among soft-water macrophyte species to acidification. Can. J. Bot. 68: 2127–2134.

Braun-Blanquet, J., 1964. Pflanzensoziologie. Grundzüge der Vegetationskunde. Springer, Wien, 865 pp.

Carbiener, R., M. Trémolières, J.-L. Mercier & A. Ortscheidt, 1990. Aquatic macrophyte communities as bioindicators of eutrophication in calcareous oligosaprobe stream waters (Upper Rhine plain, Alsace). Vegetatio 86: 71–88.

Dambrine, E., M., Bonneau & G. Nourrisson, 1994. Apports de pollution et de nutriments aux peuplements forestiers par l'atmosphère: intensité, variations spatiales et historiques et conséquences sur la nutrition des forêts. Ann. Sci. Rés. Bios. Vosges du Nord 3: 7–24.

Farmer, A. M., 1990. The effect of lake acidification on aquatic macrophytes: a review. Envir. Pollut. 65: 219–240.

Grahn, O., 1977. Macrophyte succession in swedish lakes caused by deposition of airborne acid substances. Wat. Air Soil Pollut. 7: 295–305.

Guérold, F., D. Vein, G. Jacquemin & J-C. Pihan, 1995. The macroinvertebrate communities of streams draining a small granitic catchment exposed to acidic precipitations (Vosges Mountains, Northeastern France). Hydrobiologia. 300/301: 141–148.

Haury, J. & S. Muller, 1991. Variations écologiques et chorologiques de la végétation macrophytique des rivières acides du Massif armoricain et des Vosges du Nord (France). Rev. Sci. de l'Eau 4: 463–482.

Lambinon, J., E. De Langhe, L. Delvosalle, J. Duvigneaud & C Vanden Berghen, 1992. Nouvelle Flore de la Belgique, du Grand-Duché de Luxembourg, du Nord de la France et des Régions Voisines (Ptéridophytes et Spermaphytes). Patrimoine du Jardin botanique national de Belgique, 4th edn: 1092 pp.

Lee, G. F., 1973. Role of phosphorus in eutrophication and diffuse source control. Wat. Res. 7: 111–128.

Massabuau, J-C., B. Fritz & B. Burtin, 1987. Mise en évidence de ruisseaux acides (pH < 5) dans les Vosges. C.r. hebd. Séanc. Acad. Sci., Paris 305: 121–124.

Muller, S., 1990. Une séquence de groupements végétaux bioindicateurs d'eutrophisation croissante des cours d'eau faiblement minéralisés des Basses Vosges gréseuses du Nord. C.r. hebd. Séanc. Acad. Sci., Paris 310: 509–514.

Phillips, G. L., D. Eminson & B. Moss, 1978. A mechanism to account for macrophyte decline in progressively eutrophicated freshwaters. Aquat. Bot. 4: 103–126.

Probst, A., B. Fritz & D. Viville, 1995. Mid-trends in acid precipitation, streamwater chemistry and elements budgets in the Strengbach catchment (Vosges mountains, France). Wat. Air Soil Pollut. 79: 39–59.

Robach, F., G. Thiébaut, M. Trémolières & S. Muller, 1996. A reference system for continental running waters: plant communities as bioindicators of increasing eutrophication in alkaline and acidic waters in North East of France. Hydrobiologia 340: 67–76.

Roberts, D. A., R. Singer & C. W. Boylen, 1985. The submersed macrophyte communities of Adirondack lakes (New-York, U.S.A.) of varying degrees of acidity. Aquat. Bot. 21: 219–235.

Roelofs, J. G. M., J. A. A. R. Schuurkes & A. J. M. Smits, 1984. Impact of acidification and eutrophication on macrophytes communities in soft waters in the Netherlands- II. Experimental studies. Aquat. Bot 18: 389–411.

Satake, K., M. Nishikawa & K. Shibata, 1989. Distribution of aquatic bryophytes in relation to water chemistry of the acid river Akagawa, Japan. Arch. Hydrobiol. 116: 299–311.

Smith, A. J. E., 1992. The Liverworts Flora of Britain and Ireland. Cambridge University Press. 362 pp.

Stephenson, S. L., S. M. Studlar, C. J. Mc Quattie & P. J. Edwards, 1995. Effects of acidification on bryophyte communities in West Virginia Mountain Streams. J. envir. Qual. 24: 116–124.

Thiébaut, G. & S. Muller, 1998. Nouvelles données relatives à la séquence de bioindication de l'eutrophisation dans les cours d'eau faiblement minéralisés des Vosges du Nord. Acta bot. Gallica 142: 627–638.

Thiébaut, G., F. Guérold & S. Muller, 1996. Impact de l'acidification des eaux sur les macrophytes aquatiques dans les eaux faiblement minéralisées des Vosges du Nord- Premiers résultats. Acta bot. Gallica 142: 617–626.

Thiébaut, G. & S. Muller, 1998. The impact of eutrophication on aquatic macrophyte diversity in weakly mineralized streams in the Northern Vosges mountains (N-E, France), Biodivers Conserv. 7: 1051–1068.

Trémolières, M., R. Carbiener, A. Ortscheidt & J-P. Klein, 1994. Changes in aquatic vegetation in Rhine floodplain in Alsace in relation to disturbance. J. Veg. Sci. 5: 169–178.

Tremp, H. & A. Kohler, 1991. Die Auswirkungen von Säure und Aluminiumbelastung auf Submerse Macrophyten. Verhandlungen Gesellschaft für Ökologie 20: 569–573.

Tremp, H. & A. Kohler, 1995. The usefulness of macrophyte monitoring-systems, exemplified on eutrophication and acidification of running waters. Acta bot. Gallica 142: 541–550.

Hydrobiologia **410:** 25–37, 2000.
J. Garnier & J.-M. Mouchel (eds), Man and River Systems.

Use of oligochaete communities for assessment of ecotoxicological risk in fine sediment of rivers and canals of the Artois-Picardie water basin (France)

J. Prygiel[1,2], A. Rosso-Darmet[3], M. Lafont[4], C. Lesniak[1], A. Durbec[3] & B. Ouddane[5]
[1]*Agence de l'Eau Artois-Picardie, 200 rue Marceline, 59508 Douai Cedex, France*
[2]*Université des Sciences et Technologies de Lille, Laboratoire d'Ecologie Numérique, Bat. SN3, 59655 Villeneuve d'Ascq Cedex, France*
[3]*BURGEAP Region Centre-Est, 19 rue de la Villette, 69425 Lyon Cedex, France*
[4]*Cemagref, Division Biologie des Écosystèmes Aquatiques, 3 bis Quai Chauveau, C. P. 220, F-69336 Lyon Cedex 09, France*
[5]*Université des Sciences et Technologies de Lille, Laboratoire de Chimie Analytique Marine, Bat. C8, 59655 Villeneuve d'Ascq Cedex, France*

Key words: aquatic oligochaetes, sediment quality, micropollutants, biomonitoring, ecotoxicological risk

Abstract

This study had a purpose to validate different tools based on the oligochaetes for the evaluation of the biological quality of sediments within a surveillance network. Biological data (IOBS index, percentage of Tubificidae with and without hair setae, worm densities), physical data (granulometrics) and chemical data (Cd, Cu, Cr, Hg, Ni, Pb, Zn, PCBs, PAHs) have been studied with co-inertia analysis. These analyses show an opposition between IOBS index and metals, a correlation between metals and percentage of Tubificidae without hair setae, and an association between PAHs, low index values, high densities and high percentage of Tubificidae with hair setae. For 24 sites, biological and chemical data gave similar information. In 11 cases on 36 they are conflicting. The use of metal speciations, according to the method of Teissier et al. (1979), show that the organic matter content seems to have an important role by sorbing metals leading to lower availability. Macrophyte cover could also explain some conflicting results. Granulometrics does not play a determinating role in the present case. As a conclusion, a table allowing the classification of river sediments is proposed for monitoring assessment. This table takes into account IOBS index values, the worm densities and the percentage of Tubificidae without hair setae.

Introduction

Problems linked to the toxicity of freshwater sediments are the subject of increasing attention on the part of water quality managers (Literathy et al., 1994; Santiago et al., 1994; Seymore et al., 1994; Johansson et al., 1995; Lim & Kiu, 1995; Teixeira et al., 1997). In the case of the Artois-Picardie water basin, where the geographical relief is low and where canalized environments compose an important part of the hydrographic network, management of contaminated sediments becomes a crucial problem due to past and present industrial activities. For various reasons (navigation, risks linked to sediment contamination, flooding, eutrophication and river maintenance), numerous river cleaning operations are envisaged in this basin.

An assessment of *in situ* sediment quality must, therefore, be made in both chemical and biological terms. Chemical analyses have long been the preferred choice of managers, but even though they provide an estimation of the concentrations and fractions of pollutants present (Belzunce-Segarra et al., 1997; Gouws & Coetzee, 1997), they do not indicate the *in situ* toxic risk. Laboratory tests have been developed for this (Brouwer et al., 1990; Reynoldson et al., 1991; Traunspurger & Drews, 1996), but extrapolation of the results in the aquatic environment has proven difficult (Seaward, 1995). In addition, measurements of the contamination in the invertebrate macrofauna have also been proposed (Duzzin et al., 1988; Bervoets et al., 1997), as well as chironomid larval deformities (Dickman & Rygiel, 1996).

It is the ecotoxicologists themselves who have been most critical of their own tests (Slooff, 1983; Petersen & Petersen, 1988; Chapman, 1989) and who have recommended a return to the field. In the scope of a management policy of the sediments of their water basin, the managers must be capable of implementing *in situ* bioindication methods which are capable of assessing biological quality, measuring the specific impact of the contaminants, defining an ecological quality objective to be preserved or restored, and predicting the risks and consequences of the presence of toxic pollutants for the aquatic environment. These objectives are part of European efforts for protection of aquatic environments (EC, 1997) and help in public decision making processes.

In this context, we have tested a biological method based on the qualitative and quantitative composition of fine sediment oligochaete communities in the Artois-Picardie water basin. The oligochaetes have been studied as indicators of the biological quality of the sediments and toxic effect of metal and organic micropollutants in different French water basins (Lafont, 1989; Rosso et al., 1994; Rosso, 1995; Lafont et al., 1996b; Rosso-Darmet et al., 1997a, b). The purpose of this study is to validate different sediment bioindication tools developed using oligochaetes, for integration in the biomonitoring network of the Artois-Picardie water basin, and to compare these with methods which are already routinely used for monitoring water quality (diatom indices, Prygiel & Coste, 1993; Prygiel, 1996).

Study area, material and methods

Thirty-six sites which are part of the different rivers and canals of the Artois-Picardie water basin monitoring network were prospected (Figure 1). Field sampling of the fine sediments was performed at 32 sites in June 1997 and at 4 sites in August 1997. Sediment samplings intended for chemical analyses were carried out at the same time and at the same locations.

For sampling sediments and oligochaetes under a water depth of 1.5 m, we used a Surber type net (with 100 cm^2 opening and with a 0.160 mm mesh size) for sandy sediments, or a core sampler (Rofes & Savary type, 1981) with a 25 cm^2 opening or transparent plastic tubes (20 cm^2 opening). When the water depth exceeded 1.5 m, a Friedinger grab with 350 cm^2 opening was used. Only the sediment top layer of 10 cm thickness is collected.

Each sampling operation is composed of a minimum of 3 samples (one sampling made at approximately every 20 steps), grouped together and preserved with 5% formalin in the same collection jar.

In the laboratory, the sediment is sieved through a 0.500 mm mesh-size sieve. The sieving residues are poured into a gridded sub-sampling tray. One hundred and thirty oligochaete specimens were extracted with forceps using a binocular magnifier from the different boxes sampled at random in the tray, using a pipette with 10 mm opening. The worm specimens are mounted between slide and coverglass in a mounting medium composed of a mixture of equal parts of glycerin and pure lactic acid. The preparations are heated on a hot plate at 60 °C for approximately 30 min. Species determination is made under microscope and the abundances of worms are converted to numbers per 0.1 m^2.

The biological quality of the sites was evaluated by examination of the IOBS index values in the fine sediment (Oligochaete Index of Sediment Bioindication; Rosso et al., 1994; Rosso, 1995). This index has the following form:

$$\text{IOBS} = 10ST^{-1},$$

where S is the total number of species per sample, and T is the percentage in the total number of oligochaetes of the dominant Tubificidae group, with or without hair setae, for all adult and immature worms combined. The index varies from 0 (maximum toxicity) to 3 or more (good to high biological quality). If for example, 10 species are present and if the proportions of Tubificidae with and without hair setae in the total number of oligochaetes are, respectively, 30% and 50%, the corresponding index IOBS value is: 100/50 = 2.

This analysis is completed by examination of the percentages of Tubificidae without hair setae, considered as describing a micropollutant effect (metals and PCBs) for values > 60% (Rosso et al., 1994), and by that of the densities of the oligochaete communities.

Metals of sediments collected for chemical analyses were analyzed by Varian Liberty 2 Inductively Coupled Plasma-Atomic Emission Spectrometry (ICP-AES). The speciation of heavy metals in sediments was examined using a sequential extraction method in order to obtain the concentration

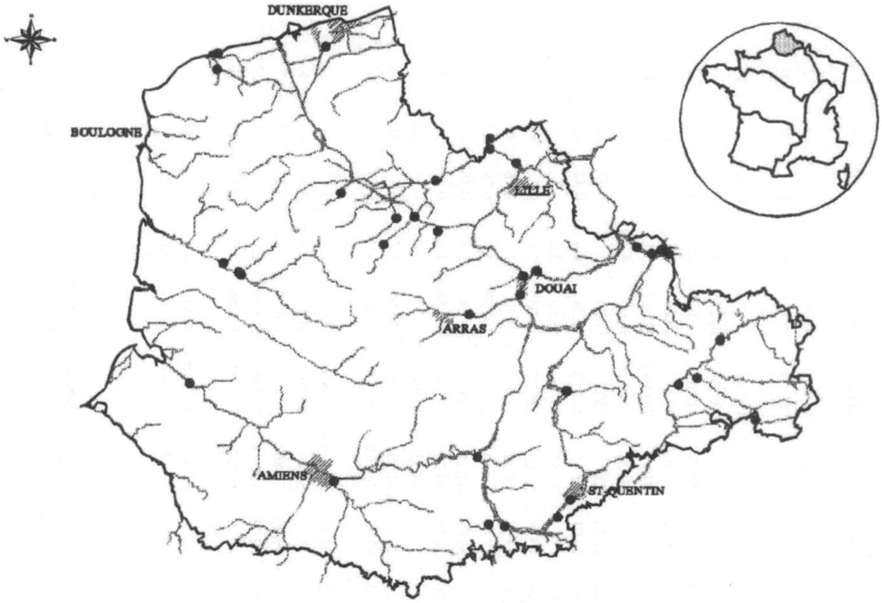

Figure 1. Location of the 36 sampling sites in the Artois-Picardie basin.

of metals bound in cation exchange positions, carbonates, Fe-Mn oxides, organic matter and sulfides, and crystal lattice (Tessier et al., 1979). Exchange and carbonate fractions have been gathered as carbonate fraction. PAHs (Polyaromatic Hydrocarbons) underwent chromatographic analysis in liquid phase with fluorimetric detection. They include fluoranthene, benzo(b)-benzo(k)-fluoranthenes, benzo(a)pyrene, benzo(ghi)perylene and indene(1,2,3-cd)pyrene. PCBs underwent chromatographic analysis in gaseous phase with electron capture detector. They include 028 trichlorobiphenyle, 052 tetrachlorobiphenyle, 101 and 118 pentachlorobiphenyles, 138, 153 and 180 hexachlorobiphenyles. The granulometrical spectrums were made by CILAF HR 850-B type laser granulometry and the total organic carbon content by Leco CHNS-932 type elementary analyzer.

The chemical data were processed by standardized Principal Component Analysis after data transformation into $\log_{10}(x + 1)$. A co-inertia analysis was used to describe the common structure of the physico-chemical and biological data. This made it possible to pair the chemical and biological tables (Chessel & Mercier, 1993; Dolédec & Chessel, 1994). The table resulting from this pairing of data is formed by co-variations between the biological variables and the chemical variables. In co-inertia analysis, the projected inertia is partially maximized because we wish to maximize the co-variation of the coordinates of the projections. The A.D.E. data analysis program (Chessel & Dolédec, 1993) was used for the two analyses.

Results

Sediment contaminations

The micropollutant analyses show a wide diversity of pollutant substances and contamination levels (Table 1). To try and demonstrate the tendencies in this set of physico-chemical data, a standardized principal component analysis was performed on the 36 samples and 9 active variables: Cd, Cr, Cu, Hg, Ni, Pb, Zn, PCBs, PAHs. The total organic carbon as well as the median value of grain-size were projected as supplementary variables.

Four principal components explain 89.5% of the inertia (Figure 2). The first (65.6% of the total variance) is positively correlated with the metal concentrations (in decreasing order of contribution: Cu, Pb, Zn, Cr, Hg, Ni and Cd) and with PCBs. The second

Table 1. Biological and chemical characteristics of sediments at the 36 sites; RNB/RC: site numbers in the Artois-Picardie basin; IOBS: IOBS index; NSPS: number of species; NC: not calculated; EFBR: worm abundances per 0.1 m^2; TUSP: percentages of Tubificidae without hair setae; Cd, Cr, Cu, Ni, Pb, Zn: $\mu g\ g^{-1}$ dry substance; Hg, PAHs, PCBs: $ng\ g^{-1}$ dry substance; grain-size: μm median values; total organic carbon (TOC): percent. Reference values for trace metals in the sediments of the Artois-Picardie basin (same units as above): Cd: 1; Cr: 30; Cu: 15; Ni: 12.5; Pb: 40; Zn: 100; Hg: 200.

Water courses	Sites	RNB/RC	IOBS	NSPS	EFBR	TUSP	Cd	Cr	Cu	Ni	Pb	Zn	Hg	C org.	PAHs	PCBs	Median
Canal de Bourbourg	Grande-Synthe	109 000	3.4	13	2781	38.5	0.36	32.76	10.20	7.36	42.08	111.22	16.90	1.2	0.052	<1	106
Torrent d'Esnes	Crèvecoeur	022 000	2.6	12	6417	46.9	0.03	32.38	19.19	11.77	37.79	80.46	40.00	0.8	3.6	<1	32
Scarpe	Courchelettes	038 000	2.3	10	2623	43.1	2.92	51.11	72.80	21.47	205.31	585.59	189.05	3.5	1.5	20	20
Jard	Hergnies	033 000	1.8	9	4963	36.9	1.43	58.60	45.35	19.36	47.42	179.69	53.03	3.5	0.21	7	32
Canche	Aubin-Saint-Vaast	094 000	1.7	9	12 594	52.3	0.10	25.84	24.29	7.17	16.76	73.60	50.22	1.2	0.41	<1	40
Escaut	Crèvecoeur	010 000	1.6	10	16 254	36.9	0.07	40.08	21.10	11.38	27.58	113.51	19.80	1.5	1.9	<1	32
Planquette	Contes	097 300	1.6	11	7709	12.3	2.23	27.89	31.61	18.41	51.36	47.89	30.26	1.3	-	<1	32
Somme	Offoy	119 000	1.4	10	7187	71.5	1.38	91.52	75.81	35.12	87.95	2415.81	273.02	5.1	1	42	32
Avre	Cagny	135 000	1.4	10	8330	73.8	0.23	22.26	25.79	9.43	30.24	134.51	54.90	4.7	1.3	14	40
Somme	Blaches	119 500	1.3	9	116	70.2	0.24	14.78	17.48	11.14	22.54	480.77	17.86	3.1	0.47	20	40
Sambre	Pont/Sambre	002 000	1.3	11	2310	84.6	0.19	54.51	31.89	20.01	35.41	142.75	41.38	1.4	0.32	<1	20
Lawe	Bruay-la Buissière	171 000	1.3	7	15 506	55.4	0.13	44.18	26.63	15.12	33.55	153.51	63.69	2.0	1.3	<1	32
Somme	Cambron	130 000	1.2	9	4110	74.6	0.51	31.72	23.48	9.51	35.65	126.37	73.06	2.6	0.22	<1	50
Rivièrette	Landrecies	009 100	1.2	7	1548	33.1	0.12	44.87	14.86	13.36	23.27	78.27	10.70	1.7	0.76	<1	32
Lawe	Essars	072 000	1.2	6	9445	50	0.48	52.31	58.91	17.85	69.59	282.67	216.03	3.5	0.78	<1	32
Courant de Bernissart	Condé-sur-l'Escaut	001 148	1.1	8	3947	71.5	0.22	71.25	55.31	33.83	41.21	165.76	46.24	8.3	0.36	<1	40
Helpe Mineure	Maroilles	006 000	1.1	8	2503	70	0.84	76.70	27.26	20.75	34.20	223.11	115.58	1.6	0.4	3	20
Créquoise	Beaurainville	097 500	1.1	8	3312	23.8	2.08	29.53	32.09	17.26	43.33	46.39	33.33	1.6	-	<1	32
Clarence	Chocques	069 000	1.0	5	6200	52.3	1.03	39.90	46.84	18.18	51.93	311.99	153.06	3.2	3.7	18	32
Scarpe	Fampoux	036 000	1.0	8	7412	77.7	0.71	59.59	75.33	32.82	166.18	369.29	58.18	4.3	1.1	6	63
Canal de al Deûle	Flers-en-Escrebieux	076 000	0.8	7	967	87.7	5.51	107.87	102.23	22.49	1291.63	1148.74	586.67	5.3	1.4	6	2 0
Canal du Nord	Rouy-le-Petit	131 000	0.8	7	388	93.1	0.21	33.31	34.97	13.71	59.49	196.91	61.86	2.9	1.5	4	40
Helpe Mineure	Rocquigny	005 000	0.7	5	23 084	21.5	0.83	77.36	59.51	19.29	77.86	474.43	113.36	3.1	3.1	17	20
Lys	Warneton	058 000	0.7	4	20 130	57.7	7.10	86.12	82.89	32.03	135.11	672.68	213.99	2.7	1.7	31	32
Canal de Mons	Saint Aybert	030 500	0.7	6	279	92.3	2.60	177.16	547.93	40.09	639.09	1979.40	142.86	5.9	2.1	15	20
Laquette	Witternesse	060 000	0.7	5	5917	75.4	1.82	34.69	21.47	21.31	9.91	95.45	26.67	1.4	-	<1	20
Canal de Marck	Calais	114 600	0.7	3	462	46.2	0.42	41.39	31.21	14.55	67.40	1925.08	20.41	4.7	4.3	<1	32
Lys	Estaires	055 000	0.6	3	25 676	47.7	0.80	42.50	32.63	40.27	37.94	205.66	286.64	5.6	0.29	<1	32
Jard	Condé-sur-l'Escaut	017 100	0.6	5	3056	82.3	0.20	60.45	31.81	21.81	37.07	96.39	233.18	2.4	0.48	<1	20
Somme	Séracourt	117 000	0.4	4	2360	1.5	4.35	259.92	334.61	108.60	278.12	2735.04	195.35	12.6	4.9	155	40
Scarpe	Râches	039 000	0.4	4	228	98.5	274.57	87.51	305.83	36.36	1287.53	5680.27	1800.00	7.6	1.6	9	40
Canal de Calais	Coulogne	114 000	0.4	4	922	93.1	1.30	66.65	35.41	22.74	144.01	443.19	173.55	4.3	1	<1	20
Canal de la Deûle	Quesnoy/Deûle	-	0.3	3	198	94.6	26.37	146.48	230.43	26.52	359.75	2359 .51	431.19	5.5	-	61	32
Somme	Gauchy	116 500	0.2	2	14 602	96.9	1.81	95.52	139.05	45.12	197.62	859.65	208.33	5.4	4.3	113	20
Canal de la Deûle	Wambrechies	081 000	0.1	1	244	100	9.43	84.19	83.63	18.85	148.90	680.58	418.08	0.8	0.64	13	32
Surgeon	Cuinchy	064 000	NC	1	14	-	0.78	83.99	88.71	25.39	104.29	410.61	50.00	5.2	2	7	50

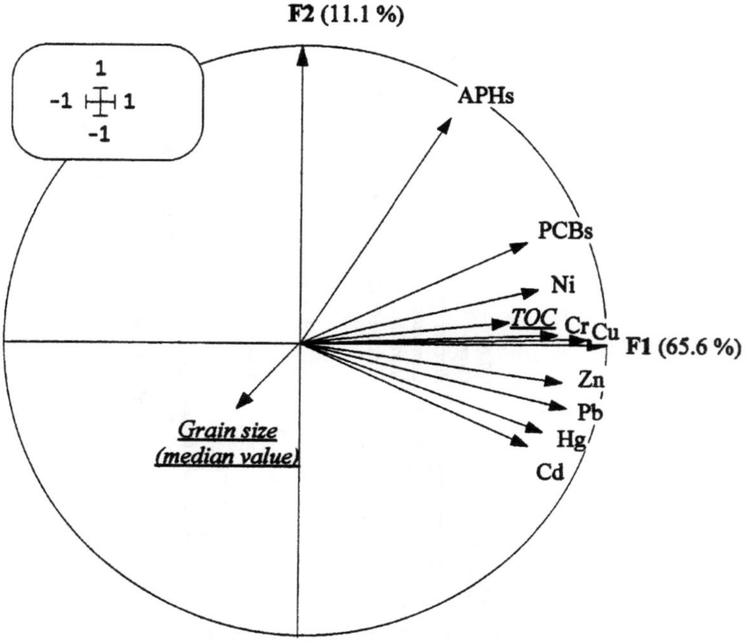

Figure 2. Correlation circle of the Principal Component Analysis: correlations of the first two factors with active and supplementary variables (italics); TOC – total organic carbon.

component (11.1% of the total variance) is positively correlated with the PAHs and the PCBs, and negatively correlated with the cadmium content. The organic load of the sediments is significantly correlated with the metal contamination. No correlation is observed between the grain-size and the F1 and F2 axes. Contamination by PAHs appears to be more independent than the other forms of contamination.

Oligochaete communities

Classification of the biological quality of the sediments was evaluated by a graphic table containing the IOBS index values and the percentages of Tubificidae without hair setae (Figure 3). Of the 36 sites studied, only one presented a fair to good biological quality (the Canal de Bourbourg at Grande-Synthe) and two showed a fair quality (Torrent d'Esnes at Crèvecoeur and Scarpe at Courchelettes). The biological quality of the sediments remains poor for 16 sites and bad for 16 other sites.

The most pronounced toxic effect is observed at Cuinchy where the sediment is found to be almost totally devoid of oligochaetes (the IOBS index can not be calculated). The biological quality of the sediments is also greatly degraded on the canal of the Deûle at Wambrechies and at Quesnoy-sur-Deûle, on the Somme at Gauchy and at Séraucourt, on the Scarpe at Râches and on the canal of Calais at Coulogne (IOBS < 0.5). The effect of the micropollutants on the worm communities is seen at 18 sites (the percentages of Tubificidae without hair setae $> 60\%$, Figure 3).

Relationships between sediment contamination and oligochaete community structure

We have tried to verify here the significance, in terms of sediment contamination, of the predictions obtained from the examination of the oligochaete communities, especially the 'micropollutant' effect prediction. A co-inertia analysis was used in selecting 4 biological variables (IOBS index, percentages of Tubificidae with and without hair setae, and densities) and 9 chemical variables (Cd, Cr, Cu, Hg, Ni, Pb, Zn, PAHs and PCBs concentrations). The inertia projected on the first two axes represents 98.3% of the total inertia (Figure 4), with 87.4% on the F1 axis and 10.9% on the F2 axis. The essential part of the co-structure is thus associated with the first analysis axis. The chemical variables structuring the F1 axis have negative coordinates (in

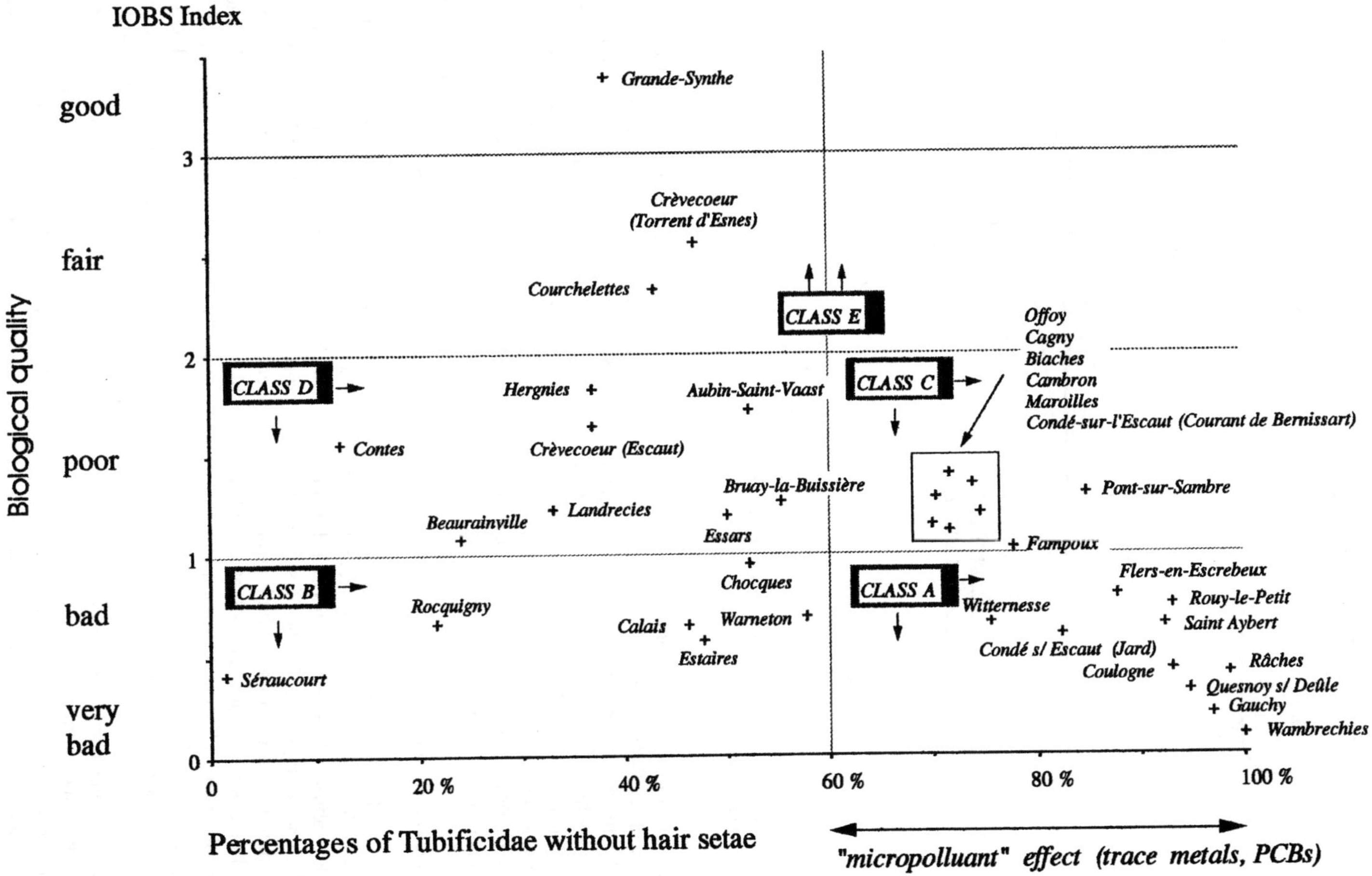

Figure 3. Graph of biological quality and micropollutant incidence assessments; the five classes (A, B, C, D, E) were defined according to a decreasing gradient of suspected micropollutant effects (metals and PCBs) and from the experience acquired in different water basins (Rosso, 1995; Rosso-Darmet et al., 1997b).

Table 2. Analysis of 3 types of discrepancy between biological and chemical diagnoses; F1, F2, F3, F4: metal fractions ($\mu g\ g^{-1}$) according to Tessier et al. (1979); F1: carbonate; F2: oxyde; F3: organic; F4: residual.

Séraucourt		F1	F2	F3	F4		
	Cd	0	3.9	1.2	0	PCBs	155
	Cr	11.3	51.1	183.9	40.4	APHs	4.9
	Cu	0.1	5.9	337.5	25.9	C%	12.6
	Ni	33.6	32.7	40.2	13	IOBS	0.4
	Pb	61	109.5	108	4	TUSP	1.5
	Zn	607.9	1883.5	572.7	58.6	densities	2360
Couchelettes							
	Cd	0.9	1.8	0.5	0	PCBs	20
	Cr	1.7	5.2	8.6	36.5	APHs	1.5
	Cu	0.3	7.4	55.4	13.9	C%	3.5
	Ni	2.3	4.5	4.2	11.5	IOBS	2.3
	Pb	83.4	61.4	51.1	11	TUSP	43.1
	Zn	177.7	254.7	138.2	52.3	densities	262.0
Rouy-le-Petit							
	Cd	0.1	0.1	0.1	0	PCBs	4
	Cr	0.7	2.1	4.6	27.2	APHs	1.5
	Cu	1.4	2.6	22.3	7.3	C%	2.9
	Ni	1.9	2.1	2	0.2	IOBS	0.8
	Pb	14.6	10.4	17.7	20	TUSP	93.1
	Zn	66.5	72	49.5	24.5	densities	388

decreasing order of contribution: Zn, Hg, Pb, Cu, Cd and Cr) (Figure 4a). The F2 axis is essentially structured by the PAHs and nickel. The PCBs have very little effect on the first factorial plane.

The biological variables which structure the F1 axis are mainly the IOBS index (positive coordinates) and the percentages of Tubificidae without hair setae (negative coordinates) (Figure 4a). The F2 axis principally separates the IOBS index from the percentage of Tubificidae with hair setae and to the densities. Simultaneous reading on the first factorial plan of the chemical and biological variables shows a discrimination between the IOBS index and contamination by heavy metals, a correlation between the heavy metals and the percentages of Tubificidae without hair setae, an association between the PAHs, the low IOBS index values, the high worm densities and the communities dominated by Tubificidae with hair setae.

The biological and chemical views of the site positions were obtained from standardized coordinates of the projections on the co-inertia axes (Figures 4b, c). The length of the arrow illustrates the amplitude of the differences between the two tables. This amplitude is seen as being more significant on the F1 axis (87.4% of the total inertia) than on the F2 factor (10.9% of the total inertia). On the F1 axis, the sites are divided into two large, very distinct groups: 24 have close biological and chemical coordinates (within less than 1 factorial coordinate amplitude unit) indicating that the biological and chemical results give similar indications with respect to the level of contamination (Figure 4b). On the contrary, eleven sites have chemical and biological coordinates which are farther apart (more than 1 amplitude unit separating them) (Figure 4c) and offer results which can be considered as being relatively conflicting. These differences are found to be of three principal types (Table 2 and see 'Discussion'):

1. a toxic effect is predicted, but the micropollutant effect (percentage of Tubificidae without hair setae $> 60\%$) is not shown despite a strong sediment contamination (the case of Séraucourt);
2. the biological results do not predict the toxic effect even though the micropollutants are relatively abundant (the case of Courchelettes);
3. the biological results predict a toxic effect even though the micropollutants are not very abundant (the case of Rouy-le-Petit).

Discussion, conclusion

This study is based on 36 sites in the Artois-Picardie basin and had for purpose to validate different tools based on the oligochaetes for the evaluation of the biological quality of sediments within a surveillance network. Biological data (IOBS index, percentage of Tubificidae with and without hair setae, worm densities), physical data (granulometrics) and chemical data (metals, PCBs, PAHs) have been studied with co-inertia analysis. These analyses show an opposition between IOBS index and the metals, a correlation between metals and percentage of Tubificidae without hair setae, and an association between PAHs, low index values, high densities and high percentage of Tubificidae with hair setae. Granulometrics does not seem to play a determinating role in the present case.

One of the main problems for this type of study comes from the fact that managers need rapid responses, while researchers need time to establish their diagnosis and verify their hypotheses (Lafont et al., 1996a). Any response in terms of environmental management involves compromises. For example, the time factor is not taken into consideration here (with only a

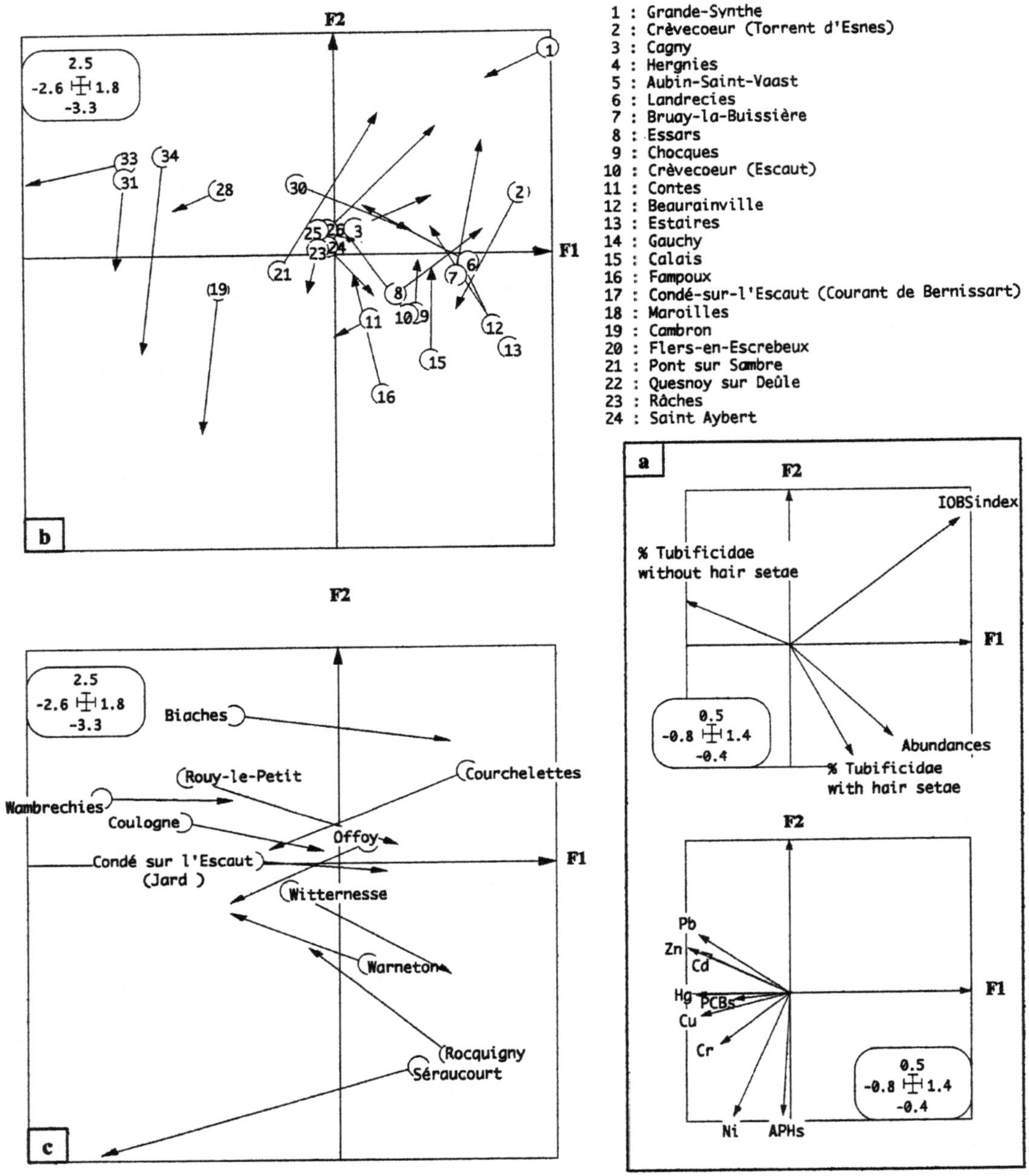

Figure 4. Co-Inertia Analysis; (a) plots of chemical and biological variables on the first plane; (b, c) plots of sites on the first plane; (b) sites where chemical and biological variables are strongly related; (c) sites where the relationships are lower; the labels of sites indicate their positions related to biology, and the end of the arrow their position related to chemical analyses.

single occasion of samplings), not making it possible to deepen the understanding of the complex mechanisms and phenomena existing in the field. Two parallel and complementary objectives exist:

1. a short term objective of predictive and operational management of the evolution of sediments in the water basin;

2. a more long term objective of research to try and understand the relationships observed in the field.

In the operational management context of this study, all of the tendencies recorded in this study have been illustrated by a predictive diagram of the biological quality of the sediments and the incidence of toxic pollutants from observation of the oligochaete communities (Figure 5). This diagram is also supported by the results of similar works carried out in other French water basins (Lafont, 1989; Rosso et al., 1994; Rosso, 1995; Lafont et al., 1996b; Rosso-Darmet et al., 1997b).

When the value of the IOBS index is near 3, the environment may still be stressed (fair water quality and/or physical disturbances, etc.), but metals or organic pollutants are non-existent or not very abundant at all.

In terms of ecological quality objectives, the canal of Bourbourg at Grande-Synthe constitutes the best station encountered and could serve as a model of ecological quality to be preserved or restored. This quality objective must however be considered as a temporary one in this hydrological basin as IOBS index values significantly higher than 3 have been encountered in reference sediments of various water basins (IOBS = 9 in the Rhin-Meuse water basin, 7 in the Adour-Garonne basin and in the Rhône basin).

For IOBS values situated between 2 and 3, the environment must not be considered *a priori* as being free of pollution, micropollutants can be found in a transitional manner in a biologically unavailable form.

IOBS index values < 1 describe a bad to very bad biological quality, whatever the incriminated pollutants may be. IOBS value $= 1$ is here confirmed as a threshold of the incidence of industrial and/or toxic wastes (Lafont, 1989; Rosso et al., 1994; Rosso, 1995; Lafont et al., 1996a; Rosso-Darmet et al., 1997b). This threshold value can be associated too with a very marked state of 'organic pollution' and/or a 'PAH effect', with, generally, the dominance of Tubificidae communities with hair setae and the existence of considerable numbers of oligochaetes (sometimes $> 2.10^4$ $(0.1\ m^2)^{-1}$.

In ultimate cases of pollution (metals + PCBs + PAHs + organic pollution), as on the river Somme at Séraucourt, Tubificidae with hair setae are found to be the predominant group.

The necessity of developing research in ecotoxicology has recourse, in particular, to the examination of the three types of differences shown in the set of data (Table 2).

The Séraucourt site on the Somme River shows a very strong contamination by both metals and by organic micropollutants. The IOBS index is low (IOBS = 0.4) and a marked toxic effect has been revealed. Nevertheless, the percentage of Tubificidae without hair setae is very low (1.5%, Table 2) and it is the Tubificidae with hair setae which largely predominate, not making it possible to predict a 'micropollutant effect'. Examination of the different metal fractions shows that the main part of the metals present are found to be linked to organic matter (Cr, Cu, Ni, and to a smaller degree Pb). Only two metals (Cd and Zn) are found in high proportions in the oxide fraction. This distribution corresponds to the concentrations in organic matter, which reach their highest value in the data set (12.6%).

In the same way as the PAHs concentrations, the metal concentrations are relatively high on the Scarpe at Courchelettes. However, a toxic effect is not predicted by the oligochaetes. The IOBS index reaches a value of 2.3, high densities of oligochaetes are recorded and the percentage of Tubificidae without hair setae remains $< 60\%$. The study of the distribution of the metals in the different fractions does not give an explanation. The metals are distributed in all 4 fractions: Cr and Ni are essentially found in the residual phase while Pb, as well as a large part of Cd and Zn, are found in the oxide and exchangeable phases. It should be noted that this station is located immediately upstream from a lock provided with an automatic spillway whose functioning depends on the water level of the canal, and under a large cover of aquatic macrophytes. The renewal of the well oxygenated section of water covering the sediments is therefore frequent and a purifying effect due to the aquatic macrophytes is suggested.

In the case of Rouy-le-Petit on the Canal du Nord, the opposite situation is observed (Table 2). The IOBS index is low (IOBS = 0.8), as are the densities of oligochaetes, and the percentage of Tubificidae without hair setae is high (93.1%). A pronounced toxic effect due to the metal micropollutants and to the PCBs is therefore predicted by the oligochaete communities. However, with the metals essentially being found in the residual and organic matter fractions, we cannot conclude that a larger bioavailability exists, but per-

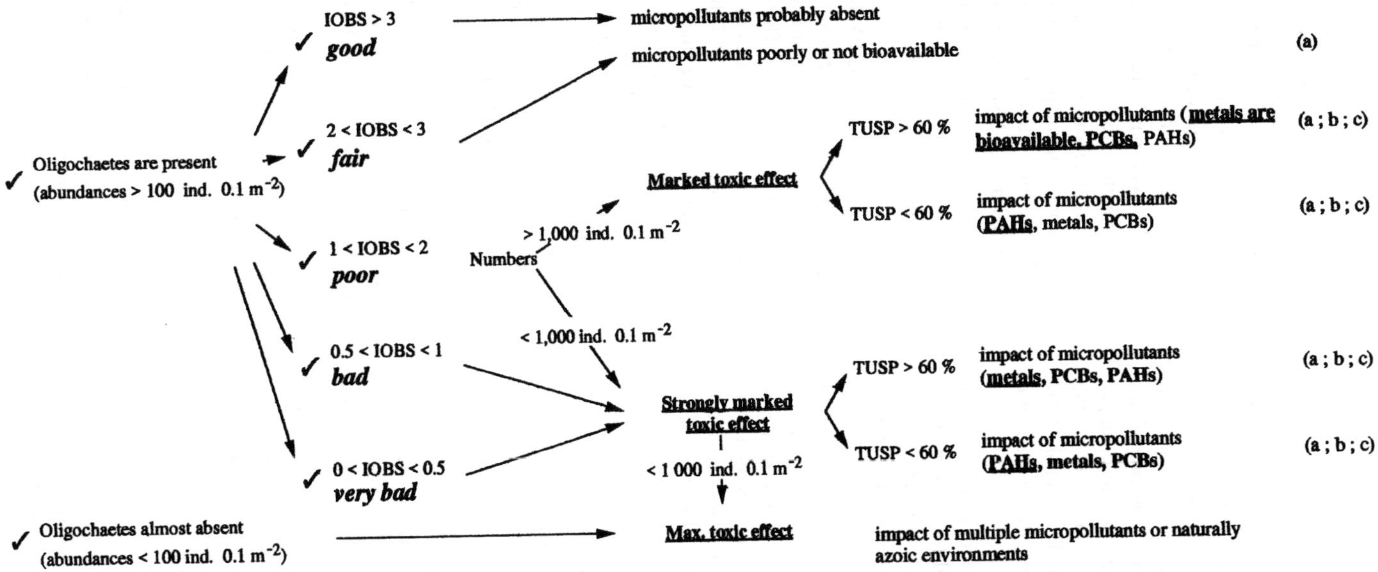

a) Abundances > 10,000 ind. 0.1 m^{-2}. Polluting effect of organic matter ; this effect can come on top of other types of contaminations ;
b) If there is a proliferation of a species not belonging to the Tubificidae, the suspected cause of deterioration is identified via examination of the Tubificidae ;
c) If metals, PCBs and PAHs are absent, non-analyzed micropollutants , high organic pollutions, physical disturbances and chlorides are the suspects.

Figure 5. Use of oligochaete criteria for predicting the incidence of pollutants, including micropollutants, in the fine sediments *in situ*.

haps that other pollutants which are not analyzed here are also present.

These apparent contradictions between the pollutants and the biocenoses of the oligochaetes emphasize the extreme complexity of the phenomena which exist in the field. For example, the ecotoxicological significance of a metal is not only determined by its concentration in the environment, but also by the form in which it is found, and the bioavailablity of the toxic elements depends on multiple factors (Calmano et al., 1993; Warren & Zimmerman, 1994). The appearance of bioavailable forms of pollutants obeys complex, and still badly understood, biological and chemical interrelationships, where the pH, the redox potential, the sorbents and the flow of pollutant matter, play an important role (Campbell & Stokes, 1985; Tessier & Campbell, 1987; Wang, 1987; Gunn et al., 1989; Campbell, 1995).

It should be added that the relationships between pollutants also play a role as well. For example, toxic synergism relationships exist between Cd and Zn (Le Du et al., 1990; Huebert & Shay, 1992), as well as antagonistic relationships between Ni and Cd and between Ni and Zn (Le Du et al., 1990).

Most particularly, nothing is known about the seasonal dynamics of the appearance of the bioavailable form of a pollutant and the exposure mechanisms of organisms to the pollutants. It is possible, for example, that the appearance of a toxic effect without associated micropollutants is the result of earlier or relatively recent bioavailable contaminations, with the absence of pollutants in the sediments being only temporary.

Furthermore, organic matter is too often underestimated as a pollutant substance. Numerous products coming from its degradation are toxic (NH_3, NO_2, phenols, etc.) and prolonged deficits in dissolved O_2 can have harmful effects. For example, the Canal du Nord at Rouy-le-Petit and the Somme River at Séraucourt present high mean concentrations of ammonium chlorides in the water column (respectively, 1.2 and 4.2 mg l^{-1}, Agence de l'Eau Artois-Picardie, 1996) and prolonged deficits of dissolved O_2 occur at these sites. The organic matter therefore has a beneficial effect, because it offers large sorbing capacities with respect to the metals which then become less bioavailable (Duzzin et al., 1988; Coquery & Welbourn, 1995; Fernandes, 1997), or a harmful effect which can interfere with that of the micropollutants.

It can be pointed that metals associated to organic matter may become available through the digestive process. Honeycutt et al. (1995) have shown that metals could be stored mainly in the digestive tractus and in the body walls of oligochaetes. In *Limnodrilus hoffmeisteri*, which was very common in this study, metals are stored in rich sulphur granules in particular cells called chloragocytes which constitute the chloragogen tissue covering the digestive tractus (Klerks & Levinton, 1989; Klerks & Bartholomew, 1991). This mecanism also exists for terrestrial oligochaetes (Morgan & Morgan, 1989). An other way of detoxification occur in terrestrial oligochaetes of the genus Tubifex (Lucan-Bouché et al., 1998). In metal contaminated environments, metals are stored in the caudal part of the worm which is finally lost. This amputation is then followed by a regeneration. So we considered that bioaccumulation studies was not suitable for this kind of biomonitoring and that biocenotic approach was more useful for field evaluation.

In the same way, the role of the aquatic macrophyte cover also deserves to be discussed. It can offer beneficial effects by participating in the purification of nutrients and the detoxication of toxic substances (Hammer & Bastian, 1989; Behrends et al., 1994), and this seems to be the case at Courchelettes.

Nevertheless, the understanding of the complexity of the interrelationships encountered in the aquatic environment must not be considered *a priori* as an objective which is impossible to reach. In particular, we are currently developing multi-disciplinary researches concerning the concept of the 'ecotoxicological ambience of sediments' (Rosso-Darmet & Lafont, 1996), researches which make use of field experiences in biology and chemistry, associated with laboratory experimentations, in the context of the 'triad' paradigm (Chapman, 1986, 1990).

In terms of operational management, the different studies conducted in the Rhin-Meuse, Rhône-Méditerranée-Corse, Adour-Garonne and Artois-Picardie water basins have led to selecting the IOBS index and the percentage of Tubificidae without hair setae as indicators of the biological quality of sediments in the biological portion of the river Quality Assessment System of French Water Agencies.

Work remains to be done in order to achieve a perfectly operational tool on a national level. Examples are the verification of the relevance of the predictive diagram (Figure 5) in the new situations, and notably for reference sites that are untouched by any contamination, the development of a sampling methodology and the adaptation of the oligochaete indices for the study of very particular environments, such as deep rock-gravel substrates and the rocks along the water banks. These habitats often constitute the dominant ones over long distances in some canalized environments.

References

Agence de l'Eau Artois-Picardie, 1996. Qualité des eaux de surface, schéma directeur. Document Agence de l'Eau Artois-Picardie, Douai.

Behrends, L. L., E. Bailey, M. J. Bulls, H. S. Coonrod, & F. J. Sikora, 1994. Seasonal trends in growth and biomass accumulation of selected nutrients and metals in six species of emergent aquatic macrophytes. Proc. 4th Int. Conf. on wetlands systems for water pollution control, Guangzhu, Pop. Rep. China, 6–10th November 1994: 274–289.

Belzunce-Segarra, M. J., J. R. Bacon, R. Prego & M. J. Wilson, 1997. Chemical forms of heavy metals in the surface sediments of the San Simon inlet, Ria de Vigo, Galicia. J. Envir. Sci. Health A32: 1271–1292.

Bervoets, L., R. Blust, M. de Wit & R. Verheyen, 1997. Relationships between river sediment characteristics and trace metal concentrations in Tubificid worms and Chironomid larvae. Envir. Pollut. 95: 345–356.

Brouwer, H., T. Murphy & L. McArdle, 1990. A sediment contact bioassay *Photobacterium phosphoreum*. Envir. Chem. Toxicol. 9: 1353–1358.

Calmano, W., J. Hong & U. Förstner, 1993. Binding and mobilization of heavy metals in contaminated sediments affected by pH and redox potential. Wat. Sci. Technol. 28: 223–235.

Campbell, P. G. C., 1995. Interactions between trace metals and aquatic organisms: a critique of the free-ion. Activity model. In A. Tessier & D. R. Turner (eds), Metal Speciation and Bioavailability in Aquatic Systems. J. Wiley & Sons Ltd Publishers: 45–102.

Campbell, P. G. C. & M. Stokes, 1985. Acidification and toxicity of metals to aquatic biota. Can. J. Fish. aquat. Sci. 42: 2034–2049.

Chapman, P. M., 1986. Sediment quality criteria from the sediment quality triad: an example. Envir. Toxicol. Chem. 5: 957–964.

Chapman, P. M., 1989. Current approaches to developing sediment quality criteria. Envir. Toxicol. Chem. 8: 589–599.

Chapman, P. M., 1990. The sediment quality triad approach to determining pollution-induced degradation. The Science of the Total Environment, 97/98: 815–825.

Chessel, D. & S. Dolédec, 1993. Programmathèque ADE. Analyses multivariées et expression graphique des données environnementales. Version 3.6, Fascicules 1 à 8.

Chessel, D. & P. Mercier, 1993. Couplage de triplets statistiques et liaisons espèces-environnement. In J. D. Lebreton & B. Asselain (eds), Biométrie et Environnement. Masson: 350 pp.

Coquery, M. & P. M. Welbourn, 1995. The relationship between metal concentration and organic matter in the sediments and metal concentration in the aquatic macrophyte *Eriocaulon septangulare*. Wat. Res. 29: 2094–2102.

Dickman, M. & G. Rygiel, 1996. Chironomids larval deformity frequencies, mortality and diversity in heavy metal contaminated sediments of a Canadian riverine wetland Envir. Int. 22: 693–703.

Dolédec, S. & D. Chessel, 1994. Co-inertia analysis: an alternative method for studying species environment relationships. Freshwat. Biol. 31: 277–294.

Duzzin, B., B. Pavoni & R. Donazzolo, 1988. Macroinvertebrate communities and sediments as pollution indicators for heavy metals in the river Adige (Italy). Wat. Res. 22: 1353–1363.

E. C., 1997. European Council directive proposition establishing a framework for EEC action in the field of water quality. EC document: 73 pp.

Fernandes, H. M., 1997. Heavy metals distribution in sediments and ecological risks assessment: the role of diagenetic processes in reducing metal toxicity in bottom sediments. Envir. Pollut. 97: 317–325.

Gouws, K. & P. P. Coetzee, 1997. Determination and partitioning of heavy metals in sediments of the Vaal Dam System by sequential extraction. Water SA 23: 217–226.

Gunn, A. M., D. T. E. Hunt & D. A. Winnard, 1989. The effect of heavy metal speciation in sediment on bioavailability to tubificid worms. Hydrobiologia 188/189: 487–496.

Hammer, D. A. & R. K. Bastian, 1989. Wetland ecosystems: natural purifiers ? In D. A. Hammer (ed.), Constructed Wetland for Waste Water Treatment. Lewis publishers, Chelsea (U.S.A.): 5–19.

Honeycutt, M. E., B. L. Roberts & D. S. Roane, 1995. Cadmium disposition in the earthworm *Eisenia fetida*. Ecotox Envir. Safety 30: 143–150.

Hopkin, S. P., 1989. Ecophysiology of metals in terrestrial invertebrates. Elsevier Applied Sciences, New-York.

Huebert, D. B. & J. M. Shay, 1992. Zinc toxicity and its interaction with cadmium in the submerged aquatic macrophyte *Lemna trisulca* L. Envir. Toxicol. Chem. 11: 715–720.

Johansson, K., A. Andersson & T. Andersson, 1995. Regional accumulation pattern of heavy metals in lake sediments and forest soils in Sweden. The Science of the Total environment 160/161: 373–380.

Klerks, P. L. & J. S. Levinton, 1989. Rapid evolution in metal resistance in a benthic oligochaete inhabiting a metal contaminated site. Biol. Bull. 176: 135–141.

Klerks, P. L. & P. R. Bartholomew, 1991. Cadmium accumulation and detoxification in a Cd-resistant population of the oligochaete *Limnodrilus hoffmeisteri*. Aquat. Toxicol. 19: 97–112.

Lafont, M., 1989. Contribution à la gestion des eaux continentales: utilisation des oligochètes comme descripteurs de l'état biologique et du degré de pollution des eaux et des sédiments. Thèse de Doctoratès Sciences, Université Lyon I: 311 pp.

Lafont, M., J. C. Camus & A. Rosso, 1996a. Superficial and hyporheic oligochaete communities as indicators of pollution and water exchange in river Moselle, France. Hydrobiologia 334: 147–155.

Lafont, M., J. Mouthon & A. Rosso-Darmet, 1996b. Etude des mollusques et des oligochètes dans les sédiments de cours d'eau du bassin du Rhône. Rapport Cemagref Lyon/Agence de l'Eau RMC: 26 pp.

Le Du, A., D. Dive & A. Philippo, 1990. Intercation between components of electroplating industry wastes. Influence of the receiving water on the toxicity of the affluent. Envir. Pollut. 65: 251–267.

Lim, P-E. & M-Y Kiu, 1995. Determination and speciation of heavy metals in sediments of the Juru River, Penang, Malaysia. Envir. Monit. Assess. 35: 85–95.

Literathy, P., F. Laszlo & B. Csanyi, 1994. Approaches for sediment associated pollutant monitoring in the river Danube. Wat. Sci. Technol. 30: 157–165.

Lucan-Bouché, M. L., S. Biagianti-Risbourg, F. Arsac & G. Vernet, 1998. Sur un mécanisme original de décontamination développé par l'oligochète *Tubifex tubifex*. 101° journées annuelles de la

Société Zoologique de France, 29–30 juin et 1° juillet 1998, Université des Sciences et Technologies de Lille.
Morgan, J. E. & A. J. Morgan, 1989. The effect of lead incorporation on the elemental composition of earthworm (Annelida, Oligocheta) chloragosome granules. Histochemistry 55: 237–241.
Petersen R. C. & L. B. M. Petersen, 1988. Compensatory mortality in aquatic populations: its importance for interpretation of toxicant effects. Ambio 17: 381–386.
Prygiel, J., 1996. Utilisation de bioindicateurs pour l'estimation de la qualité des milieux aquatiques marins et continentaux du bassin Artois-Picardie. Mémoire HDR, Université de Lille: vol. 1: 117 pp. + annexes; vol. 2: publications.
Prygiel, J. & M. Coste, 1993. The assessment of water quality in the Artois-Picardie water basin (France) by the use of diatom indices. Hydrobiologia 269/270: 343–349.
Reynoldson, T. B., S. P. Thompson & J. L. Bamsey, 1991. A sediment bioassay using the tubificid oligochaete worm *Tubifex*. Envir. Toxicol. Chem. 10: 1061–1072.
Rofes, G. & M. Savary, 1981. Description d'un nouveau modèle de carottier pour sédiments fins. Bull. fr. Piscic. 283: 102–113.
Rosso, A., 1995. Description de l'impact des micropolluants sur les peuplements d'oligochètes des sédiments de cours d'eau du bassin-versant de l'Ill (Alsace). Elaboration d'une méthode biologique de diagnostic de l'incidence des micropolluants. Thèse de Doctorat, Université Lyon I: 232 pp.
Rosso, A., M. Lafont & E. Exinger, 1994. Impact of heavy metals on benthic oligochaete communities in the River Ill and its tributaries. Wat. Sci. Technol. 29: 241–248.
Rosso-Darmet, A. & M. Lafont., 1996. Ambiance écotoxicologique des sédiments: exemples concrets dans le bassin de l'Ill (Alsace); essai de conceptualisation. In Altérations Physico-Chimiques et Biologiques des Écosystèmes Aquatiques. Cemagref, Département Gestion des Milieux Aquatiques, Paris, 26–27 novembre 1996: 167–175.
Rosso-Darmet, A., M. Lafont & A. Durbec, 1997a. Synthèse bibliographique sur l'intérêt des oligochètes comme indicateurs de la qualité du milieu et de la contamination métallique et micro-organique. Acquisition des données biologiques et résultats des inventaires d'oligochètes. Rapport BURGEAP/Cemagref Lyon/AE Artois-Picardie: 14 pp.
Rosso-Darmet A., M. Lafont & J. Mouthon, 1997b. Utilisation des peuplements d'oligochètes et de mollusques comme descripteurs du degré de pollution des sédiments. Rapport Cemagref Lyon/AE Adour-Garonne: 34 p + annexe.
Santiago, S., R. L. Thomas, G. Larbaigt, C. Corvi, D. Rossel, J. Tarradellas, D. J. Gregor, L. Mccarthy & J. P. Vernet, 1994. Nutrient, heavy metals and organic pollutant composition of suspended and bed sediment in the Rhône River. Aquat. Sci. 56: 220–242.
Seaward, M. R. D., 1995. Use and abuse of heavy metal bioassays in environmental monitoring. The Science of Total Environment 176: 129–134.
Seymore, T., H. H. Du Preez, J. H. J. van Vuren,, A. Deacon, & G. Strydom, 1994. Variations in selected water quality variables and metal concentrations in the sediment of the lower Olifants and Selati rivers, South Africa. Koedoe 37: 1–18.
Slooff, W., 1983. Benthic macroinvertebrates and water quality assessment: some toxicological considerations. Aquat. Toxicol. 4: 73–82.
Teixeira, E. C., J. D. Sanchez, L. D. Fernandez, M. L. L. Formoso, G. Pegorini, & M. H. D. Pestana, 1997. A preliminary study of metals in sediments from areas influenced by coal processing and steel industry activities - Baixo Jacui region, RS-Brazil. Envir. Technol. 18: 581–592.
Tessier, A., P. G. C. Campbell & M. Bisson, 1979. Sequential extraction procedure for the speciation of particulate trace metals. Anal. Chem. 51: 844–851.
Tessier, A. & P. G. C. Campbell, 1987. Partitioning of trace metals in sediments: relationships with bioavailability. Hydrobiologia 149: 43–52.
Traunspurger, W. & C. Drews, 1996. Toxicity analysis of freshwater and marine sediments with meio- and macrobenthic organisms: a review. Hydrobiologia 328: 215–261.
Wang, W., 1987. Factors affecting metal toxicity to (and accumulation by) aquatic organisms – overview. Envir. Int. 13: 437–457.
Warren, L. A. & A. P. Zimmerman, 1994. Suspended particulate oxides and organic matter interactions in trace metal sorption reactions in a small urban river. Biochemistry 23: 21–34.

Hydrobiologia **410:** 39–46, 1999.
J. Garnier & J.-M. Mouchel (eds), Man and River Systems.

Contribution of the macrobenthic compartment to the oxygen budget of a large regulated river: the Mosel

Vincent Bachmann[1] & Philippe Usseglio-Polatera[1,2]
[1]*Equipe de Démoécologie, UPRES EBSE, 'Ecotoxicité, Biodiversité et Santé Environnementale', Université de Metz, BP 4116-1, Rue des Récollets 57040 Metz Cedex 01, France*
[2]*UPRESA 5023, 'Ecologie des Eaux Douces et des Grands Fleuves', Université LYON I-43, Bd du 11 Novembre, 69622 Villeurbanne Cedex, France*

Key words: large river, river bottom, macrobenthos, oxygen budget, respiration, filtration rate

Abstract

The contribution of the benthic compartment to the oxygen budget of three sectors of the river Mosel (France, Luxembourg, Germany) was evaluated, in terms of benthos respiration, filtration rate and sediment oxygen demand. Bottom substrates and benthic invertebrate communities were investigated using benthic dredges during six field campaigns from summer 1994 to summer 1996. Bivalvia (*Corbicula fluminalis, Dreissena polymorpha, Anodonta cygnea, Unio pictorum*), Prosobranchia (*Viviparus viviparus*), Crustacea (Gammaridae) and Diptera (Chironomidae) were the prevailing taxa of the benthic community, both in density and biomass. An original method based on multivariate analyses and a clustering approach was developed to describe the habitat mosaic in terms of substrate-type frequencies. The biomass of each taxon was recalculated for each substrate-type. The dissolved oxygen uptake and the filtration rate of the prevailing taxa of the benthic community were modelled using results of laboratory experiments. Both effects of water temperature and trophic level conditions were taken into account. Evaluations showed that benthic compartment respiration processes may have reach up to 25% of the total amount of dissolved oxygen per day, in the river Mosel. Furthermore, the macrobenthic compartment may have a great impact on phytoplankton communities, and then on primary production, by filtration activity. Indeed, the water volume daily filtered by filter-feeder macroinvertebrates potentially ranged from 26.9 to 83.8% of the total volume of a sector, in summer conditions. Consequently, respiration and filtration activities of the benthic compartment appear to be two important biological processes of the dissolved oxygen budget of a large regulated river.

Introduction

Critically low concentrations (about 2–3 mg O_2 l^{-1}) of dissolved oxygen in summer have been observed in the river Mosel since 1990s (Garnier et al., 1995). So, a pluri-disciplinary research programme has been engaged by the International Commissions for the Protection of the rivers Mosel and Saar (CIPMS) to identify, to quantify and to model the different biological processes involved in the oxygen budget of this river. Preliminary results, obtained in summer 1994, focused on the importance of the respiration of the bacterioplankton and the nitrification processes which can respectively reach up to 75.5 and 34.9% of the total dissolved oxygen availability (i.e., instream production, aerial and upstream inputs) for the downstream part of the river Mosel (Garnier et al., 1995). The respiration of the phytoplankton and the zooplankton compartments were, respectively, evaluated at about 6.8–17.7% and 0.9–4.6% of the total available dissolved oxygen (per day) in a sector. Because no information about the contribution of the benthic compartment to the oxygen budget of the river Mosel was available, we attempted to evaluate respiration processes for sediments and macrozoobenthic communities. We tried also to evaluate the grazing impact of benthic filterers on phytoplankton communities, and consequently on primary production.

The major problem was to evaluate biomasses of benthic taxa at the sector (i.e., a river section of about

20–25 km length) scale. Sectors may be regarded as hierarchical systems of patches that differ in age, size and internal composition and are characterized by different environmental conditions. Both anthropogenic and natural processes create environmental patchiness resulting in complex biotope structure (Power et al., 1988; Townsend & Hildrew, 1994). The microdistribution of benthic macroinvertebrates results from the balance of a variety of requirements for organisms. Many references have confirmed that substrate-type is a useful and convenient predictor of the abundance, biomass and diversity of benthic macroinvertebrates (Jowett & Richardson, 1990; Milner et al., 1981; Minshall, 1984; Ormerod & Edwards, 1987). Consequently, we must elucidate the link between species requirements and environmental variability to explain the local distribution of aquatic organisms. Thus, we developed a method to (1) characterize the faunal structure of samples according to the granulometric structure of substrates within the bottom mosaic, (2) define substrate-types according to both their faunal and granulometric structure, (3) evaluate the composition of the bottom mosaic in terms of relative frequencies of substrate-types, and (4) estimate the biomass of each taxon for each substrate-type. The next step was to evaluate the contribution of the macrobenthic community to the oxygen budget of each sector of the river from experiments performed in '*in situ*' conditions. These experiments allowed an appraisal of oxygen consumption and filtration rate for the prevailing benthic taxa highlighted by the faunal study.

Materials and methods

Study site and sampling procedure

The 535-km long Mosel river (Figure 1) is, for the most part, canalized in order to enable a high level of ship traffic with high loading gauge. The downstream part of the river was divided into three different sectors. Each sector was considered as a nearly homogeneous section according to its morphometrical and physical characteristics.

Granulometric and faunal samples were performed at a water depth of 1–4 m by means of two benthic dredges employed according to the particle size of the bottom substrate of each sampling site. A triangular dredge with a 39-cm side length, for coarse mineral substrates (up to cobbles), and a cylindrical dredge (Rallier du Baty's type) with an opening diameter of 30 cm were used from a boat along several transects of the three sectors of the river Mosel (see Berly, 1989, for full description of sampling methodology). Samples were taken as much as possible on the left bank, on the channel and on the right bank.

The material collected, including macroinvertebrates, were put into plastic jars and preserved in 70% ethanol. In the laboratory, macroinvertebrates were sorted, identified, counted and dried for 24 h at 105°C in order to evaluate the dry biomass of each taxon in each sample. Only soft tissues were taken into account for molluscs. On each sampling occasion, the surface of the sampled bottom area was evaluated according to (1) the length of the dredge draught estimated from the volume of dredged material, (2) the dredge section and (3) the sampling sediment depth (Berly, 1989). Biomasses of macroinvertebrates were then expressed per square metre. Only taxonomic groups occurring at dry biomass above 0.1 g m^{-2} in at least one sample were preserved in the faunal data set (biomasses of taxa transformed to $\log_2$ classes × samples). The granulometric structure of each substrate sample was also estimated by sifting and weighting particulate materials after drying for 24 h at 105°C, and then 500°C in order to define organic matter content. Ten granulometric fractions were considered: (1) silt (Ø<0.5 mm); (2) fine sand (0.5–1 mm)0; (3) coarse sand (1–2 mm); (4) fine gravel (2–4 mm); (5) medium gravel (4–8 mm); (6) coarse gravel (8–16 mm); (7) small pebbles (16–32 mm); (8) pebbles (32–64 mm); (9) cobbles (64–128 mm); (10) boulders (Ø>128 mm). For all samples, the dry masses of the 10 substrate modalities (expressed as percentages of the total substrate dry mass of a sample) were brought together in the granulometric data set.

Reconstruction of the habitat mosaic of each sector

We developed an original method (1) to examine the relationships between granulometric traits and faunal structure of samples, and (2) to efficiently describe the substrate mosaic in each sector of the river. Because of the complexity of the relationships between community structure and environmental conditions, a multivariate approach was required. First, both faunal and mesological data tables were independently processed by within-campaign (in order to eliminate the temporal effects due to seasonality and between-year variations) correspondence analysis (Dolédec & Chessel, 1987) (results of these analyses are not shown

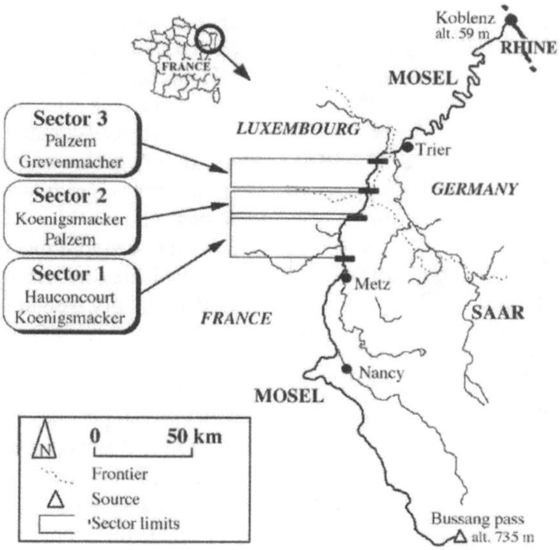

Figure 1. Map of the sectors and the sampling sites of the river Mosel. Each sampling site was representative of an homogeneous sector.

here for brevity). Then, co-inertia analysis (Chessel & Mercier, 1993; Dolédec & Chessel, 1994) which investigates the common structure between two tables sharing the same statistical units (samples in our case), was used to check for a relationship between community structure and granulometric traits of samples. Thereby, we obtained scores for the samples that had a maximal covariance. According to the definition of covariance, this maximal covariance was obtained by optimal correlation and simultaneously optimal standard deviation of these sample scores derived from each of the two arrays on taxa biomasses and granulometric frequency distributions. A good way to compare the new environmental ordination and the new faunal ordination of samples (in each sector) from co-inertia analysis, was to plot the new standardized environmental and faunal scores together on a factorial map and to link the two positions of a given sample by an arrow (Figure 2). Hence, it was possible to discuss the link between the faunal structure and the substrate characteristics of samples. The beginning and the end of this arrow indicated the positions of the sample resulting, respectively, from the new granulometric and faunal ordinations. For example, samples could simultaneously display close granulometric structure (if arrows began in the same area) but fairly different faunal contents in terms of biomasses (if arrows pointed to different directions and if their ends were well separated).

Finally, a clustering approach was used to identify substrate-types represented by samples with both close faunal contents and close granulometric descriptions. This approach was based both (1) on the euclidean distance calculated on the F_1 and F_2 coordinates of samples according to their faunal and mesological structure in co-inertia analysis and (2) on the Ward's linkage (Orloci, 1967, 1978; Pielou, 1984) recommended for river benthic community analysis (Cao et al., 1997). The number of samples for each season, in each substrate-type (i.e. each cluster) may be considered as an evaluation of the frequency of this kind of substrate-type in the bottom mosaic of a sector, for a given date. As a result, a reconstruction of the habitat mosaic was possible, in terms of substrate-type frequencies. All calculations and graphs were performed using the package ADE-4 (Thioulouse et al., 1997).

Evaluation of oxygen uptake processes at the sector level

Preliminary faunal studies conducted in 1994 on the river Mosel benthic community highlighted Bivalvia (*Dreissena polymorpha, Corbicula fluminalis, Anodonta cygnea, Unio pictorum*), Prosobranchia (*Viviparus viviparus*), Crustacea (Gammaridae) and Insecta (Chironomidae) as prevailing taxa which were retained for laboratory experiments of oxygen demand V_{O_2} and filtration rate (FR) measurements. An

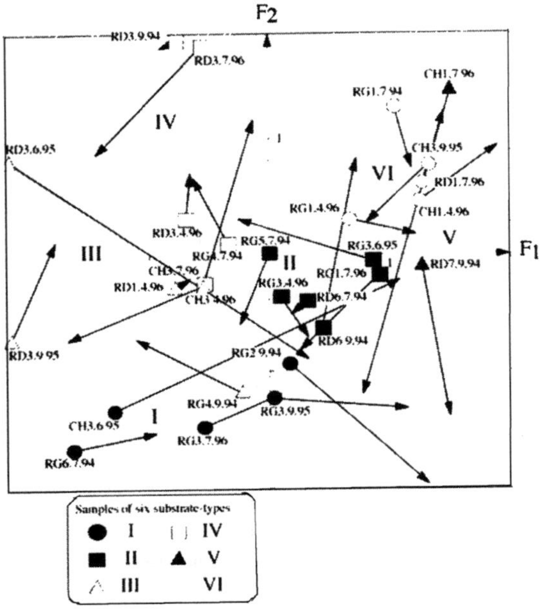

Figure 2. Co-structure between biomasses of taxa and granulometric characteristics of samples by co-inertia analysis for the Hauconcourt-Koenigsmacker sector of the river Mosel. Each sample was defined by an arrow. The beginning of the arrow (= closed/open point, square or triangle) marked the position of the sample on the $F_1 \times F_2$ co-inertia plane, according to the granulometric structure. The end of the arrow marked its position according to the faunal structure. The relative locations of sample vectors allowed to identify six substrate-types (I–VI) (see also Figure 5). Samples were labelled with letters giving information about (1) sample position (RD, right bank; RG, left bank; CH, channel), (2) sampling month (4, April, 6, June, 7, July; and 9, September), and (3) sampling year (94, 1994; 95, 1995; and 96, 1996).

enclosed method was used (for further details, see Bachmann et al., 1998). A large variety of water temperature/trophic level combinations, descriptive of '*in situ*' conditions of the river Mosel from spring to autumn, was tested. Results allowed to model V_{O_2} and FR of taxa, according to both temperature and trophic level, with non-linear multiple regressions. Studied processes were expressed per gramme of soft tissue dry weight.

For each season, the biomass of each taxon could be recalculated for each substrate-type as the weighted (by the sampling surface of each sample) average of its biomass in the different samples of this substrate-type. Then, according to (1) the biomass evaluation of the different taxa on a whole sector, (2) the oxygen consumption (and filtration rate) models, (3) the mean water temperature and food availability observed at each season, (4) the morphometric characteristics of each sector, we were able to estimate the global oxygen uptake and the filtered water volume on each sector, for each season.

The oxygen demand of sediments (DSO) was evaluated, taking into account the organic matter amount of substrates. Each sample, according to the organic matter amount (MO) of its fine fractions (Ø< 2mm), corresponded to a mean oxygen demand experimentally evaluated to 4.16 mg O_2 h^{-1} m^{-2} (if MO $<$ 3% of the total dry weight of these fractions) or 42.09 mg O_2 h^{-1} m^{-2} (if MO $\geq$ 3%). Then, the DSO of each substrate-type was recalculated as the weighted (by the sampling surface of each sample) average of estimated DSO of samples belonging to this substrate-type, for each season and each sector. As a result, the DSO may be considered in the model as the consumption of a 'particular' taxon.

Then, the global oxygen uptake on a sector was evaluated by:

$$V_{O_2} = \sum_{i=1}^{s} \left[\frac{\sum_{k=1}^{m} \left[p_k \left[\frac{\sum_{j=1}^{r_k} s_j b_{ijk}}{\sum_{j=1}^{r} s_j} \right] \right]}{\sum_{k=1}^{m} p_k} \right] x[\mathrm{Mod}_{V_{O_2}}(i)] \qquad (1)$$

with: b_{ijk} = biomass of the taxon i in the sample j of the substrate-type k; s_j = sampling area of the sample j; p_k = ratio of the substrate-type k in the substrate mosaic of the studied sector, for a given season; r_k = the number of samples in the substrate-type k; m = the number of substrate-types in the studied sector; s = the number of taxonomic groups in the faunal list of the studied sector; f = the number of filterer groups in the faunal list of the studied sector; $\mathrm{Mod}_{V_{O_2}}$(i) = predictive model of oxygen uptake, in controlled conditions (water temperature and trophic level), for the taxon i.

The global filtration rate (FR) on a sector was evaluated by replacing V_{O_2} by FR and $\mathrm{Mod}_{V_{O_2}}$(i) by Mod_{FR}(i) in the Eq. (1). Mod_{FR}(i) corresponded to the predictive model of filtration rate, in controlled conditions, for the taxon i.

Results

Fifty-six taxa were captured during the sampling programme by dredging during April, June, July and/or September of three years: 1994, 1995 and 1996. In terms of biomasses, the benthic macroinvertebrate assemblage of the lower section of the river Mosel was

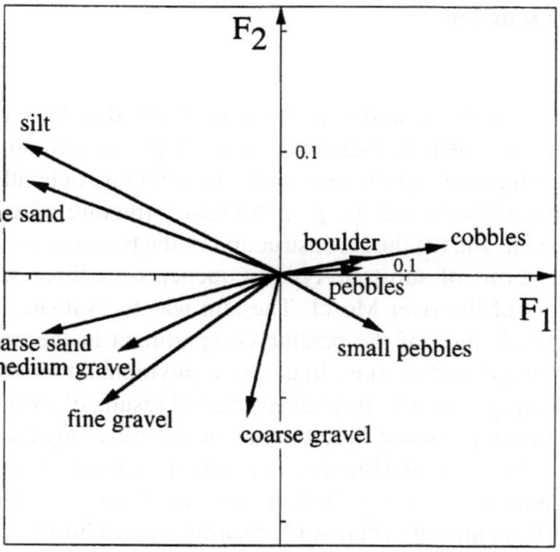

Figure 3. Co-structure between biomasses of taxa and granulometric characteristics of samples. Positions of 10 granulometric fractions on the $F_1 \times F_2$ co-inertia plane.

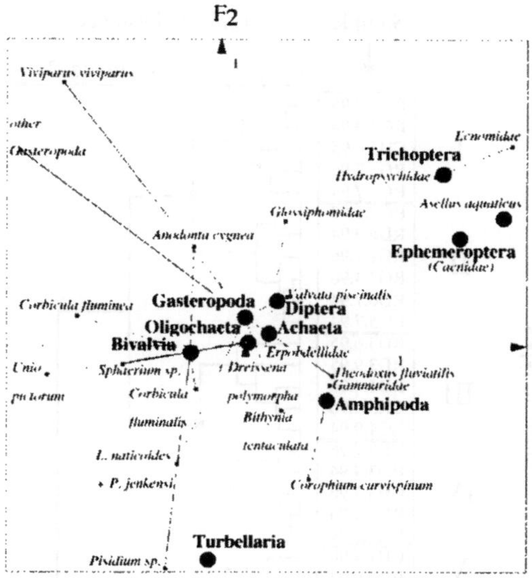

Figure 4. Co-structure between biomasses of taxa and granulometric characteristics of samples. Distribution of taxa (squares) on the $F_1 \times F_2$ co-inertia plane. Taxa were grouped according to higher taxonomic units. We plotted, for each unit, a closed circle at the weighted average of the taxon positions of that unit. Taxa were linked to their unit position by lines.

largely dominated by Bivalvia (mainly *D. polymorpha* (55.5% of the total biomass of samples), *C. fluminalis* (19.6%), *A. cygnea* (10.8%) and *U. pictorum* (3.4%)). Crustacea (mainly *Corophium curvispinum* (1.8%) and Gammaridae (3.5%)) corresponded to 5.3% of the total biomass of samples. The other groups represented less than 1% of the community biomass.

Ordination of samples

The co-inertia ordination of twenty-eight samples from the Hauconcourt-Koenigsmacker sector is shown in Figure 2 (results from the other sectors are not illustrated here for brevity). The first two axes of the co-inertia analysis explained 64.3 and 17.1% of the total variance, respectively. The first co-inertia axis represented 75% of the variability of the first axis resulting from the separate correspondence analysis of taxa biomasses and 86% of the first axis resulting from the separate PCA of granulometric structure of samples. The optimal correlation between the first axis scores of samples was equal to 0.72. Thus, the first co-inertia axis explained the major part of the variability. Comparing the positions of granulometric modalities (in Figure 3), taxa (in Figure 4) and sample vectors (in Figure 2) on the first factorial plane summarized the results of co-inertia analysis. The ordination of samples was related to their dominant particle size. Samples displaying a coarse granulometry (pebbles, cobbles) were opposed to samples having a large amount of sand and/or silt (Figure 3). On substrate of coarse particle size (at the positive side of F_1 axis), *D. polymorpha* was dominant. *C. curvispinum*, gammarids, *Asellus aquaticus*, net spinning caddisfly larvae and ephemeropteran Caenidae were specific. At the negative side of F_1 axis, on sand and silt, the faunal assemblage dominated by *C. fluminalis*, was characterized by *V. viviparus, Corbicula fluminea* and *U. pictorum.*

Classification of samples and reconstruction of the bottom mosaic

As an example, the classification of samples using Ward Linkage (Figure 5) allowed to distinguish six different substrate-types for the Hauconcourt-Koenigsmacker sector. Consequently, according to its faunal and granulometric structure, we considered the bottom mosaic of the Hauconcourt-Koenigsmacker as composed of 21.4% (=3/14 samples) of substrate-type I, 28.6% (=4/14) of substrate-type II, 7.15% (=1/14) of substrate-type III, 21.4% (=3/14) of substrate-type IV, 7.15% (=1/14) of substrate-type V, and 14.3% (= 2/14) of substrate-type VI in summer. A similar procedure was used for other seasons and other sectors.

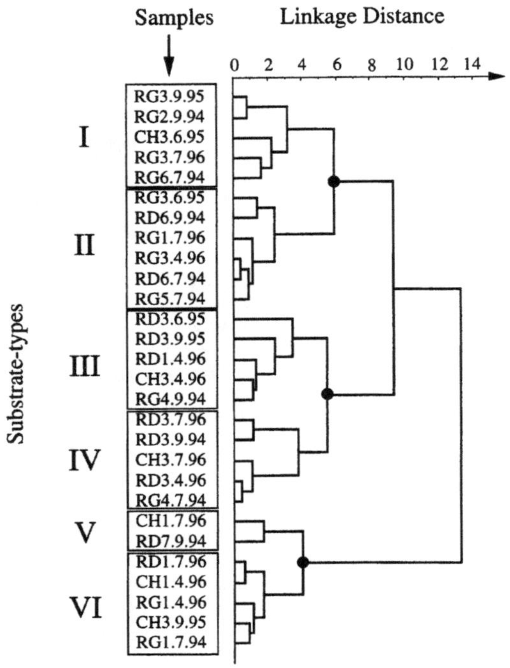

Figure 5. Co-structure between biomasses of taxa and granulometric characteristics of samples. Classification of samples from the Hauconcourt-Koenigsmacker sector using Euclidean distance and Ward's linkage method. Identification of six different substrate-types for this sector. Closed circles indicated truncation levels of the dendrogram.

Evaluation of oxygen uptake processes at the sector level

The total oxygen production and the total water volume of each sector were only available for summer conditions (Garnier et al., 1995). Consequently, the dissolved oxygen consumption (expressed as a percentage of the whole dissolved oxygen availability) and the filtered water volume (expressed as a percentage of the total water volume) could not have be evaluated for other seasons. Resulting evaluations showed that benthic compartment respiration processes may have reached up to 25.4% of the total amount of dissolved oxygen per day in summer in the lower section of the river Mosel (Table 1). Furthermore, the macrobenthic compartment may have a great impact on phytoplankton communities, and then on primary production, by filtration activity. Up to 83.8% of the total water volume of a sector might be filtered every day.

Discussion

This work allowed us to develop a method authorizing (1) the identification of substrate-types according to the relationship between the faunal structure of benthic assemblages and the granulometric structure of substrata, and (2) the reconstruction of the bottom mosaic in terms of substrate-type frequencies for three sectors of the river Mosel. The aim was to evaluate the contribution of the benthic compartment in the oxygen and carbon (i.e., impacts on phytoplankton community) budgets, from experimental results of oxygen consumption and filtration rate of the prevailing taxa.

On the map (Figure 2) on which each sample was defined as a vector, the high number of relatively long arrows indicated that most of samples had a limited co-structure between their community organization and their granulometric characteristics, because there was not often a good superposition of the samples when positioned by community structure and by granulometric composition. Indeed, samples with fairly close granulometric structure, could display quite different faunal communities (see the arrows of RD1.7.96, CH1.4.96 and CH3.9.95 in Figure 2). Conversely, samples with close faunal composition could have quite different granulometric structure (see the arrows of RD3.6.95, RG1.7.96 and RG3.4.96). These observations illustrated the complex relationship between bottom characteristics and macroinvertebrate community distribution. They justified the definition of substrate-types both on faunal and mesological traits.

Respiration and filtration activities of the benthic compartment appeared to be two important biological processes which cannot be neglected for the appraisal of the dissolved oxygen budget of the river Mosel in regards to the evaluations of other biological processes involved in this budget (phytoplankton and zooplankton respiration, bacterioplankton and nitrification activities, Garnier et al., 1995) for example in summer period.

If the method is now improved, further investigations (=samples) must be necessary to specify the oxygen demand of the benthic compartment, taking into account more accurately temporal variations of communities. We have also to specify the magnitude of potential qualitative (in terms of prevailing taxa) or quantitative (in terms of biomass) variations of the benthic community in time. Furthermore, a higher number of samples will allow to detail relative frequencies of substrate-types in each river sector. Only

Table 1. Global assessment of macrobenthos respiration (V_{O_2}) and filtration rate for three sectors of the river Mosel. Comparison of the relative contributions of the main biological compartments to the dissolved oxygen consumption in summer conditions.

Sector characteristics and associated processes	Units	Sector 1 Hauconcourt Koenigsmacker	Sector 2 Koenigsmacker Palzem	Sector 3 Palzem Grevenmacher
Length	km	27	29	17
Mean depth	m	1.95	3.38	4.1
Water column area	m^2	256	342	415
Total volume	m^3	6.90 10^6	9.92 10^6	7.06 10^6
Total bottom surface	m^2	3.54 10^6	2.90 10^6	2.90 10^6
Input discharge	$m^3 s^{-1}$	23	27	28
Output discharge	$m^3 s^{-1}$	27	28	28
Total O_2 production	tO_2 day^{-1}	63.7	39.4	19.2
Benthos V_{O_2}	tO_2 day^{-1}	11.3	5.08	4.88
Respiration ratio	%	17.7	12.9	25.4
Benthos filtration rate	m^3 day^{-1}	5.78 10^6	2.67 10^6	2.82 10^6
Filtration ratio	%	83.8	26.9	40
Sediment oxygen Demand	tO_2 day^{-1}	0.35	1.76	2.599
DSO ratio	%	0.6	4.5	13.5

an efficient modelling of benthos dynamics will enable an accurate modelling of oxygen budget processes.

It should be interesting for modelling more efficiently the benthos contribution to the oxygen budget of a large river, to evaluate more accurately the link between the oxygen demand or the filtration rate and different environmental or biological traits (reproductive stage of invertebrates, quality of available food, etc.). Further investigations have been in process to develop and to model these aspects.

Acknowledgements

This work was initiated and financially supported by: Agence de l'eau Rhin-Meuse, Ministères de l'environnement des Land de Saar (Saarland) et Rhenanie-Palatinat (Rheinland-Pfaltz), Administration de l'Environnement du Luxembourg et de l'Union Européenne (INTERREG 2).

References

Bachmann, V., P. Usseglio-Polatera, J. C. Moreteau & P. Wagner, 1998. A new method to assess the contribution of the macrobenthic compartment to the dissolved oxygen budget of a large river. In G. Bretschko & J. Helesic (eds), Advances in River Bottom Ecology. Backhuys, Leiden: 143–153.

Berly, A., 1989. Distribution spatio-temporelle des peuplements macrobenthiques prélevés par dragage dans une station du Haut-Rhône, Ph.D., Univ. Lyon I, 309 pp.

Cao, Y., A. W. Bark & W. P. Williams, 1997. A comparison of clustering methods for river benthic community analysis. Hydrobiologia 347: 25–40.

Chessel, D. & P. Mercier, 1993. Couplage de triplets statistiques et liaisons espèces-environnement. In J.D. Lebreton & B. Asselain (eds), Biométrie et environnement. Masson, Paris: 15–44.

Dolédec, S. & D. Chessel, 1987. Rythmes saisonniers et composantes stationnelles en milieu aquatique, I. Prise en compte et élimination d'effets dans un tableau faunistique. Acta Oecol., Oecol. Gener. 10: 207–232.

Dolédec, S. & D. Chessel, 1994. Co-inertia analysis: an alternative method for studying species environment relationship. Freshwat. Biol. 31: 277–294.

Garnier, J., G. Billen & L. Palfner, 1995. Bilan d'oxygène et processus biologiques dans les rivières régulées Moselle et Sarre. Rapport auprès des Commissions Internationales pour la Protection de la Moselle et de la Sarre contre la pollution, 45 pp.

Jowett, I. & J. Richardson, 1990. Microhabitat preferences of benthic invertebrates in a New Zealand river and the development of in-stream models for *Deleatidium* spp. N. Z. J. Mar. Freshwat. Res. 24: 19–30.

Milner, N. J., J. Scullion, P.A. Carling & D.T. Crisp, 1981. The effects of discharge on sediment dynamics and consequent effects on invertebrates and salmonids in upland rivers. Freshwat. Biol. Assoc. Scientific Publication: 153–220.

Minshall, G. W., 1984. Aquatic insect substratum relationships. In The Ecology of Aquatic Insects, Resh, V. H. & D. M. Rosenberg (eds), Praeger, New York: 358–400.

Orloci, L., 1967. An agglomerative method for classification of plant communities. J. Ecol. 55: 193–206.

Orloci, L., 1978. Multivariate Analysis in Vegetation Research, 2nd edn. W. Junk, Boston.

Ormerod, S. J. & R.W. Edwards, 1987. The ordination and classification of macroinvertebrate assemblages in the catchment of the river Wye in relation to environmental factors. Freshwat. Biol. 17: 533–546.

Pielou, E. C., 1984. The Interpretation of Ecological Data. Wiley, New York.

Power, M. E., R. J. Stout, C. E. Cushing, P. P. Harper, F. R. Hauer, W. J. Matthews, P. B. Moyle, B. Statzner & I. R. Wais de Badgen, 1988. Biotic and abiotic controls in river and stream communities. J. N. am. Benthol. Soc. 7: 456–479.

Thioulouse, J., D. Chessel, S. Dolédec & J. M. Olivier, 1997. ADE-4: a multivariate analysis and graphical display software. Statistics Comput 7: 75–83.

Townsend, C. R. & A. W. Hildrew, 1994. Species traits in relation to a habitat templet for river systems. Freshwat. Biol. 31: 265–275.

Hydrobiologia **410:** 47–58, 2000.
J. Garnier & J.-M. Mouchel (eds), Man and River Systems.

Contribution to the ecological quality of Aliakmon river (Macedonia, Greece): a multivariate approach

M. Lazaridou-Dimitriadou[1], V. Artemiadou[1], G. Yfantis[1], S. Mourelatos[2] & Y. Mylopoulos[3]
[1]*Department of Zoology, School of Biology, Aristotle University, 54006, Thessaloniki, Greece*
[2]*Department of Genetics, Development & Molecular Biology, School of Biology, Aristotle University, 54006 Thessaloniki, Greece*
[3]*Department of Hydraulics and Environmental Engineering, School of Civil Engineering, Aristotle University, 54006 Thessaloniki, Greece*

Key words: biomonitoring, macroinvertebrates, Aliakmon

Abstract

In the present study, an attempt was made to estimate the water quality of Aliakmon river using five biotic scores (based on benthic macroinvertebrates), physicochemical variables and multivariate data analyzes. Benthic and chemical samples were obtained from 20 sites, during the low flow season (September, 1995). Benthic macroinvertebrates were collected with the three minute kick-sweep method and several physicochemical parameters (D.O., pH, temperature, nutrients etc.) were measured. The multivariate techniques FUZZY and CANOCO were applied to the data. Aliakmon river appeared to have poorer water quality during the low flow season than during the high flow season as determined by comparison with a previous survey undertaken in April 1995. The upstream sites had better water quality than the downstream sites. The performance of the biotic scores was generally satisfactory, but in some cases they were deficient in distinguishing between running waters of good and excellent water quality. On the contrary, the multifactorial analyzes, which consider the abundance of the taxonomic groups, appeared to be more appropriate for the classification of the sites based on the structure of their biocommunities.

Introduction

Many of the streams and rivers suffer from the anthropogenic effects of abstraction and regulation. According to the proposed EU directive (C184/20/17.6.97), all Member States are obliged to establish national freshwater monitoring networks by the end of 1999. Many countries of the EU have specialized organisations or institutions to undertake the regular monitoring of running waters (e.g. Environment Agency in U.K., Agence de Bassin in France, La Direccion General de Calidad de las Aguas in Spain). Greece is one of the EU countries that does not have a national or regional network.

In this directive, one expression of the quality of running waters is the 'ecological status', refering to the biological elements of this ecosystem as well. Pollution biomonitoring of freshwater is already widespread because it provides information on the water quality and ecology before and during sampling, while the chemical approaches characterize the water system only during sampling (Metcalfe, 1989; Karr & Chu, 1999). However, the ideal method of pollution monitoring combines both physicochemical and biotic aspects.

Biomonitoring includes both sublethal changes at the cellular or tissue level (Vukmirovic et al., 1994; Cajaraville et al., 1995), and changes in community structure (Prat et al., 1984; Lafont & Juget, 1988; Barmuta, 1988; Graca et al., 1989; Quinn & Hickey, 1990; Castella et al., 1995). The use of changes in community structure to monitor pollution, commonly involves benthic invertebrates since this group is considered the most appropriate biotic indicator of water quality in EU countries (Metcalfe, 1989), including Greece (Anagnostopoulou, 1992).

A study of the applicability of two biotic indices and three scores was carried out in the Aliakmon river during the high flow season in April 1995 (Yfantis et al., 1996). In this study, the low flow season (September, 1995) was chosen in order to compare the quality of the water during a period of the year when the discharge is low and agricultural, urban and industrial activities increase. This river was chosen because it has a typical mediterranean hydrological regime and it is one of the longest rivers in Greece exclusively within Greek territory. It receives urban effluents, industrial discharges and surface runoff mainly from agricultural land. Twenty sites were sampled during the low flow season (September, 1995). For the classification of the 20 sites, both the benthic macroinvertebrates and the physicochemical characteristics of the sites were taken into consideration.

Study area description

The springs of the Aliakmon river are situated in the Grammos and Vernon mountains, in north-western Greece. It discharges into the gulf of Thermaikos. Along the river course, a series of artificial lakes has been formed following the construction of three hydroelectric dams. The catchment area is 8700 km^2 and consists of alluvial deposits, a few ophiolites, gneisses, schists and amphibolites (Bornovas & Rondogianni-Tsianbau, 1983). The physicochemical characteristics of Aliakmon river have been studied by Kouimtzis et al. (1994). The maximum discharge varies according to the year (from 3.58 m^3/s in July 1989 to 141.49 m^3/s in April 1991) and nutrients and major ions fulfill the required standards for drinking water (Kouimtzis et al., 1994). Studies have also been carried out on zooplankton abundance and taxonomic composition. Zooplankton abundance was very low along the length of the river except for the lakes of the river. It correlated with water temperature, suspended solids and NO_3-N at the upstream stations and with NO_2-N and PO_4-P at the downstream stations (Zarfdjian et al., in prep.). Fish fauna was found to be richer in the downstream Aliakmon sites than in the upstream sites (Economidis et al., 1981).

Twenty sampling sites situated upstream and downstream of the lakes were chosen in relation to point - or non-point pollution sources in September 1995 (see Figure 1). Two of the upstream sites (Melas and Nestorio) are situated on two different tributaries from which Aliakmon river originates. Four of the upstream sites (Ammoudara, Neapoli, Grevena and Panagia) are located along the main course of the river. The rest of the upstream sites belong to different tributaries: Kostarazi is on the Gioli stream that receives water from the lake overflow and the treated wastes from the waste treatment plant of Kastoria, Greveniotis site is on the homonymous tributary, downstream from the city of Grevena, and Trikomo and Eleftherohori are located on the Venetikos tributary. Trikomo is situated near the springs of the stream and Eleftherohori on its juction with Aliakmon river (see Figure 1). All the upstream sites do not receive significant amounts of pollutants apart from Kostarazi (sewage wastes) and Greveniotis (sewage and industrial wastes). The downstream sites are located on tributaries of the Aliakmon, the rivers Moglenitsas, Edesseos and the drainage channel, Canal 66 (see Figure 1). These tributaries are subject to agricultural (pesticides and fertilizers), urban (sewage and detergents) and industrial (mainly fruit and vegetable canneries) sources of pollution (Kouimtzis et al., 1992; Grammatikopoulou et al., 1992; Anagnostopoulou, 1992).

Materials and methods

Sampling of macroinvertebrates and water

Two samples of benthic macroinvertebrates were taken at each site with the 3-minute kick-sweep method (Armitage et al., 1983), using a standard pond net (surface 575 cm^2, mesh size 900 mm, depth 27.5 cm). During the three minute period, all the habitats at each site (submerged macrophytes, bankside macrophytes, bare substrate) were sampled. The samples were placed in plastic bottles and preserved in 4% formaldehyde. On return to the laboratory, the animals were sorted using a 500 mm mesh sieve and a magnifying lens and preserved in 75% ethanol. Identification of the invertebrates was made to the lowest possible taxonomic level.

Surface water was sampled using 1 l polypropylene sampling bottles. The samples were transfered in a portable refrigerator to the laboratory and the determinations were carried out in filtered water (except for total-P). Results of nutrient analyzes were provided by Dr Bobori.

Biotic scores

Having no Greek biotic score or index, five European biotic scores were applied to the samples: the Biological Monitoring Working Party (B.M.W.P.) score (Chesters, 1980), the Average Score Per Taxon (A.S.P.T.) (Armitage et al., 1983), the Lincoln Quality

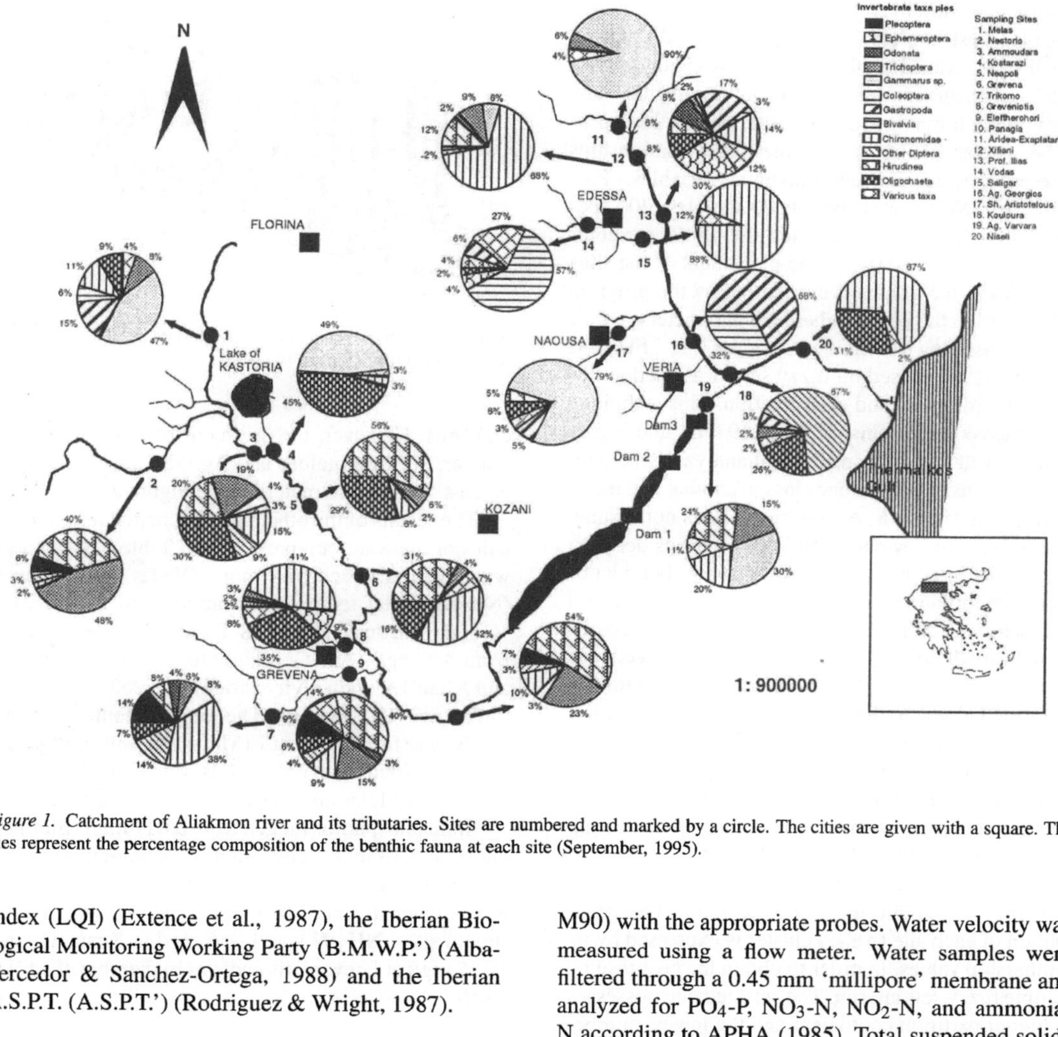

Figure 1. Catchment of Aliakmon river and its tributaries. Sites are numbered and marked by a circle. The cities are given with a square. The pies represent the percentage composition of the benthic fauna at each site (September, 1995).

Index (LQI) (Extence et al., 1987), the Iberian Biological Monitoring Working Party (B.M.W.P.') (Alba-Tercedor & Sanchez-Ortega, 1988) and the Iberian A.S.P.T. (A.S.P.T.') (Rodriguez & Wright, 1987).

Physicochemical parameters

Several physicochemical parameters (substrate, flow, pH, temperature, total suspended solids, total dissolved solids, conductivity, dissolved oxygen, PO_4-P, total-P,NO_3-N, NO_2-N, NH_3-N) were measured.

Substrate was estimated visually by percentage occurrence of the following particle categories using the Wentworth scale: clay/silt (<0.0625 mm), sand (0.0625–2 mm), gravel (2–4 mm), pebbles (4–16 mm), cobbles (16–256 mm), boulders (>256 mm). *In situ* measurements of pH, dissolved oxygen (D.O.), temperature, conductivity and total dissolved solids (TDS) were measured using a digital meter (Ciba Corning M90) with the appropriate probes. Water velocity was measured using a flow meter. Water samples were filtered through a 0.45 mm 'millipore' membrane and analyzed for PO_4-P, NO_3-N, NO_2-N, and ammonia-N according to APHA (1985). Total suspended solids (TSS) were measured as the weight of material retained by the fibre-glass filters. Total-P was carried out on unfiltered samples.

Statistical analyszs

The relationships between the macroinvertebrate taxa [transformed to ln(x+1)] and the physicochemical parameters and sampling sites were analyzed by Canonical Correspondence Analysis (program CANOCO, Ter Braak, 1988). This method escapes the assumption of linearity and is able to detect unimodal relationships between species and external variables. It is particularly suited for a forward selection of environmental variables in order to determine which variables best explain the species data. CANOCO, which is an extension of DECORANA (Hill, 1979a), selects a linear combination of environmental variables, while it maximizes the dispersion of the scores of the species.It does not assume linear relationships between species and environmental variables. Program CANOCO can analyze 1000 samples, 700 species, 75 environmental variables and 100 covariables (total data size < 80 000). The placement of the sites or species in the graphs is carried out by the program with regard to the physicochemical parameters

A clustering technique known as FUZZY, (Equihua, 1990) based on fuzzy sets, was used to obtain both ordination and classification. This technique is an extension of Twinspan (Hill, 1979b), and is particularly well suited for ecological analyzes where no clear-cut transition from one class to another and many intermediate situations may occur. It does not assume the existance of discrete benthic populations between the various stretches of a river system, but identifies the continuum and gradual change in their faunal composition. The maximum FUZZY partition coefficient is usually low (0.5–0.7) and it rarely exceeds the value of 0.9, which confirms the fact that communities are formed along gradients, without sharp boundaries. Two opposite requirements must coexist:

(a) the production of a discriminating partition enabling unambiguous feature extraction and,
(b) the intentional introduction of fuzziness to rank data memberships (Equihua, 1990).

The grouping of sampling sites is based on the assemblages of benthic macroinvertebrates. The participation of each site to each cluster is expressed by its membership values. In order to obtain the best results, multiple fuzzy sets are run. The selected one possesses the higher partition coefficient and the lowest number of clusters. The sites' membership values from the chosen fuzzy set are the ones used for their clustering.

Results

Physicochemical parameters

The upper reaches of the Aliakmon river are characterized by a substrate mainly of coarse material (cobbles, pebbles, gravel) except at Melas and Greveniotis where 50% was made of clay, (see Figure 2) whereas the majority of downstream sites were more depositional, consisting of finer particulates (gravel, sand, clay/silt). However, the sediment of Aridea, Xifiani, Saligar, Sh. Aristotelous and Ag. Varvara consisted of at least 50% coarse material (see Figure 2).

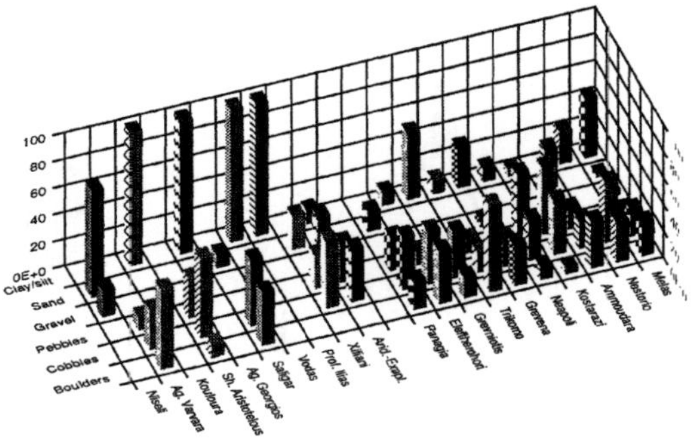

Figure 2. Percentage composition of substrate at sampling sites on the Aliakmon river (September, 1995).

The results of the other physicochemical characteristics of the water are presented in Table 1. Discharge was low and varied from none (Vodas) to 11.3 m^3/s (Niseli). Water temperature varied from 13.4 to 21.5 °C and pH from 7.11 to 8.58. Dissolved oxygen varied from 4.1 mg/l (Ag. Georgios) to 12.1 mg/l (Panagia and Xfiani). Conductivity varied from 262 mS (Melas) to 780 mS (Greveniotis). The range of total dissolved solids was from 139 mg/l (Melas) to 389 mg/l (Greveniotis). Total suspended solids were higher upstream (except for Melas and Nestorio) than downstream.

Total phosphorus ranged from 0.011 mg/l (Eleftherohori) to 3.18 mg/l. (Greveniotis). Ammonia was low, as were nitrites (NO_2-N) and nitrates (NO_3-N), which were higher in Canal 66 (Kouloura). Among the upstream sites, NH_3-N was 1.356 mg/l at Kostarazi and 1.159 mg/l at Greveniotis and was higher than the E.U. levels.

Biotic scores

Using the five biotic scores, the same pattern of results was obtained for the different sites but the A.S.P.T. and the Lincoln index were the most informative (as shown in Table 2), although the interpretation of the latter stands totally only for English running waters. According to the number of high scoring taxa and the Lincoln index, the upstream stations had excellent water quality, except for Kostarazi and Greveniotis which had good quality, whereas the downstream stations had moderate to very poor quality apart from Aridea, Xifiani and Ag. Varvara which had excellent water quality.

Table 1. Values of physicochemical parameters at Aliakmon sites (September, 1995). The values which exceed the EU suggested levels for drinking water are indicated with underlined characters and the ones exceeding the permitted levels with bold characters

Sites	D. O.%	D. O. mg/l	pH	Temp. °C	Cond. μS	TDS mg/l	TSS mg/l
Melas	82	9.2	7.28	13.4	262	139	2
Nestorio	89	8.6	8.16	18.85	276	143	7
Ammoudara	92	8.9	7.53	16.1	392	195	8.2
Neapoli	82	8.1	7.39	15.95	416	210	48.8
Kostarazi	75	7.1	7.57	16.6	628	316	5.3
Grevena	53	5.3	8.15	16.65	426	212	40.7
Trikomo	104	10	8.05	18.1	288	145	46.5
Greveniotis	75	7	7.79	19.2	80	389	10.1
Eleftherohori	105	9.2	8.57	23	348	176	51.1
Panagia	137	12.1	8.58	20.8	429	215	41.5
Arid.-Exapl.	101	9	7.76	18.7	437	217	2.2
Xifiani	139	12.1	8.33	19.8	392	196	6.4
Prof. Hlias	128	11	8.45	21.55	406	202	2.8
Bodas	111	10	7.11	18.8	501	205	1.1
Saligar	99	8.8	7.96	19.3	579	293	13.3
Ag. Georgios	45	4.1	7.46	20.7	523	262	1.8
Sch. Arist	94	9.3	7.95	14.2	506	249	15.2
Kouloura	65	5.6	7.4	20.2	510	255	4
Ag. Varvara	114	10.2	8.06	20.2	376	188	0.7
Niseli	75	6.8	7.67	20.8	471	236	4.9

Sites	Disch. m^3/s	NO_3-N mg/l	NO_2-N mg/l	NH_3-N mg/l	PO_4-P mg/l	tp mg/l
Melas	0.37	0.396	0.02	0.077	0.052	0.07
Nestorio	1.11	0.063	0.005	0.026	0.027	0.031
Ammoudara	4.31	0.968	0.034	0.177	0.095	0.3
Neapoli	8.5	1.177	0.012	0.041	0.035	0.102
Kostarasi	-	1.557	**0.293**	**1.356**	1.296	2.894
Grevena	2.1	1.112	0.006	0.039	0.090	0.096
Trikomo	1.43	0.032	0.005	0.024	0.028	0.047
Greveniotis	-	1.56	**0.37**	**1.159**	0.964	3.18
Eleftherohori	3.84	0.256	0.006	0.026	0.011	0.011
Panagia	8.6	0.471	0.011	0.028	0.047	0.06
Arid.-Exapl.	3.65	0.978	0.009	0.037	0.035	0
Xifiani	2.6	0.876	0.046	0.136	0.089	0.106
Prof. Hlias	3.5	0.827	0.033	0.065	0.074	0.078
Bodas	0	0.049	0.006	0.035	0.011	0.019
Saligar	0.83	1.862	0.046	0.061	0.128	0.135
Ag. Georgios	11.2	0.557	0.091	0.328	0.055	0.08
Sch. Arist	0.41	1.29	**0.11**	**0.909**	0.144	0.346
Kouloura	9.4	1.307	**0.103**	**1**	0.116	0.19
Ag. Varvara	-	0.535	0.011	0.051	0.019	0.044
Niseli	11.3	0.483	0.049	0.295	0.144	0.153

Table 2. Abundance of benthic macroinvertebrates, B.M.W.P. taxa and values of the biotic scores applied at Aliakmon sites (September 1995). The numbers in the parentheses show the abundance of benthic macroinvertebrates in April 1995. Wherever the abundance of macroinvertebrates is higher in September than in April, it is indicated with bold characters

Sites	Abundance of benthic macroinvertebrates	Number of BMWP taxa*	Number of high-scoring BMWP taxa (8 &10 score)	BMWP	A.S.P.T.	L.Q.I.			B.M.W.P.'	A.S.P.T.'
Melas	498 (4854)	12	5	80	6,66	5,5	A+	Excellent	77	6,42
Nestorio	378 (453)	12	7	83	6,9	6	A++	Excellent	102	6,8
Ammoudara	766 (4431)	22	9	149	6,77	7	A++	Excellent	157	6,54
Kostarazi	**5149** (4025)	14	3	71	5,07	4,5	B	Good	70	4,67
Neapoli	803 (1763)	15	7	96	6,4	6,0	A++	Excellent	93	6,2
Grevena	336 (3958)	13	4	77	5,92	5	A	Excellent	78	5,6
Trikomo	71 (617)	12	5	74	6,16	5,5	A+	Excellent	78	6
Greveniotis	444 (2320)	10	2	47	4,7	4,5	B	Good	62	4,77
Eleutherochori	97 (484)	14	7	97	6,93	6	A++	Excellent	95	6,33
Panagia	**785** (142)	15	7	103	6,87	6,5	A++	Excellent	108	6,35
Arid.-Exaplat.	**6449** (3916)	15	4	82	5,47	6,0	A++	Excellent	83	5,19
Xifiani	1267 (2840)	14	4	81	5,78	5	A	Excellent	77	5,13
Prof. Ilias	216 (561)	17	2	76	4,47	4,0	C	Good	73	4,29
Vodas	27 (49)	7	0	19	2,71	2	G	Poor	22	2,75
Saligar	24 (2406)	1	0	2	2	1	I	Very poor	6	3
Ag. Georgios	95 (365)	3	0	9	3	1,5	H	Very poor	38	3
Sch. Aristot.	**2610** (635)	10	1	38	3,8	3,0	E	Moderate	89	3,8
Kouloura	197 (3534)	5	0	17	3,4	2	G	Poor	17	2,8
Ag. Varvara	**407** (396)	8	1	43	5,37	5	A	Excellent	44	4,89
Niseli	370 (1935)	6	0	23	3,83	3	E	Moderate	29	3,63

*The BMWP taxa do not represent the total number of taxa per site, since several of them are not included in the biotic indices or scores.

Benthic macroinvertebrates

Twenty thousand, nine hundred and eighty-nine individuals were sampled and sixty-six benthic macroinvertebrate taxa were identified during this study. Several of them were not included either in the biotic indices or scores, because they are not taken into consideration in the different European indices or scores. The dominant taxa were: *Leuctra* spp. (Plecoptera), *Ecdyonurus* spp., *Ephemerella* spp., *Baetis* spp. and *Caenis* spp. (and occasionally *Ephemera* spp. and *Potamanthus luteus)* (Ephemeroptera), *Hydropsyche* spp. and *Psychomyia pusilla* (Trichoptera), Gomphidae (Odonata), *Ferrissia* spp., *Planorbis* spp., *Physa* spp., *Valvata* spp. (Gastropoda), *Gammarus* spp. (Amphipoda), Elminthidae (Coleoptera), Chironomidae and Tabanidae (Diptera), Erpobdellidae (Hirudinea) and several taxa of Oligochaeta. The abundance of benthic macroinvertebrates varied and at Kostarazi, Panagia, Arid.-Exaplatanos, and Sh. Aristotelous it was higher in September than in April, while at Ag. Varvara it was almost the same (as shown in Table 2).

Regarding the high-scoring taxa, Plecoptera were recorded at upstream sites (see Figure 1) and at Aridea-Exaplatanos, which is located in the northern part of Moglenitsas river. Among the upstream sites, Plecoptera were most abundant (highest number of individuals) at Panagia but accounted for only 7% of all macroinvertebrates found in that site. Trichoptera were recorded mainly at the upstream sites. The percentage composition of this taxon ranged from 0.06% (Kostarazi) to 48% (Nestorio). Ephemeroptera were found at all upstream and downstream sites, except for Saligar, Ag. Georgios and Kouloura. Ephemeroptera constituted from 0.8% (Niseli) to 56% (Neapoli) of the total taxa found per site. Odonata were recorded at eight upstream sites and at three downstream ones, but their percentage contribution was low (see Figure 1).

Concerning the low-scoring taxa, Diptera were present at all sites apart from Ag. Georgios, and Chironomidae were the dominant dipteran family. This order varied greatly in abundance and contributed between 0,003% (Kopanos) and 67% (Niseli) to the total taxa found per site. Amphipods *(Gammarus* spp.) were found at seven out of twenty sites and contributed between 0.7% (Ammoudara) and 90% (Arid. Expl.) of the taxa found. Oligochaeta were recorded at all sites apart from Saligar, Ag. Georgios and Ag. Varvara; at Kostarazi they accounted for 45%. Gastropoda were collected at five upstream sites and at six

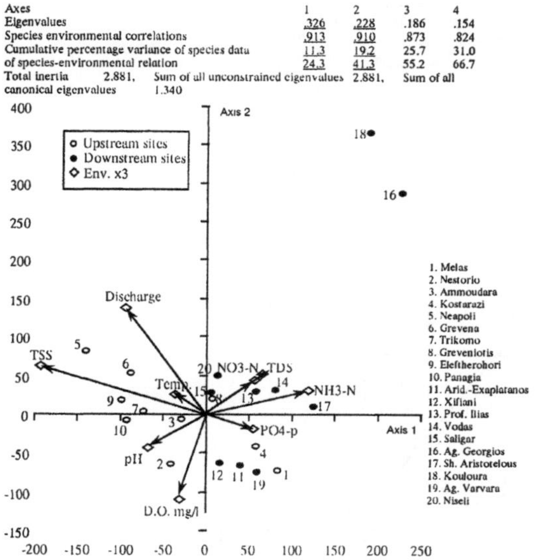

Figure 3. CCA diagram with physicochemocal variables and 20 numbered sites on the Aliakmon river, September 1995. Scale for the physicochemical variables multiplied by three. Parameters of CCA are given in the table below. The best inter set correlations of environmental variables with axes are underlined.

CorE: Inter set correlations of environmental variables with axes

Name	Axis 1	Axis 2	Axis 3	Axis 4
Extracted	0.1407	0.0925	0.1014	0.0718
D.O. mg/l	−0.1242	−0.4849	−0.2553	0.0930
pH	−0.27840	−0.1980	−0.4781	0.3150
Temperature	−0.1499	0.1061	−0.7507	0.3054
TDS	0.2629	0.2257	−0.0830	−0.1619
TSS	−0.7833	0.2691	0.1385	−0.0232
Discharge	−0.3827	0.6020	−0.0654	0.2275
NO3-N	0.2212	0.1881	0.0167	−0.5011
NH3-N	0.4746	0.1331	0.1539	−0.2373
PO4-P	0.2184	−0.0878	0.0273	−0.2437

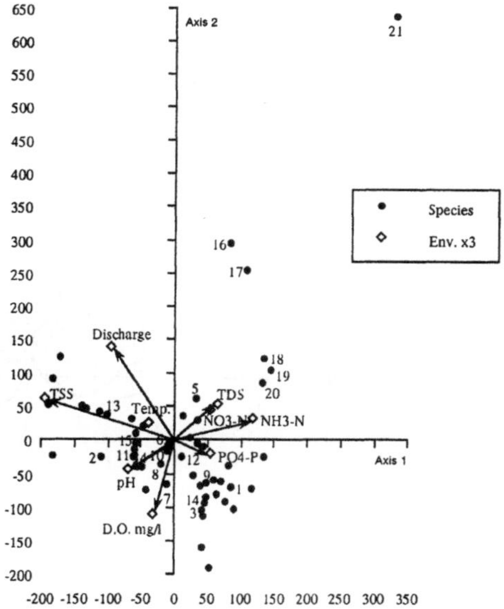

Figure 4. CCA diagram with invertebrate taxa and physicochemical variables for 20 sites on the Aliakmon river in September 1995 (invertebrate taxa are represented by numbers: 1=*Gammarus* spp., 2=Perlidae, 3=*Isoperla* spp., 4=*Leuctra* spp., 5=Oligochaeta, 6=Chironomidae, 7=Tabanidae, 8=*Hydropsyche* spp., 9=Coenagriidae, 10=Calopterygidae, 11=Gomphidae, 12=*Baetis* spp., 13=*Caenis* spp., 14=*Ephemerella* spp., 15=Elminthidae, 16=Ephydridae, 17=*Physa* spp., 18=*Valvata* spp., 19=*Planorbis* spp., 20=*Pisidium* spp., 21=Dytiscidae, Culicidae, Cyclorrhapha).

downstream ones accounting for 68% at Ag. Georgios. The rest of the taxa (Coleoptera, Hirudinea, Bivalvia, Hemiptera and Lepidoptera) never constituded more than 8% of the macroinvertebrate fauna at any site.

Statistical analyses

The successive eigenvalues of the four axes of CANOCO indicated a well-structured dataset (see Figures 3 & 4). The first two axes explained 55% of the variance. TSS (−0.7833) was the variable best correlated with the first axis, whereas the second axis was best correlated with discharge (0.6020). Interpreting the results of CANOCO site analysis, Ag. Georgios and Kouloura were set apart as outgroup sites because of their high discharge (see Figure 3). The upstream sites (except for Melas, Kostarazi and Greveniotis) were placed together in relation to TSS (see Figure 3). The downstream stations (including the three upstream stations mentioned above) were put (see Figure 3) on the opposite side of the TSS variable, which is on the positive side of axis 1. The nutrient parameters (NO_3-N, NH_3-N and PO_4-P) were the next variables best correlated to the first axis (see Figure 3). As to the macroinvertebrates, CANOCO placed the rare and unique taxa (e.g. Dytiscidae, *Physa* spp. etc.) at the edges of the diagram and the taxa characteristic of the upstream stations at the negative part of the first axis (see Figure 4) in relation to TSS.

With FUZZY analysis, the sampling sites were grouped into four clusters. (see Figure 5). The composition of the benthic macroinvertebrate communities was so different in Kouloura and Ag. Georgios as to justify the separation of these sites from all others as

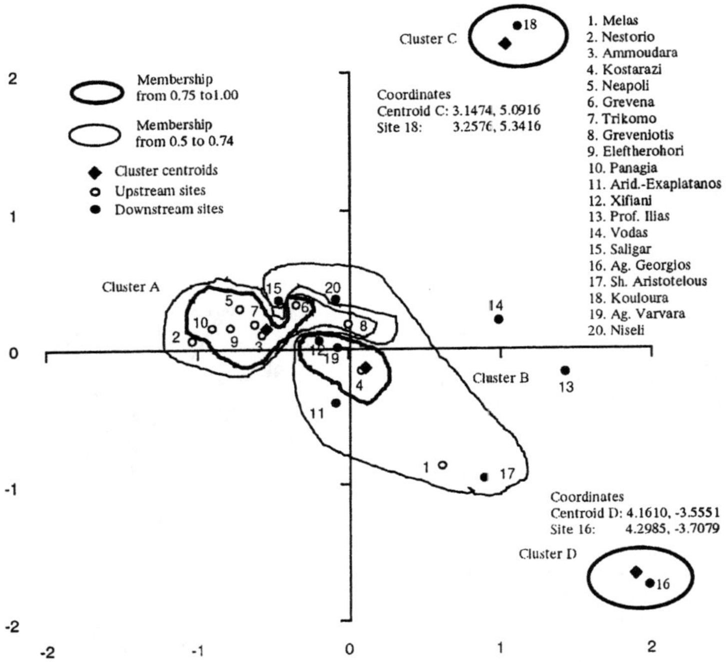

Figure 5. Classification of the Aliakmon river sites with the FUZZY method based on the invertebrate fauna. The scale of the two axes has been enlarged. The correct coordinates of clusters C anb D with their sites are given below and above their graphical representation. The membership of each site to each cluster is given below. The membership values which are between 0.5 and 0.75 are indicated with underlined characters whereas the ones 0.75 and more with bold characters in the table below. The convergence criterium and other parameters about the analysis are also indicated below the membership values.

Membership of the upstream sites					Membership of the downstream sites				
Sites/Clusters	A	B	C	D	Sites/Clusters	A	B	C	D
1	0.2355	<u>0.6736</u>	0.0465	0.044	11	0.2952	<u>0.6783</u>	0.0137	0.0128
2	<u>0.7116</u>	0.2206	0.0348	0.0330	12	0.1940	**0.7916**	0.0082	0.0063
3	**0.9365**	0.0586	0.0026	0.0022	13	0.3936	0.4137	0.0951	0.0976
4	0.0719	**0.9172**	0.0061	0.0048	14	0.3254	0.3346	0.1784	0.1615
5	**0.9210**	0.0667	0.0067	0.0057	15	0.3964	<u>0.5494</u>	0.0314	0.0228
6	**0.7801**	0.1972	0.0126	0.0100	16	0.0021	0.0020	0.0012	**0.9946**
7	0.9598	0.0362	0.0021	0.0018	17	0.2897	<u>0.6312</u>	0.0354	0.0437
8	<u>0.5171</u>	0.4335	0.0272	0.0222	18	0.0031	0.0031	**0.9923**	0.0015
9	**0.9286**	0.0612	0.0054	0.0047	19	0.1826	**0.7729**	0.0263	0.0182
10	**0.8663**	0.1107	0.0121	0.0108	20	0.2507	0.7306	0.0110	0.0076

The convergence criterium of the analysis and the description of the method	
Partition coefficient	0.5525
Membership Sum of Squares	Between: 0.02096 Within 1.65077
Eigenvalues	0.453/0.386/0.351/0.275/0.274
Number of sites	20
Number of taxa	66
Number of clusters	4
Transformation	$\ln(x+1)$
Axes	c-means (5)
Norm	Diagonal
Convergence Criterium	0.00010000

different clusters (C & D). Cluster A comprised the upstream sites of Aliakmon river apart from Melas and Kostarazi, whilst cluster B included the downstream stations (except for Ag.Georgios, Kouloura and Prof. Ilias and Vodas from which the first two were differentiated as seperate clusters and the third as outgroup) with Kostarazi and Melas.

Discussion

River management (e.g. damming), point sources of pollution and diffuse loads of agricultural runoff all affect macrobenthos communities. In the case of Aliakmon, the human interference in the natural flow of the river is evident since three artificial lakes divide the river into two parts. The first one flows through a mountainous and woodland area with small towns, receiving little sewage (at Kostarazi and at Greveniotis) and little industrial (at Greveniotis) wastes. The second part, after the lakes, flows through a plain area with cities, extended agricultural land and industrial activities (Kouimtzis et al., 1992; Grammatikopoulou et al., 1992). The upstream part of the river (except for the stations Melas, Kostarazi and Greveniotis) was rather homogeneous having more sensitive and thus high scoring taxa than the downstream stations (as shown in Table 2), whereas the low scoring taxonomic groups were almost equally present in all of the sites (Figure 1). Slight differences, among the rest upstream sites, as to their fauna composition, were probably due to hydrological or physicochemical differences (as shown in Table 1).

Melas, despite its physicochemical characteristics (as shown in Table 1) which indicated good water quality, appeared to have similarly poor fauna with dowstream sites in September. This was not in agreement with the results of the servey in the high flow season (April, 1995), where this site was very characteristic for the richness and the sensitivity of its fauna (Yfantis et al., in press). In this study, although the temperature was low enough to permit the presence of Plecoptera, the most sensitive taxon, (Hynes, 1970; Hellawell, 1986), only one specimen was captured. In contrast, at Kranionas, a site near Melas in the catchment area of Aliakmon, with similar physicochemical characteristics, Copeland et al. (1997) found a lot of Plecoptera in September 1995 (75 individuals were captured). Since there aren't any known intensive human activities near Melas, the absence of Plecoptera at this site was probably due to: a) the erosion effect which caused a change in Melas substrate (almost 50% clay/silt) inhabited by low-scoring taxa (whereas in April it had no clay/silt) and b) the decreased discharge in all of the sites (in April the mean discharge was 16.6±3.5 m^3/s whereas in September (d)=4.3±1 m^3/s) and especially in Melas station (2.19 m^3/s in April and 0.37 m^3/s in September). Additionally, in September 1995, at Kranionas, the substrate comprised less silt/clay (38%) and the dicharge was higher than at Melas (0.59 m^3/s) and so the Plecoptera were abundant.

The performance of the biotic scores (as shown in Table 2) also showed 'excellent' water quality at the upstream sites with the exception of Kostarazi and Greveniotis. However, Melas, despite having 'excellent' water quality according to the indices, was placed with the downstream stations by CANOCO and FUZZY, too, for the reasons mentioned above. This proves that the European indices do not totally match to the Greek hydrological peculiarities, and that they have to be modified or new ones must be created. The deficiencies observed in their performance are probably due to the fact that they do not include several taxa found in the samples (e.g. families of Diptera and Ephemeroptera). In addition, the biotic scores used are mainly based on the presence-absence of specific taxonomic groups and do not take their abundance into account, thus they are not quite as effective in distinguishing running waters of moderate, good and/or excellent quality. In April, all the biotic indices showed 'excellent' water quality at Melas and its benthic fauna was rich and sensitive.As a result both statistical analyzes (CANOCO and FUZZY) regarded this site as an outgroup. Kostarazi and Greveniotis, being located downstream of important point sources of pollution (Kastoria town and Grevena town) had their fauna composition affected by them. The performance of the biotic indices showed 'good' quality of water and consenquently they were both placed with the downstream sites by both statistical methods, while in the April 1995 survey, only Kostarazi was placed with the downstream stations.

Concerning the physicochemical parameters, the fact that total suspended solids were higher in the upstream sites is probably due to the high gradient of upstream sites. During the high and the low flow season, at Aliakmon river conductivity was higher than the E.U. suggested levels (400 mS) (EU directive 80/778/15.07.80) for drinking water in most of the stations. Eleftherohori had the highest suspended solids, because of erosion effect. Kouloura (Canal 66) from the downstream sites, and Kostarazi and Greveniotis from the upstream ones, had the highest NH_3-N, NO_2-N and NO_3-N (higher than E.U. levels) because they receive sewage as well as industrial wastes, whereas Canal 66 (Ag. Georgios, Kouloura) also receives a lot of agricultural wastes (Grammatikopoulou et al., 1992). The concentrations of the nutrients at the above mentioned three sites were higher during the low flow season than in the high flow season due to the lower discharge. Additionally, Kouloura site was more affected during the low flow season than in the high flow one because of the seasonal industrial activities in the cathment area of Canal 66 (Grammatikopoulou et al., 1992).

Comparing the results of this study with the survey undertaken at Aliakmon river in April 1995 (Yfantis et al., in press), the benthic fauna was less diverse in September (113 taxa found in April, while only 66 in September) and less abundant (20 989 individuals captured in September and 39 684 individuals in April) (as shown in Table 2). At Kostarazi, Aridea-Exaplatanos and Sh. Aristotelous though, the total number of individuals was higher in September than in April due to the great abundance of Ampipoda, while at Panagia the total abundance was higher in September because of the Ephemeroptera *Baetis* sp. and *Ecdyonurus* sp. Additionally, high-scoring taxa were less in September than in the high flow season. Hence, the biotic indices and scores had lower values indicating that the water quality of Aliakmon was poorer during the low flow season than during the high. It seems that in the mediterranean ecosystems, the water quality becomes worse during the low flow season, especially in the rivers where seasonal industrial activities coincide with the low discharge (Anagnostopoulou, 1992).

According to the statistical analysis, CANOCO, the upstream sites (except for Melas, Kostarazi and Greveniotis) were placed at the negative part of axis 1, which was mainly correlated with TSS (see Figure 3). Melas, Kostarazi and Greveniotis were set together with the downstream sites at the positive part of axis 1, which was well correlated with the nutrient parameters (NO_3-N, NH_3-N and PO_4-P). Melas was placed with the downstream sites because of the similarity of its benthic fauna with theirs, whereas Kostarazi and Greveniotis because of their similar fauna with the downstrean stations and their high values of nitrates, ammonia and orthophosphates. Kostarazi was comprised mainly of Oligochaeta and Gammaridae and Greveniotis of Oligochaeta and Diptera. Their water quality was poorer since the levels of NO_2-N and NH_3-N were much higher than the other upstream sites and higher than the EU permitted levels as a result of the sewage wastes from Kastoria in the case of Kostarazi and sewage and industrial wastes at Greveniotis. Although Melas was the first site in the system and the chemical caracteristics of its water were almost similar to that of the other upstream sites, it was placed with the downstream stations because of its lower TSS and discharge and the fact that it was mainly comprised of low-scoring taxa (Gastropoda 15%, Gammaridae 47% and Diptera 11%). Among the downstream sites, some were differentiated because of their fauna (Ag. Georgios) or/and water quality (Kouloura). Comparing with the results of CANOCO analysis in April (Yfantis et al., in press), the downstream sites were differentiated from the upstream stations as did in this study, with two exceptions: Kostarazi and Niseli. The former, Kostarazi, affected by the sewage of Kastoria, was placed with the downstream sites with poorer water quality for the same reasons as in September. The latter, although it was the last sampling site on the river body, it was placed with other upstream sites which had very good to excellent water quality according to the indices (the fauna) and the physicochemical parameters, showing the high purification capacity of the system. However, this was not possible to be detected chemically in the surface water of this station (Niseli) in April. The deficiency to detect chemically the purification capacity of a river was also observed in other Greek rivers as Aggitis (Tsiaousi, 1995) and Maurolakkas (Ford et al., 1998). On the contrary, in September when the discharge was very low, the purification capacity of the river was very low and Niseli was characterized by poor fauna and hence was placed with the downstream sites in CANOCO analysis.

FUZZY functioned satisfactorily. According to this analysis, four clusters with different characteristic assemblages of benthic invertebrates were formed. Most of the upstream sites were grouped together indicating that the upper part of the system is differentiated from most of the downstream stations because of the artificial lakes that divide the continuity of the river. The only exceptions were Melas and Kostarazi which were grouped with the downstream sites for the reasons mentioned above. Greveniotis was clustered with the upstream stations as in April 1995, although it was placed with the downstream sites according to CANOCO. The downstream sites though, seemed to be less similar among them than the upstream stations. Two downstream sites (Prof. Ilias and Vodas) were not grouped with any of the clusters and Ag. Georgios and Kouloura formed two different clusters by themselves. These two stations were differentiated because of their fauna (Ag. Georgios) or/and water quality (Kouloura) (as shown in Table 1 & Figure 1). The major discontinuity provoked by the presence of the artificial lakes, was accentuated , especially in the downstream sites, by the hydrological extremes which characterize the mediterranean and by the seasonal industrial activities. In the survey carried out in April though, three sites from the upstream stations having the best biological water quality formed a seperate group from the rest upstream stations. Additionally, in April, Kostarazi was clustered with the downstream sites as in September and Niseli was grouped with the upstream ones (with very good water quality) indicating the purification capacity of the system. Among the downstream stations, Kouloura formed a seperate cluster by itself as in September.

A FUZZY analysis using all the stations studied in the Aliakmon river in both April and September also grouped the upstream stations together apart from Melas-September, Kostarazi-April & September, and Greveniotis-September. This clustering shows that Aliakmon river is divided in two parts in relation to the artificial lakes and the affect of the seasonal industrial activities which take place downstream of the lakes.

The above multifactorial analyzes that take into account the abundance of each taxonomic group, with or without the inclusion of physicochemical parameters, proved to be more appropriate than the European biotic indices or scores for the classification of sampling sites based on the structure of their biocommunities. The same conclusion was reached when a similar study was carried out in Maurolakkas river (Chalkidiki, Greece) where pollution was inorganic and not organic (Ford et al., in press). It seems that in our case, a creation of a Greek biotic index or score based on the abundance too, as Chandler's Score System (Cook, 1976) does, would overcome the problems and dangers appearing with the use of indices or scores developed elsewhere, particularly when they have labels attached (e.g. 'excellent', 'good' etc.). One of the best ways, though, for monitoring the ecological quality of running waters is the use of predictive models like RIVPACS (Wright et al., 1989; Wright, 1995) which use multifactorial methods.

Acknowledgements

This study was carried out as part of the Environmental and Climate Research Program (BINOCULARS) (DG XII) and PENED 96–98 program. The latter was fiananced by the Greek Ministry of Development (Direction of Research and Technology). We would like to thank Parathyra Natasa, Ganias Kostas and Katsaris Nikos for carrying out the samplings for this study, and the unknown referee for the valuable comments on a previous draft of this paper.

References

Alba-Tercedor, J. & A. Sanchez-Ortega, 1988. Un metodo rapido y simple para evaluar la calidad biologica de las aguas corrientes basado en el de Hellawell (1978). Limnetica 4: 51–56.

American Public Health Association, 1985. Standard methods for the examination of water and wastewater. 16th edn. American Public Association Inc., Washington D. C.

Anagnostopoulou, M., 1992. The relationship between the macroinvertebrate community and water quality, and the applicability of biotic indices in the River Almopeos system (Greece). M. Sc. Thesis., University of Manchester: 133 pp.

Armitage, P. D., D. Moss, J. F. Wright & M. T. Furse, 1983. The performance of a new biological water quality score system based on macroinvertebrates over a wide range of unpolluted running water sites. Wat. Res. 17: 333–347

Barmuta, L. A., 1988. Habitat patchiness and macrobenthic community structure in an upland stream in temperate Victoria, Australia. Freshwat. Biol. 21: 223–236.

Bornovas, J. & Rondogianni-Tsiambau, 1983. Geological map of Greece. Scale 1:500.000. IGME.Greece

Cajaraville, M. P., Y. Robledo, M. Etxeberria & I. Marigomez, 1995. Cellular biomarkers as useful tools in the biological monitoring of environmental pollution: Molluscan diagestive lysosomes. In M. P. Cajaraville (ed.), Cell Biology in Environmental Toxicology. University of the Basque Country Press Service, Bilbao: 29–55.

Cajaraville, M. P., I. Abascal, M. Etxeberria & I. Marigomez, 1995. Lysosomes as cellular markers of environmental pollution: time and dose-dependent responses of the diagestive lysosomal system of mussels after petroleum hydrocarbon exposure. Envir. Toxicol. Water Qual.: 10: 1–8.

Castella, E., M. Bickerton, P. D. Armitage & G. E. Petts, 1995. The effects of water abstractions on invertebrate communities in U.K. streams. Hydrobiologia 308: 167–182.

Chesters, R. K., 1980. Biological Monitoring Working Party. The 1978 National testing exercise. Departement of Environment. Water Data Unit. Technical memorandum 19: 1–37.

Cook, S. E. K, 1976. Quest for an index of community structure sensitive to water pollution. Envir. Pollut., 11: 269–288.

Copeland, R. S., M. Lazaridou-Dimitriadou, V. Artemiadou, G. Yfantis, K. N. White & S. Mourelatos, 1997. Ecological quality of the water in the catchment of river Aliakmonas (Macedonia, Greece). Proceedings of the 5th Conference on Environment Science and Technology, Molyvos 2: 27–36.

Economidis, P. S., M. E. Kattoulas & A. Stefanidis, 1981. Fish fauna of the Aliakmon river and the adjacent waters (Macadonia, Greece). Cybium, 3e série 5: 89–95.

Equihua, M., 1990. FUZZY clustering of ecological data. J. Ecol. 78: 519–534.

European Union Commission, 1978. Directive on the quality of fresh water for the protection and improvement of fish life. Official Journal of the European Communities, No 222/1/ 14.8.78.

European Union Commission, 1980. Directive on the quality of drinking water. Official Journal of the European Communities, No 80/778/15.07.80

European Union Commission, 1997. Proposal for a Council Directive establishing a framework for Community action in the field of water policy. Official Journal of the European Communities, No C 184/20, 17.6.97.

Extence, C. A., A. J. Bates, W. J. Forbes & P. J. Barham, 1987. Biologically based water quality management. Envir. Pollut. 45: 221–236.

Ford, J., G. Yfantis, V. Artemiadou, M. Lazaridou-Dimitriadou & K. N. White, 1998. Ecological evaluation of water quality in a Greek river (river Mavrolakkas Olympiada, Halkidiki) during the months of May – August 1997. Proceedings of International Conference 'Protection and Restoration of the Environment IV', Sani Halkidiki 1: 144–152.

Grammatikopoulou, N., S. Rekkas, G. Darginis, C. Pitsiavas, D. Mandalis, N. Papadopoulos, A. Papadimos & G. Papaioannou, 1992. Pollution of Canal 66 and River Loudias from the industrial wastewater and the urban sewage of the OTA of Pella and Hemathia provinces. Greece Republic. Thessaloniki (in Greek).

Graca, M. A. S., D. M. Fonseca & S. T. Castro, 1989. The distribution of macroinvertebrate communities in two Portuguese rivers. Freshwat. Biol. 22: 297–308.

Hill, M. O., 1979a. DECORANA – A Fortran program gor detrended correspondence analysis and reciprocal averaging. Ecology and Systematics, Cornell University, Ithaca, New York.

Hill, M. O., 1979b. TWINSPAN – A Fortran program for arranging multivariate data in an ordered two-way table by classification of individuals and attributes. Ecology and Systematics, Cornell University, Ithaca, New York.

Hellawell, J. M., 1986. Biological indicators of freshwater pollution and environmental management. Elsevier Applied Science Publisher, London.

Hynes, H. B. N., 1970. The Ecology of Running Waters. Liverpool University Press.

Karr, J. R. & E. N. Chu, 1999. Restoring Life in Running Waters. Better Biological Monitoring. Island Press, U.S.A.

Kouimtzis, T., C. Samara, S. Sklavounos, P. Albanis, D. Voutsa & G. Zachariadis, 1992. Analytical determinations and characterization of the quality of surface water: Aliakmon case. Greece General Secretary of Research and Technology (in Greek).

Kouimtzis, T., C. Samara, D. Voutsa & G. Zachariadis, 1994. Evaluation of chemical parameters in Aliakmon river, northern Greece Part I: Quality Characteristics and Nutrients. J. Envir. Sci. Health A29: 2115–2126.

Lafont, M. & J. Juget, 1988. Les Oligochètes: utilisation pratique. In D. Champiat & J. P. Larpent (eds), Biologie des Euax. Méthodes et Techniques, Masson, Paris: 255–259.

Mason, C. F., 1991. Biology of Freshwater Pollution. Longman Publishers Ltd, Singapore.

Metcalfe, L. J., 1989. Biological water quality assessment of running waters based on macroinvertebrate communities: history and present status in Europe. Envir. Pollut. 60: 101–139.

Prat, N., M. A. Puig, G. Gonzalez, M. J. Tort & M. Estrada, 1984. The Liobrebregat: a mediterranean river fed by the Pyrenées. In B. A. Whitton (ed.) Blackwell. 527–552.

Quinn, J. M. & C. W. Hickey, 1990. Magnitude of effects of substrate particle size, recent flooding, and catchment development on benthic invertebrates in 88 New Zealand rivers. New Zealand J. Mar. Freshwat. Res. 24: 411–427.

Rodriguez, P. & J. F. Wright, 1987. Biological evaluation of the quality of the three Basque water courses. Proceedings of the II International Basque Congress. Leioa (Spain) 2: 223–243.

Ter Braak, C. J. F., 1988. CANOCO – a FORTRAN program for canonical community ordination (version 2.1).Tecnical report: LWA-88-02.

Tsiaousi, V., 1995. Assessment of the water quality and of the applicability of biotic indices with the use of benthic macroinvertebrates in the river Agitis system. M. Sc. Thesis, Aristotle University of Thessaloniki. Master's dissertation . Aristotle University of Thessaloniki (in Greek).

Vukmirovic, M., N. Bihari, R. K. Zahn, W. E. G. Muller & R. Batel, 1994. DNA damage in marine mussel *Mytillus galloprovinciallis* as a biomarker of environmental contamination. Mar. Ecol. Prog. Ser. 109: 165–171.

Woodiwiss, F. S., 1964. The biological system of stream classification used by the Trent River Board. Chem. Industry 11: 443–447.

Wright, J. F., 1995. Development and use of a system for predicting the macroinvertebrate fauna in flowing waters. Austr. J. Ecol. 20: 181–197.

Wright, J. F., P. D. Armitage & M. T. Furse, 1989. Prediction of invertebrate communities using stream measurements. Regulated Rivers, Res. Mgmt. 4: 147–155.

Yfantis, G., V. Artemiadou, M. Lazaridou-Dimitriadou & S. Mourelatos, 1996. Ecological evaluation of water quality in the river Aliakmonas (Macedonia, Greece). Proceedings of 7th International Congress 'Zoogeographical and Ecology of Greece and adjacent regions'. Athens, (in press).

Zarfdjian, M., D. Bobori & S. Mourelatos. Zooplankton and abiotic factors in the Aliakmon river (in prep.).

Hydrobiologia **410:** 59–68, 1999.
J. Garnier & J.-M. Mouchel (eds), Man and River Systems.

Multiple human impacts by the City of Paris on fish communities in the Seine river basin, France

Philippe Boët, Jérôme Belliard*, Romuald Berrebi-dit-Thomas* & Evelyne Tales
Cemagref, *Water Quality and Hydrology Research Unit, Parc de Tourvoie, F-92163 Antony, France*
**Present address: CSP, French National Fishery Council, 3 Rue Sainte-Marie, F-60200 Compiègne, France*

Key words: fish communities, Seine river, human impacts, watershed management, navigated rivers, regulated rivers, water quality

Abstract

The Seine, which flows through greater Paris whose population is nearly 10 million, is one of the most severely damaged European rivers. Considering several published studies, this paper examines the consequences for the fish communities of the development of Paris. Three major impacts are studied: river modifications for navigational purposes, control of river flow by storage reservoirs and degradation of water quality downstream of the City. At the basin scale, river modifications to improve navigation greatly affect the species composition of the fauna. Construction of weirs and locks along all the wider reaches of the river and its major tributaries have progressively caused the extinction of seven migratory species. Conversely, links with other catchments through waterways have favoured a colonization of the Seine river basin by several species. Locally, at the reach scale, channelling leads to homogenization of habitat conditions, which results in a general decrease of specialized species. Several storage reservoirs regulate the upper Seine basin to control floods and maintain minimum low-water flow in the lower reaches. Their overall effects on fish communities are still poorly documented. However, a detailed study reveals their negative consequences for the pike populations. Under natural flow conditions, one year out of two was suitable for pike reproduction. Since the development of flow regulation, suitable conditions for reproduction are only observed in one year out of five. Because of its very large population, the city of Paris dramatically affects the water quality of the Seine. In the last 30 years, organic pollution has been strongly reduced. However, at present, water quality degradation still causes noticeable decrease in fish species richness downstream from Paris. In summer low-flow conditions, storm events lead to sudden decreases in dissolved oxygen, causing massive fish mortality downstream from the city. In such situations, tributaries, side-channels, sandpits and areas just below weirs, constitute valuable shelters for many species.

Introduction

Paris and the Seine are intimately linked. The prosperity and expansion of Paris are due to the Seine. Likewise, without the development of human activities in the city, the river would never have become what it is today. To this long story, the present fish communities are reliable witnesses.

At the scale of the basin, the direct or indirect perturbations caused by the megalopolis in the river system and, thus, in the organization and functioning of the fish communities are numerous.

In this paper, we first briefly relate the history of the fish fauna of the Seine and summarize the major consequences of the basin development. At this time–space scale, the situation is similar to that of the majority of rivers in developed European countries (e.g. Dodge, 1989; Petts et al., 1989). We then focus on three main factors affecting fish populations, which are chiefly due to the presence of Paris and its more than ten million inhabitants on the river, which in view of its discharge is relatively small: first, the physical modifications of the aquatic ecosystem due to navigational requirements; second, the flow regulation caused by storage reservoirs aimed both at supplying water resources and preventing flooding of Paris and its suburbs; third, the water quality degradation caused by the outflow from the sewage treatment plant

Table 1. Global human effets on the fish fauna.

		Number of species	
Original fauna		27–33	
Introduction	Europe & Asia	13	40% of the actual fauna
	North America	6	
Extinction		7	20% of the original fauna
Actual fauna		46	

downstream from Paris. This work summarizes several detailed studies carried out on the fish communities of the entire Seine basin (Belliard, 1994; Belliard et al., 1995, 1997, Boët et al., 1998) as part of the 'Piren-Seine' scientific programme (Meybeck et al., 1998).

Historic background

At the historical scale, human intervention in the basin has resulted in a global enrichment of the fauna (see Table 1, and Annex).

The original fish fauna, comprising about 27–33 native species, was first confronted with introduction and colonization of exotic species from Eastern Europe and Asia. The common carp (*Cyprinus carpio*) was the first introduced species and it was spread over the whole European continent by the Roman Legions during Antiquity (Balon, 1974; Spillmann, 1961).

Later, in the Middle Ages, noblemen and religious communities helped to scatter many cyprinids from fish-farming ponds over hydrographic basins, for example tench (*Tinca tinca*) and rudd (*Scardinius erythrophtalmus*). Then, interconnections between adjacent and distant catchments through canals facilitated successful colonizations by several species from Eastern European basins, such as nase (*Chondrostoma nasus*) or pikeperch (*Stizostedion lucioperca*).

At the end of the 19th century, these colonizations were followed by voluntary introductions of North American species: rainbow trout (*Oncorhynchus mykiss*), black bass (*Micropterus salmoides*), pumpkinseed (*Lepomis gibbosus*), and black bullhead (*Ameiurus melas*).

Species extinctions are recent and date from the beginning of this century. With the exception of the eel (*Anguilla anguilla*), all the migratory species were eliminated: sturgeon (*Acipenser sturio*), salmon (*Salmo salar*), marine lamprey (*Petromyzon marinus*) and shads (*Alosa alosa* and *Alosa fallax*). This was due to the construction of weirs and locks along the river, which definitively closed the passage to their upstream spawning grounds.

Today, 46 species are found across the basin (see Annex). The longitudinal organization of the communities still conforms to the theoretical zonation described by Thienemann (1925), Huet (1949, 1959), Verneaux (1973, 1977). Local species richness increases progressively along the upstream–downstream gradient, at least up to stream-order five (Figure 1), and the trout, grayling, barbel, and bream zones are well marked (Belliard, 1994; Belliard et al., 1997; Oberdorff et al., 1993). After stream-order five, the decrease of the species richness is due to the intensification and cumulative effects of human impacts. However, disturbances induced by human activities are evident both in headwaters and lower reaches. They are due to physical modifications of the stream habitats or to pollution (Figure 1). Most of them relate to the multiple demands of past or present human activities in Paris and its suburbs.

The effects of navigation

As both its emblem and its motto — *Fluctuat nec mergitur* — indicate, Paris owes its prosperity and its expansion to navigation. Indeed, from the Celtic *Parisii* tribe to the contemporary bargemen, this has always been a flourishing activity. The communities of river traders have always been powerful, thanks to the development of commerce and to the growing needs of the City (Manéglier, 1992; Velay, 1992).

For a long time, navigation adapted itself to the imposed river conditions. Then, with the increase of human technical capacities to transform the environment, rivers were progressively adapted to navigational requirements. This resulted in great physical modifications of the whole fluvial ecosystem with severe detrimental effects on the fish populations and communities (Table 2).

As already mentioned, at the basin scale, navigation has led to a loss of integrity of the original fish fauna (Belliard, 1994).

On the one hand, the canal links with waterways in adjacent and distant catchments are directly responsible of the colonization by species from Eastern European riverine systems, such as ruffe (*Gymnocephalus cernua*), nase (*Chondrostoma nasus*) and pikeperch (*Stizostedion lucioperca*) (Goubier, 1975; Mor-

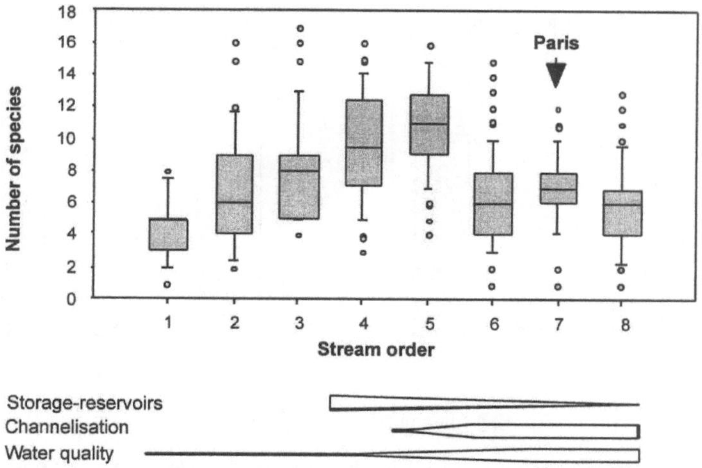

Figure 1. Longitudinal evolution of the local species richness in the Seine river basin (from Belliard 1994). Bold lines indicate the major human perturbations.

Table 2. Historical effects of navigation.

Scale	Intervention	effect	Fish changes	
Basin	Canal links between catchments	Modifications of the fauna	Colonization	3–6 species
	Weirs and locks		Extinction	7 migratory species
Reach	Channelization	Modifications of the communities	General decrease of the rheophilous component	

eau, 1881; Nelva-Pasqual, 1985; Spillmann, 1961), while probably facilitating that by barbel (*Barbus barbus*) (Persat & Berrebi, 1990), French nase (*Chondrostoma toxostoma*) and Crucian carp (*Carassius carassius*). On the other hand, the construction of weirs and locks along the river definitively closed the passage to the upstream spawning grounds and caused the progressive extinction of seven migratory species: sturgeon (*Acipenser sturio*), salmon (*Salmo salar*), sea lamprey (*Petromyzon marinus*), sea trout (*Salmo trutta trutta*), the two shads, allis shad (*Alosa alosa*) and twaite shad (*Alosa fallax*), and European smelt (*Osmerus eperlanus*) (Lavollée, 1902; Magnin, 1959; Moreau, 1937; Persat & Keith, 1997; Roule, 1920; Spillmann, 1961).

In this regard, the situation of the Seine basin resembles that of other European basins (Petts et al., 1989), particularly the Meuse basin where three species were introduced through canals, and seven migratory species became extinct (Micha & Borle, 1989).

Besides these global modifications of the Seine river fauna, the intensifying human pressure since the mid-19th century has resulted, at the reach scale, in a general decrease of the rheophilous component of the fish community (Belliard et al., 1995).

The primary mechanism of channelization consists in simplifying the channel topography by widening and deepening the channel, removing meanders and eliminating instream riparian features. This leads to a homogenization of the flow and thus to a dramatic loss of habitat diversity. A comparison with a sub-natural bypassed reach shows that this destruction of habitats in the modified sections results in strong structural alterations of the local communities. In terms of both ecological and reproductive groups, the most specialized species decrease while the most tolerant ones increase (Figure 2).

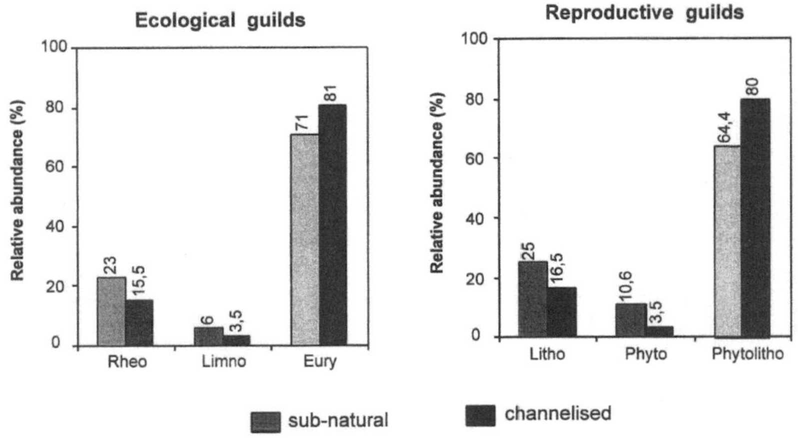

Figure 2. Structure of the fish communities between sub-natural and channelized reaches in the Seine. Fish species are classified according to Balon (1975) and Schiemer & Waidbacher (1992). (Rheo, rheophilic species; Limno, limnophilic species; Eury, eurytopic species; Litho, lithophils; phyto, phytophils; phytolitho, phytolithophils).

Another important consequence of the channelization is the isolation of the backwaters and floodplains from the main channel (Holcik, 1990; Schiemer et al., 1991). In natural situations, these backwaters provide spawning grounds and nursery habitats for many species (Berrebi-dit-Thomas et al., 1999; Tales et al., 1996). The disappearance of such off-channel areas drastically reduces the fish-breeding value of the channelized reaches. In the Seine, the channelization is often accompanied by the creation of new artificial backwaters, such as sandpits or cut-off meanders. However, the conditions of these new environments are not as favourable to reproduction as natural backwaters. Indeed, in these artificial backwaters, the young-of-the-year (0+) fish assemblages are less dense (Figure 3) and less diversified than in natural backwaters (Berrebi-dit-Thomas & Boët, 1997; Tales et al., 1996).

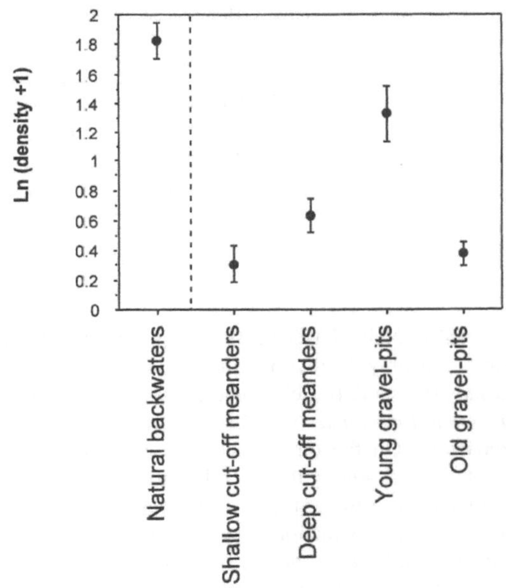

Figure 3. Young-of-the-year densities in 29 different types of backwaters within a sub-natural and a navigated fifth-order reach of the Seine River.

Storage reservoirs

The water supply of Paris has always been an important concern and has led to modifications of the Seine river morphology and functioning. The Romans were the first to divert streams to bring water into the City by way of aqueducts (Husson, 1996).

Nowadays, dams and reservoirs regulate the upper reaches of the Seine River and its major tributaries, the Aube and the Marne. These works have a two-fold objective: flood control and maintenance of minimum low-water flow. However, the predominant concern is to supply high-quality drinking water to Paris. The dams store water from December to June (by topping eventual flood peaks), and then release it from July to November.

The impacts of the flow regulation on the ecosystem are manifold. Depending on the operating mode, they greatly modify the natural flow characteristics of the river by reducing the periods and duration of recurring floods. The discharges to which biological components have adapted themselves are disturbed. This changes the structure mostly of the downstream fish communities, particularly running-water cyprinids which tend to decrease in number (Mann, 1988).

In the Seine, even still-water fish such as pike (*Esox lucius*) are influenced. As this species prefers flooded-natural grassland for spawning, it is a reliable indicator of good lateral water exchanges with the main riverbed.

Modelling the reproduction habitat of pike helps to quantify the effects of the hydraulic works (Boët & Monfort, 1998; Monfort et al., 1996). This model is an adaptation of the Instream Flow Incremental Methodology (Bovee, 1982) to a river and its floodplain using pike as the reference species. It consists in quantifying the potential carrying capacity of a stream section in the case of a given species at different stages of its development, as a function of discharge values, by coupling a hydraulic and a biological model (Bovee, 1982; Souchon et al., 1989). In this case, the potential areas suitable for pike reproduction are studied.

It provides quantitative assistance in selecting an optimal reference flow that satisfies pike requirements, while protecting the interests of riverside residents. For instance, a discharge of 175 m^3/s was considered sufficient for a site upstream from Paris, where the river remains largely natural despite hydraulic works and farming, with a main riverbed that is still a flood expansion zone, and where functional lateral components, such as natural cut-off meanders, oxbows, and flood grasslands, are abundant. Indeed, the increase in surface area of grassland flooded by greater discharges does not imply a proportional increase in carrying capacity for spawning pike (Boët & Monfort, 1998; Monfort et al. 1996).

Moreover, a 33-year compilation of discharge time-series (1960–1993) allows comparisons between different types of simulated discharge series: natural flow, flow regulated by a single dam, and flow regulated by two dams. They show a significant and general decrease of the area useful for pike reproduction in the case of flow with one and two dams. In addition, instead of one year out of two being suitable for pike reproduction under natural conditions, the regulation by two dams results in one suitable year out of five and, more importantly, in four unsuitable years that did not exist before (Table 3; Boët & Monfort 1998).

Table 3. Number of optimal years (optimal flow conditions for 30 consecutive days at least) or unsuitable years (optimal flow conditions for less than 10 consecutive days) of habitat conditions for pike reproduction simulations based on a 33-year time-series of discharge (from Monfort et al., 1996).

	Flow regime simulation		
	Natural	One dam	Two dams
Optimal years	15	10	5
Unsuitable years	0	2	4

Water quality

Release into rivers is a traditional way of removing human sewage and wastes products where the rainfall is sufficient. While in most traditional agricultural societies, human (as well animal) excrement and domestic wastes are recycled onto cropland, the release of domestic waste into surface water is a consequence of urbanization. As early as the Middle Ages, a primitive system of wastewater collection, often using small streams, was organized even in small towns to evacuate domestic waste to the rivers.

In the case of Paris, the degradation of the Seine water quality due to discharge of domestic sewage is not new. By the beginning of the 14th century, King Jean Le Bon (1325) prohibited street cleaning in Paris when it rained and ordered the waste to be dumped outside of the city (Lacordaire, 1985).

With the industrial revolution, the deterioration of water used for disposal of waste and pollutants became locally severe. During the last century, large pipe systems were constructed to collect sewage. Part of the collected wastewater was spread on cropland, but direct discharge into the Seine dominated. The process accelerated from the end of the 19th century until the 1960s, and anaerobic or near anaerobic conditions persisted for most of the year over a 100-km long section downstream from Paris (Figure 4). In particular, the conditions downstream from the Achères sewage treatment plant were, at times, totally anoxic and virtually devoid of aquatic life.

Nowadays, although much domestic sewage is treated to reduce the oxygen demand, there is still

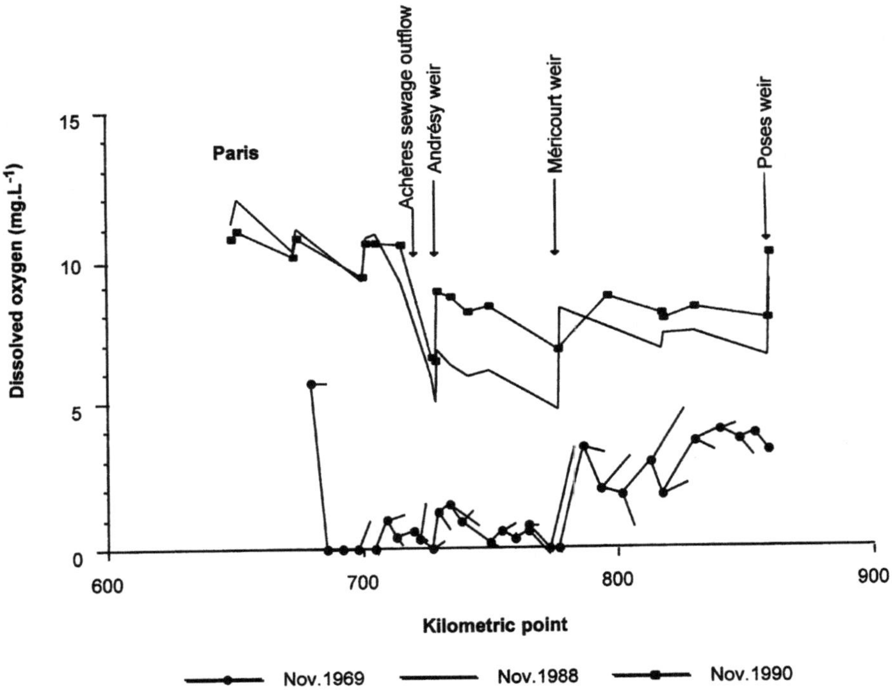

Figure 4. Longitudinal evolution of the dissolved oxygen concentrations in the Seine River downstream from the City of Paris (Cemagref, unpublished data).

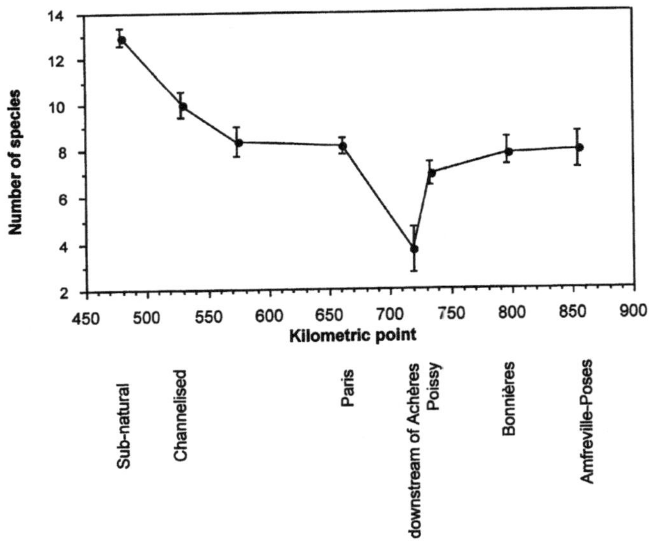

Figure 5. Longitudinal evolution of the mean number (± standard error) of fish species caught by electric fishing in Seine River reaches upstream and downstream the City of Paris (Cemagref, unpublished data).

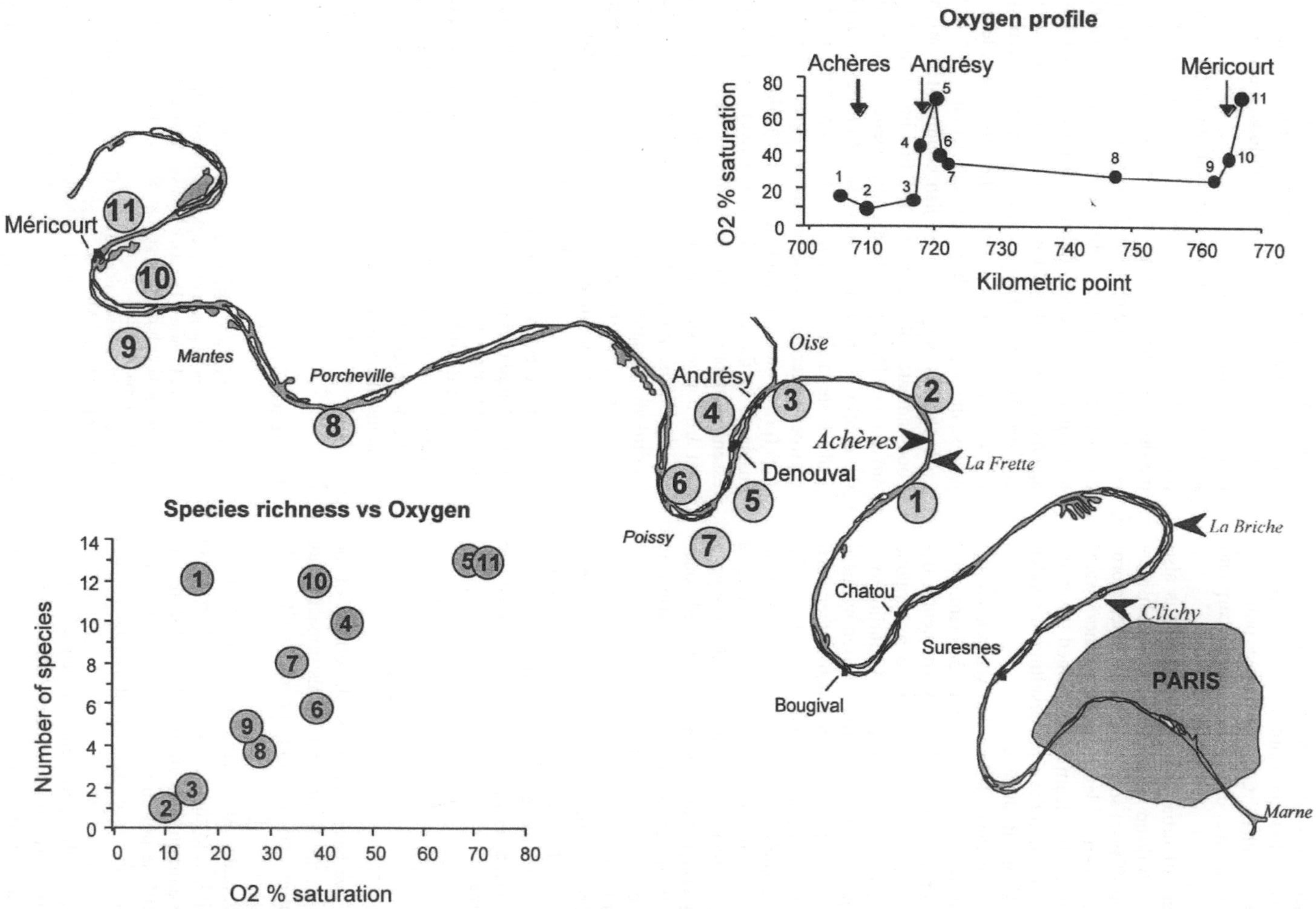

Figure 6. Relation between the number of fish species caught and the mean dissolved oxygen (expressed as percentage of saturation), after a summer storm event downstream of Paris (Boët et al., 1993). The river is severely affected just below the outflow of the sewage treatment plant of Achères (stations 2 and 3). Further downstream, stations 8 and 9 are not partially re-oxygenated by the weir, as stations 6 or 7 are. Fish find shelter in side-channel (station 4) or sandpit (station 10), and congregate below the weirs (station 5 and 11). Upstream of Achères (station 1), it seems that fish forced back by a second pollution flow from Paris were stopped by the 'chemical barrier' of the anoxic sewage outflow.

a noticeable decrease of fish species numbers downstream of Paris, below the outflow of the Achères sewage treatment plant. Only the most pollution-tolerant species survive, e.g., roach, bream, carp and perch (Figure 5).

The problem of urban stormwater runoff is also a major issue. Built-up urban areas have considerable rainfall runoff from sealed surfaces. When storm events exceed a certain level, surface runoff water plus wastewater flushed from the combined sewer are discharged directly into the river, resulting in sudden discharges peaks and flash floods.

Downstream of Paris, below the Achères sewage treatment plant, massive fish mortality frequently occurs. This reach is characterized by frequently occurring low oxygen conditions, caused by organic pollution brought by stormwater runoff during low-flow summer periods. The instability caused by these extreme events is reflected by a much higher variability in the fish community (Allardi et al., 1993). Survival during such events requires mobility. Fish are less sedentary in these perturbed environments (Karr & Dudley, 1981), and during periods of oxygen depletion they congregate in more oxygenated zones below weirs or at the confluence of tributaries, in side-channels or sandpits (Figure 6).

This phenomenon illustrates the major role played by these natural or artificial areas in the long-term equilibrium of fish populations. Such off-channel areas also serve as a refuge in periods of extreme conditions, and thus they ensure fish survival and facilitate rapid recolonization of the river.

Conclusions

In this paper, we have focused on the effects of the City of Paris on the Seine Basin. This impact is not only local but extends into the upper as well as the lower reaches.

Obviously, many other human impacts, which are not directly linked to the development of Paris, also affect the fish communities all over the basin, such as agricultural diffuse pollution or urban wastes, drainage, siltation, etc. (see Berrebi-dit-Thomas et al., 1998).

Nevertheless, the perturbations caused by the megalopolis of Paris alone strongly affect the fish communities of the Seine at the basin scale. Some thresholds have already been exceeded, since the migratory species are extinct. All the other species present in the last century are still there, but there is strong evidence of a homogenization of the fish communities, among which the rheophilous species are particularly threatened.

Until now, efforts have been concentrated on improving the water quality, mainly in order to ensure the drinking-water supply. These efforts have greatly benefited the whole biological ecosystem, particularly downstream from Paris. Obviously, it is still essential to maintain these de-pollution measures, particularly against the toxic substances, which remain of major concern. However, there is now a growing awareness of the importance of considering the habitat in order to reverse the decline of the ecosystem health.

From this point of view, fish communities are a good indicator of the global functioning of the hydrosystem. They provide a comprehensive view of the quality of the aquatic ecosystems at the regional scale. Fish data can be helpful when choice are made between watershed management strategies and priorities (preservation, rehabilitation, or remediation). For instance, in the case of the Seine river basin, it is essential to preserve the upstream diversified areas, that are not yet strongly affected by development. In these reaches, protection efforts should be directed toward discouraging physical modifications. In the already modified intermediate sections, the challenge is to recreate the diversity of the natural conditions. However, in both cases the most important problem will be to re-establish, or simulate, the variability of the discharge regime. On the other hand, in the most perturbed sites downstream from Paris, a return to natural pre-disturbance conditions appears unrealistic. Therefore, reclamation, habitat creation, or mitigation can be achieved by managing site-specific or isolated elements of ecosystems.

References

Allardi, J., J. Belliard, P. Boët & L. Boughida, 1993. Influence des actions anthropiques sur le peuplement ichtyologique du bassin de la Seine. In Fustec, E. & G. de Marsily (eds), La Seine et son bassin: de la recherche à la gestion, Symp. Piren-Seine, 29–30 juillet, Paris: 221–237.

Balon, E. K., 1974. Domestication of the Carp, *Cyprinus carpio*. R. Ont. Mus. Life Sci. Misc. Pub. 37 pp.

Balon, E. K., 1975. Ecological guilds of fishes: a short summary of the concept and its application. Ver. int. Verein Limnol. 19: 2430–2439.

Belliard, J., 1994. Le peuplement ichtyologique du bassin de la Seine: rôle et signification des échelles temporelles et spatiales. Thèse Doct. Paris VI, 197 pp.

Belliard, J., P. Boët & J. Allardi, 1995. Evolution à long terme du peuplement piscicole du bassin de la Seine. Bull. fr. Pêche Piscicult. 337/338/339: 83–91.

Belliard, J., P. Boët & E. Tales, 1997. Regional and longitudinal patterns of fish community structure in the Seine River basin, France. Envir. Biol. Fish. 50: 133–147.

Berrebi-dit-Thomas, R., J. Belliard & P. Boët, 1998. Caractéristiques des peuplements piscicoles sensibles aux altérations du milieu dans les cours d'eau du bassin de la Seine. Bull. fr. Pêche Piscicult. 348: 47–64.

Berrebi-dit-Thomas, R. & P. Boët, 1997. Biodiversity of young-of-the-year fish assemblages from natural and artificial backwaters in the Seine River floodplain (France). Ninth Int. Congress of European Ichtyologists (CEI9) 'Fish Biodiversity', 24–29 August, Trieste, Italy.

Berrebi-dit-Thomas, R., P. Boët & E. Tales, 1999. Macrohabitat characteristics influencing young-of-the-year fish assemblages in connected lentic backwaters in the Seine River (France). Arch. Hydrobiol. (accepted).

Boët, P., M. Akopian, J. Belliard, R. Berrebi-dit-Thomas, R. Pourriot, E. Tales & P. Testard, 1998. Une faune aquatique sous pressions multiples. In Meybeck, M., G. de Marsily & E. Fustec (eds), *La Seine en son bassin. Fonctionnement écologique d'un système fluvial anthropisé.* Elsevier, Paris: 627–678.

Boët, P. & O. Monfort, 1998. Gestion des inondations et reproduction du brochet. In Le Coz, C., B. Tassin & D. Thévenot (eds), Actes des Journées Sciences et Techniques de l'Environnement, Paris, 15–16 mai 1997, Presses de l'Ecole Nationale des Ponts et Chaussées: 125–142.

Bovee, K.D., 1982. A guide to stream habitat analysis using the Instream Flow Incremental Methodology. Instream Flow Information paper n° 12, U.S. Fish Wild. Serv., FWS;OBS-82;86: 248 pp.

Dodge, D.P. (ed.), 1989. Proceeding of the International Large River Symposium (LARS). Can. spec. pub. Fish. aquat. Sci. 106: 629 pp.

Goubier, J., 1975. Biogéographie, biométrie et biologie du Sandre, *Lucioperca lucioperca* (L.), Osteichthyen, Percidé. Doc. Etat Univ. Claude Bernard, Lyon I, 259 pp.

Holcik, J., 1990. Effects of hydraulic engineering on habitat and fish community in river anabranches of the middle Danube. In Van Densen, W.L.T., B. Steinmetz & R.H. Hughes (eds), Management of freshwater fisheries, Proc. Symp. EIFAC, Göteborg, 31 May–3 June 1988, Sweden, Pudoc Wageningen: 14–24.

Huet, M., 1949. Aperçu des relations entre la pente et les populations piscicoles des eaux courantes. Schweiz. Z. Hydro. 11: 332–351.

Huet, M., 1959. Profiles and biology of western European streams as related to fish management. Trans. am. Fish. Soc. 88: 155–163.

Husson, G.-P., 1996. Historique de l'alimentation en eau potable de la ville de Paris. J. Europ. Hydrol. 27: 97–108.

Karr, J.R. & D.R. Dudley, 1981. Ecological perspective on water quality goals. Envir. Mgmt 5: 55–68.

Lacordaire, S., 1985. Les inconnus de la Seine. Paris et les métiers de l'eau du XIIIe au XIXe siècle. Hachette, Paris, 306 pp.

Lavollée, M.G., 1902. Le Saumon en Seine. Bull. Soc. centr. Aquic. Pêche 14: 221–234.

Magnin, E., 1959. Répartition actuelle des Acipenseridés. Rev. Trav. Inst. (Sci. Tech.) Pêch. Mar.: 277–285.

Manéglier, H., 1992. Une grande ville et un bassin, Paris et la Seine. Rev. Géogr. Lyon 67: 299–303.

Mann, R.H.K., 1988. Fish and Fisheries of regulated rivers in the UK. Regulated Rivers 2: 411–424.

Meybeck, M., G. de Marsily & E. Fustec (eds), 1998. La Seine en son bassin. Fonctionnement écologique d'un système fluvial anthropisé. Elsevier, Paris, 749 pp.

Micha, J.C. & M.-C. Borle, 1989. Recent historical changes on belgian Meuse. In Petts, G.E., H. Möller & A.-L. Roux (eds), Historical change of large alluvial rivers: Western Europe, John Wiley & Sons, Chichester: 269–295.

Monfort, O., P. Breil & P. Boët, 1996. Selection of an ecological reference flow: support by modeling the habitat of a key species, the northern pike. In Leclerc, M., H. Capra, S. Valentin, A. Boudreault & Y. Côté (eds), 'Ecohydraulics 2000' 2nd IAHR Int. Symp. on hydraulics and habitats, Quebec City, June 11–14, 1996, INRS-Eau, B: 133–146.

Moreau, E., 1881. Histoire naturelle des poissons de la France. n°. 3 (3) Masson, Paris, 697 pp.

Moreau, E., 1937. Histoire naturelle des poissons de la France. Masson, Paris, 478, 572, 697, 144 pp.

Nelva-Pasqual, A., 1985. Biogéographie, démographie et écologie de *Chondrostoma nasus nasus* (L., 1758) (Hotu, Poisson, Téléostéen, Cyprinidé). Thèse Doc. Etat Univ. Lyon I, 362 pp.

Oberdorff, T., E. Guilbert & J.C. Lucchetta, 1993. Patterns of fish species richness in the Seine River basin, France. Hydrobiologia 259: 157–167.

Persat, H. & P. Berrebi, 1990. Relative ages of present populations of *Barbus barbus* and *Barbus meridionalis* (Cyprinidae) in southern France: preliminary consideration. Aquat. Living Resour. 3: 253–263.

Persat, H. & P. Keith,1997. La répartition géographique des poissons d'eau douce en France: qui est autochtone et qui ne l'est pas? Bull. fr. Pêche Piscicult. 344/345: 15–32.

Petts, G.E., H. Möller & A.-L. Roux (eds), 1989. Historical change of large alluvial rivers: Western Europe. John Wiley & Sons, Chichester, 355 pp.

Roule, L., 1920. Les espèces d'aloses du bassin de la Seine. Bull. Soc. cent. Aquicult. Pêche 27: 124–125.

Schiemer, F., T. Spindler, H. Wintersberger, A. Schneider & A. Chovanec, 1991. Fish fry associations: important indicators for the ecological status of large rivers. Ver. Int. Ver. Limnol. 24: 2497–2500.

Schiemer, F. & H. Waidbacher, 1992. Strategies for conservation of a danubian fish fauna. In Boon, P.J., P. Calow & G.E. Petts (eds), River conservation and management, John Wiley & Sons, Chichester: 363–382.

Souchon, Y., F. Trocherie & E. Fragnoud, 1989. Les modèles numériques des microhabitats des poissons: application et nouveaux développements. Sci. Eau 2: 817–841.

Spillmann, C.J., 1961. Poissons d'eau douce. Faune de France n°. 65 Féd. fr. Soc. Sci. Nat., Paris, 303 pp.

Tales, E., P. Boët & R. Berrebi-dit-Thomas, 1996. Les peuplements de poissons de l'année de quelques types d'annexes fluviales dans la plaine de la Bassée (Seine). Bull. fr. Pêche Piscicult. 343: 189–202.

Thienemann, A., 1925. Die Binnengewässer Mitteleuropas. Eine limnologische Einführung. Die Binnengewässer, Stuttgart, 1: 1–255.

Velay, P., 1992. De Lutèce à Paris — L'île et les deux rives. Coll. Patrimoine au présent Caisse Nationale des Monuments Historiques et des Sites, Presses du CNRS, Paris, 128 pp.

Verneaux, J., 1973. Cours d'eau de Franche-Comté (massif du Jura). Recherches écologiques sur le réseau hydrographique du Doubs. Essai de biotypologie. Thèse Doc. Ann. Fac. Sci. Univ. Besançon, 84: 257 pp.

Verneaux, J., 1977. Biotypologie de l'écosystème 'eau courante'. Détermination approchée de l'appartenance typologique d'un peuplement ichtyologique. C. r. Acad. Sci., Paris, 284 (sér. D): 675–678.

Annex. List and status of fish species in the Seine River (from Beillard 1994). (A) Native; (D) extinct; (I) exotic; (V) voluntary introductio; (C) colonization via canal links.

Family	Species	Common name	Status	Origin and way of introduction		Date
Petromyzonidae	*Petromyzon marinus*	Sea lamprey	D			1970
	Lampetra fluviatilis	River lamprey	A			
	Lampetra planeri	Brook lamprey	A			
Acipenseridae	*Acipenser sturio*	Atlantic sturgeon	D			1917
Anguillidae	*Anguilla anguilla*	Eel	A			
Clupeidae	*Alosa alosa*	Allis shad	D			1930
	Alosa fallax	Twaite shad	D			1970
Cyprinidae	*Alburnus alburnus*	Bleak	A			
	Alburnoides bipunctatus	Stream bleak	A			
	Barbus barbus	Barbel	? I		?C	?
	Abramis brama	Common bream	A			
	Blicca bjoerkna	White bream	A			
	Carassius carassius	Crucian carp	I	Europe	V + ? C	?
	Carassius auratus	Goldfish	I	Asia	V	18th
	Cyprinus carpio	Common carp	I	Europe	V	Antiquity
	Hypophthalmichtys molitrix	Silver carp	I	Asia	V	1980
	Chondrostoma nasus	Nase	I	Europe	C	1860
	Chondrostoma toxostoma	French nase	I	Europe	? C	1990
	Gobio gobio	Gudgeon	A			
	Leucaspius delineatus	Sun bleak	I	Europe	?	First mid 20th
	Leuciscus cephalus	Chub	A			
	Leuciscus leuciscus	Dace	A			
	Leuciscus soufia	Soufie	I	Europe	?	1985
	Phoxinus phoxinus	Common minnow	A			
	Scardinius erythrophtalmus	Rudd	A			
	Rutilus rutilus	Roach	A			
	Rhodeus amarus	Bitterling	A			
	Tinca tinca	Tench	A			
	Pseudorasbora parva	Stone moroco	I	Asia	?	1990
Cobitidae	*Cobitis taenia*	Spined loach	A			
	Barbatula barbatula	Stone loach	A			
Siluridae	*Silurus glanis*	Danube catfish	I	Europe	1990	
Ictaluridae	*Ameiurus melas*	Black bullhead	I	North Am.	?	1871
Salmonidae	*Salmo trutta trutta*	Sea trout	D?			
	Salmo trutta fario	Brown trout	A			
	Salmo salar	Atlantic salmon	D			1900–1910
	Oncorhynchus mykiss	Rainbow trout	I	North Am.	V	1880
	Salvelinus fontinalis	Brook trout	I	North Am.	V	1880
	Thymallus thymallus	Grayling	I	Europe	V	1960
Osmeridae	*Osmerus eperlanus*	Smelt	?(D)			?
Esocidae	*Esox lucius*	Pike	A			
Umbridae	*Umbra pygmaea*	Eastern mudminnow	I	North Am.	V	?
Gadidae	*Lota lota*	Burbot	A			
Gasterosteidae	*Gasterosteus aculeatus*	Three-spined stickleback	A			
	Pungitius pungitius	Nine-spined stickleback	A			
Cottidae	*Cottus gobio*	Sculpin	A			
Percidae	*Perca fluviatilis*	Perch	A			
	Stizostedion lucioperca	Pike-perch	I	Europe	V + C	1960
	Gymnocephalus cernua	Ruffe	I	Europe	C	1800
Centrarchidae	*Lepomis gibbosus*	Pumpkinseed	I	North Am.	V	1885
	Micropterus salmoides	Largemouth bass	I	North Am.	V	1890
Pleuronectidae	*Platichthys flesus*	Flounder	A			

Hydrobiologia **410**: 69–77, 1999.
J. Garnier & J.-M. Mouchel (eds), Man and River Systems.

Rehabilitation efforts for Atlantic salmon in the Meuse basin area: Synthesis 1983–1998

C. Prignon[1], J.C. Micha[1], G. Rimbaud[2] & J.C. Philippart[2]
[1]*Unité de Recherches en Biologie des Organismes, Facultés Universitaires Notre-Dame de la Paix, Rue de Bruxelles No. 61, B-5000 Namur, Belgique*
[2]*Laboratoire de Démographie des Poissons et d'Aquaculture, Université de Liège, Chemin de la Justice No. 10, B-4500 Tihange, Belgique*

Key words: Atlantic salmon, rehabilitation, Meuse, restocking, weir, fish pass

Abstract

In the 1950s, Atlantic salmon had become extinct in the river Meuse basin, mainly owing to the construction of navigation weirs. In 1987 started a rehabilitation program of Atlantic salmon in Belgium. Stockings are being performed in six tributaries of the river Meuse, using eyed eggs, parr and pre-smolts obtained from foreign eggs reared in a hatchery of the 'Service de la Pêche'. By the end of 1998, 896033 young salmon have been released in the wild, especially in the river Ourthe basin. In most of the cases, the evaluation of stocking operations indicates good adaptation of parr in the tributaries where they are stocked, with autumnal 0+ parr densities up to 20–30 ind 100 m^{-2}. However, fish mortality may occur when smolts pass through the turbines of the hydropower stations in the river Meuse (5% per power plant), and most of the fish passes located downstream of Namur are not suitable for anadromous fish migration. The paper reports the results of stocking experiments, the evaluation of mortality rates in turbines and the advancement of programmes aiming at the restoration of downstream and upstream migration (construction of fish migration facilities).

Introduction

Until the end of the 19th century, the Atlantic salmon (*Salmo salar* Linné) was widespread in the river Meuse basin and salmon fisheries were prosperous in France, Belgium and The Netherlands. The number of Atlantic salmon and sea trout (*Salmo trutta* Linné) captures reached 57 000 ind./year in the Dutch Rhine–Meuse delta (Philippart, 1985; Philippart, 1987).

Between 1840 and 1950, the building of navigation weirs on the river Meuse and its tributaries, as well as the increase of industrial water pollution and the overexploitation of fish stocks, led to the extinction of all anadromous fish species (Atlantic salmon, sturgeon *Acipenser sturio* Linné, houting *Coregonus oxyrhynchus* Linné, allice shad *Alosa alosa* Cuvier, twaite shad *Alosa fallax* Cuvier, sea lamprey *Petromyzon marinus* Linné and river lamprey *Lampetra fluviatilis* Linné) from this ecosystem (Philippart, 1987; Philippart et al., 1988; Philippart & Vranken, 1983).

The capture of sea trout in 1983 in the lower river Berwinne (Philippart, 1983), at the Belgian–Dutch border, and later in many sites of the river Meuse, brought the salmon problem in the Belgian river Meuse back to the surface. This major ecological event triggered the idea of rehabilitating the Atlantic salmon in the river Meuse basin (Philippart, 1985). Based on feasibility studies carried out from 1983 to 1986, the program 'Meuse Salmon 2000', aiming at the restoration of the complete life cycle of migratory salmonids (Atlantic salmon and sea trout) in Belgium (Wallonia) and the reestablishment of unimpeded fish circulation for other amphibiotic migratory species, was drafted in 1987 (Philippart & Gillet, 1990; Philippart et al., 1994).

The project covers four main aspects: (1) rearing salmon from eggs originating from rivers where salmon populations are still present; (2) restocking ecologically valuable rivers with reared fish; (3) studying the different phases of the salmon life cycle, including downstream migration in tributaries, to increase restocking efficiency; (4) define constructive proposals aiming at the improvement of downstream migration of smolts at the hydropower stations and the reestablishment of upstream facilities (construction of fish passes and reduction of water pollution).

This paper synthesizes rehabilitation efforts for Atlantic salmon in the Belgian Meuse basin area. With respect to the publication of Philippart et al. (1994), new and more precise data have been added and information has been updated, with particular emphasis on construction of fish passes, planned or in progress, to allow unimpeded upstream and downstream circulation of fish.

Description of sites studied

The river Meuse rises in the east of France and flows through Belgium and The Netherlands, where it meets the lower Rhine, forming the Dutch Delta, which opens into the North Sea (Figure 1). The total length of the river is 925 km and its catchment area is 36 011 km^2, 40.7% of which lies in Belgium (Descy et al., 1987). The main tributaries are the Chiers and the Semois in France, the Lesse, Sambre and Ourthe in Belgium, and the Roer and Niers in The Netherlands.

The Belgian stretch of the river Meuse had an initial slope of 0.23 ‰. During the last 150 years, the river has become increasingly channelized (Philippart et al., 1988). Fifteen navigation and locks weirs are located on the Belgian course of the river Meuse. Six of these weirs (ranging from 4 to 6 m high) are located downstream of Namur and are associated with hydropower stations. The rest (ranging from 1.5 to 2.3 m high) are located upstream of Namur and lack hydropower facilities.

The annual mean flow rate is about 280 $m^3\ s^{-1}$ upstream Liège (Micha & Borlée, 1989). The mean annual water temperature is about 16°C, ranging seasonally from 3–4°C in January to 25–26°C in August (Micha, 1988).

The tributaries studied in the context of the program 'Meuse Salmon 2000' are the river Ourthe (catchment area, 3672 km^2, length, 188 km; low water discharge, 3–10 $m^3\ s^{-1}$) and some of its tributaries (river Amblève, river Aisne), the river Samson (catchment area, 120 km^2; length, 21.4 km; discharge, 0.3–0.4 $m^3\ s^{-1}$), the river Lesse (catchment area, 2740 km^2, length, 83 km; low water discharge, 10 $m^3\ s^{-1}$) and the river Semois (catchment area, 1358 km^2, length, 208 km; low water discharge, 17.3 $m^3\ s^{-1}$).

Material and methods

Rearing and stocking program

Wild salmon eggs are purchased every year in Scotland (river Connon), Ireland (Lough Corrib and river Screebe), and France (river Nive).

Salmon are stocked in six tributaries of the river Meuse: the river Ourthe, the lower river Amblève, the river Aisne, the river Samson, the lower river Lesse and the river Semois (Figure 1). These tributaries were selected on the basis of habitat availability and suitability, and on water quality criteria established by regional surveys.

Three developmental stages have been tested: eyed eggs (1988–1990), parr (1988–1998), pre-smolts (1990–1998).

Parr and pre-smolts are reared in a fish farm facility of the 'Service de la Pêche' in Emptinne, near Namur (Conjaerts & Frank, 1995). Eggs are incubated on horizontal trays, in spring water at 8–10°C. After adaptation to artificial feeding, the fry (2 cm) are grown in 200–1500 l tanks, supplied with pond water at 16–20°C in summer.

In June–July, most young salmon (3–6 cm) are released in tributaries, with different stocking densities (40–120 ind. 100 m^{-2}) being tested in some rivers (e.g. Samson).

The remaining fish are reared up to a length of 10–13 cm and stocked as pre-smolts until March of the following year.

Many salmon stocked in the river Samson and in the river Lesse are microtagged, with a different code for each river (Prignon & Micha, 1996).

Evaluation of stocking operations

Young salmon were captured for demographic, growth, habitat preference and behavioural ecology analyses using electrofishing (DEKA; DC; 150–450 V, two runs) during spring and autumn. Fish sampled in their suitable habitat (rapids and riffles) were measured to the nearest mm, weighed to 0.1 g and aged. Parr population density was estimated using either the method of removal (two successive runs) (Seber & Lecren, 1967) or the mark-recapture method (Seber, 1973). Until 1996, the aim of these inventories was to assess the efficiency of stocking operations for different developmental stages and stocking densities.

Production potentialities of young salmon in the river Ourthe have been assessed from a theorical potential production of smolts of 10 ind 100 m^{-2}

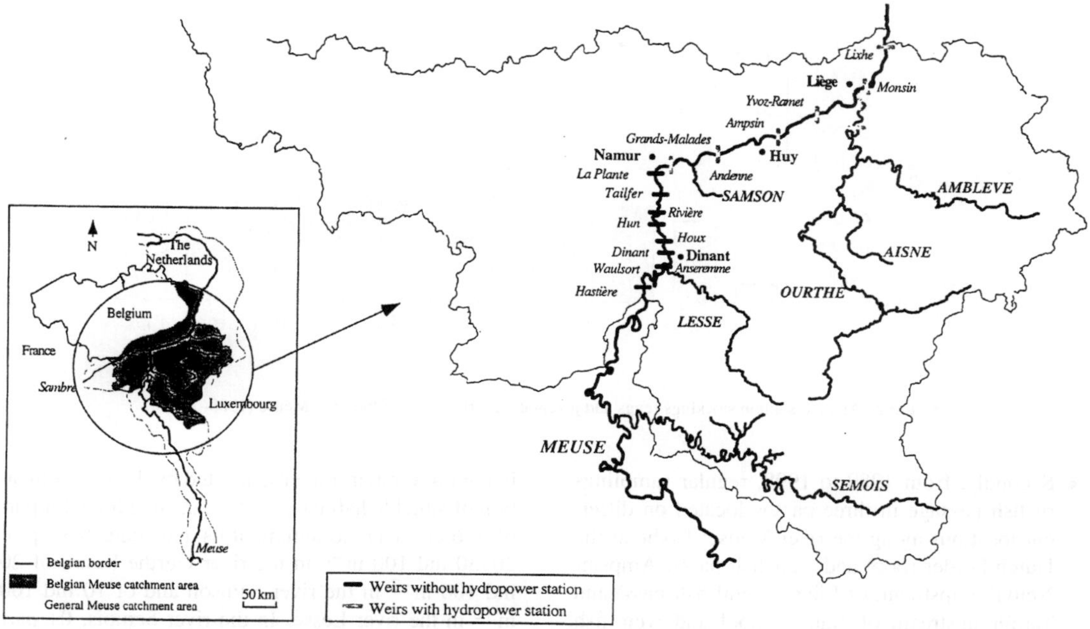

Figure 1. River Meuse hydrographic basin, with its weirs from the French border to the Dutch border, and its tributaries where young Atlantic salmon are restocked.

(Symons, 1979) multiplied by the percentage of habitats suitable for parr estimated using the photo-interpretation method described by Boudreault (1984) and by Côté et al. (1987). The carrying capacity in the river Semois have been calculated in a section of 60 km on the basis of 0+ parr densities obtained from electrofishing in suitable habitats.

Downstream migration in the river Samson

Since 1990, downstream migration of smolts has been studied in the lower river Samson, equipped with a trap installed from March to May. This trap is composed of a grid (space between bars: 1 cm) placed across the river (12 m wide), and of a box (± 1 m^3) where fish are trapped. Salmon migrating to the sea are counted every day in order to assess the efficiency of stocking, expressed as the number of smolts migrating on the number of salmon stocked in this river.

Mortality caused by hydroelectric power plants in the river Meuse

Six hydropower stations (Kaplan turbines; equipment discharge, 170–450 m^3 s^{-1}; wheel diameter, 2.60–5.60 m; four blades; rotation speed, 65.2–132.0 t min^{-1}) are located in the river Meuse between Namur and the Dutch border.

For each hydropower station, potential mortalities of smolts have been statistically assessed, in accordance with the model of Larinier and Dartiguelongue (1989), based on species, fish length and turbine characteristics:

$$P = \sin(13.41 + 42.8(TL/esp))^2 (R = 0.59),$$

where P is mortality rate; *TL* is mean total length of fish; *esp* is space between blades calculated at mid-blade.

Efficiency of fish passes

Except in Waulsort and Hastière, all mosan weirs of the river Meuse are equipped with a pool and weir fish pass or Denil fish pass.

The quality of these fish passes has been assessed using two methods.

- First, a preliminary investigation taking in consideration their situation in the weir, structure and attractivity and based on observations of the use of these devices by fish.

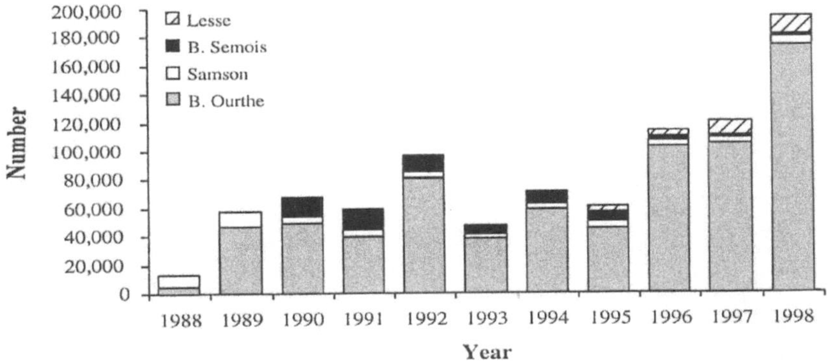

Figure 2. Atlantic salmon stockings (eggs and juveniles) in tributaries of the river Meuse (1988–1998).

- Secondly, from 1989 to 1998, regular samplings of fish passage in three passes located on different locations along the river Meuse: Lixhe at the Dutch border (pool and weir fish pass), Ampsin-Neuville upstream of Liège (Denil fish pass) and Tailfer upstream of Namur (pool and weir fish pass). Daily fish passage is monitored using a trap placed in the upper pool of the pass (Baras et al., 1994, 1996; Philippart et al., 1998; Prignon et al., 1998).

To allow adults to migrate and to spawn in tributaries, new facilities fitted to the characteristics and location of weirs have been planed in 1992 (MET, 1992).

Results

Rearing and stocking program

The hatchery run by the 'Service de la Pêche' produces about 800 parrs from 1000 salmon eggs. From 1988 to 1998, 896 033 Atlantic salmon (86 520 eyed eggs and 809 513 parr and pre-smolts of 3–16 cm) have been stocked in tributaries of the river Meuse, with the following distribution: 83% in the river Ourthe basin, 8% in the river Semois basin, 6% in the river Samson and 3% in the river Lesse (Figure 2). Stockings have been intensified the last years, especially in 1998 when they reached an amount of 193 861 parr.

Biology of young salmon

With the exception of the river Semois, the study of salmon demography reveals good adaptation of parr in their new environments, attested by dense colonization of suitable habitats. Each year, stocking with parr of 3–6 cm leads to autumnal 0+ parr densities up to 20–30 ind 100 m^{-2} in the river Ourthe basin, of 20 ind 100 m^{-2} in the river Samson and of 10 ind 100 m^{-2} in the river Lesse. In the river Semois, 0+ parr densities range from 0.2 to 30.6 ind 100 m^{-2}, with high between-year fluctuations and a tendency towards low densities (Philippart et al., 1990, 1994; Prignon & Micha, 1995b).

In the river Samson, a stocking density of 70 ind 100 m^{-2} in July produces an autumnal 0+ parr density of 20 ind 100 m^{-2}, which means a minimal survival rate of 28.6%. Indeed, we observe a parallel increase of autumnal 0+ parr density and percentage of recapture with stocking densities up to 80 ind 100 m^{-2}. At higher stocking density, parr density remains stable (optimal parr density) and stocking efficiency falls.

Production potentialities reach 100 000 smolts in the river Ourthe (Philippart, 1995; Philippart et al., 1990, 1994). The carrying capacity in the River Semois is about 40 000 parr 0+ for a 60-km reach studied (unpublished data).

Most salmon stocked in large rivers (rivers Ourthe, Semois and Lesse) reach 10–13 cm at the end of the first year (growth, 20 mm per summer month) and enter smoltification during the following spring. In colder tributaries (rivers Samson and Aisne), the size at smoltification (growth, 12 mm per summer month) is reached at the end of the second year, due to less favourable thermal regimes (Figure 3) (Philippart et al., 1994).

According to trapping experiments carried out in the river Samson, the downstream migration of smolts

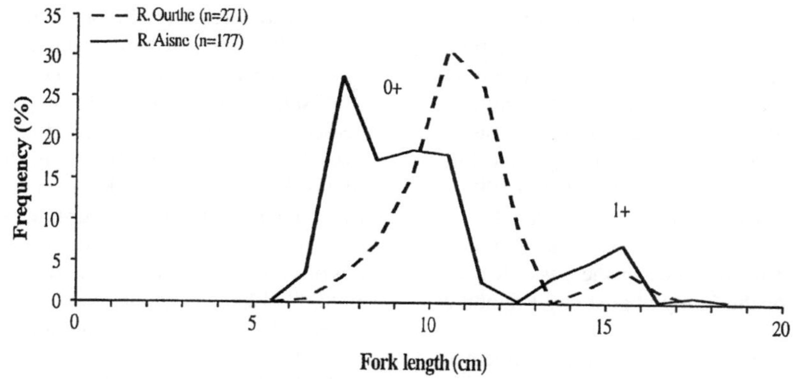

Figure 3. Length–frequency distribution of salmon parr populations in the river Ourthe and in the river Aisne in October 1992. Average water temperature in July–August: r. Ourthe, 19.3°C; r. Aisne, 14.6°C.

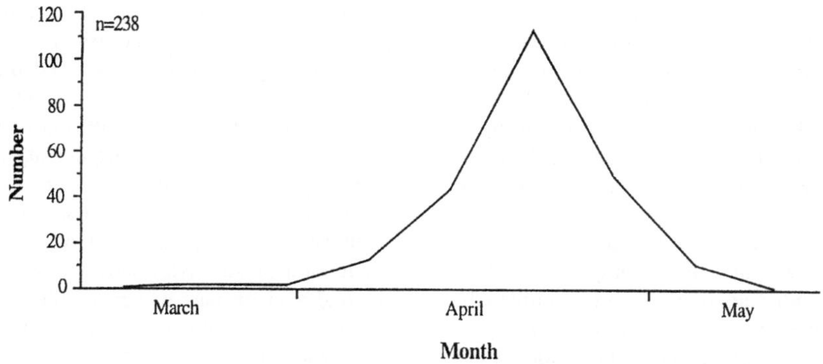

Figure 4. Downstream migration periodicity of smolts recorded in 1992 in the river Samson.

is mainly nocturnal and takes place from late March to early May, with a maximum intensity on the third and fourth weeks of April (Figure 4). Each spring, about 250 smolts are counted in the trap.

The efficiency of stockings, assessed by downstream migration rate, is estimated at a minimum of 0.2% for eggs, 3.0% for parr and 19.0% for pre-smolts. Most smolts migrate at the age of 2 years (Prignon & Micha, 1995a; Prignon et al., 1997).

Mortality caused by hydroelectric power plants in the river Meuse

Damages to fish may result from smolts passing through the turbines in the river Meuse. Potential mortality rate is estimated to be 7–9%, depending on the hydropower station. In the Belgian river Meuse, cumulative mortality rate should reach 39% for smolts migrating from the river Semois, 33% for smolts coming from the river Samson, and 14% for smolts from the rivers Ourthe – Amblève (Prignon & Micha, 1995a).

Based on these data, the energy producer Socolie, which manages the hydropower stations on the river Meuse, has decided in 1997 to install an experimental surface 'bypass' system (Travade & Larinier, 1992) at the hydropower plant of Lixhe. Smolts are attracted by a surface gate (750 l s^{-1}) and collected in a channel which bypasses the installation without passing through turbines.

Upstream migration of fish and building of new fish passes

Fish passes located downstream of Namur are not suitable for anadromous fish migration because they are inadequate either because of their location (e.g. Monsin), structure (e.g. Lixhe) and/or attractivity (e.g.

Andenne). All these fish passes must be replaced by new facilities. Upstream Namur, fish passes are functional. Problems about attractivity and structure are not so acute and may be resolved by some improvements at existing fish passes (MET, 1992; Roenen et al., 1995). For example, regular monitoring at the Tailfer fish pass indicate that, between 1989 and 1994, a total of 157 897 individuals of 23 species, included sea trout, have migrated through the fish pass. No adult salmon have been documented in the pass. A complete description of this study is given by Prignon et al. (1998).

In January 1996, a general written agreement relating to fish migration was drafted between the 'Direction Générale des Ressources Naturelles et de l'Environnement' of the 'Ministère de la Région Wallonne' and the 'Administration des Voies Hydrauliques' of the 'Ministère de l'Equipement et des Transports'. It established the opening of the migration route for migratory salmonids for 2002 in the river Ourthe and for 2007 in the entire river Meuse up to France. The building of a new pool and weir fish pass on the first Belgian weir in Lixhe, at the Dutch border has been completed by October 1998. The works on the second (Monsin) and on the third weir (Yvoz-Ramet) started in 1998 and will be completed by 1999. Radiotelemetry studies are carried out to optimize the structure and location of these new fish passes, in agreement with the requirements and behaviour of fish (Ovidio et al., 1996; Philippart et al., 1998).

Discussion

The objective of the program 'Meuse Salmon 2000' is to restore the complete life cycle of Atlantic salmon and sea trout in the Belgian river Meuse basin, ensuring the development of juveniles in river water, and to allow downstream migration of smolts and upstream migration of spawners.

In most cases, results about Atlantic salmon rehabilitation in the river Meuse basin are promising (salmon survival, habitat colonization, growth rates, smoltification and patterns of downstream migration). Except in the river Semois, results of stockings indicate good adaptation of young salmon in tributaries where they are implanted. Parr densities at the end of the first summer (up to 20–30 parr per 100 m^2 of suitable habitat) and growth rates agree with those expected in the type of environment where fish are stocked. In the French river Rhine basin, Roche (1994) and Roche & Gerlier (1996) obtained comparable results: in 1991–1992, the autumnal parr densities ranged from 5 to 70 parr 0+ per 100 m^2 (mean, 30 parr per 100 m^2) of suitable habitat where feeding fry (4–5 cm) are released. In 1995, stockings led to salmon densities of 0.8–28.7 parr 0+ per 100 m^2 in the river Bruche and 3.6–30.8 parr 0+ per 100 m^2 in the Old Rhine. In this river, adult salmon are returning again (Gerlier & Roche, 1998). In the river Adour, where native populations of salmon are still present, 0+ parr density (mean length, 11–12 cm) is about 10 ind 100 m^{-2} (Bousquet & Marty, 1987). In wild populations of Atlantic salmon in the rivers of Brittany, fish densities reach up to 24 parr 0+ 100 m^{-2} and two age classes are observed (Baglinière et al., 1987). Similarly, we observe relatively low densities and high growth preformances in larger tributaries and higher densities with slower growth in smaller and colder tributaries.

According to electrofishing and trapping experiments in the river Samson, parr stocking leads to better results than egg stocking, with higher smolt production. In this river, a stocking density of 70 parr per 100 m^2 guarantees an autumnal density of 0+ parr and an efficiency of stocking near the optimal. Indeed, in freshwaters, parr density is a key factor affecting the production of smolts (Chadwick & Randall, 1990). Beyond a given parr or egg number, the number of smolt reaches a plateau which corresponds to the maximum carrying capacity of the river (Heland & Dumas, 1994).

Rapid growth of juveniles in the tributaries of the river Meuse enables a large number of them to migrate to the sea as 1+ or 2+ smolts, with fairly high smolt production as a result. The estimate of 100 000 smolts production in the river Ourthe was confirmed by results of experimental stockings.

These studies indicate that water quality is good and that suitable habitats remain in some tributaries of the river Meuse, where salmon could potentially reproduce and grow. However, since many rivers are threatened by human engineering works (dredging of stream bed and bank rectification), protection and restoration of habitats, and improvement of water quality in some points, are a necessity to guarantee long-term success of salmon rehabilitation. It is the typical case of the river Semois, whose carrying capacity is largely reduced by the effects of high eutrophication in the summer. Indeed, critical values of NH_3 and O_2 detected in the summer months (unpublished data) could explain low densities of parr and more unpredictable results in this river. For the same reasons, salmon stockings have been interrupted in the French river Semoy (Terrier et al., 1997).

Moreover, the remaining suitable habitats in the river Meuse basin will not be available to reproduction without the building of efficient fish migration facilities for smolts and adults on the weirs of the river Meuse and the river Ourthe, and without the improvement of water quality (Admiraal et al., 1993) in the river Meuse downstream Liège.

The hydroelectric power plants in the Belgian river Meuse would cause limited losses of smolts (7–9% per installation). These relatively low mortality rates might be attributed to the large diameter of turbines, with relatively large space between the blades. Monten (1985) and Larinier & Dartiguelongue (1989) found highest mortality rates at turbines with high speed and small diameter. On the Dutch river Meuse, the injury rate of brown trout (20 mm) injected in turbines at the hydropower station of Linne (horizontal bulb; three blades, 45 m diameter; 88 t min^{-1}) and collected in a net at the outlet of turbine ranged from 3–4% (30 m^3 s^{-1}) to 8–13% (50 m^3 s^{-1}) (Hadderingh & Bakker, 1998). These values are comparable with mortality rates calculated in Belgium. The differences between the results obtained using the two methods are probably due to the differences in the turbine characteristics (in particular their diameter and the number of blades) and to parameters not included in the equation (e.g., rotation speed). However, statistical methods give a good overview of the situation. The mortality rates statistically assessed are probably over-estimated and must be reduced to 5% because of the slow rotation of turbines, the site configuration and the hydrology. The problem is that smolts must pass through several turbines and that cumulative mortalities could seriously reduce smolt production and thus return of adults (Prignon & Micha, 1995a). Downstream migration facilities and particular management measures during the downstream migration season would significantly reduce smolt losses in hydropower stations. From 1999, the efficiency of the surface 'bypass' at Lixhe will be assessed by trapping.

Until now, no capture of adult salmon has been documented in the Belgian river Meuse, mainly because of the man-made obstructions. Record from The Netherlands reveal the capture, between 1993 and 1996, of 23 adult salmons at the level of the weir of Lith, equipped with a new fish pass (de Jong & Cazemier, 1997; Lanters, 1994).

The key of the success of salmon rehabilitation in the river Meuse basin undoubtedly lies in the opportunity for spawners to successfully overcome the multiple obstacles spread along their migration route. In The Netherlands, five of the seven weirs are equipped with new efficient fish passes (Muyres, 1994; Raat, 1994). The building of new facilities on the two last weirs is scheduled for the coming years, so that in 2002 the migration route should be open for salmon between the North Sea and the Walloon part of the river Meuse. In Belgium, three weirs will be equipped with a new fish pass by 1999 and spawning areas for salmon will be accessible when the complete migratory route will be opened, in conformity with the program established . This program is supported by an international written agreement ('Décision Benelux') signed in April 1996, which forces the governments to open the migration route for migratory fish in hydrographic networks of Benelux within the next 10 years.

Conclusion

The program 'Meuse Salmon 2000' brings solid preliminary results, notably with the equipment of weirs on the river Meuse. Salmon spawning again in the tributaries in the Ardennes will obviously be dependent on the advancement of programmes aiming at the complete restoration of upstream migration from the North Sea, the reduction of smolt losses in the turbines, and the improvement of water quality in the river Meuse downstream Liège. The current and future studies will mainly focuse on the problems concerning upstream migration of spawners and on the impact of hydropower stations. It will be desirable that rehabilitation efforts for salmon in Belgium will be supported by the CIPM (Commission Internationale pour la Protection de la Meuse) which group together France, Belgium and The Netherlands, as the program of rehabilitation of salmon in the river Rhine has been supported by the CIPR (CIPR, 1994, 1996). However the case of the river Meuse is more difficult that the case of the river Rhine, mainly owing to the number of weirs between the North Sea and the Belgian and French Ardennes.

Acknowledgments

The authors wish to thank the Ministry of Agriculture, Nature Conservancy and Environment of the 'Region Wallonne' (Convention 'Meuse Saumon 2000') which financed this study.

References

Admiraal, W.G., G. van der Velde, H. Smit & W.G. Cazemier, 1993. The rivers Rhine and Meuse in the Netherlands present state and signs of ecological recovery. Hydrobiologia 265: 97–128.

Baglinière, J.L., P. Prouzet, J.P. Porcher, A. Nihouarn & G. Maisse, 1987. Caractéristiques générales des populations de saumon atlantique (*Salmo salar* L.) des rivières du Massif armoricain. In Thibault, M & R. Billard (eds), La restauration des rivières à saumons, INRA, Paris: 23–37.

Baras, E., H. Lambert & J.C. Philippart, 1994. A comprehensive assessment of the failure of *Barbus barbus* (L.) migrations through a fish pass in the canalized River Meuse (Belgium). Aquat. Living Resour. 7: 181–189.

Baras, E., J.C. Philippart & B. Salmon, 1996. Estimation of migrant yellow eel stock in large rivers through the survey of fish passes: a preliminary investigation in the River Meuse (Belgium). In Cowx, I.G. (ed), Stock Assessment in Inland Fisheries. Fishing News Books, Blackwell Science, Oxford: 82–92.

Boudreault, A., 1984. Méthodologie utilisée pour la photointerprétation des rivières à saumon de la Côte-Nord. Ministère du Loisir, de la Chasse et de la Pêche: 26 p.

Bousquet, B. & A. Marty, 1987. Le point sur les connaissances des populations naturelles de saumons dans le bassin de l'Adour. In Thibault, M. & R. Billard (eds), La restauration des rivières à saumons. INRA, Paris: 65–79.

Chadwick, E.M.P. & R.G. Randall, 1990. Facteurs dépendants ou indépendants de la densité qui affectent la production de saumon atlantique (*Salmo salar*) en milieu marin ou dulcicole. In Samson, N. & J.P. le Bel (eds), Compte rendu de l'atelier sur le nombre de reproducteurs requis dans les rivières à saumon, Ile aux Coudres, February 1988, Québec: 25–40.

CIPR, 1994. Salmon 2000. Commission Internationale pour la Protection du Rhin (CIPR), Koblenz, Germany, 30 pp.

CIPR, 1996. Salmon 2000. Etat d'avancement des projets début 1996. Commission Internationale pour la Protection du Rhin (CIPR), Koblenz, Germany, 48 pp.

Conjaerts, C. & V. Frank, 1995. Elevage des jeunes saumons atlantiques à la pisciculture du Service de la Pêche. Journée d'Information Internationale 'Meuse Saumon 2000', Université de Liège, 13 September 1995: 4 pp.

Côté, Y., D. Clavet, J.M. Dubois & A. Boudreault, 1987. Inventaire des habitats à saumons et estimation de la production par photographie aérienne. In Thibault, M. & R. Billard (eds), La restauration des rivières à saumons, INRA, Paris: 85–94.

de Jong, H.B.H.J. & W.G. Cazemier, 1997. Onderzoek naar de Salmoniden migratie via de grote rivieren in 1995. RIVO-dlo Rapport C011/97: 24 pp.

Descy, J.P., P. Servais, J.S. Smitz, G. Billen & E. Everbecq, 1987. Phytoplankton biomass and production in the river Meuse (Belgium). Wat. Res. 21: 1557–1566.

Gerlier, M. & P. Roche, 1998. A radio telemetry study of the migration of Atlantic salmon (*Salmo salar*) and Sea trout (*Salmo trutta*) in upper Rhine. Hydrobiologia 371–372: 283–293.

Hadderingh, R.H. & H.D. Bakker, 1998. Fish mortality due to passage through hydroelectric power stations on the Meuse and Vecht rivers. In Jungwirth, M., S. Schmutz & S. Weiss (eds), Fish Migration and Fish Bypasses. Fishing News Books, Blackwell Science, Cambridge: 315–328.

Heland, M. & J. Dumas, 1994. Ecologie et comportement des juvéniles. In Gueguen, J.C. & P. Prouzet (eds), Le saumon atlantique. Ifremer, Plouzané: 29–46.

Lanters, R.L.P., 1994. Het belang en de efficiënte van de vistrap Lith voor zeeforel (*Salmo trutta trutta* L.) en zalm (*Salmo salar* L.) in 1993. RIVO Rapport 94.002: 32 pp.

Larinier, M. & J. Dartiguelongue, 1989. La circulation des poissons migrateurs: le transit à travers les turbines des installations hydroélectriques. Bull. fr. Piscic. 312–313: 94 pp.

MET, 1992. Evaluation des travaux nécessaires à la libre circulation des poissons sur la Meuse et l'Ourthe. Rapport de la Commission 'Echelles à Poissons', Ministère de l'Equipement et des Transports, Ministère de la Conservation de la Nature: 26 pp.

Micha, J.C., 1988. La Meuse et les objectifs du colloque. In Micha, J.C. & S. Pilette (eds), L'impact de l'homme sur l'écosystème Meuse, Actes du colloque, Presses Universitaires de Namur, Namur, Belgium: 9–11.

Micha, J.C. & M.C. Borlée, 1989. Recent historical changes on the Belgian Meuse. In Petts, G.E. (ed), Historical change of large alluvial rivers: Western Europe: 269–295.

Monten, E., 1985. Fish and turbines. Fish injuries during passage through power station turbines. Vattenfall, Stockholm: 111 pp.

Muyres, W.J.M., 1994. Vistrappen in het stroomgebied van de Maas. In Raat, A.J.P. (ed.), Vismigratie, visgeleiding en vispassages in Nederland. Journée d'étude sur la migration des poissons, OVB, Utrecht, Nieuwegein: 187–199.

Ovidio, M., C. Birtles, E. Baras, & J.C. Philippart, 1996. A preliminary telemetry investigation on the obstacles to anadromous salmonids migration in spawning streams of the Belgian Ardennes (River Meuse Basin), in Leclerc, M., H. Capra, S. Valentin, A. Boudreault & Y. Côté (eds), Proceedings of the second IAHR Symposium on Habitat Hydraulics, Ecohydraulics 2000. INRS-Eau, Québec (Canada), Vol. A: 83–88.

Philippart, J.C., 1983. Note sur la redécouverte de 'truites de mer' dans un affluent de la Meuse liégeoise en 1983. Cah. Ethol. Appl. 3: 105–114.

Philippart, J.C., 1985. Reverrons-nous des saumons dans la Meuse? Cah. Ethol. Appl. 5: 31–68.

Philippart, J.C., 1987. Histoire de l'extinction et problématique de la restauration des Salmonidés migrateurs dans la Meuse. In Thibault, M. & R. Billard (eds), La restauration des rivières à saumons. INRA, Paris: 125–137.

Philippart, J.C., 1995. Observations sur l'habitat et la biologie des jeunes saumons réimplantés dans le bassin de l'Ourthe. Journée d'information Internationale 'Meuse Saumon 2000', Université de Liège, 13 September 1995, 7 pp.

Philippart, J.C. & M. Vranken, 1983. Protégeons nos poissons. Collection 'Animaux menacés en Wallonie'. Région Wallonne et Duculot: 206 pp.

Philippart J.C. & A. Gillet, 1990. La réintroduction du saumon atlantique dans le bassin de la Meuse. Etat d'avancement du projet en juin 1990. Tribune Eau 43: 19–26.

Philippart, J.C., A. Gillet & J.C. Micha, 1988. Fish and their environment in large european river ecosystems. The River Meuse. Sci. Eau 7: 115–154.

Philippart, J.C., E. Baras & G. Rimbaud, 1990. Premières observations sur une population de juvéniles de saumons atlantiques (*Salmo salar* L.) reconstituée dans l'Ourthe en 1989. Cah. Ethol. Appl. 10: 421–440.

Philippart, J.C., J.C. Micha, E. Baras, C. Prignon, A. Gillet & S. Joris, 1994. The Belgian project 'Meuse Salmon 2000'. First results, problems and future prospects. Wat. Sci. Tech. 29: 315–317.

Philippart, J.C., G. Rimbaud, E. Baras, M. Ovidio & D. Parkinson, 1998. Les migrations de remontée des poissons dans la Meuse liégeoise et l'Ourthe. Bilan des connaissances et perspectives.

Journée d'études Benelux 'La migration des poissons dans le bassin de la Meuse', Wandre, 21 October 1998, 21 pp.

Prignon, C. & J.C. Micha, 1995a. Etude de la dévalaison et des effets potentiels du turbinage des centrales hydro-électriques sur les saumoneaux. Journée d'Information Internationale 'Meuse Saumon 2000', Université de Liège, 13 September 1995: 4 pp.

Prignon, C. & J.C. Micha, 1995b. Synthèse des résultats 1988–1995 sur la biologie des jeunes saumons réimplantés dans le Samson et la Semois. Journée d'Information Internationale 'Meuse Saumon 2000', Université de Liège, 13 September 1995, 6 pp.

Prignon, C. & J.C. Micha, 1996. Effets du micromarquage magnétique nasal sur des juvéniles de saumon atlantique (*Salmo salar*). Aquat. Living Resour. 9: 107–112.

Prignon, C., J.C. Micha & J.C. Philippart, 1997. La réhabilitation du saumon atlantique dans le bassin de la Meuse. Projet 'Meuse Saumon 2000'. Probio Rev 2-3: 111–120.

Prignon, C., J.C. Micha & A. Gillet, 1998. Biological and environmental characteristics of fish passage at the Tailfer dam on the River Meuse, Belgium. In Jungwirth, M., S. Schmutz & S. Weiss (eds), Fish Migration and Fish Bypasses, Fishing News Books, Blackwell Science, Cambridge: 69–84.

Raat, A.J.P., 1994. Vismigratie, visgeleiding en vispassages in Nederland. Journée d'étude sur la migration des poissons, OVB, Utrecht, Nieuwegein: 279 pp.

Roche, P., 1994. Habitat availability and carrying capacity in the French part of the Rhine for Atlantic salmon (*Salmo salar* L.). Wat. Sci. Tech. 29: 257–265.

Roche, P. & M. Gerlier, 1996. Suivi annuel des peuplements de juvéniles de Salmonidés migrateurs en Alsace. Résultats 1995. Rapport CSP: 17 pp.

Roenen, F., N. Schiepers & A. Gillet, 1995. Les travaux nécessaires à la libre circulation des poissons sur la Meuse et l'Ourthe. Journée d'Information Internationale 'Meuse Saumon 2000', Université de Liège, 13 September 1995: 6 pp.

Seber, G.A.F., 1973. The estimation of animal abundance and related parameters. Griffin, London: 506 pp.

Seber, G.A.F. & E.D. Le Cren, 1967. Estimating population parameters from catches large relative to the population. J. anim. Ecol. 36: 631–643.

Symons, P.E.K., 1979. Estimated escapement of Atlantic salmon (*Salmo salar*) for maximum smolt production in rivers of different productivity. J. Fish Res. Bd Can. 36: 132–140.

Terrier, O., J.P. Raulin, F. Guidou & P. Roche, 1997. Capacité d'accueil de la Semoy française pour le saumon atlantique. Département des Ardennes. Rapport CSP: 16 pp.

Travade, F. & M. Larinier, 1992. La migration de dévalaison: problèmes et dispositifs. Bull. fr. Piscic. 326-327: 165–176.

Hydrobiologia **410:** 79–86, 2000.
J. Garnier & J.-M. Mouchel (eds), Man and River Systems.

Biophysical interactions and the structure and dynamics of riverine ecosystems: the importance of biotic feedbacks

Robert J. Naiman[1], Scott R. Elliott[2], James M. Helfield[2] & Thomas C. O'Keefe[1]
[1]*University of Washington, School of Fisheries, Box 357980, Seattle, WA 98195-7980 U.S.A.*
[2]*University of Washington, College of Forest Resources, Box 352100, Seattle, WA 98195-2100 U.S.A.*

Key words: Biotic feedbacks, riparian, microclimate, streams, animal-ecosystem interactions

Abstract

Characteristics of streams and rivers reflect variations in local geomorphology, climate, natural disturbance regimes and the dynamic features of the riparian forest. Hierarchical interactions between these components result in a rich variety of distinct stream communities which, when considered in combination with strong biotic feedbacks to the physical environment, present formidable challenges in discovering and understanding fundamental, system-level characteristics of natural rivers. The objectives of this article are to briefly review the traditional view of hierarchical physical controls on stream structure and dynamics and to show how this viewpoint is changing as recognition of strong biological influences on physical structure are emerging. In combination, identifying natural stream characteristics and the interactions among individual components, as well as recognizing the importance of biotic feedbacks on physical structure, form the basis for establishing effective conservation strategies.

Introduction

Much like other terrestrial and aquatic systems, flowing waters are structured and influenced by various biophysical agents. Variations in local geomorphology, climate, natural disturbance regimes and riparian corridor dynamics are all reflected within specific characteristics of lotic ecosystems. Interactions between these various biophysical components result in a continuum of in-stream conditions. These interactions, when coupled with the many human uses of the landscape, present formidable challenges in themselves for discovering and understanding fundamental, system-level characteristics of streams and rivers.

In essence, the key elements regulating watershed vitality in the temperate regions of North America and Europe are the delivery and routing of water, sediment, chemicals (ions and nutrients) and organic material to rivers; the temperature regime; and the characteristics of the riparian forest. These environmental features, in combination with the spatial position of a stream reach within the larger drainage network, are traditionally thought to exert hierarchical controls on the biotic character of stream ecosystems (Figure 1). However, recent research suggests that there are strong feedbacks from the biotic components that also act to shape the physical environment.

The purpose of this article is to briefly review some of the more traditional viewpoints on physical control of the biotic community and then to discuss some of the more recent advances in river and stream ecology addressing biotic feedbacks shaping the character of the physical system. Some of these advances may be 'discoveries' in a more pure, scientific sense, while others may simply represent new perspectives in how one views the flowing water ecosystems of the world.

Traditional viewpoints on physical controls

It is widely recognized that there are strong and continuous interactions between geomorphology, hydrology, water chemistry and temperature. For example, consideration of both geomorphology and hydrology is still central in the debate about an acceptable worldwide classification of rivers (Rosgen, 1994; Miller & Ritter, 1996), while hydrology and temperature are linked in their effects on the life histories of many

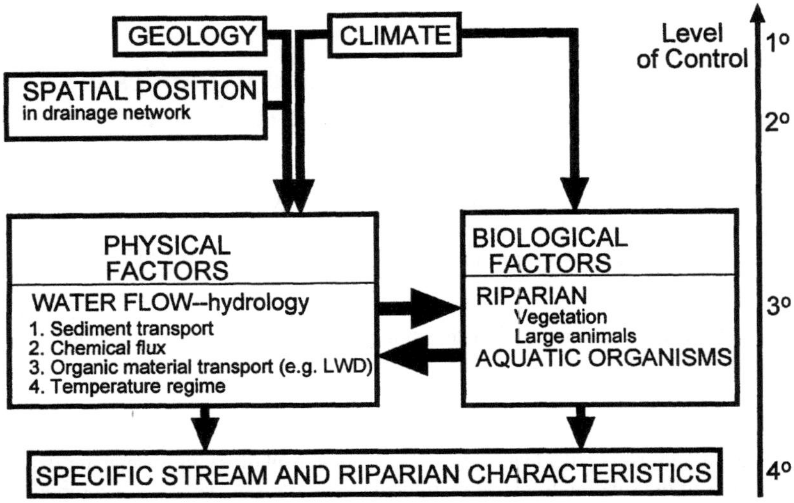

Figure 1. Shown is the concept of hierarchical controls over stream ecosystems, specifically the relative strength of factors (denoted by the width of the arrows) influencing stream characteristics, as well as some principal feed-back loops between components (from Naiman et al., 1992). Traditionally, components have been ranked hierarchically to show dominance of one component over another. However, it is now know that there may be strong feedbacks from lower components to ones that are ranked higher.

aquatic organisms (Ward, 1989). Here we provide a brief overview of each topic, setting the stage for a discussion of biotic feedbacks.

Channel geomorphology

Physical characteristics and processes related to landforms are of primary importance in understanding the structure and dynamics of lotic ecosystems. This fact ultimately provided the basis for the River Continuum Concept (Vannote et al., 1980) and related perspectives (Naiman et al., 1988a). Geologic landforms typically reflect:

1. interactions among tectonic processes that govern the development of topographic relief,
2. processes of erosion,
3. structure and lithology of the substrate, and
4. influences of climate (Montgomery, 1997).

Geomorphic processes in riverine corridors typically are affected by the type, frequency and intensity of natural disturbance. Disturbance characteristics typically reflect the spatial position of channel segments within the drainage network, and geomorphic processes (e.g. material delivery, routing and deposition) are highly interactive with hydrologic patterns and riparian vegetation, acting to shape the biotic characteristics (Naiman et al., 1992).

Hydrologic patterns

Streamflow can be considered a 'master variable' shaping in-stream conditions, given that it is correlated with many physicochemical variables, including water temperature, water quality, channel geomorphology and habitat diversity (Resh et al., 1988; Power et al., 1995). Lotic flow regimes show regional patterns that are determined primarily by river size and by geographic heterogeneity in climate, geology, topography and vegetative cover.

Poff et al. (1997) characterized five critical components of streamflow regimes that regulate ecological processes: magnitude, frequency, duration, timing and rate of change (i.e. flashiness). The *magnitude* of discharge refers simply to the amount of water moving past a fixed location in a given unit of time. The *frequency* of occurrence refers to the relative frequency of a streamflow event of a given magnitude (i.e. the concept of a 50- or 100-year flood). The *duration* refers to the period of time associated with a specific flow condition. The *timing* of a flow of a given magnitude refers to the regularity of occurrence (i.e. its *predictability*). Annual peak flows, for instance, may occur with low or high seasonal predictability. The *rate of change* or *flashiness* refers to how quickly flow changes from one condition to another.

These flow-regime components all vary within and

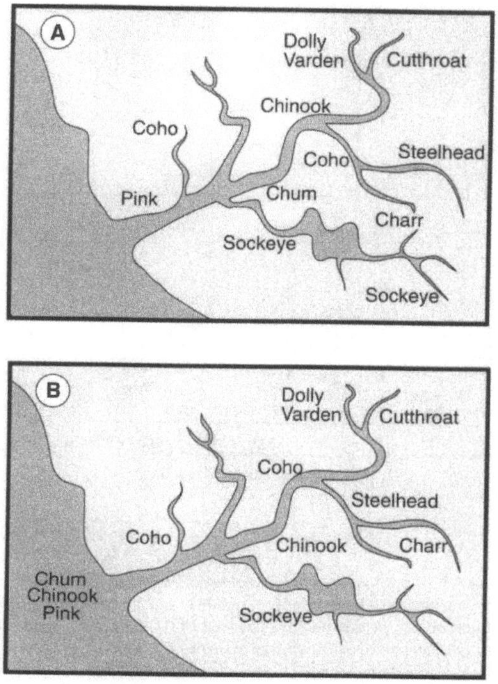

Figure 2. Shown are the relative positions in which selected salmonid species typically (a) spawn and (b) rear. Note that chinook (*Oncorhynchus tshawytscha*) are shown in two different rearing positions; this species exhibits two distinct life history strategies, one in which fry rear in the stream, and another in which emerging fry immediately migrate to sea and rear in saltwater.

between watersheds in an array of spatial and temporal patterns. A mosaic of habitat types result from these hydrologic variations; a single river may provide ephemeral, seasonal and persistent habitat patches which range from free-flowing, to standing, to no water. And for many lotic species, completion of the life cycle requires an array of habitat types, whose availability over time is regulated by the natural flow regime (e.g. Reeves et al., 1996; Naiman & Anderson, 1997). Consider, for instance, salmonids, which often require different habitat across life history stages (Figure 2). Salmonid species vary in their need for spawning habitat; different species ultimately utilize different spatial positions within the watershed. Coastal cutthroat trout (*Oncorhynchus clarki*) generally prefer cool, heavily-shaded headwater reaches, while chum salmon (*O. keta*) typically spawn lower in the watershed, often within the main channel. Rearing habitat is often different from spawning locales, and it, too, typically varies between species. Coho salmon (*O. kisutch*) normally spawn in the main channel, while emerging fry may move into off-channel habitats to feed and rear. Natural flow regime dynamics are largely responsible for the creation and maintenance of areas like off-channel rearing areas. Migratory access to these areas is subject to the timing of inundation, and many fish species have adapted to the temporal phases of off-channel access. The flow regime regulates the degree of lateral connectivity a given lotic system experiences.

The magnitude and frequency of flow extremes regulate numerous ecological processes; they in fact create 'bottlenecks' which present critical stresses and opportunities for a wide array of riverine species (Poff & Ward, 1989). Organic detritus and attached algae are transported by flow, and periods of high flow remove and transport fine sediments that clog the interstitial spaces in otherwise productive gravel habitats (Beschta & Jackson, 1979). Floods also regulate other vital riverine processes, such as the delivery of LWD, the establishment of disturbance-adapted riparian species, and the development of aquatic habitat in the floodplain. The timing of flood events are also of critical importance to many aquatic and riparian species; life history events such as egg hatching, spawning and migration are often highly flow-dependent (Ward, 1989; Naiman & Anderson, 1997).

Chemistry and temperature

Dissolved nutrients and ions have a large influence on the kinds, amounts and activities of organisms present in streams. Inorganic ion and nutrient concentrations often reflect the erodeability and solubility of the underlying geology, whereas the type and concentrations of organic materials often reflect watershed vegetation, soil processes, water residence time on the land and in the drainage network, and the ability of microbes to decompose specific materials (Welch et al., 1998).

Water temperatures and the type and duration of ice cover also have strong controls on the biotic community. Water temperatures control and synchronize most physiological functions of ectothermic life forms (Ward, 1989; Naiman & Anderson, 1997). Temperature is a crucial determinant of the incubation rate of eggs and influences the growth rates of all aquatic organisms. Where seasonal water temperatures are extreme, patchiness in temperature throughout the drainage network becomes important in the persistence of some species as well as in maintaining a diversity of life history strategies in the same species.

Ice poses special challenges for aquatic organisms (Prowse & Gridley, 1993). Each of the basic types of ice – surface ice, anchor ice and frazil ice (i.e. ice particles which may be present in the flowing water column) – has different consequences for the organisms. Anchor ice is probably the most disruptive to bottom-dwelling organisms because of its tendency to exclude organisms from the sediments and to raft sediments if the ice mass is large enough to float. Rivers with substantial ice formation each year tend to be dominated by low-diversity communities and contain numerous insect species that complete their life cycles in less than one year. In contrast, rivers with less regular ice formation tend to have higher diversity and to contain insect species that may take several years to complete their life cycles.

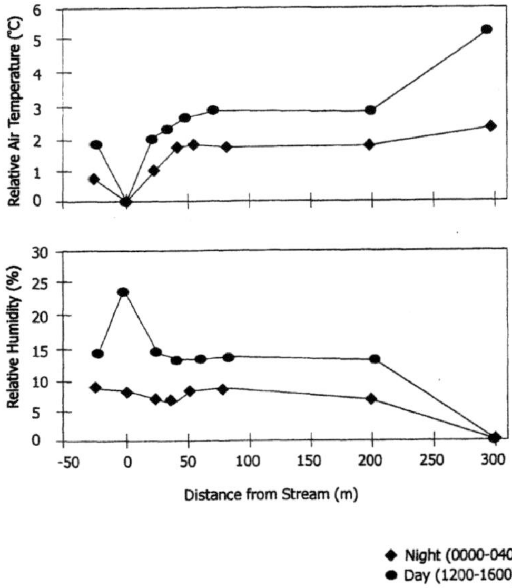

Figure 3. Shown are microclimatic data for riparian forests in western Washington. Both relative humidity and air temperature are influenced by the stream for distances of 31–62m away from the channel (adapted from Brosofske et al., 1997).

The importance of biotic feedbacks

Characteristics of the riparian corridor

The riparian perspective toward river ecology is of a relatively recent genesis, especially when compared with other advances in limnology (Hynes, 1975; Naiman & Décamps, 1990, 1997). Clearly, there are strong effects of hydrology and landform on the development of riparian forests, and these have been well-documented (Gregory et al., 1991), as have the influences of the riparian forest on the trophic character of streams (via the type and amount of allochthonous inputs; Fisher & Likens, 1973) and on water quality (Haycock et al., 1996). Although the notion that riparian forests shape the physical structure of the lotic system (although recognized some time ago by geomorphologists; Keller & Swanson, 1979), it has been only recently quantified ecologically in terms of the spatial and temporal dynamics of large woody debris (LWD).

LWD in the many of the stream ecosystems of North America is a primary determinant of channel morphology where it forms pools, regulates transport of sediment and particulate organic matter, and provides habitat and cover for fish and other biota (Bilby & Bisson, 1998). Deposited LWD often forms a low gradient area upstream of the debris jam, providing a locale for sediment and organic matter deposition. In some instances, LWD may initiate the formation of islands where vegetation colonization, growth and island coalescence is realized (Fetherston et al., 1995). Streams that have seen significant anthropogenic disturbance in the riparian zone often suffer from a lack of LWD recruitment, as well as LWD being cleared from channels (Piégay & Gurnell, 1997). The replacement of riparian forests by early-successional, easily-decayed litter may result in a long-term degradation of the quantity and quality of deposited LWD, ultimately compromising factors such as the proportion of retained organic matter and a decline in habitat value for aquatic organisms (Ralph et al., 1994). LWD in tropical regions, however, appears to play a lesser role in shaping the physical environment due to the widespread distribution of termites (Naiman & Rogers, 1997).

LWD decay rates are of interest in some ecosystems where debris may last an especially long time. Surface LWD on the Queets River, Washington (U.S.A.), has been dated to over 1500 years before present (ybp; most dates are <300 ybp, T. Hyatt & R. J. Naiman, unpublished data), while a stream in western Tasmania has significant surface LWD which dates 2000 to 4000 ybp (Nanson et al., 1995). Not only do the riparian forests and associated LWD shape the function and structure of forested streams, often LWD continues to influence the physical character of the streams for centuries to millenia.

Microclimate gradients

The notion that the riparian forest influences stream temperatures via shading of solar radiation has been known for some time (Beschta et al., 1987). Brosofske et al. (1997) have shown that small streams (2–4 m width) and their riparian zones modify air temperature and humidity for distances ranging from 31–62 m away from the channel in temperate rain forest streams (Figure 3). These results are especially important for resource management since more conservative riparian buffer widths of 10–25 m may not be adequate for preserving an unaltered microclimate or adequate growing conditions for a typical vegetative community.

In addition, the role of riparian vegetation in retaining water and attenuating the downstream effects of floods is an important landscape-scale function (Décamps, 1996). Water retention by plants and LWD in the riparian zone increases local relative humidity affecting the pattern and diversity of the vegetative community (Pollock et al., 1998).

The effects of large animals

Historically, animals have been viewed as passive components of ecosystems, responding merely to the conditions to which they are subjected. In many cases, however, large animals are responsible for biogeochemical, successional and landscape alterations that may persist for centuries (Naiman, 1988; Butler, 1995).

Early accounts of North American and European landscapes indicated a seemingly unlimited supply of migratory birds, fishes, fur-bearing mammals and large assemblages of mammalian herbivores. Some early North American accounts documented the strong interactions between these animals and the ecosystems supporting them (e.g. Bartram, 1791; Morgan, 1868; Hays, 1871). The extirpation of a number of these megafauna have eliminated their functional role from selected ecosystems, but evidence of the long-term effects of animal influences still remains (e.g. mima mounds, beaver meadows). In addition to this, a number of large animals, such as beaver (*Castor canadensis*) are now recolonizing much of their former range (due to factors such as the relative lack of predators, laws regulating take and increasing forage and habitat) and once again demonstrating some influences on the physical characteristics of streams.

Riverine systems are among the ecosystems in which animal influences have played a dominant role, although their role largely has been omitted from discussions of biophysical influences on flowing waters. This omission is particularly unfortunate, given that their influence is second only to the aforementioned primary physical factors (Naiman & Rogers, 1997). Large animals in riverine ecosystems can significantly alter both the structure (channel geomorphology, vegetative characteristics, biodiversity) and function (productivity, connectivity, resistance and resilience to disturbance) of river corridors (Figure 4).

Many ecologists and resource managers have maintained a single-species approach to the analysis of animal impacts on rivers and streams. Typically, aquatic mammal species (e.g. beaver) and riparian browsers (e.g. moose, *Alces alces*, or elk, *Cervus canadensis*) have been studied and managed separately. This focus has failed to recognize the synergistic forces that result from the interactions among riverine megafauna; indeed community interactions among species have long-term, complex ecosystem-level consequences (Butler, 1995; Naiman & Rogers, 1997). A striking example of how large animals modify the physical environment is provided by beaver. Through the cutting of wood and the construction of dams, beaver retain sediments and organic matter, create and maintain wetlands, modify nutrient cycling and decomposition dynamics, modify riparian structure and dynamics, influence the character of materials transported downstream, and ultimately influence plant and animal community composition and diversity (e.g. Naiman & Melillo, 1984). With regard to the latter point, the activities of beaver have been shown to replace invertebrate taxa normally adapted to running-waters with pond taxa (McDowell & Naiman, 1986), generate a striking increase in the biomass of lotic invertebrates per unit area (Naiman et al., 1986), and influence fish assemblages (Snodgrass & Meffe, 1998).

Interactions between species are of particular importance in influencing the physical structure of riverine ecosystems. Many Pacific species of anadromous salmon and trout (*Oncorhynchus* spp.) are semelparous (i.e. they die after spawning) and their carcasses represent a significant seasonal nutrient contribution to the system in which they perish. Epilithic organic matter, many aquatic macroinvertebrates and rearing salmonids have all been shown to be enriched with marine-derived nitrogen (^{15}N) in systems supporting anadromous salmonids (Bilby et al., 1996; Larkin & Slaney, 1997). Riparian vegetation adjacent to these streams are also ^{15}N enriched, indicating

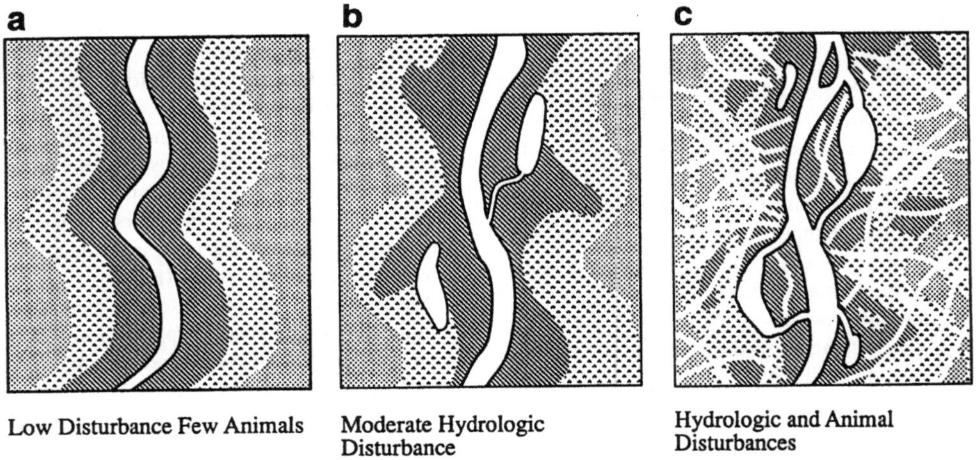

Figure 4. When channel processes are constrained by management, the result is a vegetation mosaic in a simple linear form (a). Under the influence of natural hydrologic and channel processes, heterogeneity is improved (b). Patch heterogeneity is optimized, however, only with the feeding and movements of large animals. Patches are smaller, more numerous, and more spatially dispersed under the influence of large animals (from Naiman & Rogers, 1997).

that vectors (such as grizzly bear, *Ursus arctos*) may be important in distributing limiting nutrients within the system (J. Helfield & R. J. Naiman, unpublished data). Given that most temperate ecosystems are N-limited, this contribution of marine-derived nutrients, including the distribution by various vectors, is a significant process for both the riparian vegetation (which produces LWD) and for in-channel production.

In river corridors, numbers of animals and abundance (and quality) of food also vary constantly, and the variations are irregular in time, space and ecological amplitude. These variations are interconnected; variations in animal abundance indirectly effect the abundance of others, and may indirectly affect the trophic dynamic of the system in question. Selective foraging by some large mammals (e.g. moose, *Alces alces*) can change ecosystem properties such as plant species composition, nutrient cycling rates and soil fertility. For example, selective browsing by moose on hardwood species allows unbrowsed or lightly browsed conifers to dominate the boreal landscape (McInnes et al., 1992); increasing relative abundance of conifer litter may depress the availability of soil N, often the limiting nutrient in boreal forest ecosystem productivity. Foraging strategies of individual moose may result in the creation of riparian and terrestrial landscapes which strongly affect both the vegetation (browse density and distribution) and, in turn, moose population density and survival (Moen et al., 1998).

Synthesis

The traditional view of top-down, hierarchical control of stream environments by geomorphic and hydrologic regimes remains valid. The purpose of this article is not to challenge this widely accepted and robust concept but to show that the concept, to be fully descriptive, requires modification to include biotic influences on the physical components as well. Geomorphic and hydrologic processes continue to provide strong influences on the physical template of streams and continue to form the basis for the development and maintenance of biotic communities and processes. Nevertheless, recent advances in stream ecology have demonstrated that certain biotic components also exhibit strong influences on geomorphology, hydrology and microclimate, helping shape the physical environment.

Human intervention can seriously disrupt important natural feedbacks between the biotic and physical environments. By altering natural flow regimes, modifying the riparian corridor, and directly and indirectly introducing or removing various plant and animal species, human beings have significantly altered the majority of the world's riverine ecosystems. For example, LWD deposition, forest establishment, and subsequent LWD deposition represents a process 'loop' vital to the long-term integrity of a natural stream system. Riparian modifications that result from forest management alter the LWD cycle, disrupting vital feedbacks to the physical environment (Naiman et al., 1992, 1997; Ralph et al., 1994). Replacement of mature riparian forests with stands dominated by early successional species alters the quality and quantity of recruitable LWD. Many early successional species are smaller (and therefore less likely to remain within the system through a flood season) and decay faster than late-successional species. Invasive plant species can have long-term effects on the geomorphology of channels, as seen in the Platte (Johnson, 1994), Colorado and Green (Graf, 1978) rivers in the United States. The net consequence is that long-term system-level modifications result from direct or indirect modifications to the riparian forest or to other sources of channel-shaping LWD.

Likewise, many large animal species that exert considerable influences on the structure and function of rivers and streams have been significantly affected by man's activities. Migrating salmon and trout are important components of many coastal systems, for example, yet suffer from the effects of various land use practices as well as from near-sighted fishery resource exploitation. The long-term effects of eliminating the seasonal 'pulse' of marine-derived nutrients in naturally oligotrophic coastal streams and rivers are not known precisely but are suspected to be significant. In addition, the extirpation of beaver from much of their former range (as well as their recolonization of this habitat) fundamentally altered the character of streams throughout the upper third of the world (Naiman et al., 1988b, 1994).

The perspective addressed in this article, that specific biotic components have strong feedbacks to the physical system, is not new. However, as new data are generated and new viewpoints toward streams are explored, biotic feedbacks appear to be more prevalent than originally expected. Indeed, there are strong long-term linkages between the physical and biological components and those linkages have influences and strengths in both directions, possibly with equal intensity.

Acknowledgments

Research supported by grants from the Andrew W. Mellon Foundation, the Pacific Northwest Research Station of the U.S. Forest Service and the National Science Foundation. We thank Henri Décamps and another referee for comments that substantially improved the presentation.

References

Bartram, W., 1791. The Travels of William Bartram (1928 ed.) Dover Publications, New York.

Beschta, R. L. & W. L. Jackson, 1979. The intrusion of fine sediments into a stable gravel bed. J. Fish. Res. Bd Can. 36: 207–210.

Beschta, R. L., R. E. Bilby, G. W. Brown, L. B. Holtby, & T. D. Hofstra, 1987. Stream temperature and aquatic habitat: fisheries and forestry interactions. In E. O. Salo & T. W. Cundy (eds), Streamside Management: Forestry and Fishery Interactions. University of Washington, Institute of Forest Resources Contribution No. 57: 191–232.

Bilby, R. E., B. R. Fransen & P. A. Bisson, 1996. Incorporation of nitrogen and carbon from spawning coho salmon into the trophic system of small streams: evidence from stable isotopes. Can. J. Fish. aquat. Sci. 53: 164–173.

Bisson, P. A. & R. E. Bilby, 1998. Organic matter and trophic dynamics. In R. J. Naiman & R. E. Bilby (eds), River Ecology and Management: Lessons from the Pacific Coastal Ecoregion. Springer-Verlag, New York: 373–398.

Brosofske, K. D., J. Chen, R. J. Naiman & J. F. Franklin, 1997. Harvesting effects on microclimatic gradients from small streams to uplands in western Washington. Ecol. Appl. 7: 1188–1200.

Butler, D. R., 1995. Zoogeomorphology: Animals as Geomorphic Agents. Cambridge University Press, New York.

Décamps, H., 1996. The renewal of floodplain forests along rivers: a landscape perspective. Verh. int. Ver. Limnol. 26: 35–59.

Fetherston, K. L., R. J. Naiman & R. E. Bilby, 1995. Large woody debris, physical process and riparian forest development in montane river networks of the Pacific Northwest. Geomorphology 13: 133–144.

Fisher, S. G. & G. E. Likens, 1973. Energy flow in Bear Brook, New Hampshire: an integrative approach to stream ecosystem metabolism. Ecol. Monogr. 43: 421–439.

Graf, W. L., 1978. Fluvial adjustments to the spread of tamarisk in the Colorado Plateau region. Geol. Soc. Am. Bull. 89: 1491–1501.

Gregory, S. V., F. J. Swanson, W. A. McKee & K. W. Cummins, 1991. An ecosystem perspective of riparian zones. BioScience 41: 540–551.

Haycock, N., T. P. Burt, K. W. T. Goulding & G. Pinay (eds), 1996. Buffer Zones: Their Processes and Potential in Water Protection. Quest Environmental. Harpenden, Hertfordshire, U.K.

Hays, W. J., 1871. Notes on the range of some animals in America at the time of the arrival of the whitemen. Am. Nat. 5: 25–30.

Hynes, H. B. N., 1975. The stream and its valley. Verh. int. Ver. Limnol. 19: 1-15.

Johnson, W. C., 1994. Woodland expansion in the Platte River, Nebraska: patterns and causes. Ecol. Monogr. 46: 59–84.

Keller, E. A. & F. J. Swanson, 1979. Effects of large organic material on channel form and fluvial processes. Earth Surf. Process. 4: 361–380.

Larkin, G. A. & P. A. Slaney, 1997. Implications of trends in marine-derived nutrient influx to south coastal British Columbia salmonid production. Fisheries 22: 16–24.

McDowell, D. M. & R. J. Naiman, 1986. Structure and function of a benthic macroinvertebrate stream community as influenced by beaver (*Castor canadensis*). Oecologia 68: 481–489.

McInnes, P. F., R. J. Naiman & J. Pastor, Effects of moose browsing on vegetation and litter of the boreal forest, Isle Royale, Michigan, U.S.A. Ecology 73: 2059–2075.

Miller, J. R. & J. B. Ritter, 1996. An examination of the Rosgen classification of rivers. *Catena* 27: 295–299.

Moen, R., Y. Cohen & J. Pastor, 1998. Linking moose population and plant growth models with a moose energetics model. Ecosystems 1: 52–63.

Montgomery, D. R., 1997. The influence of geological processes on ecological systems. In P. K. Schoonmaker, B. von Hagen & E. C. Wolf (eds), The Rainforests of Home: Profile of a North American Bioregion. Island Press, Washington, D.C.: 43–68.

Morgan, L. H., 1868. The American Beaver and his Works. J. B. Lippincott, Philadelphia.

Naiman, R. J. & E. C. Anderson, 1997. Streams and rivers: their physical and biological variability. In P. K. Schoonmaker, B. von Hagen & E. C. Wolf (eds), The Rain Forests of Home: Profile of a North American Bioregion. Island Press, Washington, D.C.: 131–148.

Naiman, R. J., P. A. Bisson, R. G. Lee & M. G. Turner. 1997. Approaches to management at the watershed scale. In K. Kohm & J. F. Franklin (eds), Creating A Forestry for the 21st Century: The Science of Ecosystem Management. Island Press, Washington, D.C.: 239–253.

Naiman, R. J., T. J. Beechie, L. E. Benda, D. R. Berg, P. A. Bisson, L. H. MacDonald, M. D. O'Connor, P. L. Olson, & E. A. Steel, 1992. Fundamental elements of ecologically healthy watersheds in the Pacific Northwest Coastal Ecoregion. In R. J. Naiman (ed.), Watershed Management: Balancing Sustainability with Environmental Change. Springer-Verlag, New York: 127–188.

Naiman, R. J. & H. Décamps (eds), 1990. Ecology and Management of Aquatic-Terrestrial Ecotones. UNESCO, Paris and Parthenon Publishing Group, Carnforth, U.K.

Naiman, R. J. & H. Décamps, 1997. The ecology of interfaces – riparian zones. Annual Rev. Ecol. Syst. 28: 621–658.

Naiman, R. J., H. Décamps, J. Pastor & C. A. Johnston, 1988a. The potential importance of boundaries to fluvial ecosystems. J. n. am. Benthol. Soc. 7: 289–306.

Naiman, R. J., C. A. Johnston & J. C. Kelley, 1988b. Alteration of North American streams by beaver. BioScience 38: 753–762.

Naiman, R. J. & J. M. Melillo, 1984. Nitrogen budget for a subarctic stream altered by beaver (*Castor canadensis*). Oecologia 62: 150–155.

Naiman, R. J., J. M. Melillo & J. E. Hobbie, 1986. Ecosystem alteration of boreal forest streams by beaver (*Castor canadensis*). Ecology 67: 1254–1269.

Naiman, R. J., G. Pinay, C. A. Johnson & J. Pastor, 1994. Beaver-induced influences on the long-term biogeochemical characteristics of boreal forest drainage networks. Ecology 75: 905–921.

Naiman, R. J. & K. H. Rogers, 1997. Large animals and system-level characteristics in river corridors – implications for river management. BioScience 47: 521–529.

Nanson, G. C., M. Barbetti & G. Taylor, 1995. River stabilisation due to changing climate and vegetation during the late Quarternary in western Tasmania, Australia. Geomorphology 13: 145–158.

Piégay, H. & A. M. Gurnell, 1997. Large woody debris and river geomorphological pattern: examples from S.E. France and S. England. Geomorphology 19: 99–116.

Poff, N. L. & J. V. Ward, 1989. Implications of streamflow variability and predictability for lotic community structure: a regional analysis of streamflow patterns. Can. J. Fish. aquat. Sci. 46: 1805–1818.

Poff, N. L., J. D. Allan, M. B. Bain, J. R. Karr, K. L. Prestegaard, B. D. Richter, R. E. Sparks & J. C. Stromberg, 1997. The natural flow regime. BioScience 47: 769–784.

Pollock, M. M., R. J. Naiman & T. A. Hanley, 1998. Plant species richness in riparian wetlands: a test of biodiversity theory. Ecology 79: 94–105.

Power, M. E., A. Sun, M. Parker, W. E. Dietrich & J. T. Wootton, 1995. Hydraulic food chain models: an approach to the study of food web dynamics in large rivers. BioScience 45: 159–167.

Prowse, T. D. & N. C. Gridley, 1993. Environmental aspects of river ice. National Hydrology Research Institute, NHRI Science Report No. 5, Saskatoon, Saskatchewan, Canada.

Ralph, S. C., G. C. Poole, L. L. Conquest & R. J. Naiman, 1994. Stream channel condition and in-stream habitat in logged and unlogged basins of western Washington. Can. J. Fish. aquat. Sci. 51: 37–51.

Reeves, G. H., L. E. Benda, K. M. Burnett, P. A. Bisson & J. R. Sedell, 1996. A disturbance-based ecosystem approach to maintaining and restoring freshwater habitats of evolutionarily significant units of anadromous salmonids in the Pacific Northwest. Am. Fish. Symp. 17: 334–349.

Resh, V. H., A. V. Brown, A. P. Covich, M. E. Gurtz, H. W. Li, G. W. Minshall, S. R. Reice, A. L. Sheldon, J. B. Wallace & R. Wissmar, 1988. The role of disturbance in stream ecology. J. n. am. Benthol. Soc. 7: 433–455.

Rosgen, D. L., 1994. A classification of natural rivers. *Catena* 22: 169-199.

Snodgrass, J. W. & G. K. Meffe, 1998. Influence of beavers on stream fish assemblages: effects of pond age and watershed position. Ecology 79: 928–942.

Vannote, R. L., G. W. Minshall, K. W. Cummins, J. R. Sedell & C. E. Cushing, 1980. The river continuum concept. Can. J. Fish. aquat. Sci. 37: 130–137.

Ward, J. V., 1992. Aquatic Insect Ecology, 1. Biology and Habitat. John Wiley and Sons, Inc., New York.

Welch, E., J. M. Jacoby & C. W. May, 1998. Stream quality. In R. J. Naiman & R. E. Bilby (eds), River Ecology and Management: Lessons from the Pacific Coastal Ecoregion. Springer-Verlag, New York: 67–92.

Hydrobiologia **410:** 87–96, 2000.
J. Garnier & J.-M. Mouchel (eds), Man and River Systems.

Biogeochemical nutrient cycles and nutrient management strategies

Daniel J. Conley
Department of Marine Biology and Microbiology, National Environmental Research Institute, P.O. Box 358, DK-4000 Roskilde, Denmark
E-mail: dco@dmu.dk

Key words: eutrophication, nutrient limitation, nitrogen, phosphorous, silicate

Abstract

Nutrient loading by riverine input into estuarine systems has increased by 6–50 times for the N load from pristine conditions to present, whereas a 18–180 times increase has been observed in the P load. Reductions in the ratio of N to P delivery has also occurred with time. In a review of nutrient limitation in estuarine systems, it is shown that many estuarine systems display P limitation in the spring, switching to N limitation in the summer with some estuaries displaying dissolved silicate limitation of the spring diatom bloom. Historical and recent changes in nutrient loading and their effect on nutrient limitation have intensified the debate on the control of nutrient delivery to estuaries from both agricultural and point sources, and as to what nutrient (N or P) should be managed for in estuarine systems. It is hypothesized that potential reductions in P may help oxygen depletion especially in deep estuaries and reduce fast growing macrophytes such as *Ulva* sp., although P reductions probably will have little effect on summer chlorophyll concentrations, an important recreational management goal. Reductions in N loading should reduce summer chlorophyll concentrations and improve the conditions for submerged aquatic vegetation and thus improve ecosystem functioning. Finally, if only P reductions are pursued, that is if we are able to reduce P such that it is limiting year around in estuarine systems, it is likely that the export of N from estuarine systems would increase to the bordering N-limited marine systems, thus only exporting the problem of enhanced production with eutrophication.

Introduction

The load of nutrients to aquatic systems has greatly increased with time through human activity. As a consequence of this increased nutrient loading, detrimental effects have been observed on the health of coastal ecosystems leading to, amongst other things, excess accumulations of phytoplankton biomass (Malone et al., 1986), episodes of noxious blooms, reductions in aquatic macrophyte communities (Sand-Jensen & Borum, 1991; Duarte, 1995) and the depletion of dissolved oxygen in bottom waters (Malone et al., 1986).

An important question facing us as scientists and as advisors to the management community is the role of nutrients (N, P and Si) in limiting system production, especially which nutrient limits phytoplankton production. While it is generally accepted that P limits freshwater systems (Schindler, 1974) and N limits marine systems (Nixon et al., 1996), problems arise in what nutrient to manage for in estuarine systems since these systems can show both N and P limitation at varying temporal and spatial scales (D'Elia et al., 1986; Malone et al., 1996).

In this paper, I will evaluate the changes that have occurred in nutrient loading to estuaries with time, I will review the evidence for nutrient limitation in aquatic ecosystems focusing on estuarine systems, I will examine the differences in biogeochemical cycles that may account for differences in what nutrient is limiting and finally discuss the consequences of different nutrient management strategies to estuarine systems.

Nutrient loading

It is widely accepted that increases in nutrient loading have taken place from anthropogenic activities, however, the magnitude of that increase is poorly known.

The construction of well-constrained nutrient budgets for the present day is possible for many aquatic ecosystems given the availability of information, however, modeling must be used to estimate nutrient budgets under pristine conditions. An excellent example of reconstructing historical nutrient loads is provided by Chapra (1977) where the P inputs to the North American Great Lakes were partitioned into their various sources, e.g. from the atmosphere, from land runoff, from direct inputs from anthropogenic activities, and from upstream sources, and the changes in inputs with time were estimated (Figure 1). The source proportioning method utilized by Chapra (1977) provided an important tool to evaluate historical changes in nutrient loading in the Great Lakes in order to determine the impact of man's activities on P loading through time.

Although modern day nutrient budgets are available for many estuarine and coastal systems (e.g. Nixon et al., 1996), few estimates are available to compare present day nutrient loads to those under pristine conditions. I have assembled previously made estimates of nutrient loading for four different estuarine systems (Figure 2) for pristine conditions, at the turn of the last century, and present day nutrient loads. Because of the large range in absolute nutrient loading, the nutrient loading has been normalized to watershed area to allow for the comparison of the intensity of nutrient loading from the land to different systems. It is clear that the load of nutrients to aquatic systems has increased greatly with time through human activity. The estimates presented here range from a 18 to 180 times increase in the amount of P and a 6 to 50 times increase in the amount of N from pristine conditions to present. Although determination of the exact amount of nutrients under 'pristine conditions' is problematic as most modern analogs of 'pristine' systems are perturbed by atmospheric deposition, the uncertainty in this estimate, however, is small relative to the large increases in loading observed between pristine conditions and present. Relative to conditions at the turn of the century, P loads in these systems have increased by approximately 2 – 6 times and N loads have increased 1.5 – 4.5 times (Figure 2).

It is likely that exports of N and P from pristine watersheds were also different in chemical composition and contained a different relative N:P ratio (Figure 3). In anthropogenically undisturbed systems efficient nutrient recycling processes retain most of the inorganic compounds and nutrient export is primarily in the form of organic compounds. For example, N exports

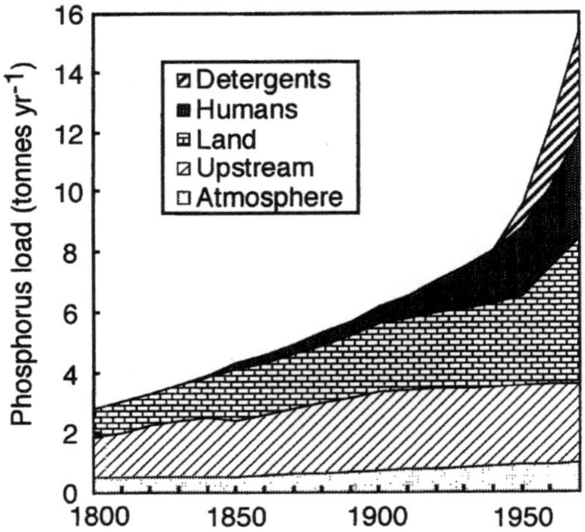

Figure 1. Historical loads of total P from various sources to the North American Great Lakes as determined from a source-partitioning model. The figure is redrawn from Chapra (1977).

from undisturbed Chilean watersheds occur primarily as dissolved organic nitrogen (Hedin et al., 1995). Alteration in the export of dissolved inorganic nutrients results in changes in nutrient ratios (Figure 3) with a lower N:P ratio observed in anthropogenically disturbed watersheds (Billen et al., 1991). Changes in the export of Si with anthropogenic activities is poorly known, especially in terms of changes in production of dissolved silicate through weathering reactions. It is clear, however, that reductions in Si delivery to the coastal zone have occurred with the enhanced retention of Si as diatoms are trapped in lakes and behind dams (Conley et al., 1993; Billen & Garnier, 1997)

Nutrient limitation

Considerable effort has been put into determining the role of nutrients (N, P and Si) in limiting system production, especially which nutrient limits phytoplankton production. A hierarchy of experimental approaches (*sensu* Hecky & Kilham, 1988) has been used to assess the response of phytoplankton to nutrient additions across scales from small-scale bioassays, mesocosms, cross-system comparisons, to whole-system fertilization. In addition, a variety of specific indicators have been used to assess nutrient limitation, including determination of the physiological state of phytoplankton, inorganic nutrient stoichiometry and

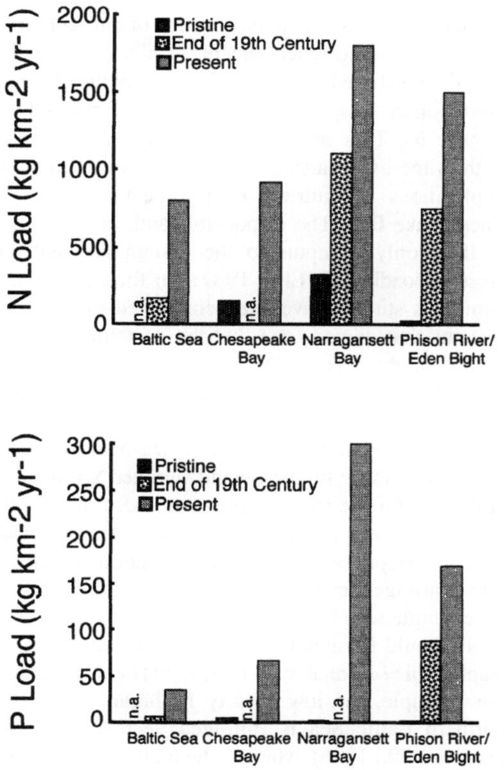

Figure 2. (A) The annual N load and (B) the annual P load per unit of watershed area (kg km^{-2} y^{-1}) for various estuarine systems under pristine conditions, at the turn of the last century and at present. Data are from Larsson et al. (1985) for the Baltic Sea with updated loading calculations of Stålnacke et al. (1998), Boynton et al. (1995) for Chesapeake Bay, Nixon (1995, 1997) for Narragansett Bay, and Billen & Garnier (1997) for the hypothetical system of Phison River/Eden Bight. All loads are for total N and total P except for Narragansett Bay which are only inorganic nutrient concentrations. n.a. is data not available.

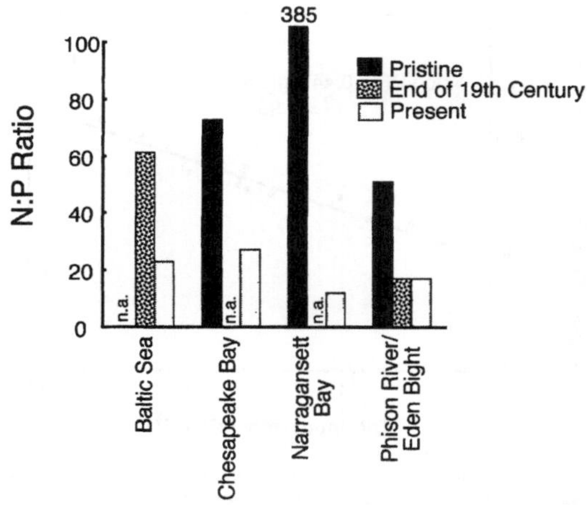

Figure 3. The N:P loading ratio for different estuarine systems under pristine conditions, at the turn of the last century and at present. The data are the same used in Figure 2. n.a. is data not available.

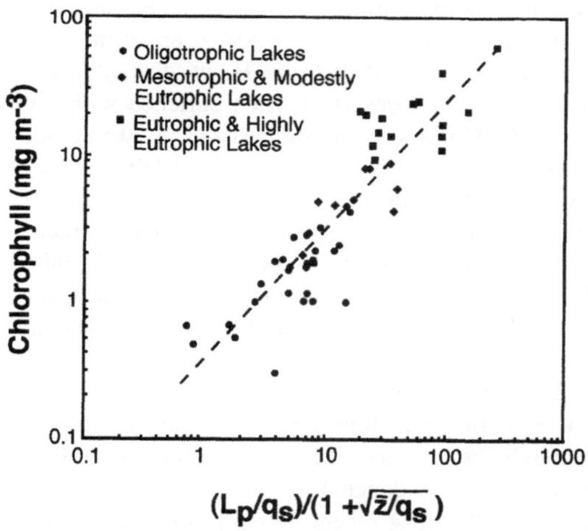

Figure 4. Statistical relationships between average epilimnetic chlorophyll concentration and phosphorus loading characteristics (from Vollenweider, 1976) in a variety of freshwater systems. L_p is phosphorus loading (g m^{-2} h^{-1}), q_s is hydraulic loading (m y^{-1} m^{-3} of lake area), and z is mean depth (m). Figure is redrawn from Hecky & Kilham (1988).

concentration, composition ratios, and various measurements of phytoplankton growth. The purpose here is to only briefly detail the evidence for nutrient limitation in freshwater and marine environments and then to focus on the results of numerous recent studies determining nutrient limitation in estuarine environments. Although the focus of this review is on nutrients, it should not be forgotten that light limitation is also one of the major factors regulating the growth of phytoplankton in aquatic systems (Cloern, 1998). In addition, the focus on a single nutrient is in some ways problematic, as the largest response to nutrient addition usually occurs when multiple nutrients are added.

In freshwaters, there is a general consensus that P is often most limiting to phytoplankton growth (Hecky & Kilham, 1988; Howarth, 1988), although during summer, when dissolved inorganic nutrients are depleted from the photic zone, a response is often elicited with N as well as P (Elser et al., 1990). The most compelling evidence that P is limiting in freshwaters came from the whole lake experiments by Schindler (1974).

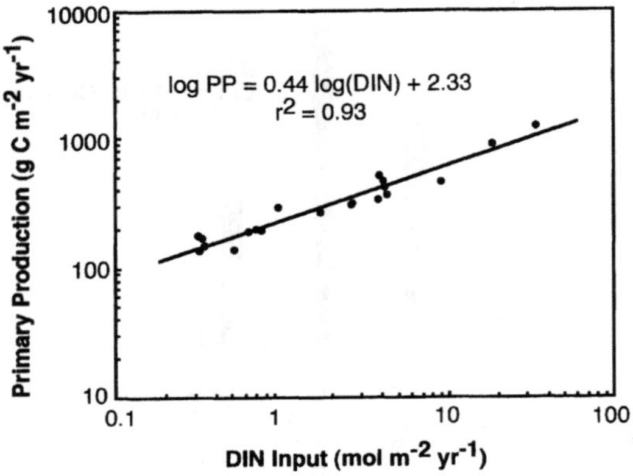

Figure 5. Primary production by phytoplankton (^{14}C uptake) as a function of the estimated rate of input of dissolved inorganic nitrogen (DIN) per unit area in a variety of marine ecosystems. Figure is redrawn from Nixon et al. (1996).

Visually dramatic differences in algal blooms were seen in lake 226 that was separated by an artificial curtain and enriched with N, C and P, but substantial algal blooms were not observed when enriched with only N and C. Cross-system comparisons, best exemplified by Vollenweider (1976), where the dependence of average epilimnetic chlorophyll concentrations were significantly related to phosphorus loading characteristics, suggested that lakes in general are P-limited (Figure 4). Similar P-chlorophyll relationships have been observed by numerous authors (see Prairie et al., 1989).

In marine waters, the general consensus has been that N is often most limiting to phytoplankton growth, although the evidence for N limitation was not as strong as the evidence for P limitation in freshwaters when the most recent reviews of nutrient limitation were written, e.g. Hecky & Kilham (1988) and Howarth (1988). The classic study demonstrating that N is most limiting in marine waters by Ryther & Dunstan (1971), although often cited, has been criticized by many (e.g. Oviatt et al., 1995). More recent, ecosystem-level nutrient limitation experiments are in support of the evidence that N is the nutrient most limiting to primary production in marine systems (Oviatt et al., 1995). In some of the early cross-system comparisons for coastal marine environments (Boynton et al., 1982), N was suggested as more important than P, although statistically significant relationships were not found. However, the recent compilation by Nixon et al. (1996) also supports N to be the limiting nutrient in marine waters (Figure 5).

In estuarine systems, the question of nutrient limitation has aroused much debate. D'Elia et al. (1986) first demonstrated that estuaries can switch from P limitation in the spring to N limitation during summer (Figure 6). This result was extremely controversial at the time as it carried with it enormous economic implications for nutrient management strategies in Chesapeake Bay. The responsible authorities wanted to limit only P inputs to the system and not address N loading (D'Elia, 1987). In the last decade, numerous studies have been completed in estuarine systems that demonstrate that the switching nutrient limitation observed by D'Elia is not a phenomena restricted to Chesapeake Bay, but is a common feature in estuarine systems. I have compiled data for different estuarine systems that exhibit seasonal switching between spring P limitation and summer N limitation (Table 1). Of the systems presented here, the P limitation observed during spring is sometimes weaker than the response observed during summer when a much stronger response is observed to N limitation, for example see Figure 6.

It should be noted that not all estuarine systems display this seasonal switching in nutrient limitation. For example, the low salinity Bothnian Bay in the northern Baltic Sea is P-limited year around (Andersson et al., 1996), whereas the higher salinity open waters of the Baltic display N-limitation throughout (Granéli et al., 1990). It is only where freshwater enters these Baltic coastal systems, as in the Himmerfjärden Estuary (Elmgren & Larsson, 1977) and in the Gulf of Riga (Maestrini et al., 1997), where seasonal switches in nutrient limitation occurs. The various systems compiled in Table 1 cover a range of salinity, nutrient loading, loading ratios and nutrient concentrations. The common factor with these various estuarine systems is that all systems have a significant quantity of seasonal freshwater input entering the system. Although the strongest P limitation tends to occur in oligohaline portions of the estuaries closer to the freshwater end-member, P limitation is also observed at higher salinities (Malone et al., 1996).

Diatom production can be seasonally limited by dissolved silicate concentrations in a number of estuarine systems (Table 1). An example is the Chesapeake Bay where dissolved silicate concentrations and loading determine the magnitude of the diatom bloom during the spring, causes the collapse of the spring diatom bloom and leads to changes in floristic composition of phytoplankton communities (Conley & Malone, 1992) allowing for diatoms to be replaced by

Table 1. Estuaries exhibiting seasonal shifts in nutrient limitation with spring P limitation and summer N limitation

Estuary	Reference
Baltic Sea	
Himmerfjärden Estuary, Sweden	Granéli et al. (1990), Elmgren & Larsson (1997)
Gulf of Riga, Latvia	Maestrini et al. (1997)
Roskilde Fjord, Denmark	Pedersen & Borum (1996)
Bay of Brest, France[a]	Del Amo et al. (1997)
Chesapeake Bay, U.S.A.[a]	
Mainstem	Malone et al. (1996)
Patuxent River Estuary	D'Elia et al. (1986)
York River Estuary	Webb (1988)
Rhode River Estuary	Gallegos & Jordan (1997)
Delaware Estuary, U.S.A.	Pennock & Sharp (1994)
Neuse River Estuary, U.S.A.	Mallin & Paerl (1994)

[a]Systems displaying seasonal dissolved silicate limitation.

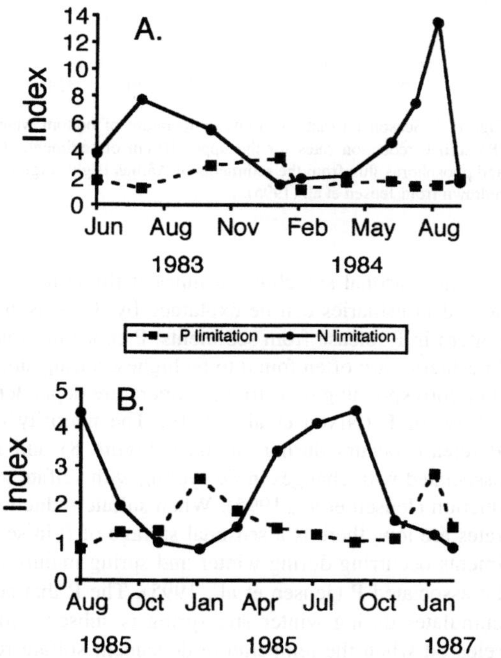

Figure 6. Nutrient limitation index (treatment chlorophyll:contol chlorophyll) for the (A) Patuxent River estuary and (B) York River estuary, tributaries to the Chesapeake Bay. Figure is redrawn from Webb (1997).

species not requiring dissolved silicate for growth. In the Bay of Brest, dissolved silicate limitation appears to be a relatively recent phenomena first observed in the 1990s, whereas prior to this time the Bay of Brest was primarily N limited (Del Amo et al., 1997).

Biogeochemical cycles

Explanations regarding why P is the primary limiting nutrient in freshwaters and N is the primary limiting nutrient in marine systems revolve around differences in the biogeochemical cycles of N and P. Three aspects of N and P biogeochemistry focusing on net gains or losses can be considered:

1. losses of N from denitrification;
2. the extent to which N deficiency can be made up for by N_2 fixation; and
3. the sediment regeneration of P.

As will be shown below, it is the sediment regeneration of P that probably accounts for the seasonal switching of limiting nutrients in estuarine systems.

In a comparison of freshwater, brackish and marine ecosystems Seitzinger (1988) noted small differences in rates of denitrification, however, there were no large differences in rates of denitrification between freshwater and marine environments. While denitrification is certainly an important loss process for N in marine systems (Nixon et al., 1996), N is not lost from marine systems at considerably higher rates than freshwater systems accounting for marine systems to be N limited. Seasonal changes in rates of denitrification are observed in estuarine systems with the highest rates often observed during winter and spring (Rysgaard et al., 1995) when water column NO_3^- concentrations are at their seasonal maxima. Thus, a major loss term in the biogeochemical cycle of N (Nixon et al., 1996) is minimized during the summer when N limitation in the water column is strongest, as NO_3^- is usually depleted from the water column and denitrific-

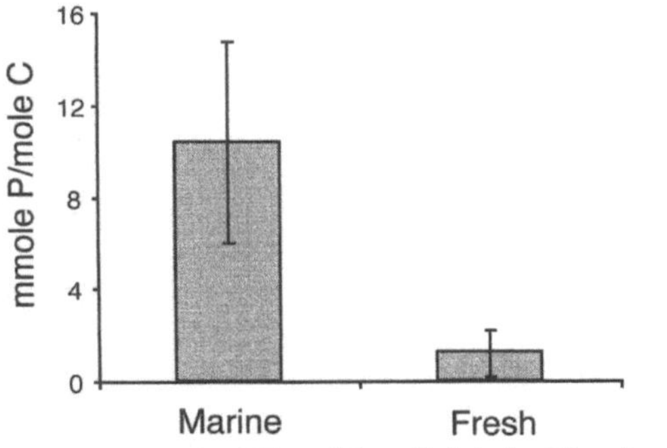

Figure 7. Average (with 95% confidence limits) of relative phosphorus release for brackish and coastal marine systems (Marine) and from freshwater systems (Fresh). The low relative release rate in freshwater systems suggests that in these systems sediments immobilize a large fraction of the P released from organics during decomposition. In most marine and brackish systems, essentially all of the remineralized P is returned to the water column. Figure is redrawn from Caraco et al. (1990).

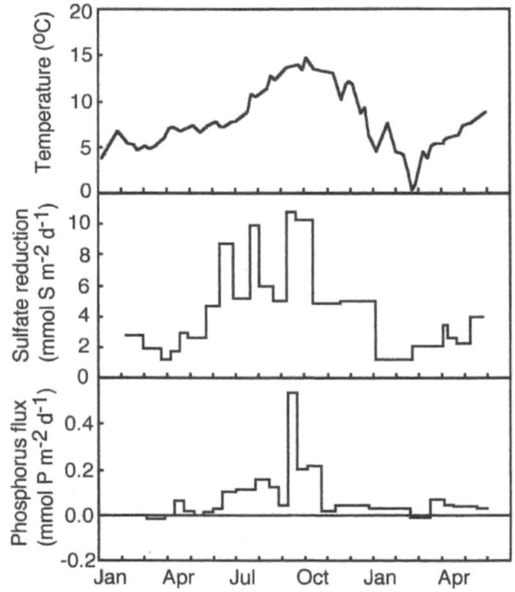

Figure 8. Seasonal variation in (A) temperature of bottom water, (B) sulfate reduction rates for the upper 10 cm of sediment, (C) and phosphorus flux from the sediments of Aarhus Bight. Figure is redrawn from Jensen et al. (1995).

ation is limited by coupled nitrification-denitrification processes.

The capacity to which deficits in N can be made up for by N_2 fixation is much greater in freshwater systems, whereas N_2 fixation is not generally important in marine systems (Howarth et al., 1988). If a lake has a low N load and concentration relative to P, deficits in N concentrations can often be made up for through N_2 fixation (Schindler, 1974) allowing for P to be controlling the accumulation of biomass even though P may initially be in excess relative to N. Although there are still many questions concerning the mechanism as to why N_2 fixation is not common in marine systems (Paerl et al., 1995; Howarth et al., 1995), this difference in biogeochemical cycling can have an important impact upon which nutrient is limiting in freshwater and marine environments.

One of the greatest differences in nutrient biogeochemical cycles between freshwater and marine systems occur in P biogeochemistry with the ability of freshwater systems to retain P in sediments through interactions with Fe. By contrast, in marine systems nearly all the P deposited in sediments is remineralized on an annual basis (Caraco et al., 1990) and returned to the overlying water (Figure 7). Caraco et al. (1989) also observed that P release from sediments is highly dependent upon sulfate concentrations, which can be used as a surrogate for salinity. Therefore, it is the preferential loss of P in freshwater systems as compared to marine environments allowing for P to be scarcest and therefore the limiting nutrient.

The seasonal switching in nutrient limitation observed in estuaries can be explained by the seasonal pattern in P release from sediments. P concentrations in estuaries are often found to be highest during summer corresponding to a strong temperature dependent release of P (Nixon et al., 1980). The majority of P release occurs during summer (Figure 8) and is associated with changes in Fe cycling with sulfate reduction (Jensen et al., 1995). When sulfate reduction rates are low, there is a seasonal storage of P in sediments occurring during winter and spring mainly as Fe-associated P (Jensen et al., 1995). The P that accumulates during winter and spring is subsequently released when the temperature dependent sulfate reduction rates increase releasing Fe bound sulfate and Fe-associated P. The release of dissolved silicate from the regeneration of diatom frustules is also a temperature dependent process with the highest concentrations of dissolved silicate often observed in estuaries during the summer (Conley & Malone, 1992). The high rates of regeneration and sediment release observed during summer for both dissolved inorganic phosphorus and dissolved silicate, brings high concentrations back into the overlying water, creating an internal source of nutrients, making it is less likely that P or Si will be limiting in estuaries during summer.

Consequences for management strategies

Given the paradigm that P is limiting in freshwaters and that N is limiting in marine waters, an important question arises as to what is the best course of action to take to reduce nutrient loads and improve the health of estuaries given that many estuaries show seasonal switches in the most limiting nutrient. Many management strategies focus primarily on P because that is the easiest nutrient to address first in high nutrient loaded systems by construction of advanced sewage treatment plants. Therefore, I will also explore the question of what are the consequences for estuarine systems by reductions in only P loading, or by reductions in only N loading.

Phosphorus

P appears to be an important limiting nutrient in many estuarine systems during spring (Table 1). It is rare that P limitation is observed year around. Year around P limitation is found to occur only in estuaries with high freshwater inputs of low dissolved inorganic phosphorus concentrations, for example in the River Göta älv estuary draining the largest freshwater lake in Sweden (Selmer & Rydberg, 1993). Reductions in P should have its largest effect on spring production. It is well established that spring is the most important time period of the year for seasonal deposition of organic matter to the sediments (Wassmann, 1991). Reductions in P loading and increasing the occurrence of P limitation, thus may lead to reductions in the seasonal deposition of organic matter. For deeper estuaries, such as Chesapeake Bay where spring P limitation is prevalent (Malone et al., 1996) and it is the organic matter produced during spring that fuels summer anoxia (Malone et al., 1986), reductions in P loading has the potential to improve summer oxygen conditions.

One of the most effective measures in reducing P loading is the control of point sources, especially the advanced treatment for the removal of P from sewage. In Danish estuaries (Odense Fjord & Roskilde Fjord) that have experienced large reductions in P loading and modest reductions in N loading with construction of advanced sewage treatment plants (Kaas et al., 1996), there has been a consequent reduction in the fast growing macrophyte *Ulva lactuca*. Although in Roskilde Fjord, *U. lactuca* was N-limited prior to the large reductions in sewage derived P (Pedersen & Borum, 1996), the spring is a critical time period for growth and the establishment of populations of *U. lactuca*. Thus, reductions in nutrient loading, especially P reductions due to better sewage treatment, may reduce the biomass of fast growing macrophytes (Duarte, 1995), such as *Ulva* and *Cladophora*, lessening the nuisance blooms of these macrophytes.

Nitrogen

As demonstrated above, marine systems are generally believed to be N limited, although as shown here, P limitation may occur during spring. However, nearly all estuaries studied to date exhibit N limitation during the summer. Thus, reductions in N loading should reduce the biomass of algae produced during the summer and possibly throughout the year. From a recreational viewpoint, summertime reductions in chlorophyll concentrations are desirable.

From an ecosystem viewpoint, reducing summer stocks of phytoplankton should have a large effect on the shading of macrophytes. For example, the depth limits of eelgrass has been shown to be related to N-loading in Danish coastal waters (Borum, 1996). A similar result was obtained in experimental mesocosms where eelgrass beds and mats of drift macroalgae declined in response to phytoplankton shading from increased water column concentrations of chlorophyll resulting from N enrichment (Taylor et al., 1995). In addition, summertime nutrient reductions should act to reduce the growth of epiphytes, which also act to shade plants (Duarte, 1995). One of the well-known ecosystem consequences of eutrophication is the replacement of submerged vegetation with planktonic communities (Sand-Jensen & Borum, 1991). The loss of submerged aquatic vegetation can cause enhanced resuspension and decreases in water column clarity with loss of macrophyte cover (Duarte, 1995). Submerged aquatic macrophytes provide important habitat for spawning and cover for fish.

Silicon

Overenrichment with N and P nutrients alters the biogeochemical cycle of Si allowing for dissolved silicate limitation to occur on a more frequent basis (Conley et al., 1993). These changes in dissolved silicate availability causes the replacement of diatoms by other algae that do not require Si for growth. The ecosystem consequences of reducing the abundance of one of the most important groups of algae on food web structure is not entirely known.

Examinations of nutrient concentrations in 'pristine' river systems yields average concentrations of 7.7 μm N and 0.32 μm P (Meybeck & Helmer, 1989) and 150 μm for the world average river dissolved silicate concentration (Tréguer et al., 1995). The resultant molar ratio is 24:470:1 for N:Si:P as compared to the Redfield ratio of 16:16:1 (Redfield et al., 1963). Unlike N or P, dissolved silicate inputs arise primarily from weathering reactions, and with our present knowledge, alteration of the weathering inputs have not been greatly impacted by mans activities. By contrast, the retention of Si has greatly increased with eutrophication (Conley et al., 1983) reducing dissolved silicate concentrations. Thus, changes in nutrient ratios may provide a sensitive indicator for eutrophication.

Conclusions

Evaluation of historical changes in nutrient loading has revealed that estuarine systems have experienced a 6–50 times increase in the N load and a 18–180 times increase in the P load from pristine conditions to present (Figure 2). Given that most management plans for reducing nutrient loads are on the order of a 50% to 80% reduction of present loads (Boynton et al., 1995; Kaas et al., 1996), while a large increase over what estuarine systems have received in the past under pristine conditions, will bring most systems near what was experienced at the turn of the century (Figure 2). In terms of what nutrient is limiting in estuaries, it was shown that many estuaries experience P limitation in spring switching to N limitation in summer (Table 1). Differences in the biogeochemical cycling of N, and especially that of P (Caraco et al., 1990) between freshwaters and marine environments contributes to the differences in limiting nutrients in the different environments. The seasonal storage of Fe-bound P during winter and spring (Jensen et al., 1995) and the subsequent temperature-dependent sediment release of P during summer can account for the seasonal switching in nutrient limitation in estuarine systems.

The construction and implementation of advanced wastewater treatment plants have resulted in reductions in nutrient loading, especially that of P, in a number of estuarine systems world-wide (Boynton et al., 1995; Kaas et al., 1996; Elmgren & Larsson, 1997). Since only modest changes have occurred in N loading, there have been large increases in the N:P ratio which have the potential to increase the importance of P relative to N in the regulation of marine ecosystem production. However, in most estuarine systems studied to date, N is nearly always limiting in summer, due to benthic-pelagic coupling and the temperature-dependent release of P from sediments (Nixon et al., 1980). It is only in a few situations that P limitation has been shown to be important year around in estuarine systems (Selmer & Rydberg, 1993). With these large reductions in P relative to N and the shift to higher N:P ratios, is it possible to make estuarine systems P limited year around? While to my knowledge there are no examples from the literature to expect this to occur, modeling studies suggest that year around P limitation is a possibility (Billen & Garnier, 1997). Evaluation of this question is relevant for example in Danish estuaries (Kaas et al., 1996) and constitute an important area for further scientific study.

Management strategies to implement reductions in P loading may help oxygen deficits in deep estuaries and reduce fast growing macrophytes, however, reductions in N are required to reduce summer chlorophyll concentrations and improve conditions for submerged aquatic vegetation. While reductions in P may limit spring production, it probably will have little effect on summer chlorophyll concentrations. Reductions in N loading should reduce summer chlorophyll concentrations and improve the conditions for submerged aquatic vegetation and thus improve ecosystem functioning. Finally, if only P reductions are pursued, that is if we are able to reduce P such that it is limiting year around in estuarine systems, it would reduce the retention of N in estuarine systems, and export N to the N-limited marine systems bordering the estuary systems, thus only exporting the problem. Such a situation occurred upon completion of the Stockholm treatment plant, where algal blooms formerly occurring near the city itself, now occur further out into the archipelago (Brattberg, 1986). These additional N loads are potentially more severe for the large enclosed coastal seas such as the Baltic. Therefore, arguments for managing only for one of the nutrients, N or P, to improve the health of coastal ecosystems is problematic.

References

Andersson, A., S. Hajdu, P. Haecky, J. Kuparinen & J. Wikner, 1996. Succession and growth limitation of phytoplankton in the Gulf of Bothnian (Baltic Sea). Mar. Biol. 126: 791–801.

Billen, G., C. Lancelot & M. Meybeck, 1991. N, P, and Si retention along the aquatic continuum from land to ocean. In R. F. C. Mantoura, J.-M. Martin & R. Wollast (eds), Ocean Margin Processes in Global Change. John Wiley & Sons: 19–44.

Billen, G. & J. Garnier, 1997. The Phison River plume: coastal eutrophication in response to changes in land use and water management in the watershed. Aquat. Microbiol. Ecol. 13: 3–17.

Borum, J., 1996. Shallow waters and land/sea boundaries. In Eutrophication in Coastal Marine Ecosystems, Coastal and Estuarine Studies, Vol. 52, Amer. Geophys. Union, Washington, D. C.: 179–203.

Boynton, W. R., W. M. Kemp & C. W. Keefe, 1982. A comparative analysis of nutrients and other factors influencing estuarine phytoplankton production. In Estuarine Comparisons, Academic Press: 69–90.

Boynton, W. R., J. H. Garber, R. Summers & W. M. Kemp, 1995. Inputs, transformations and transport of nitrogen and phosphorus in Chesapeake Bay and selected tributaries. Estuaries 18: 285–314.

Brattberg, G., 1986. Decreased phosphorus loading changes phytoplankton composition and biomass in the Stockholm archipelago. Vatten 42: 141–153.

Caraco, N., J. Cole & G. E. Likens, 1989. Evidence for a sulfate-controlled phosphorus release from sediments of aquatic systems. Nature 341: 316–318.

Caraco, N., J. Cole & G. E. Likens, 1990. A comparison of phosphorus immobilization in sediments of freshwater and coastal marine systems. Biogeochemistry 9: 277–290.

Chapra, S. C., 1977. Total phosphorus model for the Great Lakes. J. Envir. Eng. Div., ASCE, 103: 147–161.

Cloern, J. E., 1999. The relative importance of light and nutrient limitation of phytoplankton growth: A simple index of coastal ecosystem sensitivity to nutrient enrichment. Aquat. Ecol. 33: 3–15.

Conley, D. J. & T. C. Malone, 1992. Annual cycle of dissolved silicate in Chesapeake Bay: Implications for the production and fate of phytoplankton biomass. Mar. Ecol. Prog. Ser. 81: 121–128.

Conley, D. J., C. L. Schelske & E. F. Stoermer, 1993. Modification of silica biogeochemistry with eutrophication in aquatic systems. Mar. Ecol. Prog. Ser. 101: 179–192.

D'Elia, C. F., J. G. Sanders & W. R. Boynton, 1986. Nutrient enrichment studies in a coastal plain estuary: Phytoplankton growth in large-scale continuous cultures. Can. J. Fish. aquat. Sci. 43: 397–406.

D'Elia, C. F., 1987. Too much of a good thing. Nutrient enrichment of Chesapeake Bay. Environment 29: 6–11, 30–33.

Del Amo, Y., O. Le Pape, P. Tréguer, B. Quéguiner, A. Ménesguen & A. Aminot, 1997. Impacts of high-nitrate freshwater inputs on macrotidal ecosystems. I. Seasonal evolution of nutrient limitation for the diatom-dominated phytoplankton of the Bay of Brest (France). Mar. Ecol. Prog. Ser. 161: 213–224.

Duarte, C. M., 1995. Submerged aquatic vegetation in relation to different nutrient regimes. Ophelia 41: 87–112

Elser, J. J., E. R. Marzolf & C. R. Goldman, 1990. Phosphorus and nitrogen limitation of phytoplankton growth and the freshwaters of North America: A review and critique of experimental enrichments. Can. J. Fish. aquat. Sci. 47: 1468–1477.

Elmgren, R. & U. Larsson, 1997. Himmerfjärden. Changes in a nutrient loaded coastal ecosystem in the Baltic. Swedish Environmental Protection Agency, Stockholm: 197 pp. (in Swedish).

Gallegos, C. L. & T. E. Jordan, 1997. Seasonal progression of factors limiting phytoplankton pigment biomass in the Rhode River estuary, Maryland (U.S.A.). II. Modeling N *versus* P limitation. Mar. Ecol. Prog. Ser. 161: 199–212.

Granéli, E., K. Wallström, U. Larsson, W. Granéli & R. Elmgren, 1990. Nutrient limitation of primary production in the Baltic Sea Area. Ambio 19: 142–151.

Hecky, R. E. & P. Kilham, 1988. Nutrient limitation of phytoplankton in freshwater and marine environments. A review of recent evidence on the effects of enrichment. Limnol. Oceanogr. 33: 796–822.

Hedin, L. O., J. J. Armesto & A. H. Johnson, 1995. Patterns of nutrient loss from unpolluted, old-growth temperate forests: Evaluation of biogeochemical theory. Ecology 76: 493–509.

Howarth, R. W., 1988. Nutrient limitation of net primary production in marine ecosystems. Ann. Rev. Ecol. Sys. 19: 89–110.

Howarth, R. W., R. Marino, J. Lane & J. J. Cole, 1988. Nitrogen fixation in freshwater, estuarine and marine ecosystems. 1. Rates and importance. Limnol. Oceanogr. 33: 669–687.

Howarth, R. W., D. Swaney, R. Marino, T. Butler & C. R. Chu, 1995. Turbulence does not prevent nitrogen fixation by plankton in estuaries and coastal seas (reply to the comment of Paerl et al.). Limnol. Oceanogr. 40: 639–643.

Jensen, H. S., P. B. Mortensen, F. Ø. Andersen, E. Rasmussen & A. Jensen, 1995. Phosphorus cycling in a coastal marine sediment, Aarhus Bay, Denmark. Limnol. Oceanogr. 40: 908–917.

Kaas, H., F. Møhlenberg, A. Josefson, B. Rasmussen, D. Krause-Jensen, H. S. Jensen, L. Svendsen, J. Windolf, A. L. Middelboe, K. Sand-Jensen & M. F. Pedersen, 1996. Marine områder. Danske fjorde - status over miljøstand, årsagssammenhænge og udvikling. Danish Ministry of the Environment Rapport Nr. 179, Roskilde, Denmark: 205 pp. (in Danish).

Larsson, U., R. Elmgren & F. Wulff, 1985. Eutrophication and the Baltic Sea: Causes and consequences. Ambio 14: 9–14.

Maestrini, S. Y., M. Balode, C. Béchemin, I. Purina & C. Vérite, 1997. Nutrients limiting the algal growth potential (AGP) in the Gulf of Riga, eastern Baltic Sea, in spring and early summer 1996. La mer 35: 49–68.

Mallin, M. A. & H. W. Paerl, 1994. Commentary on primary productivity and nutrient limitation in the Neuse River Estuary, North Carolina, U.S.A. Mar. Ecol. Prog. Ser. 111: 311–312.

Malone, T. C., W. M. Kemp, H. W. Ducklow, W. R. Boynton, J. H. Tuttle & R. B. Jones. 1986. Lateral variation in the production and fate of phytoplankton in a partially stratified estuary. Mar. Ecol. Prog. Ser. 32: 149–160.

Malone, T. C., D. J. Conley, T. R. Fisher, P. M. Glibert, L. W. Harding, Jr. & K. G. Sellner, 1996. Scales of nutrient limited phytoplankton productivity in Chesapeake Bay. Estuaries 19: 371–385.

Meybeck, M. & R. Helmer, 1989. The quality of rivers: from pristine stage to global pollution. Palaeogeogr. Palaeoclimatol. Palaeoecol. 75: 283–309.

Nixon S. W., J. R. Kelly, B. N. Furnas, C. A. Oviatt & S. S. Hale, 1980. Phosphorus regeneration and the metabolism of coastal marine bottom communities. In K. R. Tenore & B. C. Coull (eds), Marine Benthic Dynamics. Univ. S. Carolina Press, Columbia, SC: 219–242.

Nixon, S.W. et al. (15 co-authors), 1996. The fate of nitrogen and phosphorus at the land-sea margin of the North Atlantic Ocean. Biogeochemistry 35: 141–180.

Nixon, S. W., 1995. Coastal marine eutrophication: A definition, social causes and future concerns. Ophelia 41: 199–219.

Nixon, S. W., 1997. Prehistoric nutrient inputs and productivity in Narragansett Bay. Estuaries 20: 253–261.

Oviatt, C., P. Doering, B. Nowicki, L. Reed, J. Cole & J. Frithsen, 1995. An ecosystem level experiment on nutrient limitation in temperate coastal marine environments. Mar. Ecol. Prog. Ser. 116: 171–179.

Paerl, H. W., J. L. Pickney & S. A. Kucera, 1995. Clarification of the structural and functional roles of heterocysts and anoxic microzones in the control of pelagic nitrogen fixation. Limnol. Oceanogr. 40: 634–638.

Pedersen, M. F. & J. Borum, 1996. Nutrient control of algal growth in estuarine waters. Nutrient limitation and the importance of nitrogen requirements and nitrogen storage among phytoplankton and species of macroalgae. Mar. Ecol. Prog. Ser. 142: 261–272.

Pennock, J. R. & J. H. Sharp, 1994. Temporal alternation between light- and nutrient-limitation of phytoplankton in a coastal plain estuary. Mar. Ecol. Prog. Ser. 111: 275–288.

Prairie, Y. T., C. M. Duarte & J. Kalff, 1989. Unifying nutrient-chlorophyll relationships in lakes. Can. J. Fish. aquat. Sci. 46: 1176–1182.

Redfield, A. C., B. H. Ketchum & F. A. Richards, 1963. The influence of organisms on the composition of sea-water. In M. N. Hill (ed.), The Sea. John Wiley & Sons, New York: 12–37.

Rysgaard, S., P. B. Christensen & L. P. Nielsen, 1995. Seasonal variation in nitrification and denitrification in estuarine sediment colonized by benthic microalgae and bioturbating infauna. Mar. Ecol. Prog. Ser. 126: 111–121.

Ryther, J. H. & W. M. Dunstan, 1971. Nitrogen, phosphorus, and eutrophication in the coastal marine environment. Science 171: 1008–1013.

Sand-Jensen, K. & J. Borum, 1991. Interactions among phytoplankton, periphyton, and macrophytes in temperate freshwaters and estuaries. Aquat. Bot. 41: 137–175.

Selmer, J.-S. & L. Rydberg, 1993. Effects of nutrient discharge by river water and waste water on the nitrogen dynamics in the archipelago of Göteborg, Sweden. Mar. Ecol. Prog. Ser. 92: 119–133.

Schindler, D. W., 1974. Eutrophication and recovery in experimental lakes: implications for lake management. Science 184: 897–899.

Seitzinger, S. P., 1988. Denitrification in freshwater and coastal marine ecosystems: Ecological and geochemical significance. Limnol. Oceanogr. 33: 702–724.

Stålnacke, P., A. Grimvall, K. Sundblad & A. Tonderski, 1998. Estimation of riverine loads of nitrogen and phosphorus to the Baltic Sea, 1970–1993. J. Envir. Monit. Assem.

Taylor, D., S. Nixon, S. Granger & B. Buckley, 1995. Nutrient limitation and the eutrophication of coastal lagoons. Mar. Ecol. Prog. Ser. 127: 235–244.

Tréguer, P., D. M. Nelson, A. J. van Bennekom, D. J. DeMaster, A. Leynaert & B. Quéguiner, 1995. The silica balance in the world ocean: a reestimate. Science 268: 375–379.

Vollenweider, R. A., 1976. Advances in defining critical loading levels of phosphorus in lake eutrophication. Mem. Ist. ital. Idrobiol. 33: 53–83.

Wassmann, P., 1991. Dynamics of primary production and sedimentation in shallow fjords and polls of western Norway. Oceanogr. mar. Biol. Annu. Rev. 29: 87–154.

Webb, K. L., 1988. Comment on "Nutrient limitation of phytoplankton growth in brackish coastal ponds" by Caraco, Tamse, Boutros and Valiela (1987). Can. J. Fish. aquat. Sci. 45: 380–381.

Hydrobiologia **410:** 97–102, 1999.
J. Garnier & J.-M. Mouchel (eds), Man and River Systems.

Application of the 'snapshot' methodology to a basin-wide analysis of phosphorus and nitrogen at stable low flow

M. Salvia[1,2], J.F. Iffly[1], P. Vander Borght[2], M. Sary[3] & L. Hoffmann[4]
[1]*Centre de Recherche Public-Gabriel Lippmann, CREBS, 162a, Avenue de la Faïencerie, L-1511 Luxembourg, Luxembourg*
[2]*Fondation Universitaire Luxembourgeoise, 185, avenue de Longwy, B-6700 Arlon, Belgium*
[3]*Université de Metz, Centre d'Etudes Géographiques, Ile du Saulcy, F-57405 Metz, France*
[4]*Université de Liège, Institut de Botanique (B.22), Sart Tilman, B-4000 Liège, Belgium*

Key words: 'snapshot' methodology, spatial analysis, nutrient fluxes, eutrophication, Esch-sur-Sûre basin

Abstract

The 'snapshot' sampling methodology was applied to the spatial analysis of phosphorus and nitrogen concentrations and fluxes in the transnational upper Sûre watershed (Belgium–Luxembourg) during stable low discharge summer periods in 1991 and 1992. The results show clear differences in nitrogen concentrations and specific loads between agricultural and forested sub-basins, whereas for phosphorus the dependence on land use is less evident. The simultaneous measurement of stream discharge and nutrient concentrations permitted a quantitative evaluation of the input from point respectively diffuse sources and loads from the tributaries as well as the retention capacity by self-purification or sedimentation of the different phosphorus and nitrogen forms in the various sections of the Sûre river. The results illustrate the utility of the 'snapshot' sampling as an important tool for an integrated watershed management.

Introduction

Water quality is usually monitored on a regular basis at only a small number of sites in a watershed, often only at the watershed outlet. Thus the effect of all the point and non-point sources occurring throughout the watershed are integrated. However, effective watershed management requires data which identify major sources and processes (House & Denison, 1997; Spooner & Line, 1993; Wern & Wodsak, 1995); a spatial analysis is thus also of great importance.

The nutrient transfer to water bodies can be approached using diverse spatial scales (Merot & Bruneau, 1992; Wolock et al., 1997). Plot-scale experimental studies provide accurate data on nutrient export processes but are strongly related to a specific kind of soil and crop (Edwards et al., 1996; Schepers et al., 1995; Scholefield & Stone, 1995; Young et al., 1996). Studies using the scale between field and small experimental catchments account for export rates of a more heterogenous system and for interactions between individual components. Nevertheless it is not feasible to transpose the results of both previous scales to a larger watersheds, and modelling (usually with annual resolution) must be applied (Beusen et al., 1995; Johnes & Hodgkinson, 1998; Line et al., 1998).

'Snapshot' sampling was the technique used in our study, which constitutes a part of a wider research on the transfers of phosphorus and nitrogen in the transnational Belgo–Luxembourgish upper Sûre basin in the framework of the eutrophication of the Sûre drinking water reservoir. This methodology, which has only rarely been applied (Grayson et al., 1997), consists in giving an instantaneous picture of the concentrations and fluxes of a whole watershed by sampling a river system at every confluence and discharge point during a time period that is as short as possible and during which flow is stable. This sampling technique is very useful for a better understanding of a basin; it especially allows (1) a geographical overview of stream nutrient levels in relation with watershed geology and land-use, (2) the identification of both point and non-point source contributors to nutrient load, (3) the establishment of the background loads of nutrients, thereby giving a basis for evaluating the true influence of non-point sources from agricultural areas, (4) the identification of the river reaches which have a retention capacity (by biological self-purification, sedimentation) or which are sources of nutrients.

Watershed description

The upper Sûre basin is situated on Devonian schistous substrates in the Ardennes region (49°52′ N, 5°40′ E); it has an area of approximately 428 km^2, two-thirds of which are located on Belgian territory and one-third on Luxembourgish territory. It is a rural basin with a population density of approximately 25 inhabitants/km^2, 55% of the surface being occupied by agriculture and 45% by forests. The catchment is dominated by loamy soils with a thin superficial layer. Annual rainfall varies across the basin from approximately 900 mm in the east to 1100 mm in the west, the wettest months occurring in the winter–spring period. The mean annual discharge of the Sûre at Martelange (the Belgo–Luxembourgish boundary, area = 208 km^2) is 3.9 m^3 s^{-1}, with important differences between summer (10th percentile = 0.1 m^3 s^{-1}) and winter (90th percentile = 58.9 m^3 s^{-1}). The Sûre river supplies a storage reservoir which is used mainly as a drinking-water supply. Only few watershed villages have a sewage treatment plant and most of them exclusively depend on primary treatment. Important point sources are a small slaughter-house near Vaux-sur-Sûre and the intensive farming areas of Mecher and Harlange.

Materials and methods

Sampling was undertaken within 12 h at 50 sites (Figures 1 & 2) throughout the basin in July 1991 and June 1992. In order to ensure stable flow without any influence from rain events, the sampling took place after a period of 2 weeks without rainfall. The rationale for choosing the sites was to obtain a longitudinal profile of the upper Sûre river and its major tributaries by including all tributary junctions and point sources.

At each site, discharge was measured with an Ott current meter/integration method, according to the procedure described by Decloux and Sary (1988). A 500 ml water sample was taken from mid-stream for laboratory analyzes. An unfiltered sample was used for the determination of total phosphorus (tot-P) and total suspended matter, and a filtered sample (GF/C filter) for the determination of NH_4, NO_3 and soluble reactive phosphorus (SRP). After filtration and addition of a drop of $HgCl_2$ sat/100 ml, the samples were stored at 4°C until analyzes within a few days. Tot-P was converted to SRP by sulfuric acid–sulfate digestion at 400°C. Ammonium, nitrates and SRP were analyzed by a Technicon auto-analyzer. Instantaneous fluxes (F_i) at every station i were calculated by multiplying instantaneous concentrations (c_i) with the corresponding discharge (Q_i). Specific fluxes were obtained dividing F_i by the upstream area of the watershed (S_i) at the same station. Discharge at Martelange (station 21) was 400 and 700 l s^{-1} in July 1991 and June 1992, respectively.

Results and discussion

Nutrient concentrations and specific loads

Figures 1 and 2 show the spatial distribution of the concentrations of total phosphorus, soluble reactive phosphorus, ammonium and nitrate in the upper Sûre watershed. Total P concentrations around 0.800 mg l^{-1} P were present downstream of the slaughter house at Vaux-sur-Sûre (sampling station 5); two other sites with important point sources (village + agriculture) were found near Harlange (25L12) and Mecher (29L1) with tot-P up to 0.250 mg l^{-1} P. In agro-forestry sub-basins devoid of point sources, the concentrations of total phosphorus were below 0.100 mg l^{-1} P. A similar picture was obtained for soluble reactive phosphorus which represented 40–50% of total phosphorus. The highest SRP concentrations were measured at the same stations as for tot-P: Vaux-sur-Sûre (0.400–0.450 mg l^{-1} P), Harlange and Mecher (0.150–0.200 mg l^{-1} P), whereas in the other stations, SRP concentrations were generally below 0.100 mg l^{-1} P, especially in forested basins (below 0.050 mg l^{-1} P). Tot-P and SRP were strongly correlated for both campaigns (R^2 = 0.8).

Nitrogen concentrations are generally correlated to land use. In sub-basins mainly occupied by forests and without point sources (18R, 19R, 29L2) mineral nitrogen concentrations were below 1 mg l^{-1} NO_3-N, respectively below 0.050 mg l^{-1} NH_4-N , whereas in agricultural sub-basins nitrates were in the range of 2-4 mg l^{-1} NO_3-N and ammonium mostly 0.050–0.150 mg l^{-1} NH_4-N. The highest ammonium concentrations were found at the important point sources of stations 5 and 29L1 (2–3 mg l^{-1} NH_4-N).

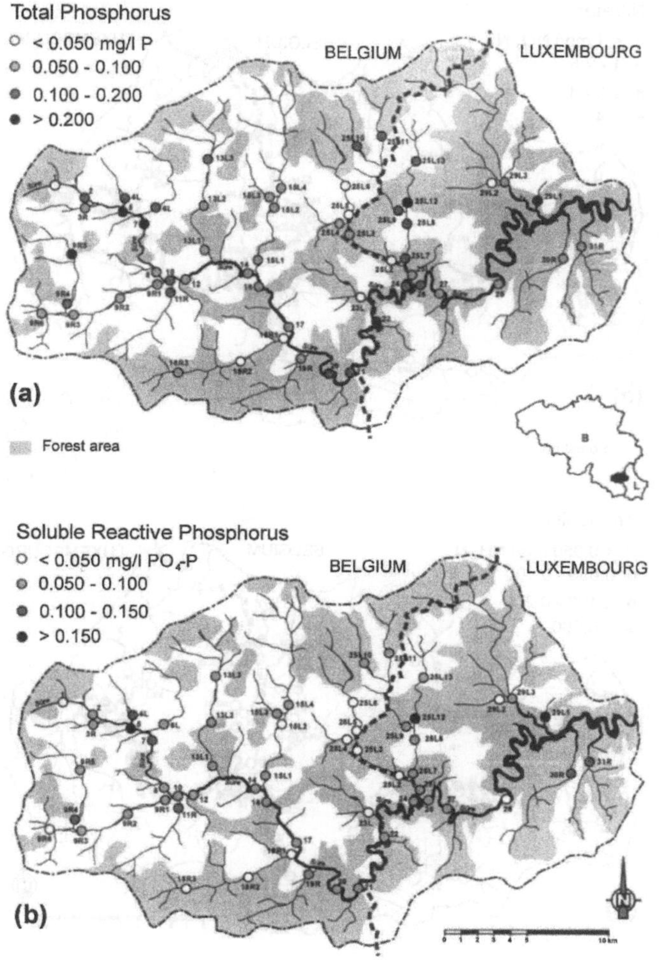

Figure 1. Mean concentrations (mg l^{-1} P) of total phosphorus (a) and soluble reactive phosphorus (b).

The results from the basin-wide sampling permit an estimate of natural water quality during low flow. The water quality at relatively unimpacted reference locations is indeed representative of regional natural water quality and may be reasonably extrapolated to other similar locations within the ecoregion (Schonter & Novotny, 1993). Hence, we can consider the forested streams 18R and 19R as characteristic of natural nutrient levels of the upper Sûre catchment.

The same distribution pattern as for nutrient concentrations is also observed for specific loads (not represented). This is a consequence of the uniformity of the specific discharges throughout the catchment, in agreement with its geological homogeneity. Specific loads of the Sûre and its tributaries (excluding important point sources) varied between 0.010 and 0.040 kg day^{-1} km^{-2} tot-P, 0.004 and 0.020 kg day^{-1} km^{-2} SRP, 0.1 and 1.2 kg day^{-1} km^{-2} NO_3-N, 0.005 and 0.050 kg day^{-1} km^{-2} NH_4-N. These daily fluxes cannot be used to calculate annual N or P export coefficients because of the large annual variability of hydrological conditions in the catchment, leading to a large increase, especially of tot-P and nitrate concentrations, during high flow periods (e.g. Emde et al., 1993; Roberts, 1977). Indeed, monthly fluxes measured at station 27 during the summer period do not

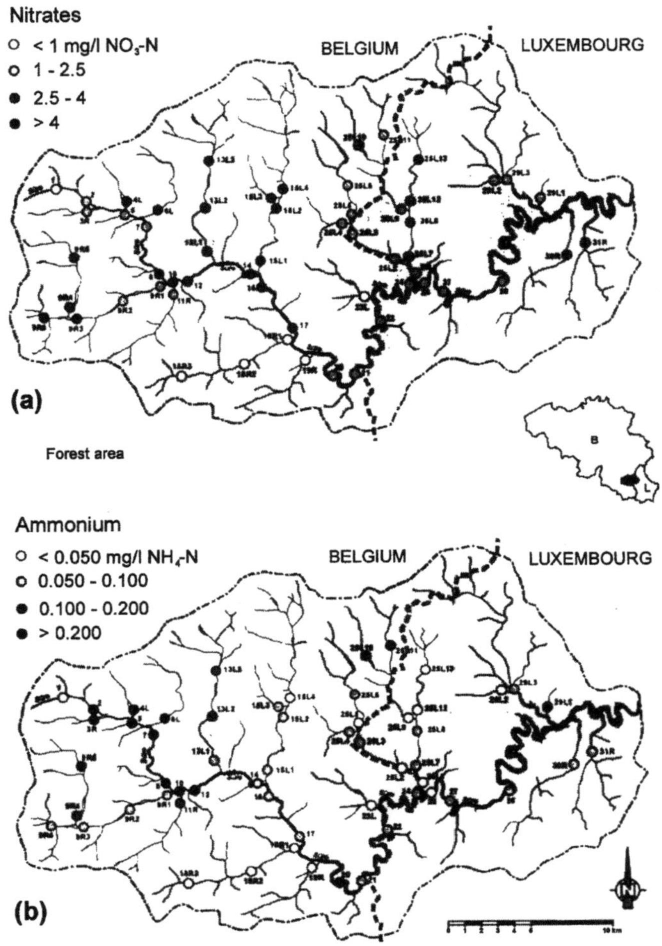

Figure 2. Mean concentrations (mg l^{-1} N) of nitrates (a) and ammonium (b).

represent 1/12 of the annual load but only 1/200 in the case of nitrate, 1/100 for ammonium and total phosphorus, and 1/40 for SRP.

Nutrient fluxes in the Sûre river

Figures 3 and 4 show the measured instantaneous concentrations and absolute fluxes of the phosphorus and nitrogen forms along the Sûre river. The measured fluxes F_i (kg day^{-1}) at every station i are compared with the expected fluxes, obtained by the sum of the measured instantaneous flux (F_{i-1}) at the closest upstream station (i–1) and the incoming fluxes by affluents and direct tributary surfaces between stations i and i–1. This allows (1) the additional inputs between successive stations by point and non-point sources to be quantified, (2) the conservative character of the nutrients to be checked, and thus the nutrients which are retained in the corresponding river section to be estimated (Dorioz et al., 1989). Furthermore, a reference line indicates the increase of the fluxes along the Sûre river assuming a constant nutrient concentration for the whole watershed; the concentration used corresponds to the mean concentration of all the sampling stations (except the important point sources 5, 7 and 29L1). This line allows the difference of the nutrient concentrations at each station with an average situation for the whole basin to be evaluated.

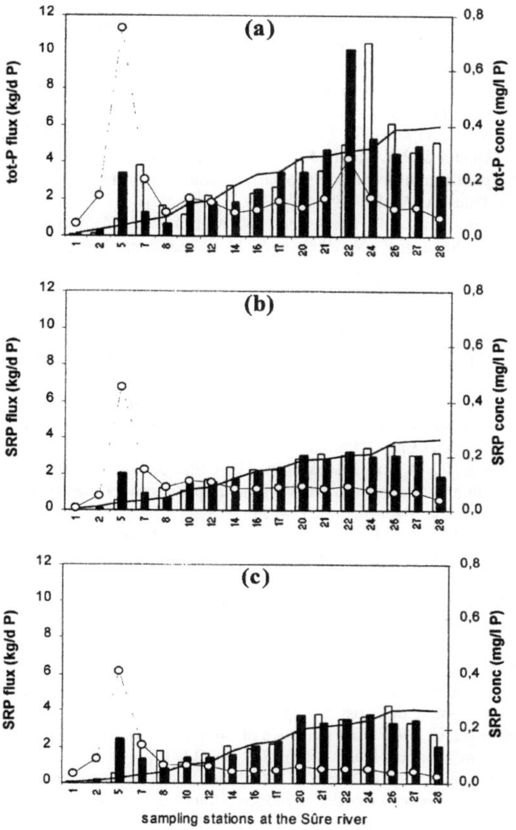

Figure 3. Concentrations (circles) (mg l^{-1} P), measured flux F_i (black bars), expected flux $[F(i-1)+\Sigma F$ tributaries] (white bars) and steady fluxes (at constant conc. = mean basin value) (line) in kg day^{-1} P of tot-P ((a) July 1991) and SRP ((b) July 1991; (c) June 1992) along the Sûre river.

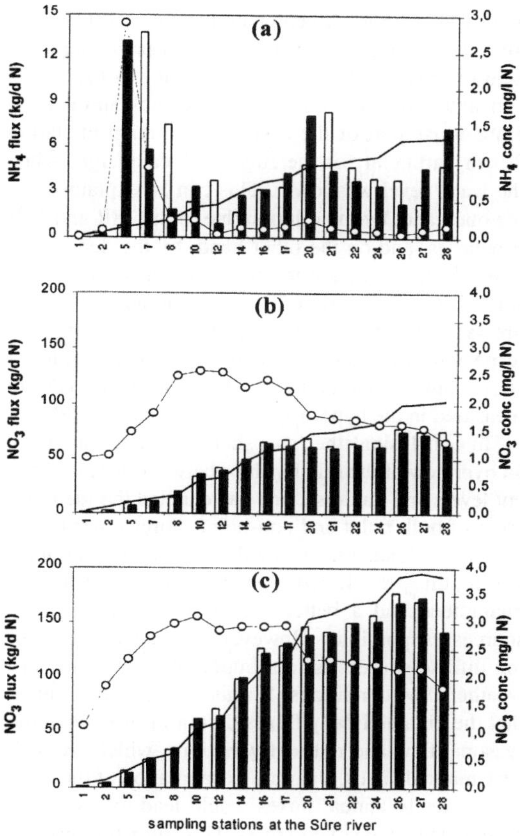

Figure 4. Concentrations (circles) (mg/l N), measured flux F_i (black bars), expected flux $[F(i-1)+\Sigma F$ tributaries] (white bars) and steady fluxes (at constant conc. = mean basin value) (line) in kg/day N of ammonium ((a) July 1991) and nitrate ((b) July 1991; (c) June 1992) along the Sûre river.

For total phosphorus (Figure 3a), the point source at station 5 (due to the village of Vaux-sur-Sûre and its slaughter-house) stands out by its high P concentration. On the basis of the difference between the measured and expected fluxes, its input can be evaluated as 2.5 kg day^{-1} tot-P in July 1991 (4 kg day^{-1} tot-P in June 1992, not represented). It is interesting to observe the substantial retention of this load by the river. Indeed, at the next station located 2 km downstream tot-P had already decreased by about 60%. Similar retention rates were obtained by Pilleboue (1987) and Taleb (1997) during low flow periods, whereas Svendsen et al. (1995) report retention rates around 30%. The important fluxes measured at station 22 are related to accidentally high suspended solids which had sedimented at the next station. They did not appear in the 1992 campaign. Figure 3b,c represents the longitudinal profiles for soluble reactive phosphorus during the 1991 and 1992 campaigns. The point source at station 5 and the rapid retention of SRP are again clearly visible. The difference observed at station 28 is due to the presence of a small predam in which SRP retention takes place in summer (unpublished results). For the other stretches of the river, no significant difference is generally observed between the expected and measured fluxes, indicating that retention is not important. It is also interesting to note that SRP fluxes are similar for both campaigns despite the different discharges (400 and 700 l s^{-1} at station 21). The situation is different for nitrate fluxes which are twice as high in

June 1992 (Figure 4c) compared to July 1991 (Figure 4b); this strong relation with discharge is characteristic for non-point sources. A reduction in the nitrate concentrations (and fluxes) is appreciable in the Luxembourgish part of the basin (downstream of station 20) in particular for the July 1991 campaign, when the water level was lower. This can be explained, on the one hand, by the dilution due to the 18R and 19R tributaries, and on the other hand, by the denitrification effect, significant in the lentic water area of the small reservoir (station 28). Ammonium loads (Figure 4a) had a more irregular pattern; several stretches (especially downstream stations 5, 20) presented a significant retention but important contributions were measured in the hypertrophic predam (station 28).

In conclusion, the 'snapshot' sampling has given an overview of the relationships between stream nutrient levels and land-use, and has enabled the identification of both point and non-point source contributors to nutrient loads under low flow summer conditions. During this period, hydrological stability is generally guaranteed and a better quantification of the point sources is possible. However, the contribution of the diffuse sources is less important during this dry weather flow conditions. It must especially be noted that the obtained daily fluxes do not inform about the seasonal flow and nutrient variations, which are both very marked. Thus, a simple extrapolation of the summer fluxes to annual fluxes would lead to a strongly biased view of the export rates. This study furthermore helped to establish background loads of nutrient levels. It thus proved to be a useful tool for an integrated watershed management in order to reduce the nutrient loads entering the drinking water reservoir.

Acknowledgements

MS acknowledges the receipt of a scholarship from the Luxembourgish government. LH is research associate of the Belgian National Fund for Scientific Research.

References

Beusen, A.H.W., O. Kleper & C.R. Meinardi, 1995. Modelling the flow of nitrogen and phosphorus in Europe: from loads to coastal seas. Wat. Sci. Tech. 31: 141–145.

Decloux, J. P. & M. Sary, 1988. Campagnes d'étiage: objectifs, traitement et valorisation des données. Mosella XVIII (No. spécial annuel – parution 1991): 121–134.

Dorioz, J.M., E. Pilleboue & A. Ferhi, 1989. Dynamique du phosphore dans les bassins versants: importance des phénomènes de rétention dans les sédiments. Wat. Res. 23: 147–158.

Edwards, D.R., C.T. Haan, A.N. Sharpley, J.F. Murdoch, T.C. Daniel & P.A. Moore Jr., 1996. Application of simplified phosphorus transport models to pasture fields in Northwest Arkansas. Trans. ASAE 39: 489–496.

Emde, W.v.d., H. Fleckseder, N. Matsché, F. Plahl-Wabnegg, G. Spatzierer, W. Stalzer & U. Wenninger, 1993. Sources of nitrogen and phosphorus in the catchment area of Neusiedlersee/Fertö, Austria/Hungary. Wat. Sci. Tech. 28: 101–110.

Grayson, R.B., C.J. Gippel, B.L. Finlayson & B.T. Hart, 1997. Catchment-wide impacts on water quality: the use of 'snapshot' sampling during stable flow. J. Hydrol. 199: 121–134.

House, W. A. & F. H. Denison, 1997. Nutrient dynamics in a lowland stream impacted by sewage effluent: Great Ouse, England. Sci. total Envir. 205: 25–49.

Johnes, P.J. & R.A. Hodgkinson, 1998. Phosphorus loss from agricultural catchments: pathways and implications for management. Soil Use Manage. 14: 175–185.

Line, D.E., R.A. McLaughlin, D.L. Osmond, G.D. Jennings, W.A. Harman, L.A. Lombardo & J. Spooner, 1998. Nonpoint sources. Wat. envir. Res. 70: 895–912.

Merot, Ph. & P. Bruneau, 1992. Échelle spatiale représentative d'un bassin versant sur le plan géochimique. C.r. Acad. Sci., Paris, sér. II, 315: 1097–1103.

Pilleboue, E., 1987. Origine, spéciation et modalités de transfert du phosphore et de l'azote dans un bassin versant. Ph.D. Thesis, Université de Paris: 251 pp.

Roberts, G., 1997. The influence of sampling frequency on streamflow chemical loads. J. CIWEM 11: 114–118.

Schepers, J.S., G.E. Varvel & D.G. Watts, 1995. Nitrogen and water management strategies to reduce nitrate leaching under irrigated maize. J. Contam. Hydrol. 20: 227–239.

Scholefield, D. & A.C. Stone, 1995. Nutrient losses in runoff water following application of different fertilisers to grassland cut for silage. Agr. Ecosyst. Envir. 55: 181–191.

Schonter, R. & V. Novotny, 1993. Predicting attainable water quality using the ecoregional approach. Wat. Sci. Tech. 28: 149–158.

Spooner, J. & D.E. Line, 1993. Effective monitoring strategies for demonstrating water quality changes from non-point source controls on a watershed scale. Wat. Sci. Tech. 28: 143–148.

Svendsen, L.M., B. Kronvang, P. Kristensen & P. Græsbøl, 1995. Dynamics of phosphorus compounds in a lowland river system: importance of retention and non-point sources. Hydrol. Process. 9: 119–142.

Taleb, A., 1997. Pollution diffuse ou localisée des eaux de la Senne en amont de Tubize (Belgique) par les nutriments. Ph.D. Thesis, Université Libre de Bruxelles: 239 pp.

Wern, W. & H. P. Wodsak, 1995. The role of non-point nutrient sources in water pollution: present situation, countermeasures, outlook. Wat. Sci. Tech. 31: 87–97.

Wolock, D. M., J. Fan & G. B. Lawrence, 1997. Effects of basin size on low-flow stream chemistry and subsurface contact time in the Neversink River watershed, New York. Hydrol. Process. 11: 1273–1286.

Young, W.J., F.M. Marston & J.R. Davis, 1996. Nutrient exports and land use in Australian catchments. J. Env. Manage. 46: 142–161.

Hydrobiologia **410:** 103–110, 1999.
J. Garnier & J.-M. Mouchel (eds), Man and River Systems.

Nutrient runoff and transfer from land and rivers to the Gulf of Riga

Per Stålnacke[1], Nils Vagstad[1], Timo Tamminen[2], Paul Wassmann[3], Viesturs Jansons[4] & Enn Loigu[5]
[1]*Jordforsk - Centre for Soil and Environmental Research, N-1432 Ås, Norway*
[2]*Finnish Environment Institute, P.O. Box 140, FIN-00251 Helsinki, Finland*
[3]*Norwegian College of Fishery Science, University of Tromsø, N-9037 Tromsø, Norway*
[4]*Latvian University of Agriculture, Dept. of Environmental Engineering and Management, LV-3001 Jelgava, Latvia*
[5]*Tallinn Technical University, Dept. of Environmental Engineering, EE-0026 Tallinn, Estonia*

Key words: nutrients, Gulf of Riga, agricultural runoff, riverine loads, marine eutrophication

Abstract

In this paper, we have synthesized and integrated results regarding nutrient loads and eutrophication of the Gulf of Riga (GoR) that were obtained in three projects that were part of a six-year research programme (1993–1998). In particular, we focused our attention on the factors that control the temporal variability in the load of nutrients in the drainage basin and rivers, as well as the effects of such nutrients on the environment of the recipient, the Gulf of Riga.

The results indicate that the rivers play a crucial role in the total input of nutrients to the GoR, and exceed the combined contribution from atmospheric deposition, point emissions from cities and industries along the coast, and nitrogen fixation by marine organisms. It was found that natural variability in water discharge was the main factor controlling the temporal variability in the riverine load of nutrients (i.e. differences in load between seasons and years). Moderate nutrient losses to surface waters, especially from agriculture, and weak riverine response to the sudden decrease in agricultural production after the disintegration of the Soviet Union in the early 1990s were also found. It is suggested that this is most likely related to the inertia in, and buffering capacity of, agricultural soils, i.e. depending of factors such as hydrological conditions, the size of ground water aquifers and water-saturated soils, creating favourable conditions for nutrient retention processes in the agricultural landscape. Regardless of the relatively low area-specific riverine inputs, the pollution loads in the rivers have contributed significantly to eutrophication of the Gulf of Riga. This seems to be true despite the relatively rapid distribution of the loads beyond the littoral zone of the gulf, resulting in a moderate all-over eutrophication of the entire recipient, characteried by a continuously shifting gradient between the southern and northern parts of the gulf. Furthermore, the marine eutrophication seems to be less prominent and less confined to the southernmost basin than previously assumed.

Introduction

Nutrient enrichment is generally regarded as the greatest threat to ecosystems of the North Sea and the Baltic Sea. The Gulf of Riga (Figure 1), which has a surface area of 19 000 km^2 and a volume of 420 km^3, has been described as one of the most polluted basins of the Baltic (Yurkovskis et al., 1993; Andrushaitis et al., 1995), and eutrophication is indicated by periodically increased abundance of algae, meiofauna and macrofauna, as well as changes in species composition. In addition, contamination with faecal micro-organisms, due to in-efficient or lack of sewage treatment, has been prevalent. Most of the pollution loads in the Gulf of Riga can be attributed to activities in the drainage basin, which covers 135 700 km^2, or more than seven times the surface area of the gulf itself.

A prerequisite for the planning of actions for water protection measures is an understanding of the forces that drive different nutrient transformation processes in seas, rivers, ground water and soils. The inter-

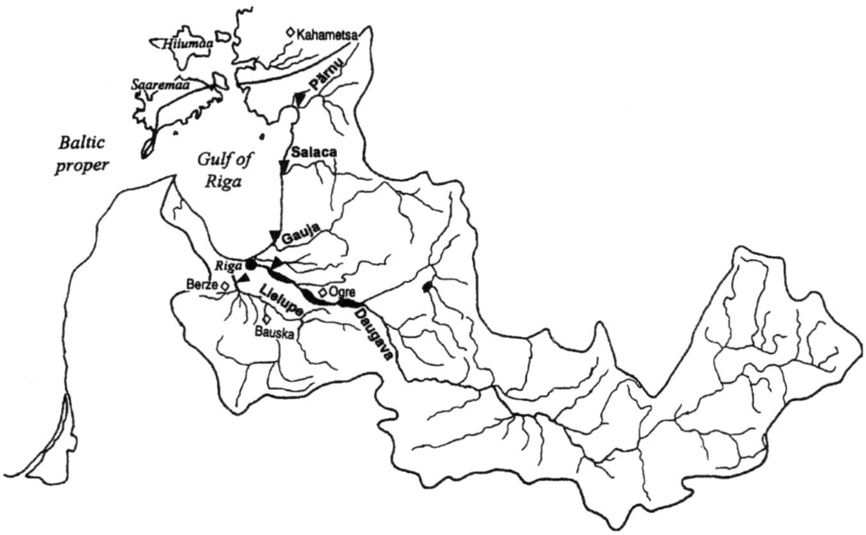

Figure 1. Map of the drainage basin of the Gulf of Riga. Sites for river water quality monitoring (▼) and monitoring of nutrient losses in agricultural catchments (◊) are marked.

linkages between the various terrestrial and aquatic media have received more attention in recent research projects. An integrated approach in this area was possible for participants in the *Gulf of Riga project*, which is a Nordic environmental research programme, that was initiated by the Nordic Council of Ministers and is part of the environmental co-operation in the Baltic Sea region. In the present paper, we have synthesized some of the findings of three of the subprojects, which are titled as follows:

(A) Soil and Nutrient Losses from Small Catchments,
(B) Riverine Transport and Source Apportionment of the Nutrient Load to the Gulf of Riga, and
(C) Pelagic Eutrophication and Sedimentation.

In particular, we addressed the following general questions:

1. To what extent can data from small catchments help us to explain and understand temporal variability in riverine loads of nutrients?
2. What factors control short-term and long-term changes in the losses from soil to rivers and the controllability of the export of nutrients to the sea?
3. To what extent can nutrient loadings in the marine environment be traced to rivers, and what are the environmental effects of the riverine loads?

The starting-point of our analysis is the recent economical recession in the former Soviet republics, which has created a unique opportunity to study riverine response to the dramatic decrease in inputs of nutrients to the agricultural sector.

Data base and study area

Forty percent of the land (mainly lowland) is used for agricultural production (Table 1). The catchment of the Lielupe River has the largest cover of farmland in the area (56%). Over 4.5 million people live in the drainage area of the Gulf of Riga (Table 1). The population is relatively uniformly distributed among and within the river basins with the exception of a few large cities (e.g. city of Riga with 900 000 inh.). Further background information on the Gulf of Riga area has been published by Tsirkunov et al. (1992) and Ojaveer (1995).

Our results and analyses are based on data obtained from the following: (A) monitoring of nutrient losses in four small agricultural catchments (0.8–8 km^2) in 1994–1997 (Figure 1); two of the catchments were characterized by cereal production (Berze & Kahametsa), and two by high livestock density and large application of animal manure (Bauska & Ogre); (B) time series of water quality data from monitoring sites in the five river basins (Daugava, Lielupe, Gauja, Salaca & Pärnu Rivers) included in Estonian and Latvian river monitoring programme (Figure 1); (C) sampling

Table 1. Land cover and population statistics for the Gulf of Riga drainage area. Source: Sweitzer et al. (1996) and Langaas (1996)

River basin	Agricultural land[a] (%)	Forested land (%)	Other land[b] (%)	Urban population (1000 inh.)	Rural population (1000 inh.)
Daugava[c]	40	35	25	1367	824
Lielupe[c]	56	25	19	207	248
Others	33	49	18	1517	475
Total	54 100 (km^2)	51 400 (km^2)	30 200 (km^2)	3091	1547

[a] Arable land and pasture.
[b] Mainly non-productive open land, urban areas and waters.
[c] Refers to area upstream of sampling site for water quality.

carried out on cruises in April, June and August–September in the Gulf of Riga in 1993–1995, which included investigations of nutrients, nutrient limitation of phytoplankton growth, distribution of suspended biomass, plankton dynamics, and the vertical flux of organic matter. Further details regarding data, study area, data analysis and analytical methods for sub-projects A, B and C have been reported by Vagstad et al. (1997, 1998, 1999), Laznik et al. (1999), and Wassmann and Tamminen (1999), respectively.

Results and discussion

Our results show that nutrients carried in the rivers are undoubtedly responsible for most of the total pollution load in the Gulf of Riga (Table 2). Approximately 113 000 tonnes of total nitrogen (79% of the total load) and 2100 tonnes of total phosphorus (68% of the total load) are annually transported by the rivers to the gulf (Laznik et al., 1999). These values correspond to unit area loads (i.e. time-averaged area-specific losses) varying from 7 to 10 kg ha^{-1} y^{-1} for total nitrogen and from 0.10 to 0.30 kg ha^{-1} y^{-1} for total phosphorus in the 5 major rivers (Table 3). The unit area loads in the areas surrounding the Gulf of Riga are far greater than those in forested and sparsely populated areas in Scandinavia and relatively modest seen in a Baltic Sea perspective (Stålnacke et al., 1999). The cited values can also be compared with the unit area loads in the Rhine basin, which are severalfold higher (van Dijk et al., 1996; Grimvall & Stålnacke, 1999). Furthermore, even higher values can be found for the nitrate nitrogen export from agricultural districts in Western Europe. In the Odense River basin in Denmark, for example, arable land covers 82% of the drainage area, and the unit area load of nitrate-N has been estimated to about 26 kg ha^{-1} y^{-1} (Grimvall & Stålnacke, 1999) as compared to 6 kg ha^{-1} y^{-1} in the most agriculturally dominated river basin in the Gulf of Riga, namely that of the Lielupe River (Table 3). These low or moderate nutrient losses may seem somewhat surprising in light of both the generally inefficient sewage treatment serving the 4.6 million inhabitants in the drainage basin and the previously intensive agriculture, which, among other things, entailed improper handling and spreading of manure and commercial fertilisers in combination with low yields (see further discussion below).

Results from a source apportionment study clearly indicate that non-point sources and, in particular, the losses from agricultural soils without any doubt were responsible the major input of nitrogen to waters in the drainage basin (Stålnacke, 1996). The agriculture was, in the time period from 1960 to 1980, characterized by marked intensification of plant and animal production, and increase in land drainage. At the end of the 1980s, the average use of N fertilisers had reached the same level or higher, compared to many Nordic countries, while crop yields were only about half (Löfgren et al., 1999). This mean that there was a relatively low nutrient utilisation efficiency which clearly indicate a high potential for losses to air and water, or an accumulation of N in the soil pool. Agriculture in the Gulf of Riga drainage basin has undergone dramatic changes during the past 8–10 years. The years 1990 and 1991 were the start of an almost total collapse in the agricultural production due to the disintegration of the former Soviet Union. The consequence of this collapse were immediate, and included dramatic decreases in nutrient inputs; mineral fertiliser use dropped with 90% and the live-stock decreased to 25–50% from it previous level (Löfgren et al., 1999). At the same time, large areas of arable land were taken out of production. Such abrupt and large land use change has hardly been recorded in modern European agricultural history. The

Table 2. Estimated annual input of total nitrogen and total phosphorus to the Gulf of Riga along the main pathways

Pathway	Total nitrogen (tonnes yr^{-1})	Total phosphorus (tonnes yr^{-1})	Reference
Riverine input	113 000	2100	Laznik et al., 1999
Daugava River	67 300	1290	
Lielupe River	16 400	290	
Gauja River	7500	80	
Pärnu River	6300	190	
Salaca River	3300	40	
Other rivers	12 500	160	
Atmospheric deposition	16–17 000	300	Yurkovskis et al., 1993, Granat, 1999
Coastal point sources	5341	684	Helcom, 1993
N_2-fixation	6900	–	Andrushaitis et al., 1995
All pathways combined	142 500	3100	

Table 3. Drainage basin area, area upstream of water quality station, time-averaged specific runoff and area-specific riverine loads of nutrients from the five major rivers of the Gulf of Riga drainage basin (study period 1977–1995). Source: Laznik et al., (1999)

River	Drainage basin area (km^2)	Area upstream of water quality station (km^2)	Specific runoff ($l\ s^{-1}\ km^{-2}$)	Nitrate-N ($kg\ ha^{-1}\ yr^{-1}$)	Total-N ($kg\ ha^{-1}\ yr^{-1}$)	Phosphate-P ($kg\ ha^{-1}\ yr^{-1}$)	Total-P ($kg\ ha^{-1}\ yr^{-1}$)
Daugava	87 900	84 100	8.4	3.10	7.60	0.11	0.15
Lielupe	17 600	16 500	5.4	6.30	9.30	0.15	0.17
Gauja	8900	8900	9.3	4.00	8.40	0.08	0.10
Pärnu	6910	5200	11.3	3.80	9.10	0.13	0.28
Salaca	3570	3570	10.8	4.60	9.20	0.09	0.11
Mean			8.5	4.00	8.30	0.11	0.15

monthly time series of various nutrient constituents for the period 1977–1995 shows perhaps somewhat surprisingly that, despite a large temporal variation (i.e. interannual and seasonal variation), the total nutrient riverine loads to the Gulf of Riga have been relatively stable (Figure 2). The decrease in loading observed in the first half of the 1990s was due almost solely to natural variability in water discharge; the period from the late 1980s to 1990 was wet, and the subsequent period (i.e. 1991–1995) was hydrologically more normal (Figure 2). In more detailed studies by Löfgren et al. (1999) and Stålnacke (1996) it was also difficult to find clear and unequivocal evidence of a large-scale decrease in the riverine loads that can be related to the prodigious economical recession in the catchment area of the Gulf of Riga after the disintegration of the Soviet Union in 1990.

Unfortunately, there have not been any direct monitoring of nutrient losses to waters from small agricultural catchments before the dramatic change in agriculture. However, there existed some sampling sites in small agricultural streams included in the national monitoring programmes for surface water quality. In 1993/94, a monitoring programme for nutrient runoff in smaller catchments was established in Estonia and Latvia.

Results from measurements in two of the small agricultural catchments dominated by cereal production with typical fertilisation have shown that the nutrient losses were roughly 10–17 kg $ha^{-1}\ y^{-1}$ for total nitro-

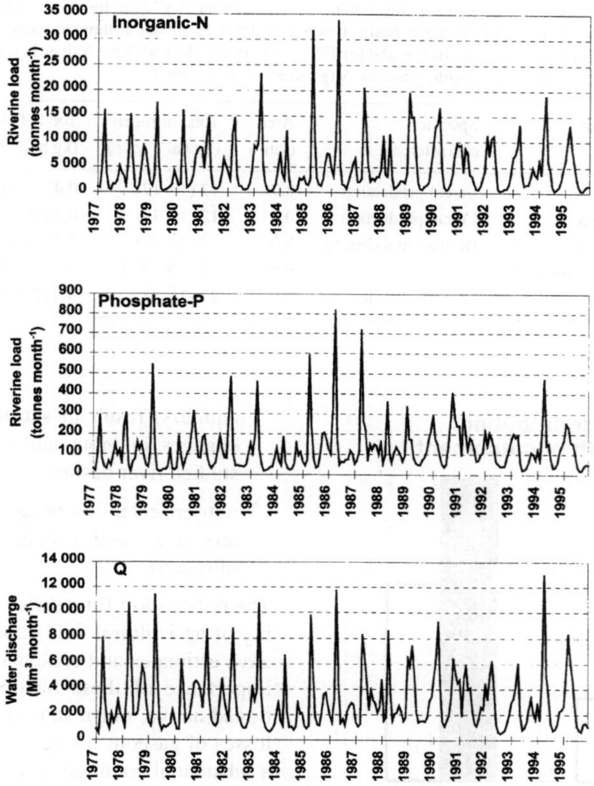

Figure 2. Monthly riverine input of inorganic nitrogen, phosphate-phosphorus and fresh water to the Gulf of Riga 1977–1995.

gen and 0.13–0.52 kg ha^{-1} y^{-1} for total phosphorus (Table 4). These values correspond to very low or moderate levels in a Nordic perspective. For example, Bechmann et al. (1998) reported values in the range 15–80 kg ha^{-1} y^{-1} for total nitrogen and 0.1–2.2 kg ha^{-1} y^{-1} for total phosphorus in three Norwegian agricultural catchments dominated by cereals with approximately the same precipitation (585–800 mm) and specific runoff (250–375 mm) as in the GoR-area.

There were also examples of tremendous nutrient losses from areas receiving large amounts of pig slurry, even though application ceased several years ago (Table 4). These high losses were related to special slurry irrigation areas, with previous application rates up to 630 kg ha^{-1} y^{-1} N. However, these sites are not regarded as typical for the farmland in general (V. Jansons, pers. comm.) and should instead be interpreted as agricultural 'hots-spots'.

Notable was also the large interannual variation in both the nitrogen and phosphorus losses (Table 4).

Some of the results reported by Vagstad et al. (1998, 1999) indicate that mineralization of soil-N, rather than application of fertilisers, is the main source of nitrogen inputs to waters. The low losses of nitrogen are probably due to the fact that water residence times in agricultural catchments are much longer in the Baltic states than for example in Norway (Deelstra et al., 1998); this prolonged water retention might lead to temporal water saturation of soils, which might enhance denitrification processes, as well as to a significant nutrient retention in small streams. Anaerobic conditions in soilwater may also inhibit nitrification of ammonia and, in combination with a very high soil pH (i.e. 7.3–7.8), promote NH_3-evaporation (Vagstad et al., 1998, 1999). In addition, it was also indications of nitrate transformation processes, presumably denitrification, in the smaller streams and channels. More precisely, it was in the main channel outlets of the catchments shown that the nitrate concentrations were 30–65% of the nitrate concentrations in the lateral tile

Table 4. Total catchment area and mean unit area loads (kg ha^{-1} y^{-1}) in four agricultural catchments in the Gulf of Riga drainage basin. Loads calculated as annual arithmetic mean value for 1994–1997 in Kahametsa and Berze, and for 1995–1997 in Bauska and Ogre. Values in parenthesis denote the range in annual loads. Sources: Vagstad et al., 1998, 1999

Catchment	Specific characteristic	Area (km^2)	Total nitrogen (kg ha^{-1} y^{-1})	Total phosphorus (kg ha^{-1} y^{-1})
Berze	Cereal production	3.7	13 (10–17)	0.47 (0.41–0.52)
Kahametsa	Cereal production	0.8	12 (10–14)	0.16 (0.13–0.18)
Bauska (main area)	High animal density	8.0	24 (23–24)	1.7 (1.0-2.3)
(slurry area)		0.5	127 (46–209)	4.8 (2.7–6.9)
Ogre	High animal density	3.0	27 (22–33)	3.9 (2.4–5.4)

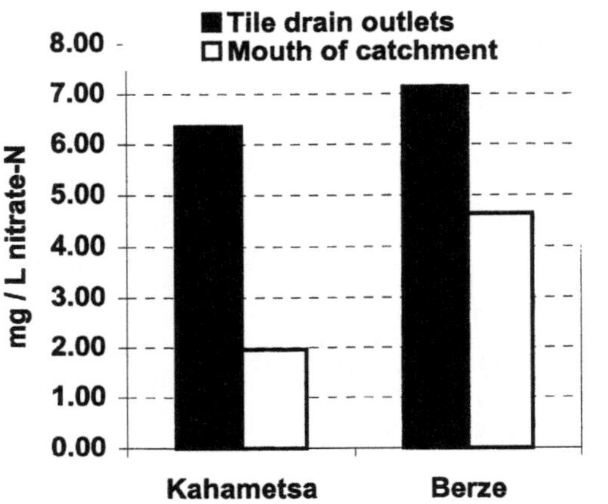

Figure 3. Mean nitrate concentrations at the outlet of the tile drains and at main channel in outlet of agricultural stream at Berze and Kahametsa.

drains within the same catchment (Figure 3). However, retention of nitrogen in the major rivers seemed to be relatively low, besides a tendency of nitrogen retention below major cities (i.e. point source discharges).

All this also indicates that the factors that control the losses of nutrients on a field scale are of significance for the dynamics on a river-basin scale and for the understanding and interpretation of long-term trends in riverine loads. More precisely, the remarkable lack of human-induced trends observed in our study areas seem to be linked to three aspects of the buffering capacities in soils and landscapes:

1. losses of nitrate from agricultural soils are maintained by mineralization of large pools of organic nitrogen that have accumulated over several years (Jenkinson, 1991);
2. long water-transit times (e.g. large ground-water aquifers) result in substantial time lags between changes in input and output of nitrogen in the studied river basins;
3. nutrient retention processes occur in soil and waters (e.g. sedimentation and gaseous losses of nitrogen).

The rather short time series of data from the monitoring of agricultural fields does not yet enable quantitative estimates and conclusions regarding the relative importance of these factors. Nonetheless, as mentioned above, our results suggest that there are large losses of gaseous nitrogen from agricultural soils and a substantial retention of nitrogen in the small agricultural streams, which might explain the low levels of N leached to waters as well as the slow response in riverine loads to the dramatic changes in agricultural practices during the 1990s. The rapid response that has been observed in some minor catchments in Estonia (Loigu & Vasilyev, 1997) and Slovakia (Pekarova & Pekar, 1996), and in the Tisza River basin in Hungary (Olàh & Olàh, 1996) has occurred under conditions which are apparently not representative for the whole drainage basin of the Gulf of Riga.

In the case of phosphorus, it must also be taken into account that emissions from point sources (i.e. municipal sewage treatment plants and large animal farms) and internal losses from the river bed may play important roles and therefore conceal some trends. Additionally, budget calculations indicate that retention in the main streams were relatively small, besides a notable retention of particulate phosphorus in the three reservoirs in the Daugava River.

Regarding specific rivers, the two largest basins, i.e. those of the Daugava and Lielupe Rivers, which cover 78% of the total drainage area, contribute substantially to the total riverine loads of nutrients to the gulf; more exactly, 74% of the nitrogen and 77% of the phosphorus load (Table 2). Spatial analysis based on monitoring of the marine environment has also shown increased concentrations of biomass and nutrients in the southernmost part of the gulf (i.e. near the outlets of the Daugava and Lielupe Rivers), as compared to the rest of the gulf (Reigstad et al., 1999; Tamminen & Seppälä, 1999). An excess of N in riverine runoff, relative to P or dissolved silica (DSi), as is the case in the Gulf of Riga, could drive the near-shore areas towards temporal limitation of P and DSi. The ultimate effects are not obvious, and examples, within this research programme indicate that a shift from the previously suggested P limitation (Yurkovskis et al., 1993) to N limitation or co-limitation has most likely occurred outside the southernmost part of the gulf. More precisely, spatial and temporal limitation patterns were studied by nutrient ratios (Tamminen & Seppälä, 1999) and experimental assays (Seppälä et al., 1999). The winter accumulation of inorganic nutrients clearly indicated basic N limitation of the basin outside the immediate influence of the River Daugava spring flood plume (Tamminen & Seppälä, 1999). Within this plume, the excess of N in riverine runoff indicated potential P limitation. After the spring flood season, the plume effect dissolutes, and horizontal variability in the basin becomes more irregular and patchy. Before late summer and autumn, N limitation of the productive layer prevailed, but towards the deep-mixing late summer season, the accumulation of mineralized nutrients, especially N, in intermediate depths provided increasing possibilities for P limited phases during periods of increased mixing of the water column (Tamminen & Seppälä, 1999). Therefore, during summertime, the spatial transport and mixing of nutrients and suspended and sinking particles in the gulf are greatly affected by wind (Reigstad et al., 1999), due to the simple and exposed topography of the basin. This implies that eutrophication is less prominent in, and less confined to, the southern part of the gulf than was previously assumed. However, clear signs of eutrophication have been seen in the littoral zone, especially on the south-east coast of the Gulf of Riga where the macrophyte communities are dominated by fast growing ephemeral algae (D. Conley, pers. comm.). This is partly due to the Coriolis-effect, which induces transport of nutrients and suspended matter from the Daugava and Lielupe Rivers along the coast in a north-easterly direction. Eutrophication in the littoral zone is also seen in Pärnu Bay attributed to direct nutrient inputs from the Pärnu River.

Moreover, the pelagic food web in the gulf plays a crucial role in modifying the traditional signs of eutrophication, such as agglomeration of biomass and vertical export of bioelements. Our results confirm previous observations in the eastern Baltic showing that, in a eutrophied environment, where nutrients are continuously supplied from both external and internal regenerative sources, C and N are lost in excess, whereas P is recycled from the pelagic system throughout the productive season (Reigstad et al., 1999). The efficiency of P retention depends on the complexity of the pelagic food web and may be enhanced with increasing numbers of trophic levels in the system. In the Gulf of Riga, this seemed to be the case, and thus the buffering capacity of the pelagic ecosystem towards external loading is relatively high.

Conclusions and concluding remarks

From a Nordic perspective, area-specific loads of nutrients in rivers and, in particular, losses from agricultural soils are moderate in the Gulf of Riga drainage basin. Furthermore, findings from the small catchments support data from the large river basins. More precisely, there is only weak evidence of a decline in the riverine load, whereas the decrease in agricultural production has been considerable, and this situation is most likely due to the inertia in, and buffering capacity of, agricultural soils. Hydrological conditions, the size of ground water aquifers and nutrient retention processes in the agricultural landscape are important for the control of long-term changes in the losses from land to the sea. This suggests that changes in the nitrogen retention capacities of the drainage basin may be of vital importance for the future load of nitrogen to the Gulf of Riga. Nevertheless, the pollution loads in the rivers, in particular those originating from the catchments of the Daugava and the Lielupe Rivers, have contributed significantly to eutrophication of the Gulf of Riga. This seems to be true despite the relatively rapid distribution of the loads beyond the littoral zone of the gulf, which gives rise to moderate all-over eutrophication of the area, with a continuously shifting gradient between the southern and northern parts of the gulf. Reduced emissions of P and organic matter from point sources may, under certain conditions, reduce the denitrification capacity in rivers (Chesterikoff et al., 1992). Therefore, it cannot be ruled out that the N loads to the Gulf of Riga may increase in the future, regardless of measures undertaken in the agricultural sector. Accordingly, it is difficult to predict changes in both the total nutrient discharge to the gulf and the N-P-DSi composition of the waters in the gulf, due to the complex buffering mechanisms of both the terrestrial and pelagic systems. As the nutrient limitation patterns of the basin are spatially and seasonally variable, and demonstrate the importance of especially N, but also P, as the main agent for the eutrophication processes, environmental management requires a wide and long-term strategy to counteract the eutrophication of the basin. In addition, it has been shown that the agricultural contribution to the nutrient loads is extremely variable over time and space (between catchments), which clearly demonstrates the complexity of the problem(s).

Acknowledgements

This study was financially supported by the Nordic Council of Ministers and the Research Council of Norway. The authors are grateful to the scientists of the '*Gulf of Riga project*' who provided data and results for this study.

References

Andrushaitis, A., Z. Seisuma, M. Legzdina & A. Lenshs, 1995. River load of eutrophying substances and heavy metals into the Gulf of Riga. In: Ecosystem of the Gulf of Riga between 1920 and 1990. E. Ojaveer, (ed.), Estonian Academy of Sciences, Tallinn, Estonia: 32–40.

Bechmann, M., H.O. Eggestad & N. Vagstad, 1998. Nitrogen balances and leaching in four agricultural catchments in South-Eastern Norway. Envir. Poll. 102: 493–499.

Chesterikoff, A., B. Garban, G. Billen & M. Poulin, 1992. Inorganic nitrogen dynamics in the River Seine downstream from Paris (France). Biogeochemistry 17: 147–164.

Deelstra, J., N. Vagstad, E. Loigu, A. Vasilyev & V. Janssons 1998. Interactions between hydrology and nitrogen runoff in small agricultural catchments. a comparative study of Estonian, Latvian and Norwegian catchments. In J. Kajander (ed.), XX Nordic Hydrological Conference, Nordic Association for Hydrology, Helsinki, 10–13 August 1998. Nordic Hydrological Programme, NHP Report No. 44: 120–128.

Granat, L., 1999. Deposition of nitrate and ammonium from the atmosphere to the Baltic. In: A Systems Analysis of the Changing Baltic Sea. F. Wulff, L. Rahm & P. Larsson (eds), Springer Verlag (accepted).

Grimvall, A. & P. Stålnacke, 1999. Nutrients with rivers to the Baltic Sea. In: A Systems Analysis of the Changing Baltic Sea. F. Wulff, L. Rahm & P. Larsson (eds), Springer Verlag (accepted).

Helcom, 1993. Second Baltic Sea pollution load compilation. *Baltic Sea Environment Proceedings*, No.45, Helsinki Commission, Helsinki: 161 pp.

Jenkinson, D. S., 1991. The Rothamsted long-term experiments: Are they still of use? Agron. J. 83: 2–10.

Langaas, S., 1998. Catchment statistics on land cover and population, Daugava and Lielupe rivers, Latvia. Technical Report 1/1996, UNEP/GRID-Arendal.

Laznik, M., P. Stålnacke, A. Grimvall & H.B. Wittgren, 1999. Riverine input of nutrients to the Gulf of Riga – temporal and spatial variation. J. Mar. Sys. (in press).

Loigu, E. & A. Vasilyev, 1997. Evaluation of water quality response to sudden changes in the amounts of fertilizers used in Estonia. In: Hydrology and Environment, Proceedings of the Baltic States Hydrology Conference, May 22–24, 1997. Kaunas.

Löfgren, S., A. Gustafson, S. Steineck & P. Stålnacke, 1999. Agricultural development and nutrient flows in the Baltic states and Sweden after 1988. Ambio 28: 320–327.

Ojaveer, E. (ed.), 1995. Ecosystem of the Gulf of Riga between 1920 and 1990. Estonian Academy of Sciences, Tallinn, Estonia

Olàh, J. & M. Olàh, 1996. Improving landscape nitrogen metabolism in the Hungarian lowlands. Ambio 25: 331–335.

Pekarova, P. & J. Pekar, 1996. The impact of land use on stream water quality in Slovakia. J. Hydrol. 180: 333–350.

Reigstad, M., A.S. Heiskanen & P. Wassmann, 1999. Seasonal sedimentation, retention and loss of particulate nutrients (C, N, P) in the pelagic system of the Gulf of Riga. J. Mar. Sys. (in press).

Seppälä, J., T. Tamminen & S. Kaitala, 1999. Experimental evaluation of nutrient limitation of phytoplankton communities in the Gulf of Riga. J. Mar. Sys. (in press).

Stålnacke, P., 1996. Nutrient loads to the Baltic Sea. Linköping Studies in Arts and Science, No. 146. Ph.D. Thesis, Linköping University.

Stålnacke, P., A. Grimvall, K. Sundblad & A. Tonderski, 1999. Estimation of riverine loads of nitrogen and phosphorus to the Baltic Sea, 1970–1993. Env. Mon. Ass. 58: 173–200.

Sweitzer, J., S. Langaas & C. Folke, 1996. Land cover and population density in the Baltic Sea drainage basin: a GIS database. Ambio 25: 191–198.

Tamminen, T. & J. Seppälä, 1999. Nutrient pools, ratios and transformations in the Gulf of Riga basin. J. Mar. Sys. (in press).

Tsirkunov, V. V., A.M. Nikanorov, M.M. Laznik & Z. Dongwei, 1992. Analysis of long-term and seasonal river water quality changes in Latvia. Wat. Res. 26: 1203–1216.

Vagstad, N., V. Jansons, E. Loigu & J. Deelstra, 1999a. Nutrient losses from agricultural areas in the Gulf of Riga drainage basin. Ecol. Engin. (in press).

Vagstad, N., E. Loigu, V. Jansons, J. Deelstra & A. Njøs, 1998. The agricultural contribution to nutrient loads in the Gulf of Riga drainage basin. Paper presented at Symposium 'The Gulf of Riga Project 1993–1998' in Jurmala, Latvia, May 10–14, 1998.

van Dijk, G.-M., P. Stålnacke, A. Grimvall, A. Tonderski, K. Sundblad & A. Schäfer, 1996. Long-term trends in nitrogen and phosphorus concentrations in the Lower River Rhine. Arch. Hydrobiol. Suppl. 113: 99–109.

Wassmann, P. & T. Tamminen, 1999. Pelagic eutrophication and sedimentation in the Gulf of Riga: an introduction. J. Mar. Sys. (in press).

Yurkovskis, A., F. Wulff, L. Rahm, A. Andruzaitis, & M. Rodriguez-Medina, 1993. A nutrient budget of the Gulf of Riga; Baltic Sea. Estuar. coast. shelf Sci. 37: 113–127.

Hydrobiologia **410:** 111–122, 2000.
J. Garnier & J.-M. Mouchel (eds), Man and River Systems.

Retention of nutrients in river systems: dependence on specific runoff and hydraulic load

Horst Behrendt & Dieter Opitz
Institute of Freshwater Ecology and Inland Fisheries, Department of Lowland Rivers and Shallow Lakes, Müggelseedamm 260, D-12587 Berlin, Germany

Key words: phosphorus, nitrogen, retention, river systems, specific runoff, hydraulic load

Abstract

The total nutrient inputs (emissions) from point and diffuse sources into 100 different river basins in Europe are compared with the measured load (transport). The catchment areas range between 121 and 194 000 km^2. Other characteristic parameters of the basins, including specific runoff, surface water area and the nutrient concentration show large differences. Independent of the different methods applied for estimating nutrient emissions, these are in general higher than the actual transport. The ratio of the transport to the emission was used to exclude the effects of basin size and to facilitate comparisons between river basins. A statistical model of the retention, including net sedimentation and denitrification, is derived for the description of the discrepancy between the measured transport and the estimated total emission of nutrients, which is assumed to reflect retention of nutrients in a river system. It was found that the load-weighted retention of phosphorus is controlled practically only by the specific runoff of the basin. The load- weighted retention of nitrogen is further dependent on the proportion of the basin area occupied by surface water, the basin size itself and the mean annual nitrogen concentration at a specific monitoring station. Application of the statistical model reduces the mean deviation between the calculated emissions and the values given by different authors from about 40% to 20%.

Introduction

Reduction of nutrient input into lakes and seas should focus especially on decreasing the inputs by rivers, and requires a knowledge of the sources and their contribution to the transport by the rivers. The number of results of analysis of the emission within the whole basin of a particular river is increasing but has been limited up to now, especially when such analysis includes different diffuse pathways. Furthermore, confusion exists with regard to a possible comparison of the measured transport (output or immission) of a substance at a monitoring station of a river with the sum of all diffuse and point sources of emission which represents the input. The common opinion is that especially in large river basins, the discrepancy between the sum of all inputs and the output is low and retention and/or loss processes play a subordinate role with exception of the process of denitrification in the case of nitrogen. Some authors take into account losses of nutrients within the river basin related lakes only, because losses in lakes are accepted and predictable since the investigations of Vollenweider (1968). Small river basins are often used to estimate the non-point source loading and it is assumed in most cases that the processes operating within the waters of such a small catchment can also be ignored.

As shown by different authors (e.g. Billen et al., 1995; Behrendt, 1996), the assumption that retention processes can be neglected in river systems is in general wrong. For nitrogen, it has been shown by many investigations that losses of nitrogen (especially denitrification) occurs also in rivers. Measured denitrification rates in rivers are often higher than in lakes (Howarth et al., 1996). From these studies, it can be concluded that the intensity of the denitrification process is similar in lakes and rivers. Lower losses in rivers are mainly caused by the lower sediment surface compared with lakes. Probst (1985) and Hilbricht-Illkowska (1989) have already pointed out qualitatively that the sum of the phosphorus inputs to a river system is higher then the observed transport.

The number of studies on P-retention is limited up to now, but indications exist also that P-retention occurs at low flow conditions as well as during floods, where the river flows partly outside of its bed. Further consumption of nutrients by benthic organisms such as submerged and emergent macrophytes and mussels (Welker & Walz, 1998) can contribute to the retention of nutrients to a large extent. The direct measurement of the different retention and loss processes is possible only at certain locations (Svendsen & Kronvang, 1993). But upscaling of such measurements requires knowledge of the distribution of the identified driving parameters along the whole river system.

The increasing knowledge of the pathways of nutrient emissions from point and diffuse sources gives the possibility to estimate the total inputs into a river system. If the emissions are known, the sum of all riverine retention can be approximately estimated from the difference between the emissions and the measured load. This approach also includes errors associated with upscaling and insufficient knowledge of the processes. Therefore, differences between the inputs and the measured load can be caused by an over-estimation of the inputs or an under-estimation of the transport by the river. But the application of different methods of emission estimations to the same river systems indicates that the standard deviation of the mean of all methods decreases significantly, if the most simple method (emission is equal to transport) is excluded (Behrendt, 1999). This behavior can be explained because neglect of retention in rivers leads to larger errors than these associated with other methods, which include retention directly or indirectly. In the following, it is assumed that the difference between the observed transport and emissions is mostly caused by retention processes. The aim is to identify the driving parameters of the nutrient retention.

Material and methods

The analysis was carried out using data on the emission and transport of nutrients of 100 different rivers basins mostly located in Central Europe, which are published by different authors. An overview of the rivers and the references is presented in Table 1. River basins smaller than 100 km^2 were not considered, because the comparison of methods shows an increase in the standard deviation of the mean of the emissions for small basins (Behrendt, 1999).

The rivers listed in Table 1 include the data sets for 15 river basins given by Behrendt (1996) and other authors. Whereas the point emissions are calculated from inventories of industrial and municipal waste water treatment plants, the methods used for the estimation of diffuse sources are different. Behrendt & Bachor (1998) estimated the diffuse emissions of nutrients according to the hydrological pathways (groundwater, tile drainage and surface runoff). Additionally the emissions from urban areas, atmospheric deposition and agricultural direct load were considered. A detailed description of this method is given by Werner et al. (1991), Kroiß et al. (1997), Prasuhn & Braun (1994) and Isermann (1997) used a similar method. Tonderski (1997) considered specific export coefficients for different land use categories and soil types. Braun (1994) used the discharge dependency of the nutrient load to estimate the relative contribution of point and diffuse sources to the measured transport. By comparing of net transport of the point sources with the gross load from the point source inventory, the retention and the total emissions are derived. Svendsen et al. (1995) applied three different methods of source apportionment. The diffuse emissions into the rivers Gudena and Vantaanjoki (Table 1) were estimated using as the residual from the balance of load plus retention for lakes minus point emissions. Arnheimer & Brandt (1998) used the model HBV-N for the estimation of the total nitrogen inputs into river systems of southern Sweden. From these results, a selected database of 10 river systems could be used by courtesy of Brandt (pers. comm.).

The parameter specific runoff (q) was calculated from the discharge for the same period as the nutrient load and is given in Table 1. The percentage of the area of the basin occupied by surface water (W) was calculated using the percentage of lake and reservoir area (W_{Lake}) taken from land use maps or statistics and the river surface area by the following formula:

$$W = W_{\text{lake}} + 0.1 \cdot A^{0.185}[\%], \quad (1)$$

where A is the total area of the river basin.

The second part of the sum is derived from analysis of different river systems according to stream order (Billen, 1993; Billen et al., 1995) and our own measurements in rivers of different size. For the rivers investigated by Prasuhn & Braun (1994) and Prasuhn et al. (1996) the given water area was used without the application of Equation (1), because these data include both the area of lakes and rivers. The lake area of the Austrian rivers (Kroiß et al., 1997) could not be con-

Table 1. Basin size, surface water area, specific runoff and nutrient transport and emission for the investigated river basins

Ref.	River basin	basin size [km^2]	surface waters [%]	specific runoff [l km^{-2} s^{-1}]	TP-transport [ktP/a]	TP-emission [ktP/a]	DIN-transport [ktN/a]	TN-emission [ktN/a]
1)	Warnow 83/87	3139	4.9	4.4	0.096	0.388	1.23	5.8
1)	Warnow 92/94	3139	4.9	3.8	0.080	0.292	1.1	4.4
1)	Mulde	7400	1.0	7.8	1.773	3.00	14.77	33
1)	Neckar	14 000	0.8	12.3	3.159	4.791	33.93	46.31
1)	Main up. Kahl	23 422	1.2	8.2	3.501	4.943	32.6	59.63
1)	Havel	24 025	3.4	4.6	1.58	5.10	9.18	46
1)	Saale	24 079	1.1	5.8	3.954	7.70	49.66	92
1)	Main	27 226	1.2	8.6	5.91	7.37	51.76	78.23
1)	Mosel	28 165	0.9	14.0	5.23	6.40	55.63	96.4
1)	Rhein up Koblenz	110 075	2.3	16.9	24.472	28.16	214.89	346.91
1)	Elbe up. Schnackenb. 83/87	132 060	2.1	4.3	5.8	16.95	102	252
1)	Elbe up. Schnackenb. 92/94	132 060	2.1	5.8	11.62	29.6	135	365
1)	Rhein up. Seltz	50 200	3.6	25.8	9.0	9.40	117	141
1)	Rhein up. Lobith	159 794	2.3	15.7	35.32	45.561	356.17	562.3
1)	Elbe czech part	51 330	1.5	4.7	1.527	5.249	47.5	87.38
2)	Elbe czech part	51 330	1.5	6.4			57.68	114.7
3)	Ruhr Villigst 1997	1988	0.9	10.5	0.110	0.176	3.34	4.42
3)	Ruhr Villigst 1990	1988	0.9	11.3	0.150	0.256	3.18	4.22
3)	Ruhr Essen 1990	4171	1.0	14.0	0.520	0.846	9.26	14.13
3)	Ruhr Essen 1997	4171	1.0	11.9	0.240	0.476	8.02	13.33
4)	Donau up. Jochenstein	77 086	1.8	18.4	8.69	10.90	109.16	123
5)	Melk	286	0.8	10.0	0.006	0.012	0.21	0.43
5)	Piesting II	288	0.8	10.4	0.009	0.011	0.14	0.32
5)	Schunter	295	0.5	5.1	0.015	0.035	0.42	0.52
5)	Pielach upper part	307	0.8	19.0	0.005	0.005	0.26	0.25
5)	Krems	392	0.8	15.0	0.022	0.046	0.98	1.65
5)	Pielach	539	0.8	14.0	0.026	0.035	0.59	0.9
5)	Ybbs	1183	0.5	26.0	0.082	0.096	2.56	2.43
5)	Leitha	2141	0.6	8.0	0.081	0.213	1.63	3.19
5)	Inn	7145	1.0	28.0	0.855	0.707	7.6	10.33
5)	Drau	12 051	1.0	23.0	0.380	0.480	9.21	9.21
6)	Vils	756	0.4	8.6	0.054	0.090	0.83	1.13
7)	Vantaanjoki	1686	2.3	14.0	0.068	0.086	1.05	1.9
7)	Gudena	2603	2.9	11.0	0.117	0.164	2.78	5.8
8)	Schelde upper part	6200	0.7	12.9	3.336	4.625	18.58	23.15
12)	Schelde total	21 000	0.8	7.1	4.693	7.629	36.79	81.91
8)	Seine	64 500	1.0	6.5	6.837	11.804	58.24	94.43
8)	Loire	116 000	1.1	6.9	3.364	5.568	79.81	95.12
9)	Hunte upper part	403	0.8	5.9	0.015	0.047	0.5	0.83
9)	Hunte	1714	1.3	5.8	0.081	0.263	1.26	3.65
10)	Oder	110 074	2.0	4.4	4.753	16.814	44.92	115.67
10)	Vistula	194 484	2.1	4.4	5.656	22.015	72	204
11)	Wallensteingraben	156	3.3	7.6	0.015	0.032	0.15	0.5
11)	Ryck	231	0.4	5.3	0.005	0.016	0.23	0.41
11)	Barthe	292	1.4	7.2	0.007	0.020	0.56	0.64
11)	Randow	668	1.5	2.2	0.008	0.042	0.18	0.35
11)	Recknitz	669	0.4	5.9	0.027	0.050	0.6	1.07
11)	Stepenitz	701	1.4	7.7	0.034	0.066	1.07	2.05
11)	Zarow	748	2.0	3.6	0.016	0.056	0.47	0.65
11)	Trebel	992	0.5	5.1	0.023	0.071	1.22	1.55
11)	Tollense	1809	2.6	4.2	0.038	0.133	1.47	2.31

Table 1. Continued

11)	Sude	2253	2.0	6.6	0.058	0.155	1.27	2.71
11)	Uecker	2401	3.5	2.5	0.046	0.158	0.83	1.87
11)	Elde	2990	8.0	3.2	0.072	0.192	0.73	2.75
11)	Warnow	3140	4.9	5.1	0.074	0.25	1.81	3.98
11)	Peene	5110	2.5	4.1	0.15	0.368	3.19	6.41
13)	Nysa	4297	0.9	6.3	0.293	0.568	3.78	5.14
13)	Nysa Klodzka	4515	1.3	5.2	0.163	0.505	2.31	4.46
13)	Odra up. Chalupki	4666	0.9	7.1	0.563	1.144	5.78	10.23
13)	Prosna	4824	0.6	2.5	0.148	0.382	2.26	3.64
13)	Notec upper part	5508	2.8	1.8	0.113	0.613	0.77	3.58
13)	Bobr	5869	1.0	5.0	0.33	1.498	1.88	8.18
13)	Odra up. Raciborz	6684	1.1	7.7	0.954	1.407	10.16	13.69
13)	Warta up. Sieradz	7929	0.8	4.2	0.222	1.261	2.67	7.45
13)	Odra up. Groszowic	10 989	1.1	6.3	1.10	13.01	13.6	27.76
13)	Notec	17 331	2.9	3.8	0.435	1.219	2.02	14.28
13)	Odra up. Wroclaw	21 572	1.3	4.9	1.483	4.702	13.53	35.83
13)	Warta up. Poznan	25 911	1.2	3.1	0.788	3.64	10.68	29.73
13)	Odra up. Scinawa	29 584	1.2	4.6	2.247	7.121	24.65	48.99
13)	Odra up. K. Odrzanskie	40 910	1.5	4.0	1.795	8.048	26.09	60.53
13)	Odra up. Kostrzyn	53 532	1.6	4.5	2.349	9.984	26.54	74.68
13)	Warta	54 518	2.0	3.2	1.851	5.122	11.99	47.99
13)	Odra up. Krajnik Dolny	110 074	1.8	4.1	5.12	14.883	44.92	133.88
14)	Stockacher Aach	213	0.3	8.9	0.035	0.041	0.27	0.34
14)	Seefelder Aach	273	0.5	13.8	0.036	0.04	0.63	0.63
14)	Argen	653	1.3	30.2	0.191	0.192	1.99	1.91
14)	Schussen	789	1.0	16.0	0.181	0.186	2.45	2.5
14)	Bregenzerach	826	1.3	50.6	0.221	0.213	1.51	1.57
14)	Ill	1282	1.2	47.4	0.165	0.128	1.85	1.61
14)	Alpenrhein	4453	1.4	35.6	0.845	0.799	4.16	4.53
14)	Rhein up. Diepoldsau	6549	1.4	39.1	1.075	1.143	6.95	8.56
15)	Gürbe	127	1.1	22.8	0.01	0.011	0.35	0.34
15)	Langete	146	0.3	14.1	0.017	0.019	0.46	0.52
15)	Birs	201	0.4	16.2	0.021	0.024	0.31	0.37
15)	Schüss	215	0.2	26.4	0.012	0.013	0.32	0.34
15)	Sense	355	1.2	24.2	0.013	0.021	0.44	0.47
15)	Lütschine	399	0.6	46.7	0.077	0.073	0.28	0.31
15)	Emme upper part	437	0.8	24.6	0.03	0.033	0.67	0.55
15)	Aare Oberhasli	556	1.9	70.1	0.151	0.108	0.38	0.35
15)	Kander/Simme	1096	0.8	37.8	0.167	0.202	0.8	0.93
16)	Stensan	285	1.3	12.3			0.17	0.3
16)	Kilaan	432	5.2	6.8			0.11	0.27
16)	Ljungbyan	758	1.1	4.8			0.27	0.4
16)	Söderköpingsan	882	9.1	5.2			0.21	0.56
16)	Rönnea	1897	3.7	12.9			1.47	3.32
16)	Eman	4472	7.0	6.7			0.82	2.24
16)	Helge a	4725	5.3	9.4			2.28	5.97
16)	Motala strom	15 480	20.3	10.9			1.1	5.91
16)	Mälaren	22 650	12.4	7.0			0.3	3.15
16)	Göta älv	50 119	19.3	12.2			1.76	4.35

[1] Behrendt (1996) [2] Blazkova et al. (1998) [3] Ruhrverband (1998) [4] Isermann (1997) [5] Kroiß et al. (1997) [6] Braun (1994) [7] Svendsen et al. (1995) [8] Billen et al. (1995) [9] Raderschall (1996) [10] Tonderski (1997) [11] Behrendt & Bachor (1999) [12] ICWS (1989) [13] Behrendt et al. (1998) [14] Prasuhn et al. (1996) [15] Prasuhn & Braun (1994) [16] Brandt (pers. comm.)

sidered up to now with exception of the River Drau. The total percentage of the area of the basins occupied by surface waters is also given in Table 1.

The mean annual hydraulic load (HL) as a further important characteristic parameter of a river system can be calculated from the data in Table 1 by the following equation:

$$\mathrm{HL} = \frac{q[\mathrm{l \cdot km^{-2} \cdot s^{-1}}] \cdot 8.64 \cdot 0.365}{W[\%]} [\mathrm{m/a}] \quad (2)$$

The hydraulic load gives the average annual value of the height of water column flowing over one square meter water surface of the river system.

Results and discussion

Table 1 and Figure 1 indicate that the observed transport of nitrogen and phosphorus is lower for most of the considered river systems than the total nutrient emissions. This means that retention or losses of nutrients can occur also in rivers. The question is what are the factors controlling this mechanism? A statistical analysis of the data was undertaken based on the mass balance equation.

The mean annual transport or load of nutrients in a river system ($L_{N,P}$) within a certain time period is the result of the sum of all point (E_{point}) and diffuse (E_{diffuse}) emissions ($E_{N,P}$) reduced by the sum of all nutrient retention and losses ($R_{N,P}$):

$$L_{N,P} = E_{N,P} - R_{N,P} = \sum E_{\mathrm{point}N,P} + \sum E_{\mathrm{diffuse}N,P} - \sum R_{N,P} \quad (3)$$

If the load, as well as the point and diffuse emissions are known, the sum of the retained or lost nutrients can be estimated according to Equation (3). Because the river basins differ in size, it is essential to eliminate the influence of catchment area when comparing river basins. Behrendt (1996) has shown that the use of the ratio of load to the sum of emissions is one possible way of normalization. The sum of the retention and losses is than linked to this ratio by the Equation (4):

$$\frac{L_{N,P}}{E_{N,P}} = \frac{1}{1 + R_{L_{N,P}}}, \quad (4)$$

where $R_{L_{N,P}}$ is the transport weighted sum of all retention and losses ($R_{N,P}$ divided by $L_{N,P}$). Equation (4) is used for the calculation of the load or emission weighted retention according to the following equation:

$$R_{L_{N,P}} = \frac{E_{N,P}}{L_{N,P}} - 1 \quad (5)$$

The load weighted nutrient retention (R_L) can be described by the following statistical model

$$R_{L_{N,P}} = a \cdot x^b, \quad (6)$$

where x is the driving force and a and b are the coefficients of the model. According to Behrendt (1996), the specific runoff is one parameter controlling the retention behavior of river systems. As known from the analysis of lakes, the hydraulic load and especially the residence time are identified as the most important driving forces (e.g. Dillon, 1975; Vollenweider & Kereekes, 1982). Kelly et al. (1987) proposed a simple mass balance model for the denitrification losses, where the nitrogen retention in lakes is dependent on the quotient between mean depth and residence time, which is identical with the hydraulic load. Howarth et al. (1996) have shown that a model of this type can also be applied for river systems. Because the estimation of the residence time of river systems is currently an open problem, the statistical analysis, according to equation (6), is focused here only on the specific runoff and the hydraulic load.

Figures 2 and 3 show that a relationship exists between the ratio of the average annual load to the sum of the emissions of nutrients and the mean specific runoff (Figure 2), as well as the hydraulic load (Figure 3). Because total nitrogen is not measured for all rivers, the nitrogen load given in the Figures 2B and 3B and Table 1 is the dissolved inorganic nitrogen load (DIN = NH_4 + NO_2 + NO_3).

The coefficients of Equation (6) were estimated using the solver unit of a spreadsheet program. The conformity between model results and original data was tested by linear regression.

The results of the models related to specific runoff and hydraulic load are presented for all river basins and for different size dependent groups in Table 2. The regression analysis shows that the specific runoff has the highest influence on the total variance for phosphorus. About 80% of the variance of the phosphorus retention is explained by the dependency on specific runoff. Compared with the results of Behrendt (1996), the coefficient for the dependency on specific runoff is reduced from about -2 to about -1.7, which is mainly caused by the inclusion of further river basins with low specific runoff into the analysis.

The dependency of the phosphorus retention on the hydraulic load shows a regression coefficient which

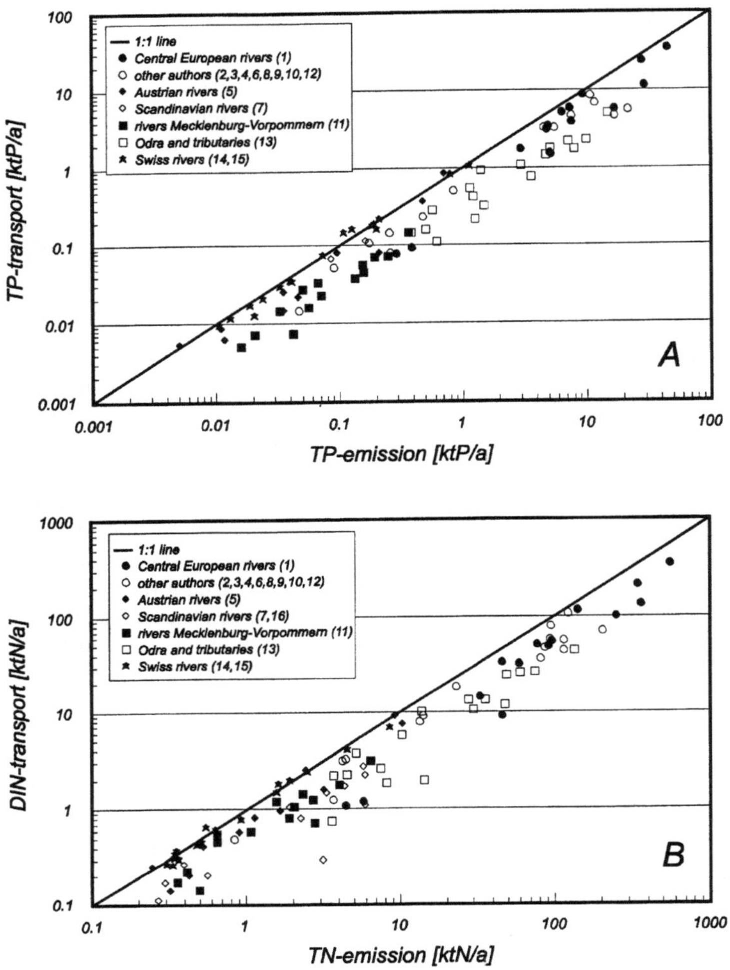

Figure 1. Relationship between observed nutrient transports and the total emissions from point and diffuse sources for different river basins (A) Phosphorus; (B) Nitrogen. The numbers in brackets refer to the reference items in Table 1.

explains only 61% of the total variance. This is also illustrated by the dotted lines in the Figures 2A and 3A, which represent the limits for a prediction of the load to emission ratio at a level of significance of 90%. The range between both limits is obviously smaller in Figure 2A than in Figure 3A.

The three separated groups of the river systems show a similar dependency on the specific runoff, whereas the regression coefficient increases with increasing size of the basin. The reason for this behavior seems to be the phenomenon that the accuracy of the measured load, as well as the estimated emissions is increasing with increasing size of the river basin. Furthermore, the model results of the considered three different groups indicate that the phosphorus retention is higher for smaller catchments, especially in the range of low specific runoff and low hydraulic load.

For nitrogen, the results are different as shown in Table 2. The relationship with specific runoff alone explains only 51% of the variance, whereas the regression coefficient of the dependency on hydraulic load amounts 65%. That means the area of surface water have a certain higher influence on the load standardized retention. Contrary to the behavior of phosphorus, the nitrogen retention seems to increase with increasing size of the catchment.

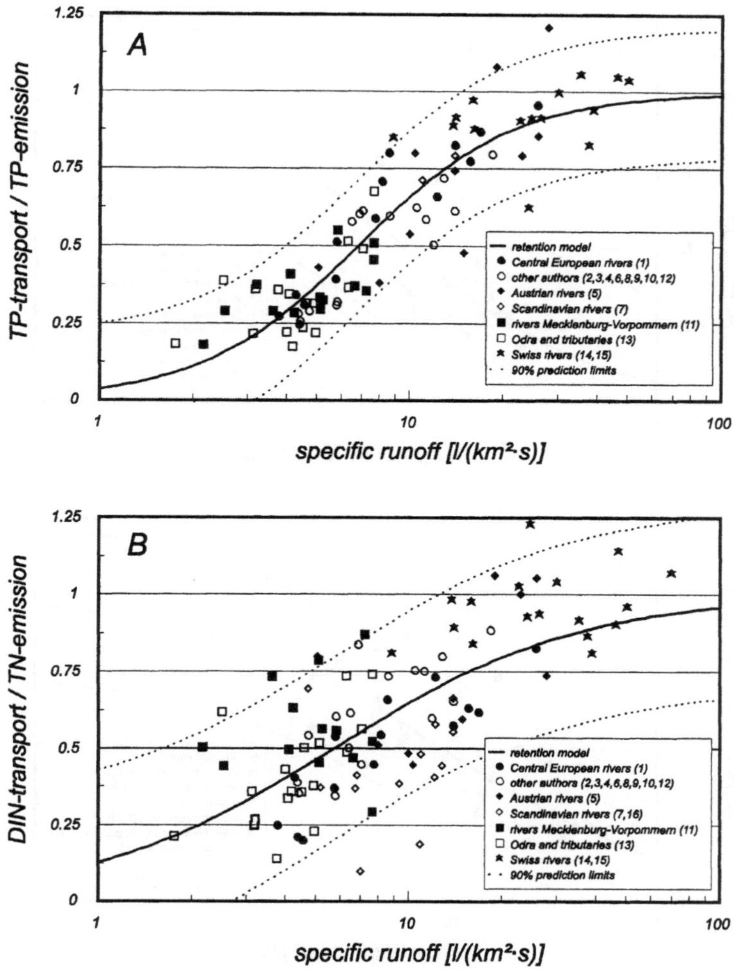

Figure 2. Dependency of the transport to emission ratio of the nutrients on the specific runoff for different river basins in Europe (A) Phosphorus; (B) Nitrogen.

The differences between the coefficients of the statistical retention model for phosphorus and nitrogen can be explained by the different processes which control the retention. The reason for the higher significance of the dependency of the P-retention on the specific runoff is not clear up to now, but it seems to be related to the main process of P-retention, the sedimentation, which is more dependent on the flow conditions than on the area of surface waters. Otherwise, the better relation between the N-retention and the hydraulic load can be caused by the denitrification process, which is related to the area of the basin occupied by surface water, because this area is nearly proportional to the area of the sediment surface. Anoxic conditions occur more likely in the upper sediment layers of larger river systems because of higher organic pollution from point sources and would so explain the increase in N-retention with increasing size of the basin.

Howarth et al. (1996) applied the mass transfer model of Kelly et al. (1987) to estimate and to compare the nitrogen retention of lakes and rivers.

The model of Kelly et al. (1996) propose the following relation between nitrogen retention and hy-

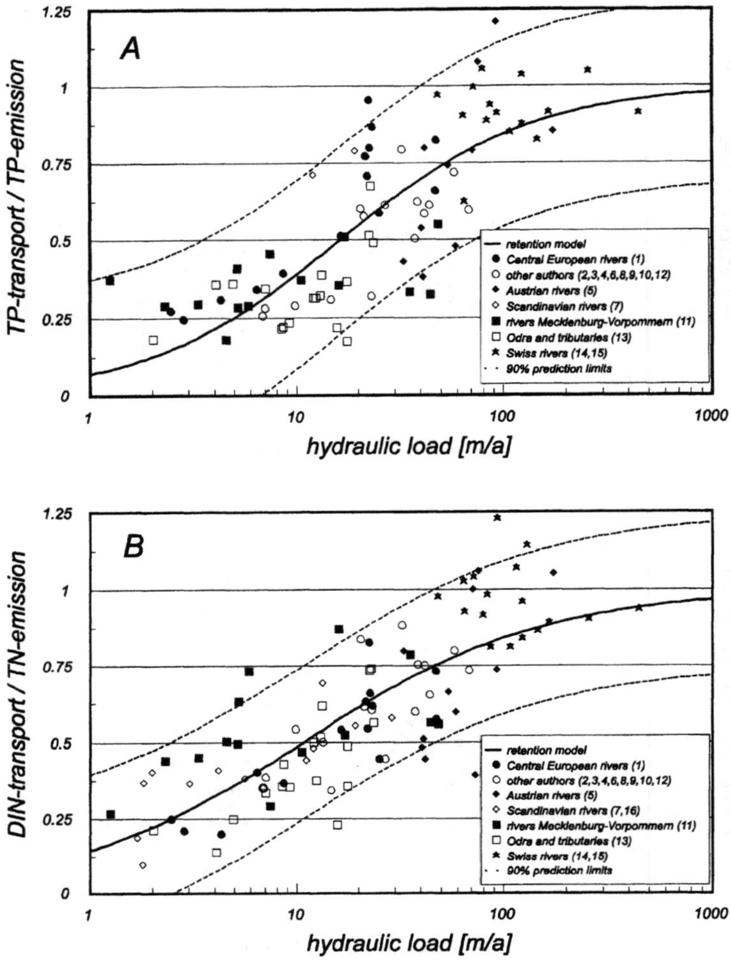

Figure 3. Dependency of the transport-emission ratio of nutrients on the hydraulic load for different river basins in Europe (A) Phosphorus; (B) Nitrogen.

draulic load:

$$R_E = \frac{S_N}{S_N + \mathrm{HL}}, \tag{7}$$

where R_E is the nitrogen retention normalized on the total emissions, HL is the hydraulic load or the ratio between mean depth of the aquatic system and the residence time and S_N an average mass transfer coefficient given in m/a. The relation between the transport weighted retention given in Equation (5) and the emission weighted retention is given by:

$$R_E = \frac{R_L}{(1 + R_L)} \tag{8}$$

If Equations (7) and (8) are compared the relation between the transport weighted retention and the hydraulic load is:

$$R_L = \frac{S_N}{\mathrm{HL}} \text{ or } R_L = \frac{a}{\mathrm{HL}} \text{ with } a = S_N \cdot \tag{9}$$

Therefore, the results of Howarth et al. (1996) can be directly compared with the estimations according to Equation (6), if the coefficient b is set to -1. The results of this analysis are given for the investigated river systems also in the Table 2 (column $\mathrm{HL}^{2)}$). For all river systems, we estimated an average value of the mass transfer coefficient for nitrogen of about 11.9 m/a, which is in the range given for lakes also by

Table 2. Results of regressions between the nutrient retention per load (R_L) of river systems and the specific runoff (q) and the hydraulic load for the investigated river systems

x	All river basins			river basins < 1000 km^2			rivers between 1000 and 10 000 km^2			river basins > 10000 km^2		
	q	HL[1)]	HL[2)]	q	HL[1)]	HL[2)]	q	HL[1)]	HL[2)]	q	HL[1)]	HL[2)]
Phosphorus												
r^2	0.8090	0.6148	0.6130	0.7529	0.5785	0.5671	0.7988	0.5884	0.5746	0.8765	0.6879	0.6699
n	89	89	89	29	29	29	32	32	32	28	28	28
a	26.6	13.3	16.6	41.4	57.6	21.1	21.7	9.3	17.2	28.9	26.9	13.9
b	−1.71	−0.93	−1	−1.93	−1.26	−1	−1.55	−0.81	−1	−1.80	−1.25	−1
Nitrogen												
r^2	0.5096	0.6535	0.6173	0.3936	0.4423	0.3647	0.5763	0.6607	0.5395	0.4548	0.7373	0.7357
n	100	100	100	33	33	33	35	35	35	32	32	32
a	6.9	5.9	11.9	3.5	3.3	9.5	5.8	4.4	12.7	7.9	10.9	12.7
b	−1.10	−0.75	−1	−1.01	−0.65	−1	−0.96	−0.62	−1	−1.03	−0.94	−1

1) Results of a model according to Eq. (6).
2) Results with a constant mass transfer coefficient - a, according to the model approach of Kelly et al. (1987).

Howarth et al. (1996) between 5 and 20 m/a and by Höhener & Gächter (1993) between 3 and 30 m/a for different lakes mostly located in the alpine region.

If the results of our analysis are compared with the range of 20 to 50 m/a suggested by Howarth et al. (1996), a large difference exists for river stretches. The reason seems to be that Howarth et al. (1996) assumed large hydraulic load for river stretches, which varies little in the range 150–300 m/a. Because the mass transfer coefficients were calculated from measured denitrification rates and the hydraulic load, they have to find for the assumed range of hydraulic load larger values of S_N as for lakes. As shown in Figure 3, the range of the hydraulic load of a river system can vary between 1 and 500 m/a. If river systems with a high proportion of lakes (W>1%) are excluded, the hydraulic load ranges between 15 and 500 m/a. For these river systems, we estimated a mass transfer coefficient of about 19 m/a, which is larger than the average of all investigated river systems but within the range given for lakes. Otherwise, we found for the river systems with lakes (W>2%) an average of the mass transfer coefficient of 8.5 m/a. Consequently, our analysis also support the suggestion that the lateral part of river systems possesses a higher mass transfer coefficient for nitrogen, but the differences seem to be lower as assumed by Howarth et al. (1996).

Whereas the differences of the mass transfer coefficients of nitrogen are low for the different groups of river systems according to Table 2, we found for phosphorus a decrease of these parameter with increasing size of basin, but the differences are not significant at a level of confidence of 90%. An explanation of this phenomenon could be that smaller river basins and especially such with low specific runoff possess larger areas where submerse and emerse macrophytes can growth. These macrophytes enlarge the area with low flow velocities and influence so directly the sedimentation of particles.

A direct comparison with results for lakes or other river basins is not possible up to now.

In general, the statistical analysis alone cannot further explain the different results for phosphorus and nitrogen retention. Therefore, it is necessary to develop a conceptual model of the Vollenweider type (e.g. Vollenweider, 1968).

For the evaluation of the results of the statistical model, the calculated emissions are compared with the estimates produced in Table 1. These comparisons are shown in Figure 4. It includes the calculated emissions using the statistical model with the highest r^2 for phosphorus (4A) and nitrogen (4B), respectively.

It is obvious that the differences associated with the model results are low. The deviation of the calculated emissions exceeds the 50% deviation level of the 1:1 line for nitrogen and phosphorus only in a few cases.

The river systems with the highest deviation from the 1:1 line are the Elde and the Uecker located in north eastern Germany. Both basins contain big lakes in the catchment area. These lakes are located in the upper part of the basin. Therefore, these lakes, or rather their sediment surface area, can only contribute

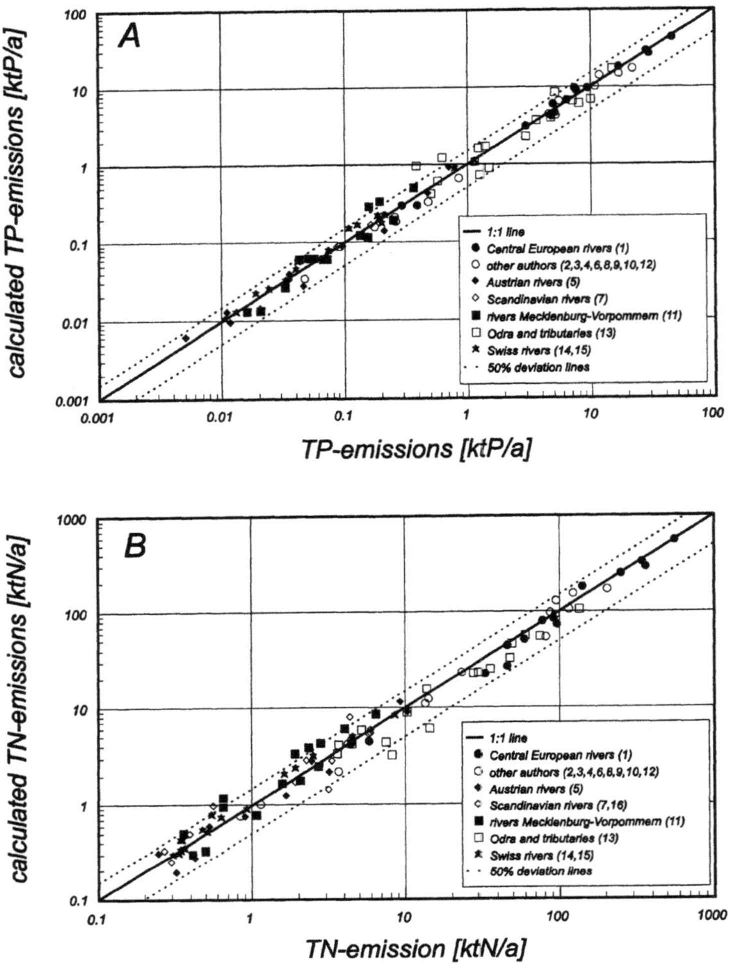

Figure 4. Relationship between the emissions of nutrients given in Table 1 and the calculated emissions according to the retention model with the highest regression coefficient for the investigated river basins (A) Phosphorus; (B) Nitrogen.

to the denitrification of nitrogen emitted from a small part of the basin. Because the model does not take into account the location of the water surface within the basin, the non homogeneous distribution of the water surface can be a source of error. For such cases, the whole basin should be divided into subbasins with more or less similar conditions. Whereas for nitrogen, the mean deviation does not show differences between certain groups of rivers, such differences were found for phosphorus.

In the case of the Odra and its tributaries, the mean deviation between calculated and estimated emissions is about twice the deviation for all other basins. That is a clear indication that the simpler model for the estimation of the diffuse phosphorus sources in the Odra basin (see Behrendt et al., 1998) possess a lower precision than the other models, but a systematic underestimation seems not to exist. The mean deviation for all river basins is about 23% and 22% for nitrogen and phosphorus, respectively. A normal log/log regression between the emissions and the transports would lead to a mean deviation of 38% for nitrogen and 48% for phosphorus. The introduction of the statistical type of model leads to a significant reduction of the variance in the relation between the transport and the nutrient emission of different river systems.

Conclusions

The derived statistical model allows the description of the retention behavior in river basins for phosphorus and nitrogen. A prediction of the total nutrient emissions in a river basin seems to be possible if the nutrient load and the main characteristic parameters of the basin such as specific runoff, area of the basin occupied by surface water and basin size are known for a certain time period at least of one year or more. However, the statistical model alone does not give an explanation of the riverine processes involved, like sedimentation and denitrification.

Indeed, the coefficients of the model are only derived from the data set employed and errors associated with the estimation of nutrient transport and especially the nutrient emission from point and diffuse sources have to taken into account. The number of rivers considered and the independence of the results for the different authors reduce the probability of systematic errors. But the possibility that the result is partly caused by a systematic over-estimation of the emissions and/or under-estimation of the transport cannot be fully excluded. Nevertheless, the statistical model represents to a certain extent the present state of our ability to estimate nutrient emissions especially from the diffuse sources of river basins larger than 100 km^2. Furthermore, the data set used for the development of the model represents different river basins and over a longer time period. Therefore, the models cannot be used for the description of interannual fluctuations within one river system. Furthermore, the investigation of river systems located outside of the Central European region is necessary, because the retention processes can depend on other parameters including interannual distribution of the discharge and drainage density.

Beside the limited data base, the estimated high correlations of the different retention models give the possibility to use these models for a control of emission calculations for river basins at least in the Central European region. Furthermore, the models seem to be applicable as an approximation of the retention in river systems, especially where detailed investigations on the retention and/or the total inputs and outputs of nutrients have not been done up to now.

References

Arnheimer, B. & M. Brandt, 1998. Modelling Nitrogen transport and retention in the catchments of Southern Sweden. Ambio 27: 471–480.

Behrendt, H., 1996. Inventories of point and diffuse sources and estimated nutrient loads – A comparison for different river basins in Central Europe. Wat. Sci. Technol. 33: 99–107.

Behrendt, H., 1999. Application of different methods of source apportionment of nutrients to river basins in Northern Germany. Wat. Sci. Technol. (in print).

Behrendt, H. & A. Bachor, 1998. Point and diffuse load of nutrients to the Baltic Sea by river basins of North East Germany (Mecklenburg-Vorpommern). Wat. Sci. Technol. 38, 10, 147–155.

Behrendt, H., R. Korol, M. Stronska-Kedzia & W.-G. Pagenkopf, 1998. Nutrient emissions of point and diffuse sources, transport and retention within the Odra Basin and its main tributaries. Proceedings, International Conference 'Management of transboundary waters in Europe', 22–25 September 1997, Mrzezyno, Poland: 485–497.

Billen, G. J., 1993. The PHISON river system: a conceptual model of C, N and P transformations in the aquatic continuum from land to sea. In R. Wollast, F. T. Mackenzie & L. Chou (eds), Interactions of C, N,P and Si Biogeochemical Cycles and Global Change. NATO ASI Series 14: 141–161.

Billen, G., J. Garnier, C. Billen & E. Hannon, 1995. Global change in nutirent transfer from land to sea: biogeochemical processes in river systems. GMMA, Free Univ. of Brussels: 99 pp.

Blazkova, S., I. Nesmerak, M. Michalova & M. Kalinova, 1998. Das nationale Projekt Elbe II in den Jahren 1996 – 1997. In W. Geller et al. (eds), Gewässerschutz im Einzugsgebiet der Elbe. Teubner Stuttgart - Leipzig: 25–28.

Braun, P., 1994. Modellhafte Erarbeitung ökologisch begründeter Sanierungskonzepte kleiner Fließgewässer- Fallbeispiel Vils/Oberpfalz. Teilbericht Chemie. Bayerisches Landesamt für Wasserwirtschaft, Materialien, Nr. 35: 239 pp.

Dillon, P. J., 1975. A critical review of Vollenweider's nutrient budget model and other related models. Wat. Resour. Bull. 10: 969–989.

Hilbricht-Illkowska, A., 1988. Transport and Transformation of phosphorus compounds in watersheds of baltic lakes. In H. Tiessen (ed.), Phosphorus Cycles in Terrestrial and Aquatic Ecosystems. Regional workshop 1: Europe, Proceedings of SCOPE workshop, University Saskatchewan, Saskatoon, Canada: 193–206.

Höhener, P. & R. Gächter, 1993. Prediction of dissolved inorganic nitrogen (DIN) concentrations in deep, seasonally stratified lakes based on the rates of DIN input and N-removal processes. Aquat. Sci. 55: 112–131.

Howarth, R. W., G. Billen, D. Swaney, A. Townsend, N. Jaworski, K. Lajtha, J. A. Downing, R. Elmgren, N. Caraco, T. Jordan, F. Berendse, J. Freney, V. Kudeyarov, P. Murdoch & Zhu Zhao-Liang, 1996. Regional nitrogen budgets and riverine N & P fluxes for the drainages to the North Atlantic Ocean: Natural and human influences. Biogeochemistry 35: 75–139.

ICWS, 1989. Scheldt river basin study. International Centre of Water Studies, ICWS-report 89.09, Amsterdam: 43 pp.

Isermann, K., 1997. Die Stickstoff- und Phosphorbilanz im Donaueinzugsgebiet von Deutschland: Ausgangssituation, Lösungsansätze und Lösungsaussichten zu ihrer nachhaltigen Gestaltung. BNLA, Bericht: 163 pp.

Kelly, C. A., J. W. M. Rudd, R. H. Hesslein, D. W. Schindler, P. J. Dillon, C. T. Driscoll, S. A. Gherini & R. E. Hecky, 1987. Prediction of biological acid neutralization in acid sensitive lakes. Biogeochemistry 3: 129–141.

Kroiß, H., M. Zeßner, K. Deutsch, W. Schaar & N. Kreuzinger,

1997. Nährstoffbilanzen der Donauanrainerstaaten - Erhebungen für Österreich. TU-Wien, Institut für Wassergüte und Abfallwirtschaft. 154 pp.

Prasuhn, V., E. Spiess & M. Braun, 1996. Methoden zur Abschätzung der Phosphor- und Stickstoffeinträge aus diffusen Quellen in den Bodensee. Ber. Int. Gewässerschutzkommission Bodensee, 45: 113 pp.

Prasuhn, V. & M. Braun, 1994. Abschätzung der Phosphor- und Stickstoffverluste aus diffusen Quellen in die Gewässer des Kantons Bern: FAC Schriftenreihe, Nr. 17, Eidgenössische Forschungsanstalt für Agrikulturchemie und Umwelthygiene, Liebfeld-Bern: 113 pp.

Probst, J. L., 1985. Nitrogen and phosphorus exportation in the Garonne basin (France). J. Hydrology 76: 281–305.

Raderschall, R., 1996. Abschätzung der diffusen Stoffeinträge in die Hunte über Ergebnisse aus Modell-Einzugsgebieten. Wasserwirtschaft 86: 14–18.

Ruhrverband, 1998. Emissions- und Immissionsbetrachtungen bezüglich der eutrophierenden Nährstoffe Stickstoff und Phosphor in der Ruhr. In: Ruhrwassergüte 1997. Ruhrverband, Essen: 66–75.

Svendsen, L. M. & B. Kronvang, 1993. Retention of nitrogen and phosphorus in a Danish lowland river system: implications for the export from the watershed. Hydrobiologia 251: 123–135.

Svendsen, L. M., H. Behrendt, H. Herata, S. Knuuttila & S. Salo, 1995. Guideline to estimate natural and anthropogenic contributions to riverine fluxes (source apportionment). TC-POLO 3/7 1995. Ad Hoc Expert Group on pollution Load to the Baltic Sea. Report of Third Meeting, Riga, Latvia, 5–8 June 1995, Helsinki Commission, Baltic Marine Environment Protection Commission, Annex 5.

Tonderski, A., 1997. Control of Nutrient fluxes in large river basins. Linköping Studies in Arts and Sciences, 157: 50 pp.

Vollenweider, R. A., 1968. The scientific basis of lake and stream eutrophication, with particular reference to phosphorus and nitrogen as eutrophication factors. Tech. Rep., OECD, Paris, DAS/CSI/68, 27: 182 pp.

Vollenweider, R. A. & J. Kerekes, 1982. Eutrophication of waters - Monitoring, assessment and control. Synthesis Report, OECD Paris: 154 pp.

Welker, M. & N. Walz, 1998. Can mussels control the plankton in rivers? – a planktological study applying a Lagrangian sampling strategy. Limnol. Oceanogr. 43: 753–762.

Werner, W., H.-W. Olfs, K. Auerswald & K. Isermann, 1991. Stickstoff- und Phosphoreintrag in Ober- flächengewässer über 'diffuse Quellen'. In A. Hamm (ed.), Studie Über Wirkungen und Qualitätsziele von Nährstoffen in Fließgewässern. Academia Verlag, Sankt Augustin: 665–764.

Hydrobiologia **410:** 123–130, 2000.
J. Garnier & J.-M. Mouchel (eds), Man and River Systems.

Modelling nutrient fluxes from source to river load: a macroscopic analysis applied to the Rhine and Elbe basins

Marcel de Wit
Utrecht University, Department of Physical Geography, P.O. Box 80115, 3508 TC Utrecht, The Netherlands
Tel: [+31]-30-2534014; E-mail: m.dewit@geog.uu.nl

Key words: Nitrogen, Phosphorus, rivers, modelling

Abstract

In many European rivers, including the major streams of the Rhine and Elbe basins, the nutrient load (N and P) still exceeds target levels. In this paper, a model is presented that describes the river nutrient load as a function of nutrient sources, runoff and lithology in the upstream basin. The model was tested with independent data on nutrient sources (input) and nutrient river load (output) for 130 specific combinations of sub-basins and 5 year periods. A wide range of river systems within the Rhine and Elbe basins were included and the analysis covers a period of 25 years from 1970 to 1995. Most of the observed spatial and temporal variation in the average annual river nutrient load was successfully described by the model. It is therefore concluded that the model can be used to predict the effect of changes in nutrient sources (e.g. reduction of livestock numbers, further improvement of waste water treatment plants etc.) on the average annual nutrient loads of the rivers Rhine, Elbe and their main tributaries.

Introduction

In many European rivers, including the major streams of the Rhine and Elbe basins, the nutrient load (N and P) still exceeds target levels. High levels of these nutrients cause eutrophication in fresh waters and coastal zones (Owens, 1993; Oslo and Paris Commission, 1993; Stanners & Bourdeau, 1995; Van Dijk et al., 1995). In order to reduce river nutrient loads, policy makers need an analysis of nutrient fluxes from all sources (e.g. agriculture, atmospheric deposition, households and industry) that help them to take the most effective measures to improve water quality. Numerous studies have analyzed nutrients and water quality, but only a few studies have dealt specifically with large river basins (e.g. Behrendt, 1993; Garnier et al., 1995; Klepper et al., 1995; Behrendt, 1996; Grimvall et al., 1996). The Rhine and Elbe basins cover respectively 180 000 km^2 and 140 000 km^2 (upstream of Hamburg) (Figure 1). The size of the basins limits the spatial and temporal resolution of their analysis. However, strategies to reduce nutrient loads of large rivers will also be made for relatively long time periods and large regions. It was therefore decided to analyze the nutrient fluxes with a macroscopic view. As such, the scale of the analysis matches with the scale of interest (river basin policies). The aim is to analyze whether the measured 5 year average annual nutrient loads of the rivers Rhine, Elbe and their main tributaries can be modelled as a function of macroscopic basin characteristics and nutrient sources in the upstream basin. The analysis is based on a large supra-national database of nutrient sources, lithology, discharge and water quality. A wide range of river systems from all over the Rhine and Elbe basins was included and the analysis covers the 25 year period from 1970 to 1995.

Materials and methods

From source to river load

The basic formulation to describe the transport of nutrients from their sources to river load is given in Equation 1. All inputs can be divided into those originating from direct and indirect sources. Direct emissions of nutrients (mainly point sources) are not in contact with the soil/groundwater system, whereas indirect nutrient emissions (diffuse sources) reach the

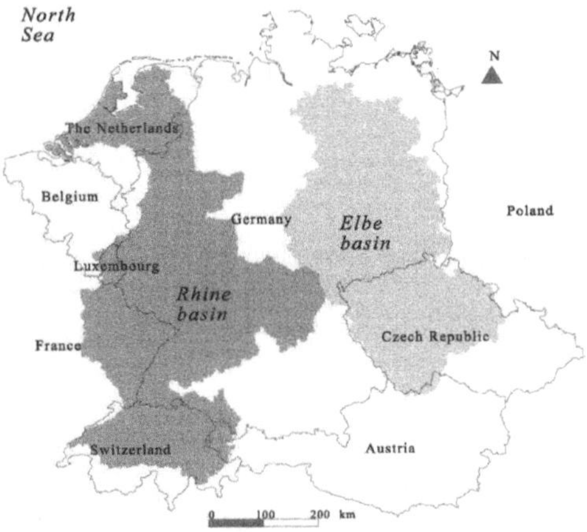

Figure 1. Location of the Rhine and Elbe basins.

surface water via the soil/groundwater system.

$$Lx = a * (DEx + IEx), \quad (1)$$

where Lx is the 5-year avg. river load at location x (kg yr^{-1}), DEx is the 5-year avg. direct inputs to the surface water upstream of x (kg yr^{-1}), IEx is the 5-year avg. indirect (via soil/groundwater) inputs to the surface water upstream of x (kg yr^{-1}) and a = parameter to account for losses in the surface water during transport (-).

Behrendt (1996) analyzed the ratio between inputs to the surface water in the upstream basin and river load (parameter a, Equation (1). He found that this ratio is strongly related to the area specific runoff in the upstream basin. This is expressed in Equation (2).

$$a = 1/(1 + \alpha^* q^{\beta}), \quad (2)$$

where q is the 5-year avg. area specific runoff (l s^{-1} km^{-2}) and α,β are the parameters

Indirect emissions *(IE)* are determined by the nutrient surplus at the land surface (the amount that can potentially runoff, leach, erode) and the conditions that influence the loss (e.g. denitrification) and retention (e.g. residence time groundwater) from the land surface to the surface water. De Wit & Behrendt (1999) have shown that the 5 year average indirect nutrient input to the surface water is more strongly related to the long term nutrient surplus at the land surface than to the nutrient surplus at the land surface averaged over the last 5 years only. Also, P is strongly bounded to the soil, whereas N is more mobile. For these reasons, it was decided to use the long term average surplus at the land surface (kg yr^{-1}) for N and the long term accumulated surplus at the land surface (kg) for P. This is expressed in Equation (3).

$$IEx = b^* \; SLSx, \quad (3)$$

where b is the parameter to account for loss/retention in soil/groundwater system N (-); P (yr^{-1}) and SLSx = N: average (last 30 years) annual surplus at the land surface (kg yr^{-1}) P: total accumulated P surplus at the land surface since 1945 (kg).

De Wit & Behrendt (1999) found that the ratio between the average indirect nutrient inputs to the surface water (IE) and the long term nutrient surplus at the land surface (SLS) appeared to be larger for regions with consolidated rocks than for regions with unconsolidated rocks. Therefore, different values for parameter b were used for consolidated rocks and unconsolidated rocks.

Equation (4) is the product of Equations (1), (2) and (3).

$$Lx = (1/(1 + \alpha^* q^{\beta}))^*(DEx + (b^* SLSx)) \quad (4)$$

Parameter α en β were derived from Behrendt (1996). Parameter b was fitted for the Rhine basin and applied to the Elbe basin. The parameter values are given in Table 1. These values were used for all sub-basins and 5 year periods.

Table 1. Parameter values

Parameter	N	P
α	12.58	45.9
β	−1.5	−2.03
b: unconsolidated rocks	0.20	0.0002
b: consolidated rocks	0.45	0.00055

A supra-national database

A supra-national database was specially created in order to compute Equation (4) for a large number of combinations of sub-basins and 5 year periods. This database contains maps with information on nutrient sources (DE and SLS), lithology (Figure 2) and area specific runoff (q). Altogether there are 287 administrative units in 9 different countries with data on population numbers, connection rates to waste water treatment plants and industrial emissions (DE), and data on agricultural land use, crop yields, livestock numbers and atmospheric deposition (SLS) (see Figure 3). The mapping of the nutrient sources (DE and

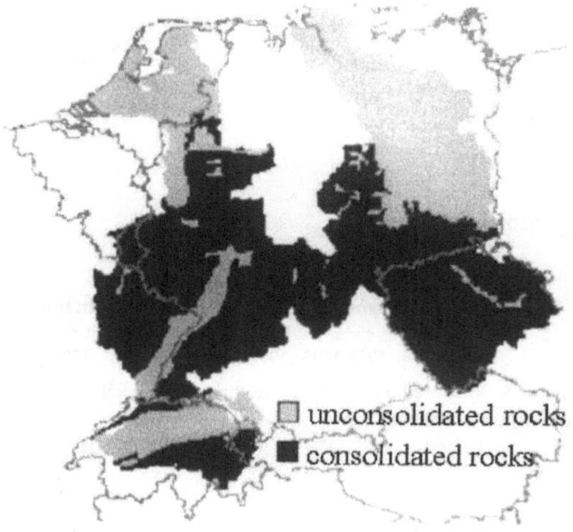

Figure 2. Major lithological subdivision Source: IAH.

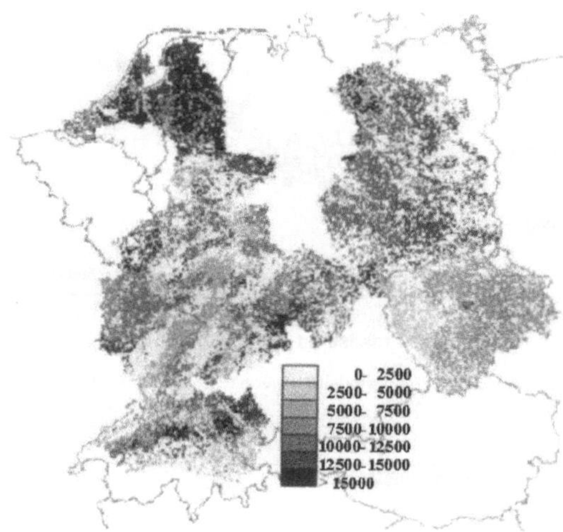

Figure 4. SLS average N: 1965-1995 (kg yr^{-1} km^{-2}) Source: De Wit (1999).

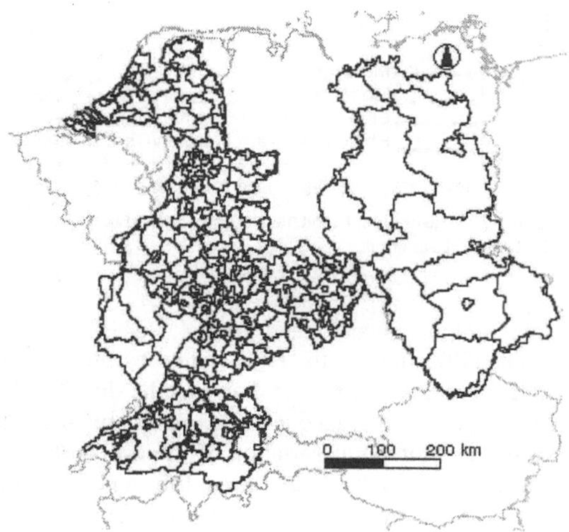

Figure 3. Administrative units.

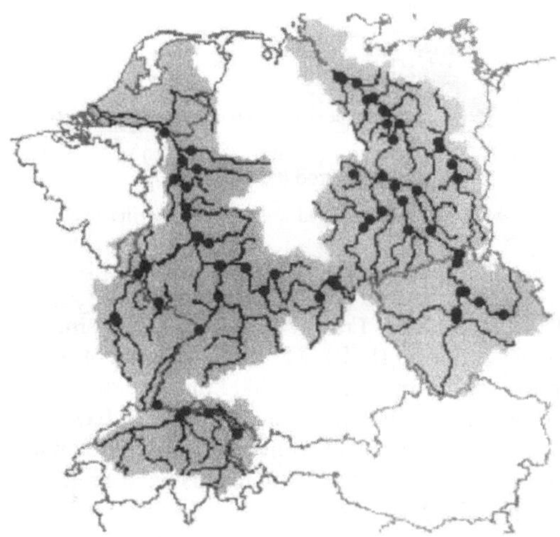

Figure 5. Monitoring stations (q and L).

SLS) in the Rhine and Elbe basins is described in detail in de Wit (1999). DE and SLS were calculated for each 5 year period (from 1970–1975 to 1990–1995) and administrative unit and were allocated to the appropriate land cover classes using a land cover map of the Rhine and Elbe basins (at a resolution of 1 km^2). Figure 4 presents an example of these maps. DE and SLS were based on information on nutrient sources only and no water quality data were used. The calculations of DE and SLS were in good agreement with emissions calculations that were reported in other regional studies (Hamm et al., 1991; IKSR, 1992; Werner & Wodsak, 1994; Brouwer et al., 1995; Frede & Bach, 1995; IKSE, 1995).

The 5 year average annual runoff (q) and river nutrient load (L) were calculated with discharge and water quality data measured at 60 different monitoring stations all over the Rhine and Elbe basins (Figure 5). These data were collected by the Dutch National Institute of Public Health and the Environment (RIVM)

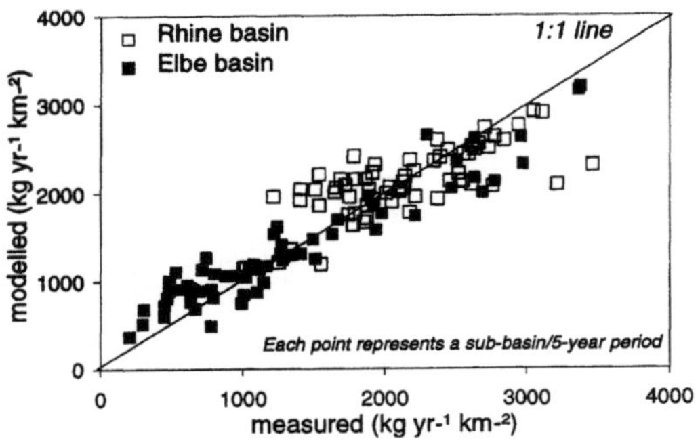

Figure 6. Measured and modelled average area specific river load (N).

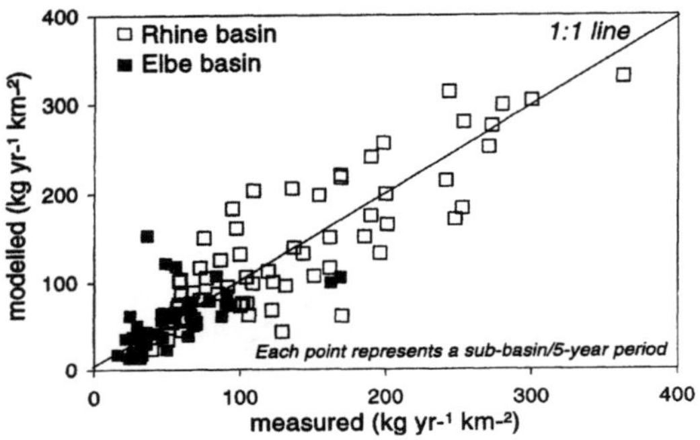

Figure 7. Measured and modelled average area specific river load (P).

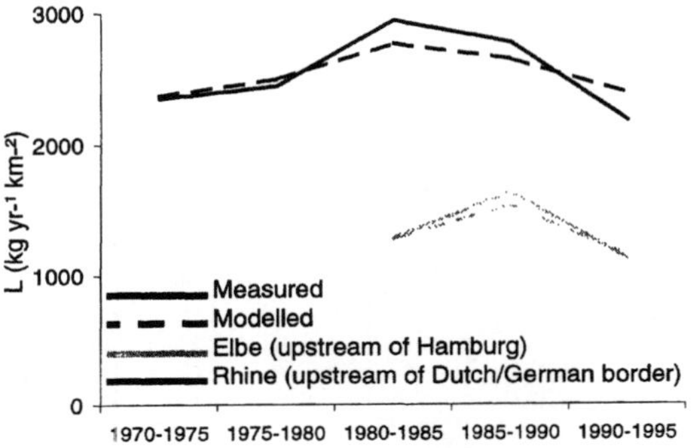

Figure 8. Measured and modelled average area specific river load (N) Temporal variation.

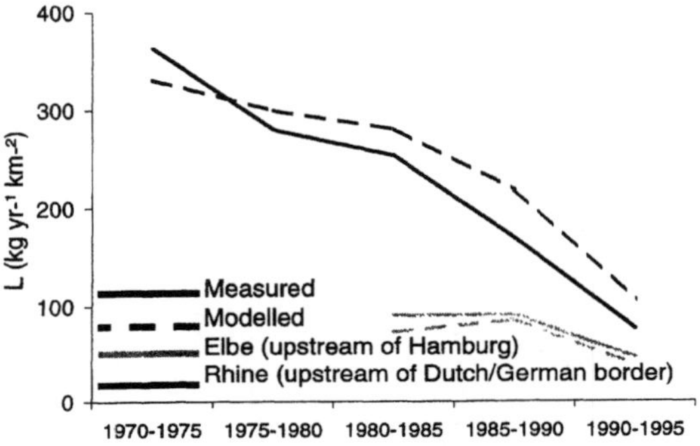

Figure 9. Measured and modelled average area specific river load (P) Temporal variation.

and the Institute of Freshwater Ecology and Inland Fisheries in Berlin (IGB). The upstream area of these monitoring stations exceeds 1000 km^2. For some stations only dissolved inorganic nitrogen (DIN) data were available. From measured ratios of Ntot with DIN it is observed that the Ntot load is on average 1.3 times the DIN load. This ratio was used to calculate the Ntot load with DIN data. Some stations had data for all 5 year periods from 1970–1975 to 1990–1995, but other stations had only data for 1990–1995.

Results

Direct inputs (DE), long term surplus at the land surface (SLS), area specific runoff (q) and load calculations (L) were available for 130 sub-basin/5-year combinations (see Tables 2 & 3). DE, SLS, q and L were calculated independently from each other and could therefore be used to test the model (Equation 4). The model results are compared with measured data in Figures 6 and 7 and Table 4. Because the sub-basins differ in size, the results are presented as area specific values. The results show that the simple model formulation used, adequately explains much of the spatial and temporal variation of the average annual nutrient loads in the rivers Rhine, Elbe and their main tributaries. Figures 8 and 9 show that the model also successfully described temporal changes in the river nutrient load.

Discussion

The nutrient load at a given location in the river is a complex function of numerous physical, chemical and biological processes that take place in the upstream basin. However, it seems possible to explain much of the large scale spatial and temporal variation in the nutrient river load without having a detailed description of these processes. Processes that determine the loss of nutrients in the surface water during

Table 2. DE, SLS, q and L. Rhine basin

River	Monitoring Station	Size basin km2	Unconsolidated Rocks % of basin	Period	q $l\,s^{-1}\,km^{-2}$	L measured N $kg\,yr^{-1}\,km^{-2}$	L measured P $kg\,yr^{-1}\,km^{-2}$	DE N $kg\,yr^{-1}\,km^{-2}$	DE P $kg\,yr^{-1}\,km^{-2}$	SLS N $kg\,yr^{-1}\,km^{-2a}$	SLS P $kg\,km^{-2b}$
Aare	Brugg	11314	69	1990/1995	27.3	2174	73	700	70	4473	50 280
Erft	Eppinghoven	1588	75	1990/1995	6.9	1003	58	1123	133	3144	42 704
Lahn	Limburg	4900	0	1990/1995	8.7	1539	57	703	88	4611	48 501
Lippe	Schermbeck	4811	37	1990/1995	7.9	2698	581	3731	318	1679	0627
Main	Bischofsheim	27070	5	1980/1985	8.7	2608	252	1057	263	4732	46 349
				1985/1990	9.0	2544	185	949	205	4971	50 819
				1990/1995	6.5	1850	68	736	90	5022	53 093
Main	Kahl	22520	3	1980/1985	10.9	2202	247	822	206	5019	49 084
				1985/1990	7.9	1749	119	745	162	5277	53 847
				1990/1995	10.6	1686	55	591	77	5339	56 325
Fr. Saale	Gemunden	2270	0	1990/1995	7.0	1548	50	291	43	3799	42 470
Regnitz	Pettstadt	7039	2	1990/1995	6.4	1787	61	699	88	5532	55 793
Main	Schermbeck	2373	0	1990/1995	10.9	2049	69	320	44	4990	52 396
Tauber	Waldhausen	1775	0	1985/1990	6.8	2523	85	575	129	6858	74 657
				1990/1995	3.4	1249	39	483	70	6982	78 467
Rodach	Marktzeuln	1019	0	1990/1995	12.3	1773	60	229	35	4196	44 422
Roter Main	Rothwind	1275	0	1980/1985	9.7	2100	162	557	142	5113	50 019
				1985/1990	11.5	2516	151	507	112	5412	55 071
				1990/1995	10.9	2455	106	408	53	5528	57 901
Mosel	Koblenz	28505	1	1970/1975	8.2	1869	143	1015	213	3629	31 941
				1975/1980	10.2	2130	161	1018	192	4081	38 309
				1980/1985	15.5	2724	201	1020	170	4500	43 945
				1985/1990	13.0	2385	196	922	140	4776	48 523
				1990/1995	10.7	2135	102	726	81	4846	50 985
Mosel	Millery	7100	0	1990/1995	16.7	1860	122	574	54	4420	44 411
Mosel	Palzem	11258	3	1990/1995	13.7	1779	75	844	77	4929	51 496
Saar	Kanzem	7425	0	1990/1995	9.8	1505	108	940	117	4307	45 599
Mosel	Liverdun	3983	0	1990/1995	15.9	1533	49	480	47	4840	50 196
Sure	Wasserbillig	4578	0	1990/1995	11.9	2589	100	452	61	6095	63 053
Saar	Saarbrucken	3733	0	1990/1995	9.6	1532	105	635	83	4478	47 590
Neckar	Mannheim	14233	1	1980/1985	12.6	3045	243	1521	372	4966	49 087
				1985/1990	12.0	3107	198	1449	235	5224	54 106
				1990/1995	9.8	2364	73	1307	140	5293	57 006
Rhine	Bad Honnef	140888	22	1990/1995	13.5	1793	91	957	97	4445	47 374
Rhine	Dipoldsau	6482	6	1990/1995	34.5	1019	129	321	36	2089	17 587
Rhine	Koblenz	109952	28	1970/1975	13.9		300	1429	358	3354	29 002
				1975/1980	15.0	1927	273	1407	312	3746	34 721
				1980/1985	18.3	2608	270	1384	266	4085	39 715
				1985/1990	16.7	2548	200	1261	167	4299	43 888
				1990/1995	15.1	2141	122	1018	100	4338	46 420
Rhine	Lobith	160084	23	1970/1975	12.7	2344	363	1670	410	3448	30 068
				1975/1980	13.1	2440	280	1639	356	3857	36 018
				1980/1985	16.8	2938	253	1607	303	4217	41 236
				1985/1990	15.0	2768	169	1465	191	4449	45 582
				1990/1995	13.9	2174	77	1184	111	4505	48 183
Rhine	Mainz	98105	31	1985/1990	17.6	2627	135	1304	168	4344	44 290
				1990/1995	15.7	1911	59	1054	101	4389	46 944
Rhine	Oehningen	11188	36	1990/1995	32.3	1213		612	63	4139	41 526
Rhine	Seltz	49408	50	1975/1980	25.4	2213		1044	238	3414	30 691
				1980/1985	27.8	3211	241	1057	213	3748	35 485
				1985/1990	26.1	2753	190	989	134	3963	39 714
				1990/1995	23.2	2370	131	853	88	4006	42 811
Rhine	Village Neuf	36190	51	1975/1980	32.2	1913	169	910	218	3568	31 906
				1980/1985	31.9	1886	154	921	193	3939	37 224
				1985/1990	29.1	2106	97	876	116	4192	42 090
				1990/1995	28.3	2009	68	789	77	4279	46 053
Ruhr	Hattingen	4531	7	1990/1995	16.0	3460	87	1323	129	3345	34 427
Sieg	Menden	2783	1	1985/1990	15.7	2833	190	1259	209	4164	41 887
				1990/1995	17.5	2652	100	1128	126	4193	43 958
Mosel	Bouxiere	3065	0	1980/1985	15.0	1720		940	118	3569	31 916
				1985/1990	14.5	2032		831	97	3812	35 327
				1990/1995	13.0	1874	170	614	57	3894	37 193
Rhine	Rekingen	15040	48	1975/1980	32.2	1401	109	846	200	3668	32 361
				1980/1985	32.3	1403	95	856	176	4030	37 565
				1985/1990	29.5	1655	75	814	106	4274	42 237
				1990/1995	29.0	1646	60	734	73	4358	45 845

[a] Average over last 30 years.
[b] Accumulated since 1945.

Table 3. DE, SLS, q and L. Elbe basin

River	Monitoring Station	Size basin km2	Unconsolidated Rocks % of basin	Period	q $l\,s^{-1}\,km^{-2}$	L measured N $kg\,yr^{-1}\,km^{-2}$	L measured P $kg\,yr^{-1}\,km^{-2}$	DE N $kg\,yr^{-1}\,km^{-2}$	DE P $kg\,yr^{-1}\,km^{-2}$	SLS N $kg\,yr^{-1}\,km^{-2a}$	SLS P $kg\,km^{-2b}$
Elbe	Zollenspieker	135024	41	1980/1985	4.1	1039		1101	204	5115	48 913
				1985/1990	5.4	1488		1037	180	5545	56 054
				1990/1995	4.3	956		769	92	5158	56 693
Ilmenau	Rote Schleuse	1545	100	1980/1985	5.0	1277	79	892	198	8712	85 492
				1985/1990	4.6	1074	53	837	171	9480	97 915
				1990/1995	4.5	801	36	611	75	9533	10 3457
Sude	Bandekow	2253	100	1980/1985	7.1	881		484	90	6414	57 497
				1985/1990	7.3	714		468	82	6991	64 854
				1990/1995	6.6	651	32	327	41	6272	65 097
Elde	Domitz	2990	100	1980/1985	5.2	571		763	124	5644	48 609
				1985/1990	3.8	449	32	739	112	6148	54 823
				1990/1995	3.2	297	24	542	61	5528	55 005
Jeetzel	Luggau	1660	100	1980/1985	4.3	483	27	539	112	7951	76 325
				1985/1990	4.4	613	22	522	101	8647	86 832
				1990/1995	3.8	632	16	400	52	8290	89 833
Elbe	Schnacklenburg	125482	35	1980/1985	4.5	1284	90	1150	212	4972	47 723
				1985/1990	5.5	1629	91	1082	187	5387	54 750
				1990/1995	4.5	1116	47	804	97	4995	55 238
Aland	Wanzer	1820	100	1985/1990	5.8	1023		639	124	6749	61 202
				1990/1995	4.5	989	32	513	70	6082	61 334
Biese	Osterburg	924	100	1985/1990	4.0	663		558	109	6119	53 932
				1990/1995	3.2	774	27	447	61	5536	54 092
Havel	Toppel	24297	96	1980/1985	5.3	530	84	1188	266	5050	40 620
				1985/1990	4.7	487	66	1102	224	5485	46 014
				1990/1995	3.8	305	33	749	83	5080	46 268
Elbe	Tangermunde	97780	19	1980/1985	4.4	1262		1161	203	4888	48 937
				1985/1990	5.8	1665		1096	181	5291	56 319
				1990/1995	4.4	1163		831	101	4913	56 874
Dahme	Neu Muhle	1362	100	1980/1985	9.9	1007	87	452	85	3684	25 773
				1985/1990	10.0	1098	68	426	75	4000	29 250
				1990/1995	6.6	446	30	294	37	3764	29 534
Bode	Wegeleben	1215	6	1980/1985	7.6	1236		680	148	4408	33 447
				1985/1990	6.4	1214		664	133	4776	37 707
				1990/1995	5.6	741	25	556	76	4392	37 960
Saale	GroB Rosenburg	23718	15	1980/1985	5.2	2683	162	1597	253	6095	52 384
				1985/1990	5.7	2627	169	1464	225	6596	59 124
				1990/1995	3.9	1344	64	971	126	6019	59 540
Spree	Leibsch	4529	83	1975/1980	5.1		47	701	146	4697	36 268
				1980/1985	4.7	693	46	720	148	5223	41 797
				1985/1990	4.8	780		615	131	5657	47 326
				1990/1995	3.9	469		843	67	5228	47 558
Mulde	Dessau	7400	22	1985/1990	7.1	2291		1813	357	6592	57 165
				1990/1995	8.6	2502	36	1169	194	6028	57 376
Schw. Elster	Gorsdorf	5453	81	1985/1990	4.4	630		678	129	6065	51 305
				1990/1995	2.1	202		519	66	5586	51 523
Ver. Mulde	Bad Duben	6171	7	1985/1990	11.5	3362		1408	282	6404	54 847
				1990/1995	9.6	2954	56	1197	150	5872	55 074
W. Elster	Ammendorf	5384	28	1985/1990	6.4	2971		1608	282	6659	58 506
				1990/1995	4.7	1934	66	1239	155	6062	58 714
Unstrut	Freiburg	6327	9	1985/1990	5.6	1882		939	186	6845	61 851
				1990/1995	4.3	1270	44	787	106	6173	62 029
Saale	Camburg	3977	0	1975/1980	8.3	1975		778	183	4252	32 950
				1980/1985	9.4	2467		827	183	4725	38 044
				1985/1990	9.2	2768		793	160	5115	43 137
				1990/1995	7.4	2211	56	636	82	4877	44 213
Fr. Mulde	Erlin	2983	0	1985/1990	12.2	3373		1236	247	6457	55 819
				1990/1995	9.7	2626	49	1042	130	5918	56 043
Elbe	Schmilka	51391	2	1990/1995	5.6	1287	68	743	79	3936	55 910
Elbe	Decin	49770	3	1990/1995	4.1	1146	32	750	79	3960	56 367
Moldau	Zelcin	28090	0	1990/1995	3.7	791	50	548	70	3823	54 601
Elbe	Obristvi	13730	6	1990/1995	4.5	1504	30	977	84	4420	72 988
Elbe	Lysa	10550	6	1990/1995	4.7	1405	36	1016	89	4375	63 908
Berounka	Mouth	8861	0	1990/1995	3.5		33	386	51	3771	53 998
Elbe	Valy	6299	5	1990/1995	6.7	1916	64	1393	107	4130	59 168
Sazava	Mouth	4349	0	1990/1995	3.5		26	371	49	4396	64 632

[a] Average over last 30 years.
[b] Accumulated since 1945.

Table 4. Average absolute error and average relative error of model results

	N Rhine basin	N Elbe basin	P Rhine basin	P Elbe basin
Avg. Relative error (%)	13	20	31	37
Avg. Absolute error (kg yr^{-1} km^{-2})	271	248	34	20

transport (e.g. denitrification, assimilation by algae and aquatic macrophytes, adsorption onto stream bed sediments and sedimentation) were described by one area specific characteristic only; runoff. Also the processes that determine the loss/retention of nutrients in the soil/groundwater system (e.g. denitrification, adsorption in the soil, erosion, leaching and groundwater flow (residence times)) were described by one area specific characteristic only; consolidated *versus* unconsolidated rocks.

The fact that lithology and area specific runoff perform so well as 'macroscopic surrogate variables' can be explained. As in many river basins, the regions with consolidated rocks are also the regions with the largest relief energy. Erosion rates and the (fast) surface runoff component are larger in these regions compared to the relatively flat regions with unconsolidated rocks. Moreover, there is in general less denitrification in the groundwater in regions with consolidated rocks, compared to regions with unconsolidated rocks (e.g. North German plain compared to regions with consolidated Mesozoic sedimentary rocks, see also Wendland, 1992). There is no doubt that the fate of nutrients in the river network is related to runoff. No runoff means that there is no transport of nutrients in the river network. Regions with a large area specific runoff have relatively large area specific flow velocities in their stream network. Processes that determine the loss of nutrients in the surface water during transport are related to flow velocities (e.g. sedimentation) or simply need time.

The loss of information and detail that is associated with aggregation makes that the results of the method presented in this study have only a limited spatial and temporal resolution. Nothing can be said about seasonal variation nor about nutrient loads in rivers that have an upstream area of less than 1000 km^2. On the other hand, it is questionable whether these issues are important for developing strategies to reduce the average nutrient load in large river basins and for long time periods. This brings us to the next question; can the model be used to predict the effect of changes in the basin (e.g. emission reductions) on the average nutrient river load? From a process point of view, one might argue that the model lacks a process-based description and it is, therefore, questionable whether it will correctly react to changing conditions. However, the analysis covers a wide range of rivers and a relatively long time period. Within the Rhine and Elbe basins, there are large regional differences in physical characteristics (e.g. precipitation, lithology) and nutrient sources. During the time period covered, there were also large temporal changes in nutrient sources. Nutrient emissions from public waste water treatment plants were strongly reduced and livestock numbers and fertiliser use dropped dramatically in Eastern Europe (Elbe basin) after the political changes in 1989/1990. These spatial and temporal differences were well reflected by the model. This suggests that the model has good predictive capabilities and can be used to estimate the effect of long term regional changes in nutrient sources on the long term regional river neutrient load.

Acknowledgements

The research described in this paper is supported by the Dutch National Institute of Public Health and Environmental Protection (RIVM). Water quality data have been derived from; -International Commission for Protection of the Rhine. -German Commission for Protection of the Rhine; -International Commission for Protection of the Mosel; -Landesamt für Wasser und Abfall Nordrhein-Westfalen; -Landeshydrologie und geologie, Switzerland; -Bayerisches Landesamt für Wasserwirtschaft; -Bundesanstalt für Gewasserkunde, Koblenz; -International Commission for Protection of the Elbe; -Landesamt für Umwelt Brandenburg, -Landesamt für Umwelt Mecklenburg-Vorpommern; -Landesamt für Umweltschutz Sachsen-Anhalt; -Sächsisches Landesamt für Umwelt und Geologie; -Thüringer Landesanstalt für Umwelt. Hans Veldkamp, Sander van Dijk, Horst Behrendt and Matthias Ley are acknowledged for their support.

References

Behrendt, H., 1993. Point and diffuse loads of selected pollutants in the river Rhine and its main tributaries. Research report-93-1. IIASA, Laxenburg, Austria: 84 pp.

Behrendt, H., 1996. Inventories of point and diffuse sources and estimated nutrient loads – a comparison for different river basins in central Europe. Wat. Sci. Tech. 33: 99–107.

Brouwer, F. M., F. E. Godeschalk, P. J. G. J. Hellegers & H. J. Kelholt, 1995. Mineral balances at farm level in the European Union. Agricultural Economics Research Institute (LEI-DLO). The Hague, The Netherlands. Report no. 137: 141 pp.

De Wit, M. J. M., 1999. Nutrient fluxes in the Rhine and Elbe basins. PhD Thesis. Utrecht University, The Netherlands.

De Wit, M. J. M. & H. Behrendt, 1999. Nitrogen and Phosphorus emissions from soil to surface water in the Rhine and Elbe basins. Wat. Sci. Tech. 39 (12): 109–116.

Frede, H. G. & M. Bach, 1995. Regional differenzierte Abschätzung des Nitrateintrags aus diffusen Quellen in das Grundwasser. Untersuchung für die Bundesrepublik Deutschlands im neuen Gebietsstand. Institut fur Landeskultur. Gießen, Germany: 126 pp.

Garnier, J., G. Billen & E. Hannon, 1995. Biogeochemical Nutrient Cycling in Large River Systems (Binoculairs). 1ste progress report: 29 pp.

Grimvall, A. & P. Stålnacke, 1996. Statistical methods for source apportionment of riverine loads of pollutants. Journal of Environmetrics 7: 201–213

Hamm A., D. Gleisberg, W. Hegemann, K. H. Krauth, G. Metzner, F. Sarfert & P. Schleypen, 1991. Stickstoff- und Phosporeintrag in Oberflächengewässer über 'punktförmigen Quellen'. In A. Hamm (ed.), Studie über Wirkungen und Qualitätsziele von Nährstoffen in Fließgewässern. Academia Verlag, Sankt Augustin: 765–798.

IAH. International Hydrogeological Map of Europe (various sheets). UNESCO, Paris.

IKSE, 1995. Bestandsaufnahme von bedeutenden punktuellen kommunalen und industriellen Einleitungen von prioritären Stoffen im Einzugsgebiet der Elbe. International Commission for the Protection of the Elbe. Magdeburg.

IKSR, 1992. Aktionsprogramm Rhein. Bestandsaufnahme der Einleitungen prioritärer Stoffe 1992. International Commission for the Protection of the Rhine. Koblenz.

Klepper, O., A. H. W. Beusen & C. R. Meinardi, 1995. Modelling the flow of Nitrogen and Phosphorus in Europe: From loads to coastal seas. RIVM report no. 461501004. Bilthoven: 32 pp.

Oslo and Paris Commission, 1993. Nutrients in the convention area 1993. The Chameleon Press Limited. London.

Owens, N. J. P., 1993. Nitrate cycling in marine waters. In T. P. Burt, A. L. Heathwaite & A. L. Trudgill (eds), Nitrate. Wiley. Chichester: 169–209.

Stanners, D. & P. Bourdeau, 1995. Europe's Environment. The Dobris Assessment. European Environment Agency (EEA). Copenhagen, Denmark.

Van Dijk, G. M., E. C. Marteijn & A. Schulte-Wülwer-Leidig, 1995. Ecological rehabilitation of the river Rhine: plans, progress and perspectives. Regul. Riv. 11: 377–388.

Wendland, F., 1992. Die Nitratbelastung in den Grundwasserlandschaften der 'alten' Bundesländer (BRD). Berichte aus der Okologischen Forschung, Band 8. KFA. Jülich: 150 pp.

Werner, W. & H. P. Wodsak, 1994. Stickstoff- und Phosphateintrag in die Fließgewässer Deutschlands unter besonderer Berücksichtigung des Eintragsgeschehens im Lockergesteinbereich der ehemaligen DDR. Agrarspektrum, 22. Frankfurt am Main: 243 pp.

Hydrobiologia **410**: 131–138, 2000.
J. Garnier & J.-M. Mouchel (eds), Man and River Systems.

Integrated modelling of hydrological processes and nutrient dynamics at the river basin scale

Valentina Krysanova & Alfred Becker
Potsdam Institute for Climate Impact Research, P.O. Box 601 203, Telegrafenberg, Potsdam, Germany

Key words: mesoscale river basin, hydrological modelling, water quality modelling, nitrogen balance in soil, nonpoint source pollution

Abstract

The paper presents a generic modelling approach for modelling nutrient transfer from soil to surface water in mesoscale river basins. The approach is tested in application of the spatially-distributed coupled hydrological / water quality model SWIM for modelling hydrological processes and nitrogen dynamics in the upper Stepenitz basin (subbasin of the Elbe, 575 km^2). Both hydrological and hydrochemical validations of the model were successful for this basin. The differences in nitrogen cycling for different soils revealed by the simulation provide a basis for further scenario evaluation and eventually for recommendations for nonpoint source pollution control at the river basin scale.

Introduction

River basins can be considered as integrators of the effects of many forces, including human use of water resources, land use and climate. Their hierarchical structure and natural boundaries represent an appropriate structure for environmental impact analysis and hydrological / water quality modelling. The functioning of a river system depends significantly on constraints imposed on hydrological, geo- and hydrochemical processes by the climate, geomorphology, soil type and anthropogenic activities (land and water use, input of chemical compounds from point and nonpoint sources).

The complexity of a specific river basin model depends on the temporal and spatial resolution and on the extent to which important hydrological and biochemical processes are considered. Progress in water quality modelling is more evident at the patch and small catchment scale, while it is for mesoscale and large river basins that decision making and water resources management is required. The objective of our work is to provide a transferable coupled hydrological / water quality model for application in mesoscale basins (from several hundreds to several thousands km^2), which can be used for land use change impact studies. This requires that a certain balance is maintained: all main processes are taken into account, but the model is not overparametrized and can be initialized using regionally-available information.

Model description

SWIM (Soil and Water Integrated Model, Krysanova et al., 1996, 1998) is a continuous-time spatially distributed model, integrating hydrology, vegetation, erosion and nutrients (nitrogen, N and phosphorus, P) at the river basin scale. It is based on two other models: SWAT (Arnold et al., 1993), and MATSALU (Krysanova et al., 1989). SWIM has an interface to the GRASS GIS, which was modified from the SWAT/GRASS interface (Srinivasan et al., 1993). The interface is used to initialize the model by extracting spatially distributed parameters of elevation, land use, soil, climate and the routing structure for the whole basin. The following spatial data are necessary: a Digital Elevation Model (DEM), a land use map, a soil map, a map delineating the subbasin boundaries (can be obtained from the DEM using the *r.watershed* program in GRASS or similar routines) and a ground water table map. All the maps have to be provided in ARC/INFO or GRASS format. For larger basins, a map of climate and/or precipitation stations can be useful. In addi-

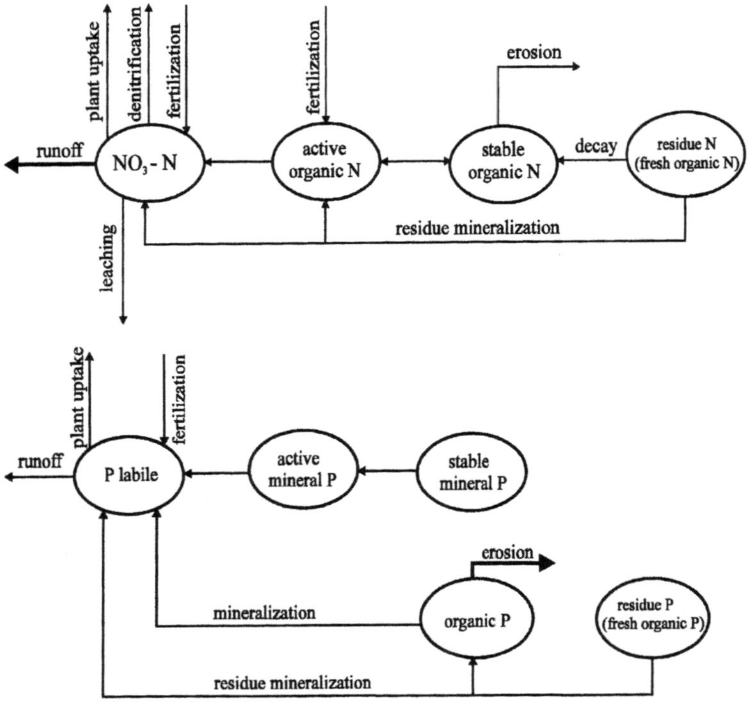

Figure 1. Flow charts for modelling the nitrogen and phosphorus cycles in soil, as implemented in SWIM model.

tion, soil parameters have to be provided for upper soil layers (root zone). Crop management data and daily meteorological time series (average, minimum and maximum temperature, precipitation, and solar radiation) are needed to initialize the model. Hydrological and water quality measurements are necessary for the validation purposes.

The simulated hydrological system consists of four control volumes: the soil surface, the root zone, the shallow aquifer and the deep aquifer. The soil column is subdivided into several layers in accordance with the soil data base. The water balance for the soil column includes precipitation, evapotranspiration, percolation, surface runoff and subsurface runoff. The percolation from the soil profile is assumed to recharge the shallow aquifer. The water balance for the shallow aquifer includes ground water recharge, capillary rise to the soil profile, lateral flow and percolation to the deep aquifer. Return flow from the shallow aquifer contributes to the streamflow. The hydrological flow chart, as well as a more full description of the model is given in Krysanova et al. (1998).

The nitrogen module includes the following nitrogen pools (Figure 1): nitrate nitrogen in soil, active and stable organic nitrogen in soil, organic nitrogen in the plant residue and the flows: mineralization, fertilization, plant uptake, runoff with surface and subsurface flows, leaching into ground water, loss with erosion, and denitrification. The phosphorus module includes the pools: labile phosphorus, active and stable mineral phosphorus, organic phosphorus and phosphorus in the plant residue and the flows: sorption / desorption, mineralization, fertilization, plant uptake, loss with erosion and runoff with lateral flow. Regarding the lateral flows of nutrients, the runoff with surface and subsurface flows are more important for nitrogen, while phosphorus is transported mainly with erosion. Amounts of nitrate-nitrogen contained in runoff, lateral subsurface flow and percolation are estimated as the products of the volume of water and the average concentration.

A three-level scheme of areal disaggregation 'basin – subbasins – hydrotopes' plus vertical subdivision of the root zone into a maximum of 10 layers are used in SWIM. A hydrotope is a set of elementary units in the subbasin, which have a uniform land use and soil type. A hydrotope can be assumed to behave in a hydrologically uniform way within the subbasin.

The main phase variables of the model are: water content in every soil layer of every hydrotope, snow content in every hydrotope, four nitrogen pools and five phosphorus pools in every soil layer of every hydrotope, total plant biomass, vegetation cover and the fraction of root weight in every hydrotope. The time step is daily. The model output is flexible, all phase variables, flows of water and nutrients, as well as some associated variables (like leaf area index) can be output on the daily, monthly, or annual basis for every hydrotope, or as averaged values for subbasins or the whole basin.

The following simulation steps are applied. At first water and nutrient balances are calculated for every hydrotope. After that the outputs from hydrotopes are integrated (area-weighted average) to estimate the subbasin outputs. Finally, the routing procedure is applied to the subbasin lateral flows, taking into account transmission and evaporation losses. Transmission losses, which are more important for semiarid regions, are estimated as a function of the effective hydraulic conductivity of the channel alluvium, the flow duration, and the channel geometry. The sediment routing considers two processes operating simultaneously: deposition in the stream channel and degradation in the routing reaches. Nitrogen retention in riparian zones can be taken into account via land use or as an additional buffering process between the subbasins and the streams. Nitrate is considered as a conservative material for the duration of an individual runoff event in mesoscale river basins and thus is routed simply by adding the inputs from all subbasins. The model is described in detail in Krysanova et al. (1996, 1998) with references to the equations in Arnold et al. (1994) and Krysanova et al. (1989).

A robust approach

There are many different complex and interlinked processes involved in the nutrient cycling in soil: physical, chemical, and micro-biological. Dealing with mesoscale watersheds with the objective of modelling the spatio-temporal dynamics of coupled hydrological and nutrient processes, we inevitably encounter many uncertainties in the input data, especially regarding soil (spatial resolution is not sufficient and / or parameters are available only for the leading profile of every soil type), crop allocation (usually only 'arable land' as a land use category) and crop management (agricultural statistics for districts only). Besides, the water quality modelling stage usually follows the hydrological modelling stage, whereby the level of uncertainty increases. As a result, the coupled hydrological / water quality modelling at the basin scale is recognized to be a very complicated task, especially as regards exact predictions of nutrient dynamics in soil and transport of nutrients. Therefore we suggest application of *a robust approach* for modelling nutrient dynamics in mesoscale river basins in general and for using SWIM particularly.

What do we understand by a robust approach? Firstly, the flow chart of nutrient dynamics implemented in the model should be as simple as possible, including only the main pools and flows (see e.g. Figure 1). This is necessary in order to be able to keep them under control. Secondly, certain requirements should be fulfilled regarding the modelling procedure, such as

(A) The hydrological module should be tested and validated in advance;

(B) The nutrient balance must be well maintained on average, in other words, the annual mineralization, N & P runoff, N leaching, crop uptake of N & P, etc. should fit into ranges specific to the region (based on data from the literature);

(C) The seasonal dynamics of nutrients in the soil must be reasonable (for example: with peaks in spring and autumn, the highest peak in spring);

(D) If conventional agriculture is practicised in a basin under study, there should be no soil degradation regarding the nutrient content in a multi-year model run.

Thirdly, the modelling results should not be interpreted as exact predictions, but in the first place qualitatively, as indicators of possible trends, as qualitative differences.

The parameter values are constrained in the following way: they are set *a priori* to some reasonable values given by process studies in the literature and then some of them are fitted by adjustment.

Model application

This approach was applied for modelling nitrogen dynamics in one of the subbasins of the river Elbe, the Stepenitz basin (gauge Wolfshagen, 575 km^2). This basin belongs to one of three different typical subregions of the German part of the Elbe basin, the

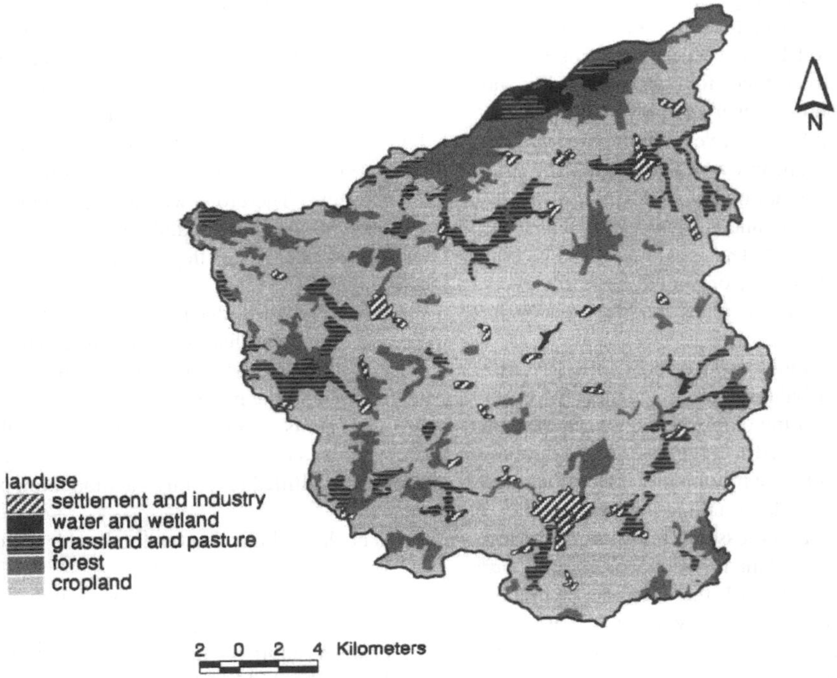

Pleistocene lowland. This watershed is predominantly agricultural, cropland and pastures occupy about 80% of the area (Figure 2). Soils are predominantly sandy with different percentages of loam. A fine disaggregation scheme was applied to yield 64 subbasins (average area 9 km^2) and 658 hydrotopes (average area 0.9 km^2).

A Digital Elevation Model with 200 m horizontal resolution (Ministerium für Umwelt, Naturschutz und Raumordnung des Landes Brandenburg) was applied for the Stepenitz basin. An available land use map with 500 × 500 m horizontal resolution, provided by the Statistisches Bundesamt, Wiesbaden, was reclassified into a new map with the following categories: 1. water, 2. settlement, 3. industry, 4. road, 5. cropland, 6. perennial grass, 7. pasture, 8. fallow, 9. forest, 10. sand dunes, 11. bare soil, 12. wetland. The digital soil map of Germany (Bodenübersichtskarte der Bundesrepublik Deutschland 1:1 000 000, generated by the Bundesanstalt für Geowissenschaften und Rohstoffe, Hannover), which provides information for 72 soil types, was used. Each soil type is characterized through a 'leading profile', which is an idealized soil profile reflecting the mean 'soil stratigraphy / composition of soil horizons' derived from bore measurements and generalized to scale 1: 1 000 000. For each horizon of each soil profile, eight attributes are specified: depth, texture class, clay content, humus content, carbon content, nitrogen content, field capacity and available water capacity.

Actual weather data obtained from German Weather Service were used for the simulation: temperature, precipitation and radiation data from one weather station and additionally precipitation data from 13 precipitation stations (on average one station per 44 km^2).

As a first step, the hydrological module was validated (requirement A). The simulated river discharge was compared with the measured discharge for an eight year period. Statistical evaluation of results was made by

(a) analyzing the simulated water balance,

(b) comparing means and coefficients of variations, and

(c) applying the common efficiency criteria after Nash & Sutcliffe (1970):

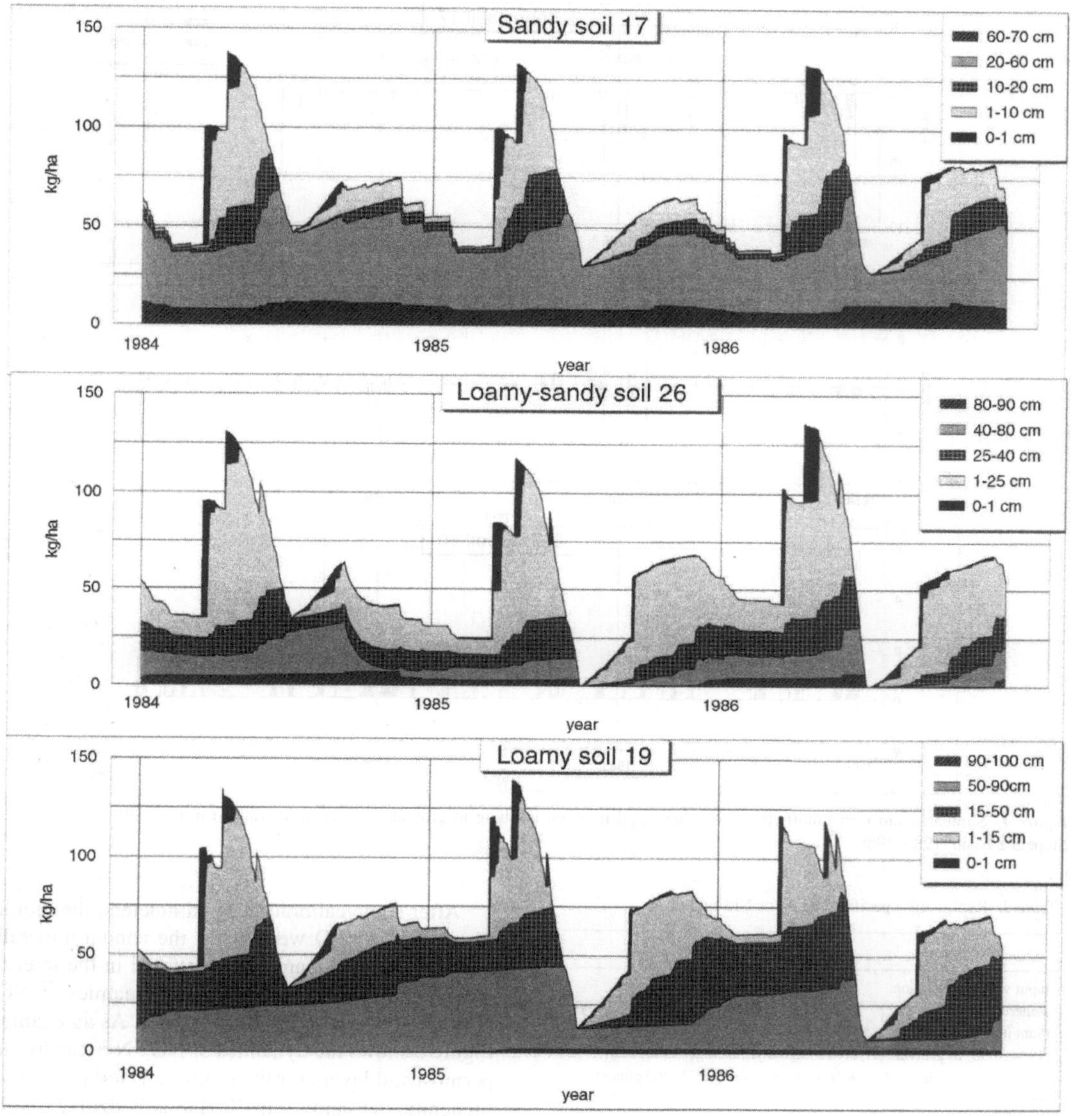

Figure 3. Simulated dynamics of nitrate nitrogen in three different soils in the Stepenitz basin, 1984–1986.

$$E = 1 - \frac{\sum(Q_{\text{obs}} - Q_{\text{sim}})^2}{\sum(Q_{\text{obs}} - \overline{Q_{\text{obs}}})^2}$$

The means and coefficients of variation of the observed and simulated time series are statistically comparable. The efficiency of runoff simulation is in a range from 0.68 to 0.85 with a daily time step. The observed and simulated river discharges for the Stepenitz are presented in Figure 5a in relation to average precipitation (given as negative values from the top of the diagram).

Before starting the nutrient modelling, the measured concentrations of NO_3-N at the gauge station were analysed in relation to the measured discharge. The positive correlation between these two time series

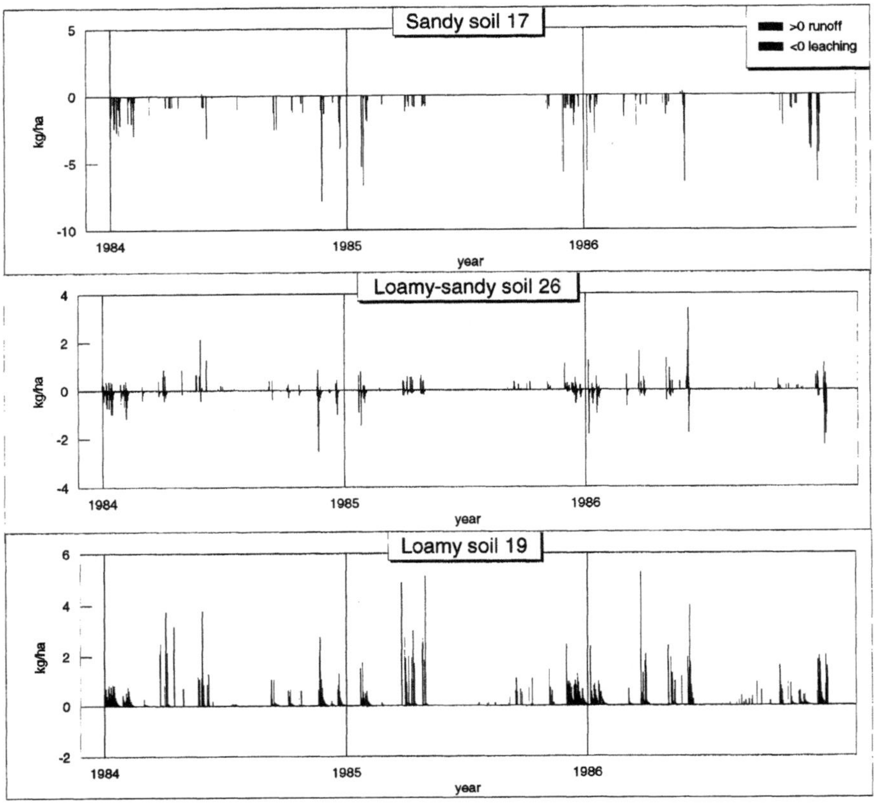

Figure 4. Simulated nitrogen runoff (positive values) and nitrogen leaching to groundwater (negative values) in three different soils in the Stepenitz basin, 1984–1986.

Table 1. Regionally-specific ranges for Nitrogen flows

Flow	*Range*
Input with precipitation:	19 – 30 kg/ha·yr
Mineralization:	1–2% org. N/yr (2% = 40 - 180 kg N/ha·yr)
Plant uptake:	50 – 150 kg/ha·yr
N - washoff, in general:	27–113 kg/ha·yr (average: 55 kg/ha·yr)
in clay soils:	9–66 kg/ha·yr (average: 21–26 kg/ha·yr)

is statistically significant. This is an indication that nonpoint source pollution prevails in the basin.

After that the regional data (Northern Germany) on nitrogen dynamics in soil and nitrogen flows were collected from the literature (Scheffer et al., 1984; Blume, 1992; McVoy et al., 1995) to obtain the regionally-specific ranges for the validation of nitrogen balance (see Table 1).

After some calibration of parameters, the requirements B, C and D were met – the annual mineralization, plant uptake and washoff fitted in the intervals indicated above and the seasonal dynamics of NO_3-N in the soil looked quite reasonable. As an example, Figure 3 shows the dynamics of NO_3-N in the five uppermost soil layers for three different soil types in the Stepenitz: the sandy soil 17 (Haplic Podzols / Cambic Podzols / Gleyic Podzols in the FAO classification), the loamy sandy soil 26 (Dystric Podzoluvisols/ Luvic Arenosols/ Dystric Cambisols), and the loamy soil 19 (Haplic Luvisols / Eutric Podzoluvisols / Stagnic Luvisols) in the period 1984 – 1986 The thickness of the upper layer is always 10 mm (the model requirement), while it varies for the other layers.

Comparison of the N-washoff for different soil types was quite reasonable: while the nitrate nitrogen from the sandy soil 17 mainly leaches into the ground water, since there is practically no surface or subsur-

Table 2. Loam and clay content and the simulated annual flows of nitrate nitrogen in three different soils in the Stepenitz basin

Soil type (texture class)	Loam content, %	Clay content, %	Mineralization, kg/ha·yr	Plant uptake, kg/ha·yr	Washoff with subsurface flow, kg/ha·yr	Leaching into ground water, kg/ha·yr
17 (sand)	5	0	67 – 111	98 – 125	0	67 – 106
26 (loamy sand)	15 – 24	5 – 10	64 – 102	120 – 142	12 – 22	48 – 64
19 (loam)	37	15 – 25	86 – 143	119 – 150	68 – 107	1 – 2

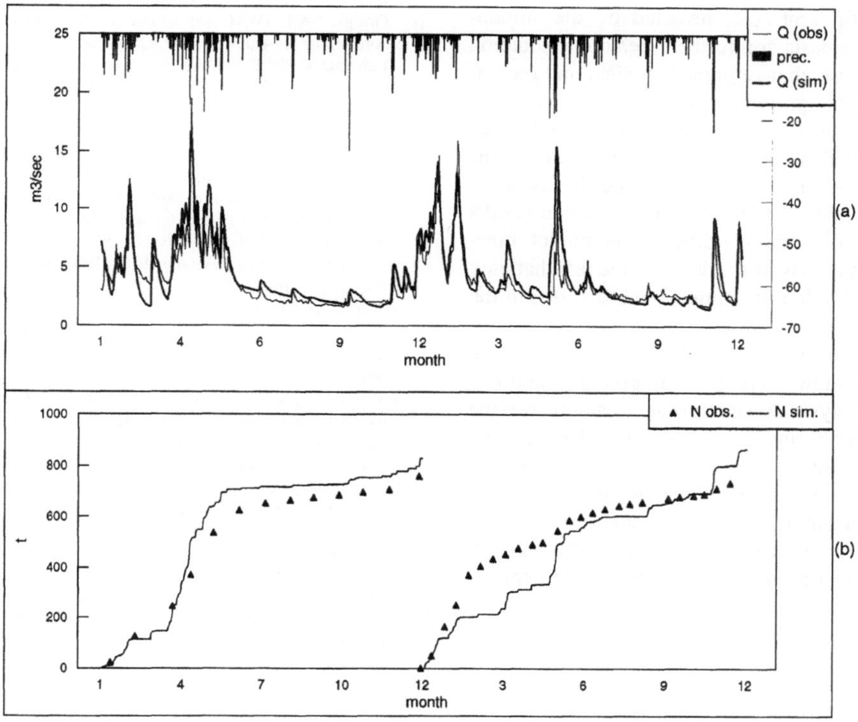

Figure 5. Simulated and observed river discharge (a) and simulated and estimated accumulated annual load (b) in the gauge Wolfshagen, Stepenitz, 1983 – 1984.

face runoff from this soil, the NO_3-N from the loamy soil 19 is mainly washed off with lateral flow and the loamy sandy soil 26 is intermediate in this respect: there are nitrogen losses in both directions (see Figure 4 & Table 2).

And finally, the accumulated simulated NO_3-N load was compared with the accumulated measured load in order to validate the nitrogen module for the whole basin. The accumulated load at the gauge station was estimated from the daily discharge and observed concentrations (the sampling frequency was once every two weeks) after linear interpolation of the concentrations for intermediate points. Figure 5b illustrates the comparison for two years, 1983 and 1984. The annual accumulated loads are close (4.9 and 8.5% difference). The seasonal dynamics looks better in 1983, while in the spring of 1984 the load is underestimated. In our opinion, this is caused mainly by the uncertainties in data (crop types and their distribution, crop management). In general, the results are satisfactory and in accordance with requirements of the robust approach the qualitative agreement is sufficient.

Conclusions

It was found that the approach was quite successful for coupled hydrological and water quality modelling in mesoscale watersheds. The differences in nitrogen cycling for different soils revealed by the simulation provide a basis for further scenario evaluation (e.g. for land use change impact studies) and recommendations for improving agricultural practices. It is expected that further testing of the model will take place in selected basins in the other two subregions of the Elbe: the mountainous area and the loess region. This would also allow to intercompare results for different basins and conditions. The model transferability to other basins results from the fact that only regionally-available information is needed to run the model.

Compared to other modelling approaches, the model SWIM allows to perform hydrological and water quality analysis at larger scale, namely for the mesoscale river basins and regions. The data requirements are modest. On the other hand, the model is quite complicated and it cannot be run as a black box, the understanding of the code and interrelations between different processes and their implementation in the model is a prerequisite for successful applications.

Acknowledgements

The authors are grateful to the German Federal Ministry for Research and Technology (BMBF) for providing the funding for this research (project 'Elbe Ecology'), and to all abovementioned agencies, which provided data.

References

Arnold, J. G., P. M. Allen & G. Bernhardt, 1993. A comprehensive surface-groundwater flow model, J. Hydrol. 142: 47–69.

Arnold, J. G., J. R. Williams, R. Srinivasan, K. W. King & R. H. Griggs, 1994. SWAT, Soil and Water Assessment Tool, USDA, Agriculture Research Service, Grassland, Soil & Water Research Laboratory, 808 East Blackland Road, Temple, TX 76502.

Blume, H. P., 1992. Handbuch des Bodenschutzes, 2nd edn, ECOMED-publisher.

Krysanova, V., A. Meiner, J. Roosaare & A. Vasilyev, 1989. Simulation modelling of the coastal waters pollution from agricultural watershed, Ecol. Modelling. 49: 7–29.

Krysanova, V., D. I. Müller-Wohlfeil & A. Becker, 1996. Integrated Modelling of Hydrology and Water Quality in Mesoscale Watersheds. PIK Report, No. 18, July 1996, Potsdam Institute for Climate Impact Research (PIK), P.O.Box 601203, D-14412 Potsdam, Germany: 32 pp.

Krysanova, V., D. I. Müller-Wohlfeil & A. Becker, 1998. Development and test of a spatially distributed hydrological / water quality model for mesoscale watersheds. Ecol. Modelling 106: 261–289.

McVoy, C. W., K. C. Kersebaum, M. Arning, P. Kleeberg, H. Othmer & U. Schröder, 1995. A data set from north Germany for the validation of agrosystem models: documentation and evaluation. Ecol. Modelling 81: 265–300.

Nash, J. E. & J. V. Sutcliffe, 1979. River flow forecasting through conceptional models, 1. A discussion of principles. J. Hydrol. 10: 282–290.

Scheffer, F. & P. Schachtschabel, 1984. Lehrbuch der Bodenkunde. 11. Edition, Stuttgart: Enke.

Srinivasan, R. & J. G. Arnold, 1993. Basin scale water quality modelling using GIS. Proc. Applic. of Adv. Inform. Technol. for Manag. of Nat. Res., June 17–19, Spokane, WA, U.S.A.: 475–484.

Hydrobiologia **410:** 139–150, 2000.
J. Garnier & J.-M. Mouchel (eds), Man and River Systems.

Nitrogen transfers through the Seine drainage network: a budget based on the application of the 'Riverstrahler' model

G. Billen & J. Garnier
UMR CNRS 7619 Sisyphe, UPMC, Boite 123, 4 place Jussieu, F-75005 Paris (France)

Key words: Nitrogen, riparian retention, river Seine

Abstract

A new version of the 'Riverstrahler' model has been developed for modelling riverine transfers of nitrogen from diffuse sources within the watershed, using land use data as a forcing function, together with an adjustable riparian transfer coefficient. Application of the model to the Seine river system (divided into four main sub-basins and one main branch) allows to establish a comprehensive budget of nitrogen transfers through the whole drainage network and to quantify the overall retention by riparian wetlands. According to the sub-basins, the figures indicate that 25 – 55% of the nitrogen coming from below the root-zone of agricultural land or from the aquifers is retained or eliminated before reaching surface water. The lower retention is found in areas where dense agricultural draining has been installed, thus by-passing riparian wetlands. At the scale of the Seine catchment, riparian retention represents 70 – 110 10^3 tonN y^{-1}, while in-stream retention accounts for only 24–32 10^3 tonN y^{-1}. The data suggest that landscape management, like restoration of efficient riparian systems in areas where they have been suppressed or by-passed, might be an efficient alternative or complimentary strategy for reducing nitrate contamination of surface water, with respect to tertiary treatment of point wastewater discharges.

Introduction

Nitrogen budgets established at the basin or regional scale have shown that a significant part of nitrogen introduced from terrestrial systems into surface water may be retained or eliminated through various processes during its down river travel before reaching the sea (Billen et al., 1991). In their analysis of the nitrogen budget of the North Atlantic catchment, Howarth et al. (1996) evaluated that the riverine nitrogen delivery to the sea only represents about 20% of the net anthropogenic inputs of this element to the terrestrial systems of the watershed, the remaining part being either stored in the biomass of growing forests and in aquifers, or eliminated in wetlands, rivers and associated stagnant systems. Behrendt (1996), compiling the nitrogen budget of a large sample of river systems in Western Europe, showed a clear relationship between the mean annual in-stream nitrogen retention and the specific runoff. In a more recent analysis, Behrendt & Opitz (1999) proposed on the same grounds an empirical relationship between nutrient retention and the estimated residence time of the water masses within the drainage network.

Deterministic models of nitrogen transfer at the scale of the whole drainage network thus have to take into account all these retention processes, along the whole continuum of aquatic ecosystems that forms a large river system. The 'Riverstrahler' model offers one of the few examples of a modelling approach allowing to represent the complex biogeochemical processes affecting nutrient and plankton dynamics at the scale of the whole drainage network of large, human impacted, river systems. The 'Seneque' model, resulting from the application of the 'Riverstrahler' approach to the Seine river and its major tributaries, has proven able to correctly simulate the seasonal and geographical variations of phytoplankton development and nutrient (N, P, Si) concentrations (Billen et al., 1994; Garnier et al., 1995; Billen & Garnier, 1996). It has also been applied to a hypothetical river system and an associated coastal marine system for studying the consequences of long term historical changes in land use of the watershed on river water quality and coastal eutrophication (Billen & Garnier, 1997).

For taking into account diffuse sources of nutrients, most budgets and models of nutrients transfer through river systems (including previous versions of the 'Riverstrahler' model) make use of loading coefficients (expressed, e.g. as kgN km^{-2} y^{-1}), which represent specific nutrient export for given land use and are obtained empirically from measurements carried out on small watersheds with various land uses (Benneton, 1986; Ryding & Rast, 1989; Quotbi, 1996). Such an approach, however, does not allow to take explicitly into account the role of riparian areas which are known to retain a significant part of the diffuse load of nutrient from agriculture (Peterjohn & Correll, 1984; Lowrance et al., 1984; Jordan et al., 1993; Haycock et al., 1993; Vought et al., 1994). We here describe an improved version of the 'Seneque' model incorporating an empirical description of the role of riparian systems at the basin scale. This version of the model is used to calculate nitrogen distribution in the Seine drainage network from watershed land use data and to establish a comprehensive budget of nitrogen transfer from soils and aquifers down to the sea.

Because of a high population density in its watershed and an intense agricultural activity, mostly based on cereals cultivation, with decreasing importance of cattle farming, the Seine river system constitutes a typical example of human impacted system. Its specific rate of nitrogen delivery (800–1400 kgN km^{-2} y^{-1}) represents about 10 times the pristine values (Howarth et al., 1996). The modelling approach proposed in this paper aims at quantifying the various sources of nitrogen to surface water and to identify and locate the processes able to reduce the nitrogen load.

The 'Seneque' model

General structure

The 'Seneque' model couples a module describing the kinetics of the various biological and physicochemical processes affecting organic matter and nutrients ('Rive': Garnier & Billen, 1995; Garnier et al., 1998), with an idealized hydrological model (Figure 1), describing the flow of water from the watershed in four sub-basins and the main branch of the Seine river (Figure 2). The 'Hydrostrahler' module, based on the concept of stream order (Strahler, 1957), represents the complex network of tributaries of each sub-basin by a regular scheme of confluence of rivers of increasing stream order with mean morphological characteristics. Using a two reservoirs description of the rain-discharge relationship, it calculates, from the seasonal variations of rainfall and evapotranspiration, the specific flow in each sub-basin as the sum of a runoff and intermediate flow contribution and a groundwater base flow. The required hydrological parameters are calibrated using pluriannual series of discharge values at the outlet of each sub-basins (Table 1). Flow velocity and depth in each stream order are then calculated by application of the Manning-Strickler formula.

Point sources of nutrients by wastewater works are taken into account, using data communicated by the Agence de l'Eau Seine-Normandie and summarized in Table 2. For the sub-basins the discharges by all wastewater works are distributed by stream order, while for the main branch they are taken individually into account at their exact kilometric position.

In-stream processes affecting nitrogen dynamics, like uptake by phytoplankton growth, ammonification of organic matter, nitrification and denitrification, are calculated by the 'Rive' module for the tributaries of each stream order in the four sub-basins, and along the main branch of the Seine river.

Definition of non-point sources and riparian retention

In a previous version of the model (Billen et al., 1994), non-point sources of nutrients were taken empirically into account by using data from a survey of the composition of first order streams sampled upstream from any point source of domestic or industrial contamination. Such data, obtained both in a period of high flow and in a period of low water, provided the required limit conditions of the model, i.e. the composition, assumed constant, respectively of the (sub)surface runoff and the base flow components of the discharge, as calculated by the hydrological module. A new approach has been developed here, in which these limit conditions are taken 'upstream' from the riparian zones. It does not rely on a deterministic model of nitrogen dynamics in soil (as e.g. in the INCA Model of Whitehead et al., 1998 a, b), but rather on the analysis of available empirical data of the mean nitrate concentration in water collected below the root zone by use of lysimeters, porous candles or agricultural drains installed under areas of different land use and agricultural practices.

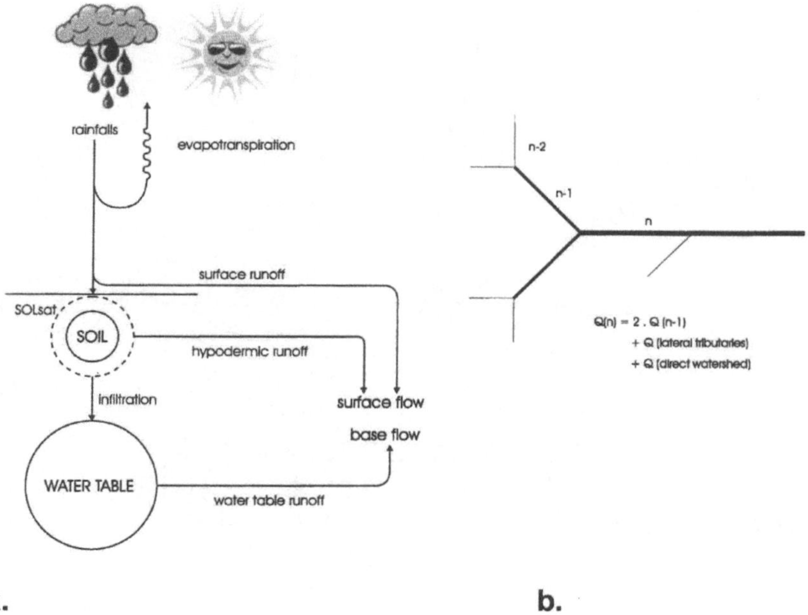

Figure 1. 'Hydrostrahler', the simple 2 compartments hydrological module describing the rainfall-discharge relationship within the 'Seneque' model. (a) Relationships between rainfall, evapotranspiration and water transfers through a soil and an aquifer compartement. Surface runoff occurs only when the soil water content is above saturation level; sub-surface (or hypodermic) runoff and infiltration are 1st order with respect to the soil water content; base flow is first order with respect to water table level. (b) The calculation of discharge of tributaries is based on the idealization of the drainage network as a regular scheme of confluence of tributaries of increasing stream order with mean morphological characteristics.

A synthesis of such data available for the Seine basin or neighbouring areas is provided in Table 3 and summarized in Figure 3. Clearly, the nitrate leaching rate is much lower from permanently vegetated soils (forests and meadows) than from arable soils. A general effect of increasing nitrogen fertilisation is also apparent for arable soils, as well as an effect of the depth of the soil horizons. The following empirical relationship (1) can be established from these data to calculate the mean nitrate concentration in water below the root zone from data on land use and agricultural practices in each sub-basin:

$$\begin{aligned} \text{NO3s (mgN/l)} &= \%\text{forest} * 0.4 \text{ mgN/l} \\ &+ \%\text{grassland} * 3 \text{ mgN/l} \\ &+ \%\text{arable} * (15 \text{ mgN/l} + 0.01 \text{ mgN/l (kgN/ha/y)}^{-1} \\ &* \text{fert} + 0.1 \text{ mgN/l (kgN/ha/y)}^{-1} * (\text{fert} - 130) \\ &(\text{fert} > 130)), \end{aligned} \tag{1}$$

where fert represents the amount of N fertilisers applied on arable land (in kgN/ha/yr) and the expression (fert>130) is zero if fert < 130 and 1 if fert >130.

The required land use data (% forest,% grassland and % arable land) for the different sub-basins were obtained from the 'Corine' data base (IFEN, 1997) and are summarized in Table 4. The corresponding calculated nitrate concentrations are used for both surface runoff and subsurface flow.

Because of the inertia of its response to diffuse pollution, the composition of phreatic groundwater often does not reflect the present land use of its catchment area. The mean nitrate concentration in phreatic water, characterizing the base flow of rivers in each sub-basin, was therefore determined empirically, not from land use data in the catchment, but from a survey of available data on the composition of groundwater pumped from the upper 10 m layer of the major aquifers of the Seine watershed (Table 5).

Using this approach, it is possible, for given hydrological conditions, i.e. for given proportions of (sub)surface and base flow, to calculate the composition of water before it crosses the riparian zones and reaches the rivers. This calculation has been carried out for a number of small streams devoid of signi-

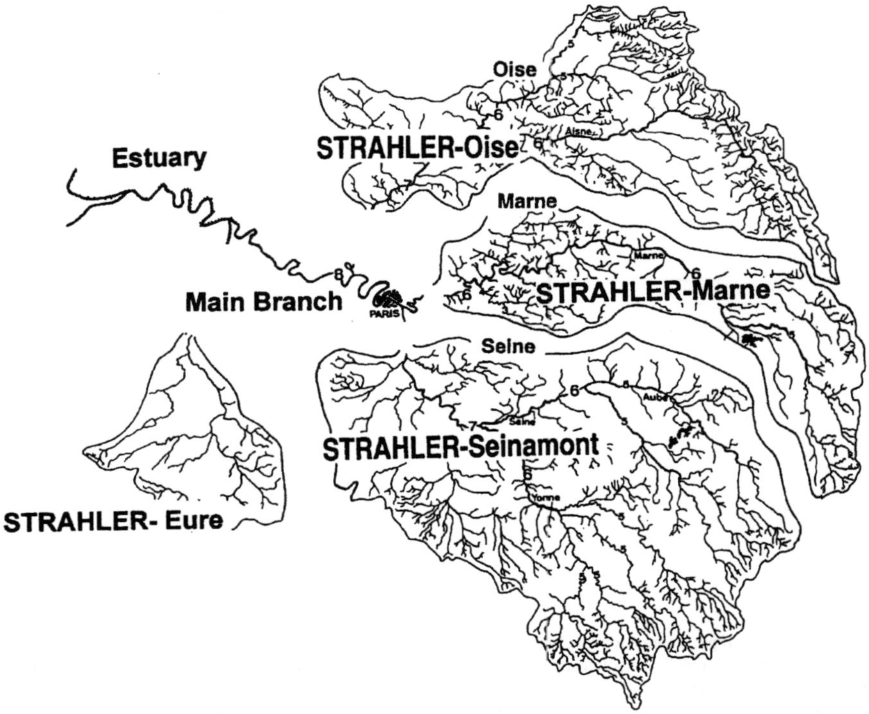

Figure 2. The Seine river drainage network, as schematised by 4 sub-basin and one branch in the 'Seneque' model. Figures indicate the stream order of some major tributaries.

Table 1. Parameters characterising the hydrology of the 4 sub-basins of the Seine catchment considered in the Seneque model. Calculated specific surface and sub-surface runoff and base flow, for two years with contrasting rainfall regime

Sub-basin	Solsat	srr	ir	arr	1991 (dry year)		1994 (wet year)	
					Surf. runoff	Base flow	Surf. runoff	Base flow
	mm	d^{-1}	d^{-1}	d^{-1}	mm y^{-1}	mm y^{-1}	mm y^{-1}	mm y^{-1}
Upper Seine	30	0.15	0.2	0.035	108	50	229	83
Marne	30	0.25	0.25	0.05	123	94	294	127
Oise	75	0.2	0.15	0.05	90	92	139	113
Eure	50	0.1	0.2	0.02	73	82	114	99

solsat: level of soil saturation, above which superficial runoff occurs.
srr: subsurface runoff rate from the soil compartment
ir: infiltration rate from the soil compartment
arr: aquifer runoff rate

ficant point contamination, for which the land use of the watershed and the mean nitrate composition of aquifers were known. The results are compared in Figure 4 with direct determinations of nitrate concentration in surface river water. As seen, the observed concentrations are most often lower than the calculated one, sometimes by a factor of five or more. We interpret this difference as resulting from the retention (denitrification) effect of riparian zones, *sensu lato*.

In the present version of the 'Seneque' model (Figure 5), land use on the one hand, and mean nitrate concentration in aquifers on the other hand, are used as the forcing functions for calculating the limit nitrogen concentration in (sub)surface and base flow

Table 2. Point sources of wastewater to the Seine system. (from data communicated by the Agence de l'Eau Seine-Normandie for the upstream basin, and by the Service de la Navigation de la Seine, Cellule Anti-Pollution de Rouen, for the estuarine zone)

	Area, km^2	Inhab. equiv., 10^6	Nitrogen inputs, ton N /day
Upper Seine sub-basin	30 796	2.44	17.2
Marne sub-basin	12 762	1.05	7.2
Oise sub-basin	16 972	1.69	12.9
Eure sub-basin	6 030	0.5	3.9
Main Branch (Paris-Poses)	5 200	8.7	97
Estuarine zone (Poses-Caudebec)	2 040	1	13
Total	73 800	15.4	139.3

Table 3. Mean annual nitrate concentration of water below the root zone for different land use. Results of measurements using drains (dr), porous candles (pc) or lysimeters (ly)

	Site	Method	NO_3, mgN/l	Authors
Forest	deciduous (Lorraine)	pc	0.45	Benoît et al. (1995)
	epicea (Vosges)	dr	0.3	Probst (1990)
Grassland	grassland without N fertilisation (Champagne)	ly	1	Ballif et al. (1996)
	pasture (Lorraine)	pc	7.0	Benoît et al. (1995)
	hay meadows (Lorraine)	pc	4.3	Benoît et al. (1995)
	temporary pasture (Lorraine)	pc	6.3	Benoît et al. (1995)
	alfa alfa (Lorraine)	pc	5.2	Benoît et al. (1995)
Cereals	winter wheat (Lorraine)	pc	10.3	Benoît et al. (1995)
	winter barley (Lorraine)	pc	14.0	Benoît et al. (1995)
	spring cereals (Lorraine)	pc	7.2	Benoît et al. (1995)
	corn (Lorraine)	pc	28	Benoît et al. (1995)
	wheat-corn (Versailles)	ly	8.5 – 23	Boniface et al. (1996)
	continuous corn (Versailles)	ly	5.3 – 8.6	Boniface et al. (1996)
	wheat-sugar beet (Champagne)	ly	23 – 29	Ballif et al. (1996)
	wheat-beets-alfa alfa (Champagne)	ly	10.6	Ballif et al. (1996)
	cereals (Beauce)	dr	11.7	Mariotti (1982)
	cereals (Brie)	dr	10.7	Belamie (1980)
other crops	peas (Brie)	dr	32	Muxart et al. (1997)
	oilrapes (Lorraine)	pc	27	Benoît et al. (1995)
	vineyard (Champagne)	ly	37	Ballif et al. (1996)

respectively. The retention effect is taken into account by a simple transfer coefficient, expressing the fraction of nitrate remaining after crossing the riparian zone and entering the drainage network. The value of this transfer coefficient is determined by adjusting the simulations to the observed nitrate concentrations.

Figure 6 illustrates this adjustment procedure on the examples of the Marne and Eure sub-basins. The values of transfer coefficient obtained by this procedure for the four major sub-basins as well as for a number of smaller sub-basins to which the model could be applied, are plotted on Figure 7. A distinct relationship appears between the riparian transfer coefficient and

Table 4. Land use data for the different sub-basins of the Seine river (Data from Corine Land Cover (IFEN, 1997) and the Recencement Général de l'Agriculture, 1988)

	area, km^2	% Urban area	% Forest	% Grass-land	% Arable land	Fertilisers on arable land (kgN/ha)
upper Seine sub-basin	30 796	3.8	28.8	8.3	58.4	136
Marne sub-basin	12 762	4.8	26.2	8.3	60.4	132
Oise sub-basin	16 972	4.4	21.7	10.2	63.0	118
Eure sub-basin	6 030	4.2	21.1	4.1	70.3	112
Main Branch	7 240	17.8	15.8	21.4	43.6	105
Total	**73 800**	**5.5**	**24.8**	**9.7**	**59.3**	-

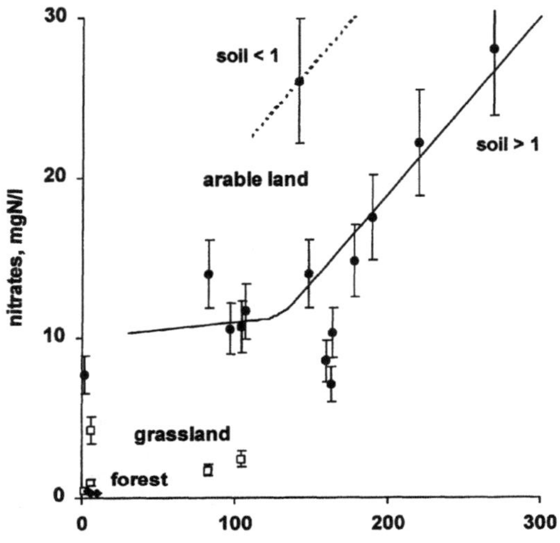

Figure 3. Relationship between land use, nitrogen inputs and the mean annual nitrate concentration in soil water below the root zone. For references, see Table 2. Inputs of nitrogen to the terrestrial systems were calculated from nitrogen fertiliser application and atmospheric deposition given by Thibert (1996).

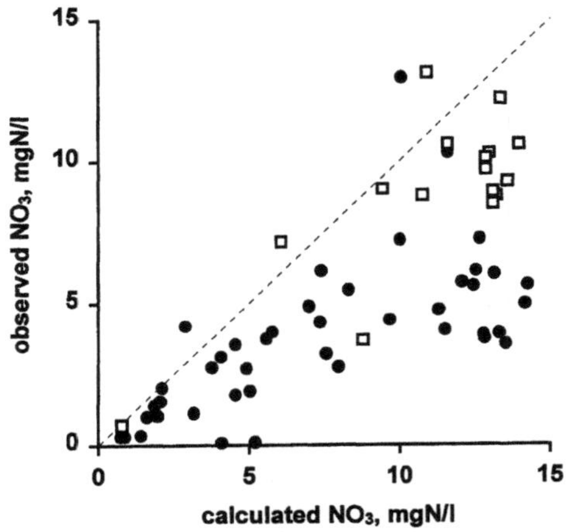

Figure 4. Comparison between mean observed nitrate concentrations in a sample of streams devoid of point nitrogen contamination and the calculated composition of soil water taking into account the land use in their watershed and the mean composition of the contributing aquifers. Open symbols refer to basins with more than 10% of the area equipped with tile-drains.

the extent of agricultural drainage in the watershed (Figure 7): basins where less than 20% agricultural area is equipped with tile-drains have transfer coefficients between 45 and 55% (implying that 55 – 45% of nitrate from soil leaching and aquifers is retained in riparian areas), while basins with more extensive tile drainage show much higher transfer coefficients. This is quite consistent with the fact that agricultural drainage most often results in the suppression or by-passing of the riparian wetlands. The same trends were observed in the data of Figure 4, where, in general, the streams draining watersheds equipped with tile drains showed less differences between the calculated and measured nitrate concentrations.

Budget of nitrogen transfers at the basin scale

The model has been used to simulate the seasonal cycle of nitrogen concentration in years 1991 and 1994, which strongly differ in hydrology (the mean annual discharge being respectively 395 and 650 m^3/s at Poses, the last gauging station). A very good agreement is found between observed and simulated nitrate and ammonium concentrations at some nodal points in the drainage network (Figure 8). A con-

Table 5. Relative area of major aquifers in the Seine river sub-basins, and their mean nitrate concentration. Data from (1) Fritz (1994), (2) AESN(1987)

		UpperSeine (1)		Marne(2)		Oise(2)		Eure(2)	
	Aquifers	%	mgN/l 1985–89	%	mgN/l 1980	%	mgN/l 1980	%	mgN/l 1980
Alluvions		10.0	5.0	12.3	3.0	10.8	5.0	5.5	5.0
Miocene/Oligocene	*calc.de Beauce, sable de Fontainebleau, calc.de Brie*	19.6	7.0	10.4	5.0	0.4		15.0	7.0
Eocene	*calc.de Champigny, de St Ouen, sable de Beauchamp*	3.9	11.3	10.7	9.0	0.7		0.6	9.0
Eocene/Paleocene	*calc.grossier, sable de Cuise, sable du Soissonnais*	4.5	13.0	16.7	8.5	29.6	8.5	3.2	10.0
Sup. cretaceous	*Craie*	18.9	6.4	14.4	4.7	44.3	6.0	64.7	6.8
Sup. cretaceous	*Sable du Perche*	0.0		0.0		0.0		11.0	1.8
inf cretaceous	*Sable albien, aptien, néocomien*	7.5	5.0	5.7	3.0	5.0	5.5		
Sup. mid jurass	*Calcaire*	22.5	10.0	22.7	2.0	5.8	3.3		
Sup. mid jurass	*Marne*	2.3	5.9	4.4	5.0	1.9	3.3		
inf jurass	*Calcaire*	5.3	5.5	2.6	3.0	0.1			
Base		5.4	1.8	0.0		1.5	1.8		
mean concentration, mgN/l			**6.7**		**4.8**		**6.3**		**5.9**

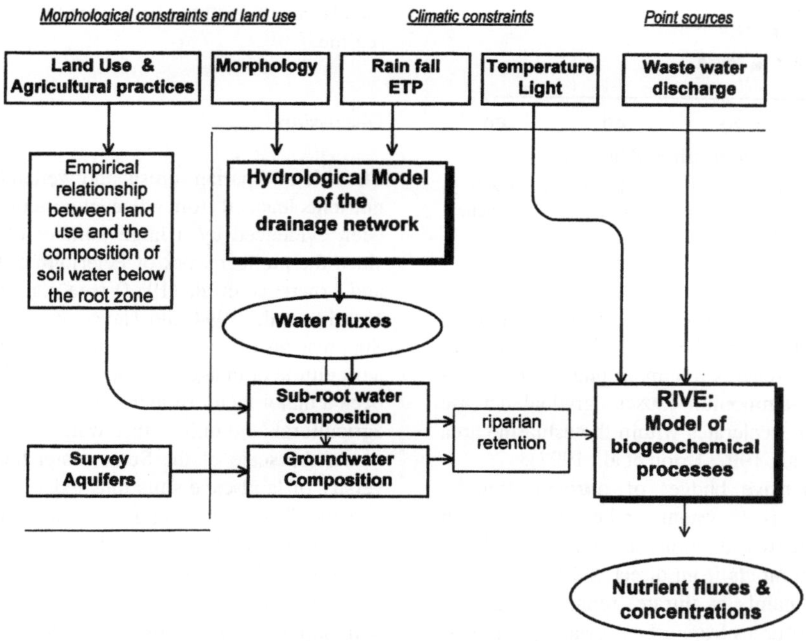

Figure 5. Structure of the 'Seneque' model explicitely taking into account land-use and mean aquifers composition as forcing functions and considering a riparian transfer coefficient.

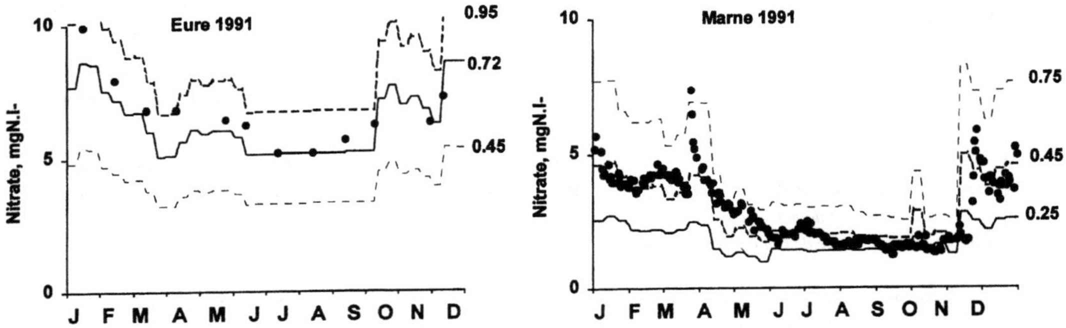

Figure 6. Illustration of the adjustment procedure of the riparian transfer coefficient on the observed variations of nitrate concentration at Neuilly-sur-Marne (outlet of the Marne sub-basin), and at Léry (outlet of the Eure sub-basin) in 1991.

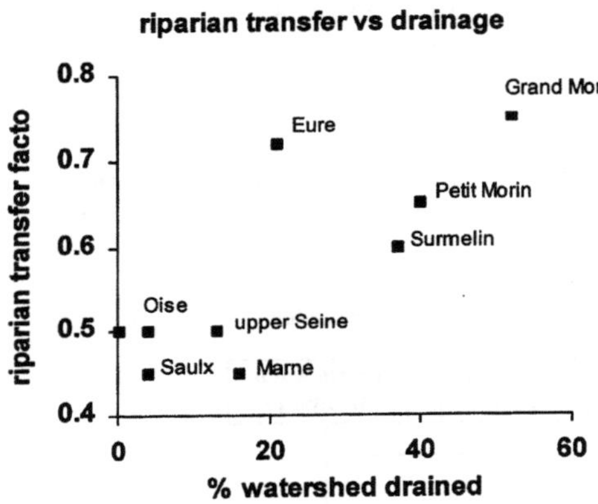

Figure 7. Relationship observed between the adjusted value of the riparian nitrate transfer coefficient and the extent of agricultural drainage in the watershed.

siderable increase in ammonium concentration at the stations below Paris, is the result from wastewater discharge at the Achères treatment plant. Further nitrification of this ammonium causes a gradual increase in nitrate which accelerates within the estuarine area (Chesterikoff et al., 1992; Brion et al., 1998).

A comprehensive budget of nitrogen transfer through the Seine river system can be calculated with the model for the two situations (Figure 9 & Table 6). Diffuse nitrogen inputs from direct soil leaching and from aquifers (mainly as nitrate) represent 2.5 to 4 times the point sources from urban sewage (mainly as organic nitrogen and ammonium). Riparian retention of these diffuse sources, in spite of large variations between sub-basins, can be estimated as about 40% at the Seine basin scale. The effective load of nitrogen in surface water thus consists of 56 – 66% from diffuse and 44 – 34% from point sources. In-stream retention (by denitrification in sediments in both river channel and associated stagnant systems like reservoirs, Garnier et al., 1999) only represents about 30 10^3 tonN y^{-1}, a rather small amount compared with riparian retention (69 – 112 10^3 tonN y^{-1}).

The budget also reveals the large seasonal and year-to-year variations in nitrogen delivery to the sea, mostly resulting from variations in the hydrological regime.

Discussion

The role of riparian forests and wetlands as sinks for nutrients leached from adjacent agricultural areas has been evidenced by a large number of local studies since the pioneer work of Peterjohn & Corell (1984) and Lowrance et al. (1984) (see e.g. the reviews by Vought et al., 1994 and Haycock and Pinay, 1993). Suppression of these areas or their short-cutting by agricultural drainage may therefore be thought to have been a major factor of increasing nutrient export from agricultural land into surface waters.

At the scale of the Seine watershed, a large diversity of landscape situations exists. In some areas, like the *Brie* region, numerous drainpipes networks have been installed during the last 50 years, transforming the landscape, once dominated by grassland and cattle farming, into an intensive crop farming area with suppression of most wetlands and extension of cultivated areas down to the very edge of rivers. In other areas, like in the *Champagne crayeuse* intensification of the agriculture has been limited to the uphill plateau, leaving large forested riparian zones as the only natural elements of the landscape. Other areas

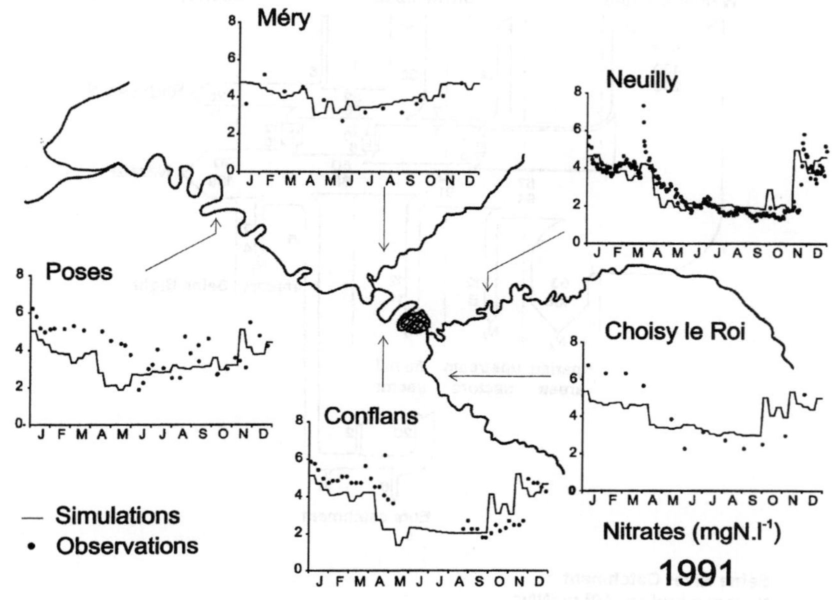

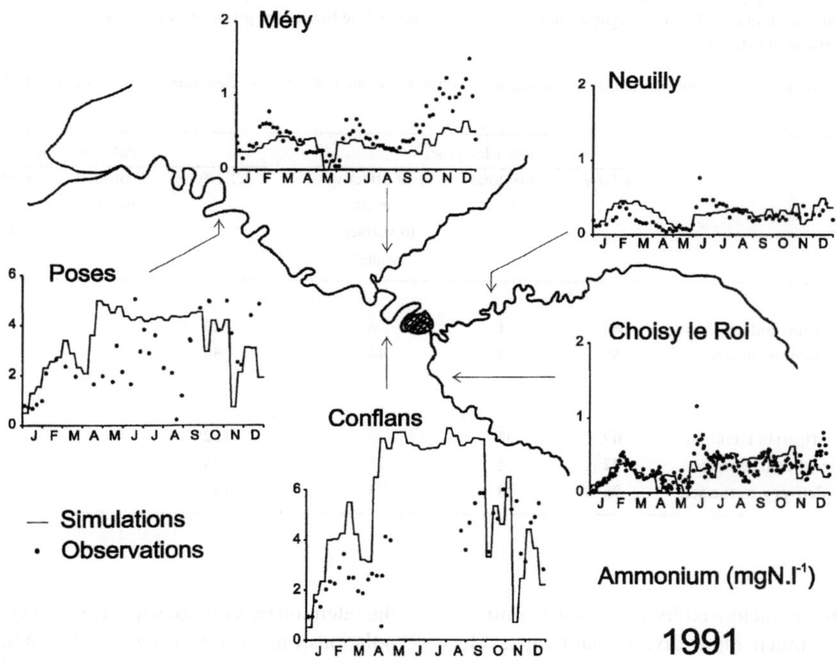

Figure 8. Simulation by the model 'Seneque' of the seasonal variations of nitrate and ammonium concentrations at some nodal points in the Seine drainage network in 1991. Comparison with observed data.

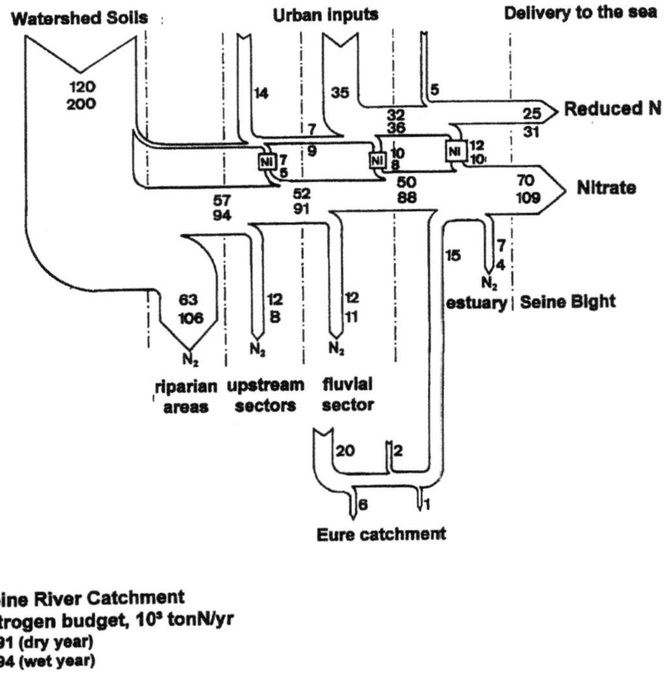

Figure 9. Budget of nitrogen transfers (in 10^3 tonsN y^{-1}) through the Seine river system, calculated with the 'Seneque' model for the hydrological conditions of year 1991 (upper figures) and 1994 (lower figures). The figures are given separately for nitrate and reduced nitrogen (ammonium and organic nitrogen).

Table 6. Input-output budget of nitrogen in the Seine drainage network, as calculated by the 'Seneque' model for a dry (1991) and a wet (1994) year

	1991 (dry year)			1994 (wet year)		
	kT N/yr	% of total inputs	% of effective inputs to surface water	kT N/yr	% of total inputs	% of effective inputs to surface water
Input						
from soil leaching and aquifers	**140**	71	56	**220**	80	66
from urban sewage	**56**	29	44	**56**	20	34
Output						
riparian retention	**69**	36	-	**112**	41	-
instream retention	**32**	16	25	**24**	9	15
delivery to the sea	**95**	48	75	**140**	50	85

in the basin are characterised by less intensive mixed farming with a much more diverse landscape. The overall effect of these various local situations is difficult to evaluate. Only very few experimental studies have been conducted to provide direct estimation of the retention capacity of small riparian systems. Extrapolation of the results of such studies at the basin scale would still require considerable efforts of wetlands typology and inventory. The approach proposed in the present paper by-passes this direct field measurement approach, but it does nor replace it. Although indirect, it provides the first quantitative estimation of riparian retention at the scale of such a large river system.

Nitrogen retention and storage at the watershed scale, as considered, e.g., in Howarth's et al (1996) overall input-output budgets, involves the following processes:

1. storage in catchment biomass as for instance in aggrading forests;
2. storage or denitrification in catrchment soils;
3. storage or denitrification in groundwater;
4. denitrification or storage in wetlands or riparian zones, i.e. at the interface between ground- and surface water;
5. in-stream processes of retention, either by benthic or planktonic denitrification or by storage of sediments.

The approach proposed in the present paper only address the evaluation of processes 4. and 5. The latter are directly calculated by the 'Rive' module of the 'Seneque' model. The former is estimated by fitting an overall transfer coefficient on observed instream nitrate concentrations, given the concentration in the sub-root zone and aquifer waters. As implicitely defined here, riparian retention (process 4) therefore includes all processes affecting nutrient concentration during the transit of water from the sub-root zone on the one hand and the aquifers on the other hand, to river channel. The processes possibly occurring in the soil below the root zone or within the aquifers themselves are not included in this estimation. Because the aquifers are not in equilibrium (their hydrographic level is fluctuating interannually and their nitrate content is regularly increasing), the approach cannot be used to estimate the rate of denitrification in aquifers (process 3). However, by multiplying the infiltration flow rate calculated by our simple hydrological model by the difference in mean nitrate concentration between water below the root zone and aquifers, it can be estimated that a total amount of about 27 and 37 10^3 tonN y^{-1} is either stored or eliminated in groundwater in 1991 and 1994, respectively.

Our figures indicate that as much as 50% of nitrogen leached from agricultural areas to surface water or flowing from aquifers are retained or denitrified within riparian areas. This estimate is very sensitive to the value taken for the mean composition of aquifers and water below the root zone. As the latter is not known with more than 20% accuracy, the estimated retention percentage can vary in the range from 40 to 60%.

Our data for different sub-basins suggest that the riparian retention is lower in areas subject to large scale agricultural drainage. It thus identifies drainage as a major factor responsible for nitrogen losses from agricultural land to surface water, as already suggested for a Swedish river system by Krug (1993). From the middle of the XIXth century and until recently, drainage of wetlands or hydromorphic soil regions was considered as an efficient way to convert marginal land into fertile cropland. The Seine watershed comprises about 5000 km^2 drained agricultural land, half of it dating back to the last 20 years. Only very recently has the rate of tile-drains installation being slowed down, owing to the sudden awareness of the ecological values of wetlands.

Our results thus confirm the importance of riparian processes in controlling nitrogen contamination of surface water and fluvial delivery to estuarine and coastal marine systems. They indicate that landscape management, like restoration of efficient riparian systems in areas where they have been suppressed or bypassed, might have a potentially more efficient impact on nitrogen loading than tertiary treatment of point wastewater discharge.

Acknowledgements

This work is a part of the French CNRS PIREN-Seine programme, and has been supported by the Commission of the European Communities (DG XII) as a part of the BINOCULARS Project. We are grateful to Dr H. Behrendt and three other anonymous reviewers for critical comments on an earlier version of the manuscript.

References

AESN, 1987. Etat des eaux en 1981: Bilans et perspectives. Oise-Aisne, Le bassin de la Marne, Seine-Aval. Agence de l'Eau Seine-Normandie. Nanterre (France).

Ballif, J. L. et al., 1996. Les lysimètres en sols de craie de Chalons-sur-Marne. In "Trente ans de lysimétrie en France 1960-1990". J. C. Muller, coordinateur. INRA editions, Paris.

Benoît, M., D. Saintot & F. Gaury 1995. Mesures en parcelles d'agriculteurs des pertes en nitrates. Variabilité sous divers systèmes de culture et modélisation de la qualité de l'eau d'un bassin d'alimentation. C.R. Acad. Agric. Fr. 81: 175–188.

Behrendt, H., 1996. Inventories of point and diffuse sources and estimated nutrient loads. A comparison for different river basins in Central Europe. Wat. Sci. Techn. 33: 99–107.

Behrendt, H. & D. Opitz, 1999. Retention of nutrients in river systems: dependence on specific runoff and hydraulic load. Hydrobiologia 410: 111–122.

Belamie, R., 1980. Influence of cropping practices and hydrological factors on the variation of nitrates in the discharges from a representative basin in the Paris region. Proceedings of the Helsinki symposium IAMS IAM publ: 130.

Benneton, J. P., 1986. Eutrophisation des plans d'eau. Inventaire des principales sources de substances nutritives azotées et phosphorées. Etude bibliographique. Partie III. Trib. Cebedeau. 39: 16–34.

Billen G., C. Lancelot & M. Meybeck, 1991. N, P and Si retention along the aquatic continuum from land to ocean. In R. F. C. Mantoura, J. M. Martin & R. Wollast, (eds), Ocean margin processes in global change. Dahlem Workshop Reports, Wiley: 19–44.

Billen G., J. Garnier & P. Hanset, 1994. Modelling phytoplankton development in whole drainage networks: The 'Riverstrahler' model applied to the Seine river system. Hydrobiologia 289: 119–137.

Billen, G. & J. Garnier, 1996. Transfert et métabolisme de l'azote et du phosphore dans l'hydrosystème Seine. In C. Le Coz, B. Tassin & D. Thévenot (eds), Transfert des polluants dans les hydrosystèmes. Presses de l'Ecole Nationale des Ponts et Chaussées: 121–140.

Billen, G. & J. Garnier, 1997. The Phison River Plume: coastal eutrophication in response to changes in land use and water management in the watershed. Aquat. Microb. Ecol. 13: 3–17.

Boniface, R. et al., 1996. Lysimètres en sols de Versailles. Bilan 1974–1990. In 'Trente ans de lysimétrie en France 1960–1990'. J. C. Muller, coordinateur. INRA editions, Paris.

Brion, N., 1998. Etude du processus de nitrification à l'échelle de grands réseaux hydrographiques anthropisés. Thesis. Université Libre de Bruxelles.

Chesterikoff, A., B. Garban, G. Billen & M. Poulin, 1992. Inorganic nitrogen dynamics in the River Seine downstream from Paris (France). Biogeochemistry 17: 147–164.

Fritz, M., 1994. Etude statistique sur la contamination en nitrates des cours d'eau et des nappes principales du bassin de l'Yonne et de la Seine en amont de Montereau. Mémoire de DEA, Université Pierre et Marie Curie, Paris.

Garnier, J., G. Billen & M. Coste, 1995. Seasonal succession of diatoms and chlorophyecae in the drainage network of the River Seine: Observations and modelling. Limnol. Oceanogr. 40: 750–765.

Garnier, J., G. Billen & L. Palfner, 1999. Understanding the oxygen budget and related ecological processes in the river Mosel: the RIVERSTRAHLER approach. Hydrobiologia 410: 151–166.

Garnier, J., Leporcq, B., Sanchez, N. & Philippon, X., 1999. Biogeochemical budgets in three large reservoirs of the Seine basin (Marne, Seine & Aube reservoirs). Biogeochemistry 47: 119–146.

Haycock, N. E., G. Pinay & C. Walker, 1993. Nitrogen retention in river corridors: European perspective. Ambio 22: 340–346.

Howarth R. W., G. Billen, D. Swaney, A. Townsend, N. Jaworski, K. Lajtha, J. A. Downing, R. Elmgren, N. Caraco, T. Jordan, F. Berendse, J. Freney, V. Kudeyarov, P. Murdoch & Z. Zhao-liang, 1996. Regional nitrogen budgets and riverine N & P fluxes for the drainages to the North Atlantic ocean: natural and human influences. Biogeochemistry 35: 75–139.

IFEN, 1997. Corine Land Cover. Base de données géographiques de l'occupation du sol. Institut Français de l'Environnement. Orléans.

Jordan, T. E., D. L. Correll & D. E. Weller, 1993. Nutrient interception by a riparian forest receiving inputs from adjacent cropland. J. Envir. Qual. 22: 467–473.

Krug, A., 1993. Drainage history and land use pattern of a swedish river system. Their importance for understanding nitrogen and phosphorus load. Hydrobiologia 251: 285–296.

Lowrance, R., R. Todd, J. FaIL, O. Hendrickson, R. Leonard & L. Rasmussen, 1984. Riparian forests as nutrient filters in agricultural watersheds. BioScience 34: 374–377.

Mariotti, A., 1982. Apports de la géochimie isotopique à la connaissance du cycle de l'azote. Thesis. Université Pierre et Marie Curie.

Muxart, T. (coord), 1996. Exportation de MES et de polluants. Chap. 7: 75–124. In 'Thème Bassins versants ruraux'. Rapport PIREN-Seine 1996/1: 181pp.

Ministère de l'Agriculture et des Forêts, 1988. Recensement Général de l'Agriculture, 1988.

Peterjohn, W. T., D. L. Correll, 1984. Nutrient dynamics in an agricultural watershed: observation on the role of a riparian forest. Ecology 65: 1466–1475.

Qotbi, A., 1996. Etude de la dynamique du phytoplancton et des nutriments dans une rivière aménagée: le Lot (France). Modélisation mathématique et simulation de scénarios. Thesis. Université Paul Sabatier, Toulouse (France).

Ryding, S. O. & W. Rast, 1989. The control of eutrophication of lakes and reservoirs. Man & the Biosphere Series, Jeffers J.N.R. (ed.), UNESCO, Paris: 314 pp.

Strahler, A. H., 1957. Quantitative analysis of watershed geomorphology. Geophys. Union Trans. 38: 913–920.

Thibert, S., 1996. Exportations naturelles et anthropiques des ions majeurs et des éléments nutritifs dans le bassin de la Seine. Approches méthodologiques. Thèse. Université de Paris VI.

Vought, L. B. M., J. Dahl, C. L. Pedersen & J. O. Lacoursiere, 1994. Nutrient retention in riparian ecotones. Ambio 23: 342–347.

Whitehead, P. G., E. J. Wilson & D. Butterfield, 1998. A semi-distributed Integrated Nitrogen model for multiple source assessment in Catchments (INCA): Part I – model structure and process equations. Sci. Tot. Envir. 210/211: 547–558.

Whitehead, P. G., E. J. Wilson, D. Butterfield & K. Seed, 1998. A semi-distributed Integrated Nitrogen model for multiple source assessment in Catchments (INCA): Part II – application to large river basins in South Wales and eastern England. Sci. Tot. Envir. 210/211: 559–583.

Hydrobiologia **410:** 151–166, 2000.
J. Garnier & J.-M. Mouchel (eds), Man and River Systems.

Understanding the oxygen budget and related ecological processes in the river Mosel: the RIVERSTRAHLER approach

Josette Garnier[1], Gilles Billen[1] & Laurent Palfner[2]
[1]*UMR CNRS 7619 Sisyphe, UPMC boite 123, Tour 26, Etage 5, 4 place Jussieu, 75005 Paris, France*
[2]*NANCIe, Centre International de l'Eau, 149, rue Gabriel Péri, B.P. 290, 54515 Vandoeuvre cedex, France*

Key words: river Mosel, ecological modelling, oxygen budget, P *vs* R ratio, phytoplankton decline

Abstract

The oxygen budget in the transboundary river Mosel was analyzed with the help of the RIVERSTRAHLER model. The model, developed for the river Seine, was used after minor modifications of the kinetics of microbial processes. The impact of benthic filter feeders (*Dreissena polymorpha*) was introduced into the model to better explain phytoplankton decline in the canalized sector of the river. Hydro-geomorphology, meteorology and point and non-point sources of nutrients were analyzed as required by the model, according to the stream order concept at the scale of the whole drainage network of the main tributaries and along the main branch of the river Mosel, from Millery to Koblenz. The model was validated on water quality data (phytoplankton biomass -Chl *a*-, nutrients, oxygen) collected at half-monthly intervals during the period 1993–1995. A reasonable agreement was found at both the seasonal and spatial scales. The validated model was used to calculate the oxygen budget that shows variations in the contributions of biological processes (net primary production, bacterial and benthic respiration, nitrification) along successive stretches of the main river branch. Bacterial respiration dominates in sectors particularly affected by effluent inputs. Benthic filter feeders colonising these canalized sectors contribute to increases in respiration and oxygen deficit through their own respiration and their impact on phytoplankton. Several possible management scenarios, aimed at improving oxygenation, were tested with the model. An 80% reduction of both phosphates (to reduce eutrophication) and organic matter (to decrease bacterial activity) restores autotrophic conditions in the 7th order sector.

Introduction

In aquatic systems, oxygenation is the result of an imbalance between the processes of photosynthesis, degradation of organic matter and reaeration. Whereas domestic and industrial organic pollution is a main factor of desoxygenation, excessive phytoplankton development followed by an abrupt decline (river Rhine: de Ruyter van Steveninck et al., 1992; Admiraal et al., 1994; river Spree: Köhler, 1995; river Seine: Garnier et al., 1995) may also be a major contributor to desoxygenation in canalized rivers (Billen & Garnier, 1996). Moreover, when these systems are polluted by ammonium input, nitrification may play an important role in controlling the oxygen budget (Brion, 1998).

The ecological model (RIVERSTRAHLER: Billen et al., 1994; Garnier et al., 1995) was recently developed for studying eutrophication and oxygen deficit problems in the river Seine (Billen & Garnier, 1996; Garnier & Billen, 1998). In contrast to already existing ecological river models (river Lot: Capblancq et al., 1982; Dauta, 1986; Admiral et al., 1993; Qotbi, 1996; river Rhine: river Mosel-Saar: Schöl et al., 1997, 1999), RIVERSTRAHLER takes into account the whole drainage network according to the concept of stream orders (Strahler, 1957). An assumption fundamental to our modelling approach is that the specific features of the ecological functioning of the river are the results rather of differences in the constraints (geomorphology, hydrology, climate and input of elements from diffuse and point sources) than of differences in the kinetics of the microbiological processes. The modeling strategy is based on the River Continuum Concept (RCC: Vannote et al., 1980) and stresses the importance of considering a drainage network as a series of interdependent systems firstly structured by longitudinal interactions changing with increasing stream orders.

With a population density of 190 inh. km^{-2}, intense industrial activity and a large proportion of arable land, the Mosel basin is subjected to considerable human pressure. The model was already taking into account all above processes identified in the oxygen deficit of the river Mosel and experimentally studied in the field; our aim was therefore to arrive to a more thorough understanding and formulation of these processes. We now have some evidence that, in rivers, as opposed to lakes, zooplankton does not control phytoplankton decline (Billen et al., 1994; Garnier et al., 1995) although it can influence the phytoplankton composition and biomass in special conditions (Gosselain et al., 1994; Gosselain et al., 1998). Whereas we searched to corroborate such results, we moreover focused on the role in phytoplankton decline by lamellibranches which are now recognized as a major controlling factor. The model was used to calculate oxygen budgets which provide information on the variations in the contribution made by the different processes. We also tested different rehabilitation scenarios and analyzed the responses in terms of autotrophy *versus* heterotrophy of the system (P *versus* R diagrams) as proposed by Vannote et al. (1980). By this modelling approach, we aimed to give quantitative support to the RCC, as the exploration of scenarios allows the concept to be tested over a gradient of human perturbations.

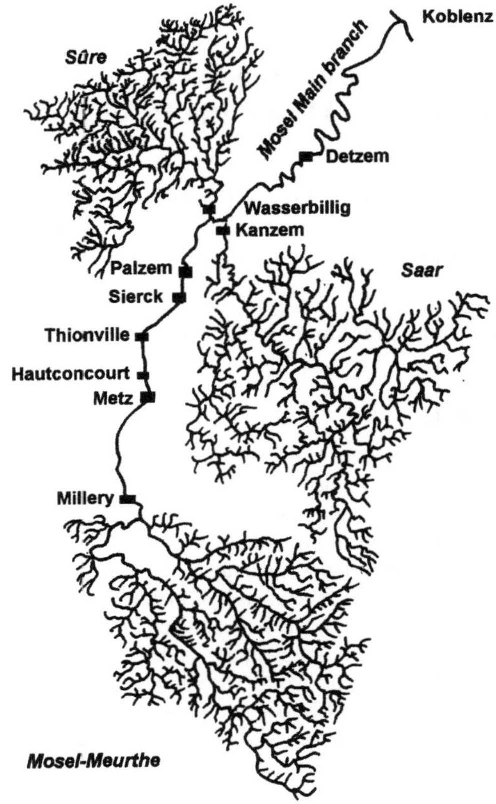

Figure 1. Map of the Mosel hydrographic network with the 3 sub-basins considered and locations of the main stations, at the outlet of the sub-basins and along the main branch of the river.

Site description

The river Mosel, a major tributary to the river Rhine, has its source in the Vosges Massif (altitude 735 m) and flows into the Rhine at Koblenz (altitude 59 m). Its hydrographic network is oriented northwards toward the schistose Massif of the Rhine. The Mosel basin (28 156 km^2) comprises three main sub-basins: the French Meurthe-Mosel (6850 km^2) upstream basin, the Saar sub-basin (7300 km^2) equally distributed between France (the upstream part) and Germany (the downstream part) and the Sure sub-basin (3740 km^2), almost completely draining Luxembourg. The rivers Saar and Sure flow into the main branch of the river Mosel in Germany (Figure 1). As early as 1956, a contract was made between France, Germany and Luxembourg for the canalization of 298 km of the river Mosel, from Metz to Koblenz and the construction of 18 weirs. In 1964, the river Mosel was opened to navigation and between 1969 and 1979 the navigable sector extended upstream to the confluence with the Meurthe. The regulation of the river Saar for navigation from Saarbrücken to the confluence with the main branch of the river Mosel was decided by Germany in 1969 (Figure 1). Regulation work really began in 1976; 91 km of river are already canalized and six weirs constructed from Saarbrücken to the river Mosel. The river Sure is not regulated for navigation.

As opposed to the Rhine, the river Mosel has a hydrological regime rainy and oceanic as snow-melting is insignificant in the Vosges Massif. Extreme discharge values at Cochem, immediately upstream from Koblenz, vary from 14 m^3 s^{-1} in dry summers to 4000 m^3 s^{-1} during winter flood events with mean discharge of about 330 m^3 s^{-1}.

Table 1. Geomorphological characteristics of the 3 sub-basins of the river Mosel as taken into account in the RIVERSTRAHLER model. Stream order: ordination according to Strahler (1957); mean for each stream order of number of rivers, length, watershed area and slope

Sub-basins	Streamorder	Number	Length (km)	Watershed area (km^2)	Slope ($m\,m^{-1}$)
Meurthe-Mosel sub-basin	1	1485	1.85	2.6	0.057
	2	333	2.7	12	0.021
	3	78	5.8	50	0.01
	4	21	13.6	199	0.0033
	5	4	63.5	1688	0.0016
	6	2	48	3416	0.0006
	7	1	(21)	6850	0.0006
Sarre sub-basin	1	400	2	2.9	0.039
	2	80	4.5	79	0.015
	3	25	9	207	0.0033
	4	8	21	632	0.0016
	5	3	38	1878	0.0011
	6	1	103	7300	0.0006
Sure sub-basin	1	950	1.3	1.85	0.039
	2	230	3	10	0.015
	3	50	6.5	45	0.0033
	4	13	16.3	203	0.0016
	5	5	17	588	0.0011
	6	1	74.5	3740	0.0006

Model description

The model calculates the seasonal and geographical variations of the main water variables for the entire drainage network, taking into account the constraints related to morphology, hydro-meteorology, and point and non-point sources of organic matter and nutrients.

Morphology and hydrology

The **MoSar** model considers the whole hydrographic network, on the basis of an idealization of the upstream basin according to the concept of stream orders (Strahler, 1957). To improve the geographical resolution of this description, the model includes three sub-basins (Meurthe-Mosel sub-basin upstream from Millery, Saar and Sure sub-basins upstream of their confluence with the river Mosel), (Figure 1, Table 1). They are connected to a main branch, with a detailed morphological description, including the wetted section and width every km and waterfall height at each weir. The corresponding values vary according to the river flow, and resulted from a detailed hydrological model set up by the BfG (Schöl et al., 1997, 1999)

An empirical model of the rain-discharge relationship (Bultot & Dupriez, 1976) and the description of the geomorphological characteristics by stream order (Strahler, 1957) make it possible to calculate the seasonal variations of discharge from all tributaries of each sub-basin and from the main branch. No limit conditions are thus required, except the composition of drainage and ground waters, resulting from diffuse sources of nutrients from soil leaching. The calculated values of discharge (Q, $m^3\ s^{-1}$) by stream order in each sub-basin for ten day periods and the morphological characteristics (width: w, m; slope: s, $m\ m^{-1}$) of each stream order, are used to compute the mean depth (d, m) and flow velocity (v, $m\ s^{-1}$) by means of rearrangement of Manning's empirical formula (Billen et al., 1994):

$$v = r^{-1}.\ d^{2/3}.\ s^{1/2} \qquad (1)$$

$$d = (r\cdot Q.\ w^{-1}.\ s^{1/2})^{3/5}, \qquad (2)$$

where r is a coefficient expressing the roughness of the channel bed varying between 0.05–0.15 (Hammer & Mc Kichan, 1981). In the case of sluice-lock regulated river stretches, a constant depth is assumed whatever the discharge.

The flow from the river catchment area of the river or from its lateral tributaries of a lower stream order, 'dilutes' the water masses flowing through the main channel. The corresponding dilution factor (dil, h^{-1}) can be calculated as follows:

$$dil = -\ (w.d)^{-1}\cdot(dQ/dx),$$

where (w.d) is the wetted section of the river and (dQ/dx) is the longitudinal change in discharge.

This dilution and its variations with the stream order and season constitutes the most important factor controlling the development of planktonic organisms in rivers.

In the main branch of the river, the calculation is similar and accounts for the contribution to the flow from both the direct watershed and the sub-basins. In regulated sectors, the values of depth and wetted section provided by the hydraulic model by Schöl et al. (1999) are used instead of the results of Manning's formula.

Meteorological constraints

Seasonal variations of water temperature, mean daily irradiance and photoperiod are represented by sine functions; during the photoperiod, the hourly irradiance is also calculated as a sine function (Billen et al., 1994).

Table 2. Non-point sources in the 3 sub-basins for each of the two components of the river flow (surface and phreatic water) NB: The values corresponding to the direct watershed of the main branch of the river Mosel are taken as identical to those of the Meurthe-Mosel sub-basin

		Meurthe-Moselle		Saar		Sure	
		surf.	phr.	surf.	phr.	surf.	phr.
Nitrate	mgN/l	3.5	1.4	3.5	2.2	4.9	4.2
Ammonium	mgN/l	.02	.02	.02	.02	.015	.015
Total phosphorus	mgP/l	.09	.03	.25	.1	.06	.03
Reactive silica	$mgSiO_2/l$	3	4.5	3	4	3	4.5
Suspended matter	mg/l	50	20	50	20	50	20

The extinction coefficient (K) is calculated from phytoplankton (Chl a) and non-phytoplankton suspended matter concentrations (SM), according to the following relationships, established empirically for the river Mosel:

$$K\ (m^{-1}) = 0.8 + 0.01\ \text{Chl}\ a\ (\mu g\ l^{-1}) + 0.05\ \text{SM}\ (mg\ l^{-1})$$

Input from the watershed

Non-point sources of nutrients and suspended matter are taken into account by associating a concentration value for each state variable of the model to the assumed two components of the river flow (superficial and ground water). These concentrations are adjusted to values close to those observed in small, unpolluted, tributaries, sampled in the drainage network of each sub-basin of the Mosel drainage network (Table 2); for the Meurthe-Mosel sub-basin, the data are those found by Kattan et al. (1986). They are considered constant in time and independent of the stream order.

Point sources of nutrients and suspended matter were arranged by order (or pK) as inhabitant-equivalents discharging wastewater and suspended matter, total organic carbon (C-Corg), ammonium (N-NH_4), nitrate (N-NO_3), phosphate (P-PO_4) in kg d^{-1}. These data were obtained from the complete record of wastewater purification plants and direct discharges from both collective housing and industries in the sub-basins, provided through the CIPMS/IKSMS by the different regional water authorities (Table 3).

The model of biological processes

The model of ecological functioning ('rive' model, Billen et al., 1994; Garnier et al., 1995) constitutes the common module for the calculation of water quality in the sub-basins and the main branch of the river Mosel. It consists of 18 variables, including nutrients (nitrate and ammonium, phosphate, dissolved silica), dissolved and particulate organic matter (as two classes of biodegradability), two taxonomic groups of phytoplankton (diatoms and non-diatoms), zooplankton and bacteria. The description of phytoplankton dynamics is based on the formulation by Lancelot et al. (1991) which distinguishes between photosynthesis, controlled by light intensity and algal growth, controlled by nutrient availability. The module was adapted to two groups of algae (diatoms and non diatoms) and a formulation for losses by excretion and grazing was added (Garnier & Billen, 1993; Garnier et al., 1995). The degradation of organic matter and bacterioplankton dynamics are described according to Billen & Servais (1989) and Billen (1991). The model also includes an original algorithm calculating nutrient exchanges across the sediment-water interface as a result of given sedimentation fluxes of organic material, taking into account organic matter degradation, associated ammonium and phosphate release and oxygen consumption, nitrification and denitrification, phosphate and ammonium adsorption on inorganic material, mixing processes in the interstitial and solid phases and accretion of the sedimentary column by inorganic matter sedimentation. Water column nitrification (Brion, 1998) and phosphate adsorption on suspended inorganic particles (and their subsequent sedimentation) are also taken into account in the model.

During the study, attention was focused on the reasons for the observed abrupt decline of phytoplankton biomass in the canalized sectors of the river Mosel, assumed to be a major factor in oxygen depletion. As hypothesized by Garnier et al. (1995) for the river Seine and shown by Caraco et al. (1997) in the Hudson River, strong benthic filter feeders such as the lamellibranch *Dreissena polymorpha* that has colonized a majority of the large rivers in Europe and in USA (Testard, 1991; Khalanski, 1997) must be taken into account in the phytoplankton dynamics. The study carried out in the river Mosel by Bachmann et al. (1997) allows us to refine the existing model by introducing a measured biomass of the benthic filter feeders, distributed along the main branch of the river. The maximum filtration rate value is closed to those found in Reeders et al. (1989), in Stanczykowska & Lewandowski (1993) and Caraco et al. (1997) and varies with temperature (Table 4).

Table 3. Actual loading in inhabitant-equivalent, and point-sources in the 3 sub-basins and main branch of the river Mosel. Total for each stream order, main branch and entire basin. SM: suspended matter; TBOC: total biodegradable organic carbon; NH_4: ammonium, NO_3: nitrates; PO_4: phosphates. Taking into account the respective drainage area for each sub-basin, the Saar is the most polluted

Sub-basins	Stream order	act. loading inhab-equ	SM kg d^{-1}	TBOC kg C d^{-1}	NH_4 kg Nd^{-1}	NO_3 kg Nd^{-1}	PO_4 kg P d^{-1}
Meurthe-Mosel sub-basin 6850 km^2	**2**	96334	4291	1060	79	88	156
	3	95726	5367	1310	202	128	144
	4	152325	5658	1377	517	152	257
	5	252660	12349	3034	1035	259	407
	6	371326	13047	2984	2797	452	792
	Total	**968371**	**40712**	**9765**	**4630**	**1079**	**1756**
Sarre sub-basin 7300 km^2	**2**	240900	14787	3327	1100	509	267
	3	627427	22338	5026	3271	284	425
	4	268174	18630	4192	1265	86	237
	5	268805	6567	1478	1011	326	142
	6	835980	36144	8132	3705	336	387
	Total	**2241286**	**98466**	**22155**	**10351**	**1542**	**1458**
Sure sub-basin 3740 km^2	**2**	24975	266	60	95	30	38
	3	8000	128	29	14	14	17
	4	398260	5647	1271	1037	585	669
	5	58525	1103	248	411	31	119
	6	78960	1316	296	312	101	165
	Total	**568720**	**8461**	**1904**	**1870**	**761**	**1009**
Mosel main Branch	**pk 0–40**	82890	225	51	211	248	58
	pk 40–80	97097	237	53	296	113	50
	pk 80–120	122170	157	36	101	165	23
	pk 120–160	185002	1228	260	341	259	55
	pk 160–200	349751	706	175	368	583	439
	pk 200–240	8394	196	41	30	9	9
	pk 240–280	591262	27720	7810	1414	1571	899
	pk 280–320	327460	11364	3190	526	1530	666
	pk 320–350	57368	6339	1426	332	5	91
	Total	**1821395**	**48172**	**10839**	**3623**	**4485**	**2293**
Entire basin 28 156 km^2	**Total basin**	**5599772**	**195811**	**44663**	**20474**	**7867**	**6516**

The kinetics of the major processes involved in the ecological functioning and the corresponding parameters are mostly derived from the experimental field work carried out in several programmes, including the CIPMS/IKSMS programme, so that no (or only minor) adjustments were required. In the model applied to the river Mosel, the kinetics formulations were essentially the same as in the original descriptions of the RIVERSTRAHLER model. (Table 4).

Results

Validations by the MoSar model in 1993, 1994 and 1995

The simulations by the model were compared to observations over the three years, for which a large data base was available (CIPMS/IKSMS, Agence de l'Eau Rhin-Meuse). Such comparisons represent a true validation of the model and not a check on the

Table 4. Kinetic formulation of the processes taken into account in the 'rive' model, and values of the corresponding parameters

Process	Kinetic expression		*Meaning*	Parameters		*Units*
phytoplankton dynamics				Diatoms	Chloro-phyceae	
Photosynthesis (phot)	kmax (1-exp-(α I/kmax))PHY	kmax*	maximal rate of photosynth.	0.2	0.5	h^{-1}
		α	initial slope of P/I curve	0.0012	0.0012	$h^{-1}(\mu E.m^{-2}s^{-1})$
reserves synthesis	srmax *M*(S/PHY,Ks) PHY	srmax*	max. rate of reserve synthesis	0.15	0.37	h^{-1}
		Ks	1/2 saturation cst	0.06	0.06	
reserves catabolism	kcr R	kcr.*	rate of R catabolism	0.2	0.2	h^{-1}
growth (phygrwth)	mufmax *M*(S/PHY,Ks) lf PHY	mufmax	max. growth rate*	0.07	0.14	h^{-1}
nutrient limitation factor	with lf = *M*(PO_4,Kpp)	Kpp	1/2 sat. cst for P uptake	15	46	$\mu g\ P\ l^{-1}$
	or *M*(NO_3 +NH_4, Kpn)	Kpn	1/2 sat. cst for N uptake	70	70	$\mu g\ N\ l^{-1}$
	or *M*(SiO_2, KpSi)	KpSi	1/2 sat. cst for Si uptake	0.42	-	$mgSiO_2\ l^{-1}$
respiration	maint PHY +ecbs phygrwth	maint*	maintenance coefficient.	0.002	0.002	h^{-1}
		ecbs	energetic cost of biosynthesis	0.5	0.5	-
excretion (phyex)	exp. phot.+ exb PHY	exp	'income tax' excretion	0.0006	0.0006	h^{-1}
		exb	'property tax' excretion	0.001	0.001	h^{-1}
lysis (phylys)	kdf + kdf (1+ vf)	kdf*	mortality rate	0.004	0.004	h^{-1}
		vf +	parasitic lysis factor	0 / 20	0 / 20	-
phyto sedimentation	(vsphy/depth).PHY	vsphy	sinking rate	.004	.0005	m/h
NH_4uptake	phygrwth /cnNH_4/(NH_4+NO_3)	cn	algal C:N ratio	7	7	$g\ C(g\ N)^{-1}$
NO_3uptake	phygrwth /cnNO_3/(NH_4+NO_3)					
PO_4uptake	phygrwth /cp	cp	algal C:P ratio	40	40	$g\ C(g\ P)^{-1}$
SiO_2 uptake	phygrwth /cSi	cSi	algal C:Si ratio	2	-	$g\ C(g\ SiO_2)^{-1}$
temperature dependency	$p(T) = p(Topt).exp(-(T-Topt)^2 / dti^2)$	Topt	optimal temperature	18	35	°C
		dti	range of temperature	13	17	°C
zooplankton dynamics				Total zooplankton		
ZOO growth	μzox.*M*(PHY-PHYo),KPHY.ZOO	μzox	max. growth rate	0.02*		h^{-1}
(zoogwth)	KPHY		1/2 sat cst to PHY	0.4		$mgC\ l^{-1}$
		PHYo	threshold phyto conc.	0.1		$mgC\ l^{-1}$
ZOO grazing	grmx.*M*((PHY-PHYo) KPHY).ZOO	grmx	max grazing rate	0.035*		h^{-1}
ZOO mortality	kdz.ZOO	kdz	mortality rate	0.001*		h^{-1}
temperature dependency	$p(T) = p(Topt).exp(-(T-Topt)^2/ dti^2)$	Topt	optimal temperature	22		°C
		dti	range of temperature	12		°C
Lamellibranchs				*Dreissena*		
Filtration rate		fmax	max filtration rate	0.01*		$m^3\ gDW^{-1}\ h^{-1}$
temperature dependency	$p(T) = p(Topt).exp(-(T-Topt)^2 / dti^2)$	Topt	optimal temperature	25		°C
		dti	range of temperature	8		°C
bacteriopankton dynamics				Bacteria		
HPi production by lysis	εpi. (phylys+bactlys+zoomort)	εp1	HP1 fraction in lysis pducts	0.2		-
		εp2	HP2 fraction in lysis pdcts	0.2		-
		εp3	HP3 fraction in lysis pdcts	0.1		
enzym. HPi hydrolysis	kib.HPi	k1b	HP1 lysis rate	0.005		h^{-1}
		k2b	HP2 lysis rate	0.00025		h^{-1}
HPi sedimentation	(vsm/depth).Hip	Vs	Hip sinking rate	0.05		$m\ h^{-1}$
Hid production by lysis	δe. (phylys+bactlys+zoomort)	εd1	HD1 fraction in lysis pdcts	0.2		-
		εd2	HD2 fraction in lysis pdcts	0.2		-
		εd3	HD3 fraction in lysis pdcts	0.1		-
enzym. HDi hydrolysis	eimax. *M(*HDi,KHi).BAC	e1max	max. rate of HD1 hydrolysis	0.75	0.75	h^{-1}
		e2max	max. rate of HD2 hydrolysis	0.25	0.25	h^{-1}
		KH1	1/2 sat cst for HD1 hydrol.	0.25	0.25	$mgC\ l^{-1}$
		KH2	1/2 sat cst for HD1 hydrol.	2.5	2.5	$mgC\ l^{-1}$
direct substr. uptake	bmax. *M(*S,Ks).BAC	bmax	max. S uptake rate	0.2	0.8	h^{-1}
		Ks	1/2 sat cst for S uptake	0.1	0.1	$mgC\ l^{-1}$
bact. growth (bgwth)	Y. bmax. *M*(S,Ks).BAC	Y	growth yield	0.25	0.25	-
bact. mortality (bactlys)	kdb.BAC	kdb	bact. lysis rate	.01	0.1	h^{-1}
bact. sedimentation	(vsb/depth).BAC	vsb	bacteria sinking rate	0	0.01	m/h
ammonification	(1-Y)/Y.bgwth/cn	cn	bact. C:N ratio	7		$gC\ gN^{-1}$
PO_4 production	(1-Y)/Y.bgwth/cp	cp	bact. C:P ratio	40		$gC\ gP^{-1}$
temperature dependency	$p(T) = p(Topt).exp(-(T-Topt)^2 / dti^2)$	Topt	optimal temperature	25	25	°C
		dti	range of temperature	15	15	°C

Table 4. contd.

nitrification and phosphorus dynamics				nitrifying bacteria	
NIT growth (nitgwth)	μnix.*M*(NH_4,KNH_4).*M*(O_2,KO_2).	μnix*	max growth rate of NIT	0.05	h^{-1}
	NIT	KNH_4	1/2 sat cst for NH_4	7	mgN l^{-1}
		KO_2	1/2 sat cst for O_2	0.6	mgO_2 l^{-1}
NH4 oxidation	nitgwth/rdtnit	rdtnit	NIT growth yield NIT	0.1	mgC $mgNH_4{}^{-2}$
NIT mortality	kdnit.NIT	kdnit*	NIT mortality rate	0.01	h^{-1}
PO4 adsorpt/desorpt.	Langmuir isotherm	Pac	SM max. adsorpt. capacity	0.0045	mgP/$mgSM^1$
(planktonic phase)		KPads	1/2 saturation ads. cst.	0.3	$mgPl^{-1}$
temperature dependency	$p(T) = p(Topt).exp(-(T\text{-}Topt)^2 / dti^2)$	Topt	optimal temperature	23	°C
		dti	range of temperature	16	°C
benthos remineralisation					
susp. matter sedim.	(vsm/depth)*MES	vsm	sinking rate	0.5	m h^{-1}
Diffusion (interstitial ph.)	Fick law	Di	app.diffusion coefficient	2 10^{-5}	cm^2 s^{-1}
Mixing (solid phase)	Fick law	Ds	mixing coefficient	2 10^{-6}	cm^2 s^{-1}
orgN mineralis. (maorg)	kib.HPi/cn				
orgP mineralis.	kip.HPi/cp	k1p*	orgP hydrolysis rate of HP1	0.05*	h^{-1}
		k2p*	orgP hydrolysis rate of HP2	0.0025*	h^{-1}
benth. nitrification	kNi*NH_4 *(in oxic layer)*	kNi	1st order nitrification cst	1	h^{-1}
NH_4adsorpt/desorpt.	1st order equilibrium	Kam	1st order adsorpt. cst for NH_4	30	-
PO_4adsorpt/desorpt.	1st order equilibrium	Kpa	PO_4 adsorpt. (oxic layer)	35	-
(in benthos)		Kpe	PO_4 adsorpt. (anoxic layer)	1.7	-
SiO_2 redissolution	kdbSi.SIB	kdbSi	silica redissolution rate	0.01	h^{-1}
temperature dependency	$p(T) = p(Topt).exp(-(T\text{-}Topt)^2\ dti^2)$	Topt	optimal temperature	25	°C
		dti	range of temperature	20	°C

*These parameters depend on temperature according to the relation mentioned.
+*M*(C,Kc) = C/(C+Kc): hyperbolic Michaelis-Menten function.
+vf: parasitic lysis amplification function. It is maintained at zero while algal density of each group remains lower than a threshold value of 65 μg Chl a / l and temperature is below 15°C.

overall outcome of the model, since the parameters are not adjusted, but determined experimentally and independently.

For the main variables of water quality (phytoplankton, nutrients) in the sub-basins and at different stations of the main branch, the comparison between observations and simulations is given for one year (1994) but for oxygen concentrations, it is three years.

On the whole, the values of the different variables and their main seasonal patterns are well reproduced by the model for both the sub-basins and the main branch. In the sub-basins, maximum phytoplankton biomass varies from 50 to 150 μg Chl *a* l^{-1}; it is highest in the upstream Meurthe-Mosel basin (Figure 2). The model tends to smooth the phytoplankton peak. The level of phosphate concentration is correctly given by the model (about 0.3 mg P l^{-1}), although the river Saar shows strong variations in winter due to high variation in simulated discharges (Figure 2). Although the model overestimates ammonium concentrations in the river Saar, the observations confirm that this river is the most polluted of the Mosel tributaries (Figure 2, see also Table 3).

The MoSar model produces satisfactory simulations of oxygen concentrations over the three years (Figure 3), but there are some discrepancies between simulations and observations, mainly for extreme values. The strongest summer oxygen deficit with respect to the saturation level within the canalized sector at Sierck is rather well given by the model.

Along the main branch of the river Mosel, comparisons are shown at three stations: Hauconcourt (pK 285) represents the beginning of the canalized sector, Sierck, 41 km downstream often shows strong oxygen deficits, and Detzem is a station downstream of the confluence with the rivers Saar and Sure, at pK 167 (see Figure 1).

Phytoplankton biomass reaches its maximum values in the canalized sector (Hauconcourt-Sierck), with seasonal variations characterised by two phases of development: first, an increase in the spring (April–May, 1994) with a maximum biomass of 80 μg Chl *a* l^{-1}, then a second phase with a sharp increase to 200 μg Chl *a* l^{-1}, followed by an abrupt decline. Unfortunately the model does not predict the true amplitude of phytoplankton biomass variations at this time. As shown by the seasonal variations of reactive silica concentration, diatoms develop during these two phases.

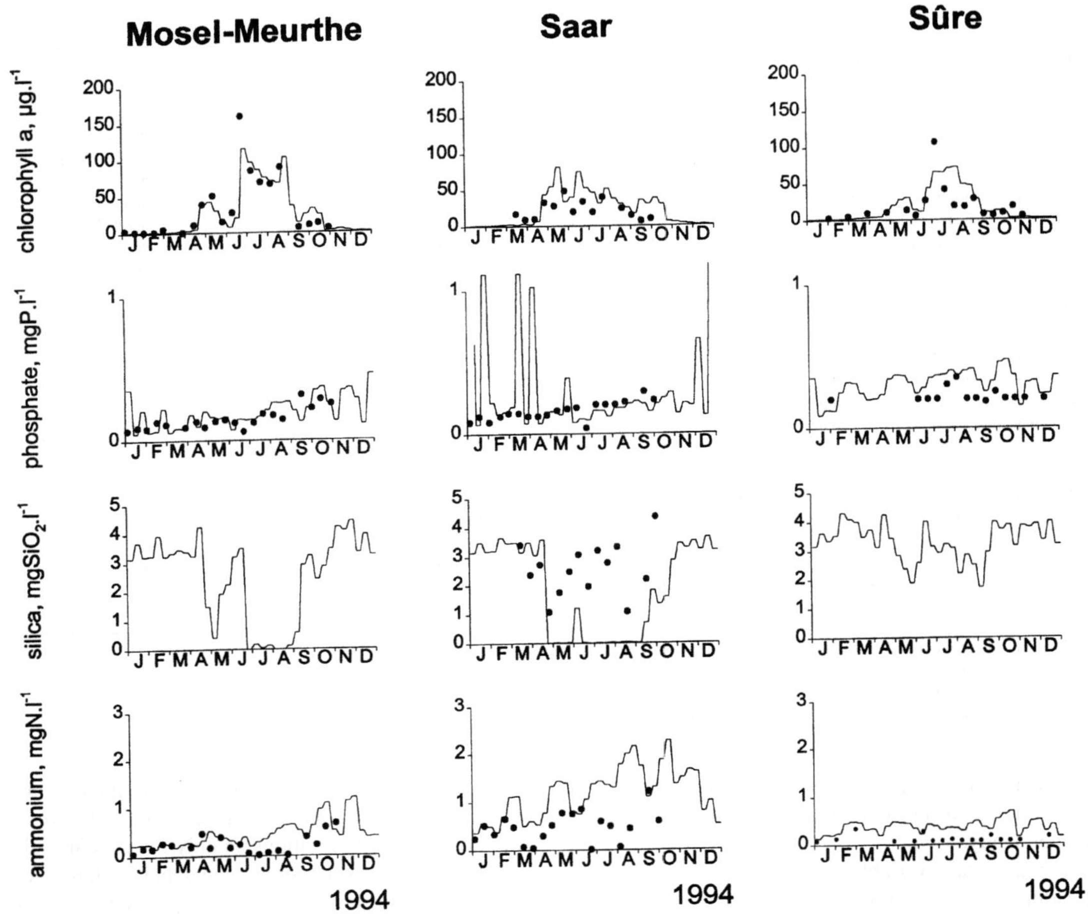

Figure 2. Sub-basins: seasonal variations during the year 1994 of chlorophyll *a* concentrations, phosphate concentrations, reactive silica and, ammonium concentration in the upstream sub-basins of the rivers Mosel, Saar and Sure.

The importance of diatoms (dominated by species of the Centrics) is confirmed by microscope observations (Gosselain et al., 1997, 1998). Phosphate concentrations vary from 0.1 to 0.4 μg P l^{-1} and are not a limiting factor for sustaining phytoplankton biomass (Figure 4). As shown by the measurements at Detzem, reactive silica is well simulated *(silica is unfortunately not systematically measured during routine sampling programmes)*. Ammonium concentrations range from 0.01 to 0.3 mgN l^{-1} and are slightly overestimated by the model (Figure 4).

Regarding year-to-year variations of oxygen concentrations, the simulations reproduce the general seasonal pattern, although the maximum amplitude of the simulations is smoothed compared to the observations. The strongest oxygen deficit in August (4 mg O_2 l^{-1}), after the phytoplankton decline, and most pronounced at Thionville and Sierck, is correctly predicted by the model (Figure 5).

Model calculations of the oxygen budget for a summer situation

The model allows the calculation of the input-output oxygen budget for specific sectors of the river. We chose four sectors that are representative of the main perturbations succeeding each other along the main branch of the river Mosel (Table 5).

For the four sectors, the oxygen budgets calculated by the model are almost balanced; however 20 tons O_2 d^{-1} ($\approx$ 20% of total) are missing in the Millery –

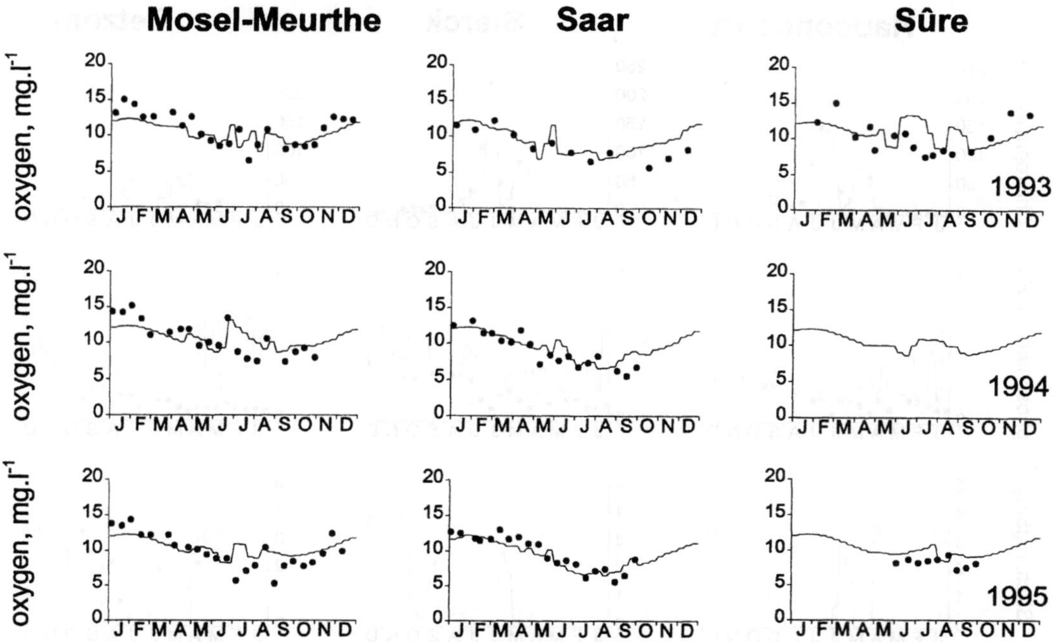

Figure 3. Sub-basins: seasonal variations in oxygen concentrations during 1993, 1994 and 1995 in the upstream sub-basins of the Mosel, Saar and Sure Rivers.

Hauconcourt sector (sector 1), for the other sectors the differences are smaller. The missing oxygen flux in the input-output balance in sector 1 corresponds roughly to an increase in oxygen concentration of 1 mg O_2 d^{-1} over the whole sector during the calculation period.

Primary production is the dominant process of oxygen input in sector 1, but it decreases regularly downstream (Table 5). In comparison, reaeration is very low. Total consumption is at its maximum in sector 2 due to a strong increase in bacterial degradation and respiration by molluscs. The increase in planktonic bacterial respiration is due to the cumulative effect of point sources of organic matter in that sector (see Table 3) whereas a high respiration rate of benthic filter feeders is the result of their large biomass (about 50 mgDW m^{-2}, Bachmann et al., 1997). Primary production exceeds total consumption in sector 1, whereas the reverse is true in sector 2. Sectors 3 and 4, have similar primary production and total consumption values.

Bacterial and benthic lamellibranch respiration, as calculated by the model is found to be the major oxygen consuming process. The contribution by planktonic nitrification and zooplankton respiration is generally low but micro-benthic respiration may become relatively significant when not masked by the one related to organic pollution and lamellibranch invasion (Table 5).

Analyses of management scenarios: autotrophic-heterotrophic sequences

Various scenarios were tested with the model to predict the effect of realistic management measures for the whole basin to improve the autotrophic conditions of the river. With its phytoplankton biomass reaching 200 μg Chl *a* l^{-1}, the river Mosel is eutrophic; phytoplankton is an autochtonous organic matter load that, in deep sectors, may consume more oxygen than it produces. An 80% phosphorus abatement of the phosphorus load discharged by waste water purification plants was tested as a mean of lowering phytoplankton biomass. An 80% reduction of the organic matter in domestic effluents, achieved by improved treatment procedures, was also tested.

The 80% phosphorus abatement scenario significantly reduced phytoplankton blooms at both Hauconcourt (pK 285) and Sierck (pK 244), whereas oxygenation was improved only at Sierck (Figure 6).

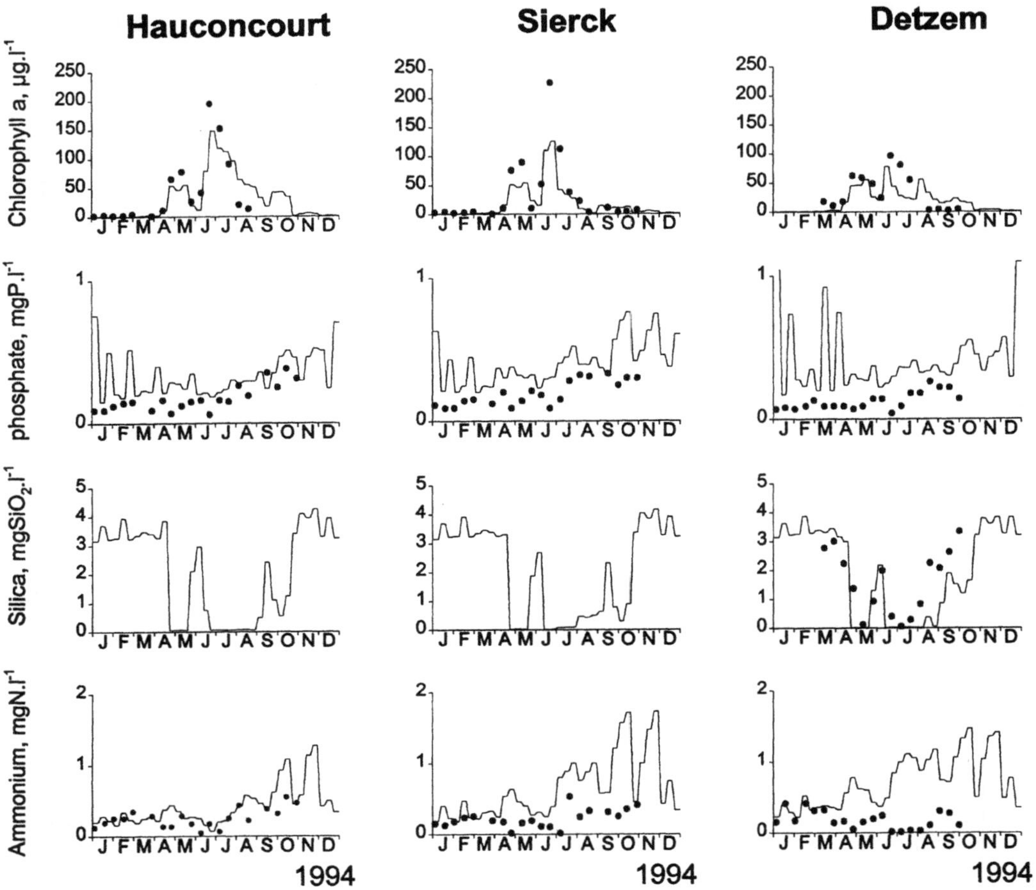

Figure 4. Main branch: seasonal variations during the year 1994 of chlorophyll *a*, phosphate, reactive silica and ammonium concentrations at Hauconcourt ((pK 285), Sierck (pK 244) and Detzem (pK 167).

This confirms that phytoplankton produces more oxygen than it consumes at the upstream station, the opposite being the case downstream. Organic matter treatment also strongly improves oxygenation, as shown at Sierck in the sector receiving maximum loads (Figure 6).

Besides seasonal variations of the variables, the model can also calculate the trophic status of the ecosystem as described in a P *versus* R diagram, where P and R are respectively the net production and total heterotrophic activity expressed as gC m^{-2} d^{-1}. Total heterotrophic activity represents the sum of bacterial and zooplanktonic activity, macro- and micro-benthic respiration. In a closed system at equilibrium, the auto- and hetero-trophic metabolisms are generally of equal intensity because autotrophy is regulated by the supply of nutrients from mineralization and heterotrophy relies on autochtonous organic matter. When the system is supplied by external sources of nutrients, primary production can exceed respiration (P/R >1), i.e. the system is autotrophic. On the contrary, an input of allochtonous organic matter leads to a mineralization greater than the production, i.e. a P/R < 1 makes the system heterotrophic. The simulation results are given for the situation of July 10–20 1994. The reference situation is characterised by an increasing autotrophy from Millery to Hauconcourt, while the system shifts to heterotrophy at Thionville and returns

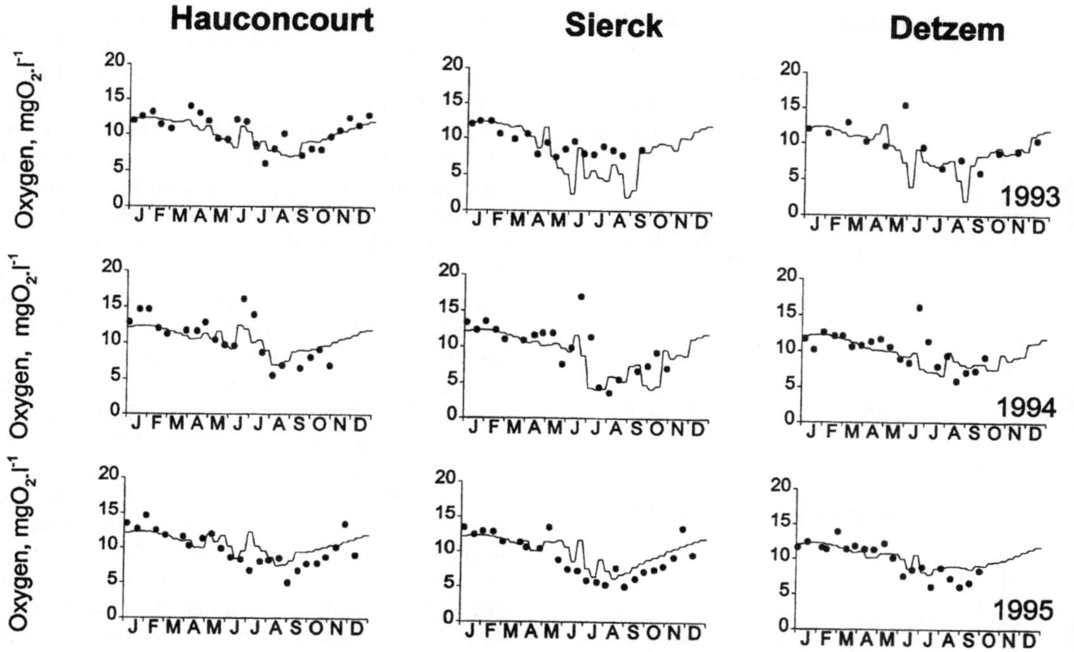

Figure 5. Main branch: seasonal variations in oxygen concentrations during 1993, 1994 and 1995 at Hauconcourt ((pK 285), Sierck (pK 244) and Detzem (pK 167).

Table 5. Oxygen budget in four sectors of the canalized river Mosel: Millery–Hauconcourt, Hauconcourt-Koenigsmacher, Koenigsmacher-Palzem, Palzem-Grevenmacher for the situation of the July 10 to 20 1994

	Sector 1 *Millery-* *Hauconcourt* pK341–285 $T\,O_2\,d^{-1}$	Sector 2 *Hauconcourt-* *Koenigsmacher* pK 341–259 $T\,O_2\,d^{-1}$	Sector 3 *Koenigsmacher-* *Palzem* pK 259-230 $T\,O_2\,d^{-1}$	Sector 4 *Palzem-* *Grevenmacher* pK 230-213 $T\,O_2\,d^{-1}$
upstream inputs &lateral inputs	46.9	49.8	23.9	29.6
reaeration	–2.3	0.6	1.9	0.9
Primary production	**65.1**	**29.2**	**18.9**	**7.7**
Total inputs & production	**109.7**	**79.7**	**44.9**	**38.2**
downstream output	45.9	19.4	26.9	31.7
Total consumption	**34.5**	**70.2**	**27.5**	**9.9**
Bacterial respiration	16.2	31.6	11.7	3.3
Zooplankton respiration	0.4	0.3	0.4	0.3
Nitrification	0.008	0.005	0.012	0.009
mollusc respiration	10.0	33.7	12.9	5.1
microbenthic respiration	7.9	4.6	2.5	1.2
Total outputs & consumption	**80.4**	**89.6**	**54.4**	**41.6**

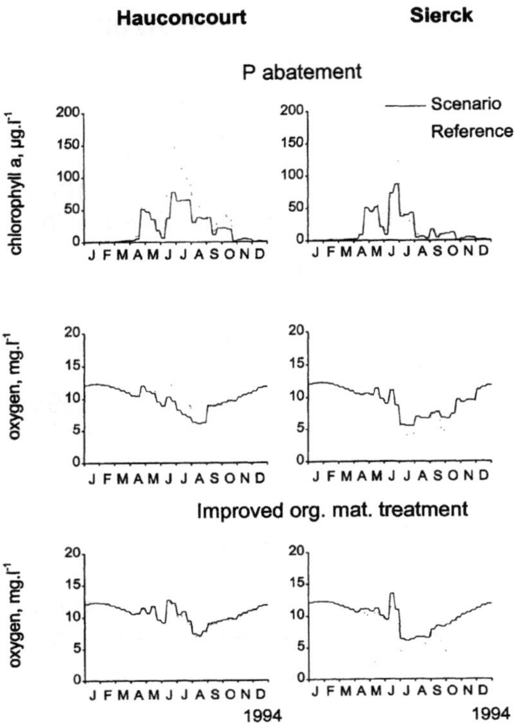

Figure 6. Seasonal variations in 1994 of chlorophyll *a* and oxygen concentrations at Hauconcourt and Sierck with an 80% phosphate abatement, and in oxygen concentrations at Hauconcourt and Sierck with an 80% reduction of organic matter.

to equilibrium downstream (Figure 7). In the scenario with improved organic matter treatment, the level of autotrophy in the sector Millery–Hauconcourt is unchanged, but the system does not become strongly heterotrophic in the lower stretches. In the scenario with phosphorus removal, the autotrophy in the upstream stretch is much less pronounced. In fact, a much slower increase in autotrophy is observed from Millery downwards, preventing the system from reaching heterotrophy in the Thionville sector, in spite of organic matter inputs. The removal of both phosphorus and organic matter greatly restores the system giving it a regular increase in autotrophy and much lower values of primary production (Figure 7).

Discussion

The effect of the invasive lamellibranchs (*Dreissena polymorpha*) is now well documented in both lakes and river systems; it does not only concern phytoplankton communities (Reeders et al., 1989; Testard, 1991; Stanczykowska & Lewandowski, 1993; Caraco et al., 1997; Boët et al., 1998) but also the zooplankton and even the entire food web (Mac Isaac et al., 1995; Pace et al., 1998). Modelling of phytoplankton development at the scale of the drainage network of the river Seine (France) led us to consider a composite factor of phytoplankton lysis to explain its abrupt decline (Billen et al., 1994; Garnier et al., 1995), which often causes strong oxygen deficits (Billen & Garnier, 1996); besides zooplankton and possibly viruses and fungii, we also suggested that lamellibranchs and protozoa might be important factors controlling phytoplankton. Based on the work by Testard (1991) in the Seine Basin and Boët et al. (1998), the role of the lamellibranchs was included in the RIVERSTRAHLER model (Garnier & Billen, 1998). The similitude between the waxing and waning of phytoplankton and the associated oxygen deficit in the rivers Mosel and Seine, together with the information available on the lamellibranch communities (Bachmann et al., 1997; Bachmann et al., 1999) allowed us to include them in the model for the river Mosel.

As opposed to bacteria, phytoplankton and zooplankton which are all known to have a physiology represented by Michaelis-Menten kinetics with experimentally measured parameters, the data on lamellibranchs are insufficient to permit modelling of their dynamics because of their complexe life cycle (larvae, adults). The factors controlling the production of larvae are not understood; year-to-year variations by a factor of 10 have been observed in both the rivers Mosel (S. Schweitzer & S. Schmitt, pers comm.) and Seine (Testard, pers. comm.; Boët et al., 1998). The growth of the larvae by cohorts and especially, the fixation rate of the adults and renewal rate of the populations are not sufficiently documented. For these reasons, the mussels were taken into account as a constraint in our RIVERSTRAHLER model. Note, however that an attempt was made by Schöl et al. (1997, 1999), to include the biomass of the mussels as a state variable in their Mosel River model and to express its variations as a function of food intake, from an arbitrarily set low initial level at the beginning of the year. It seems that a correct level of lamellibranch biomass can be simulated by the authors, but with very low values of maximum mussel filtration rates. Therefore, due to the increasing lamellibranch invasion in aquatic systems and their considerable effect on the trophic structure, we are aware that further

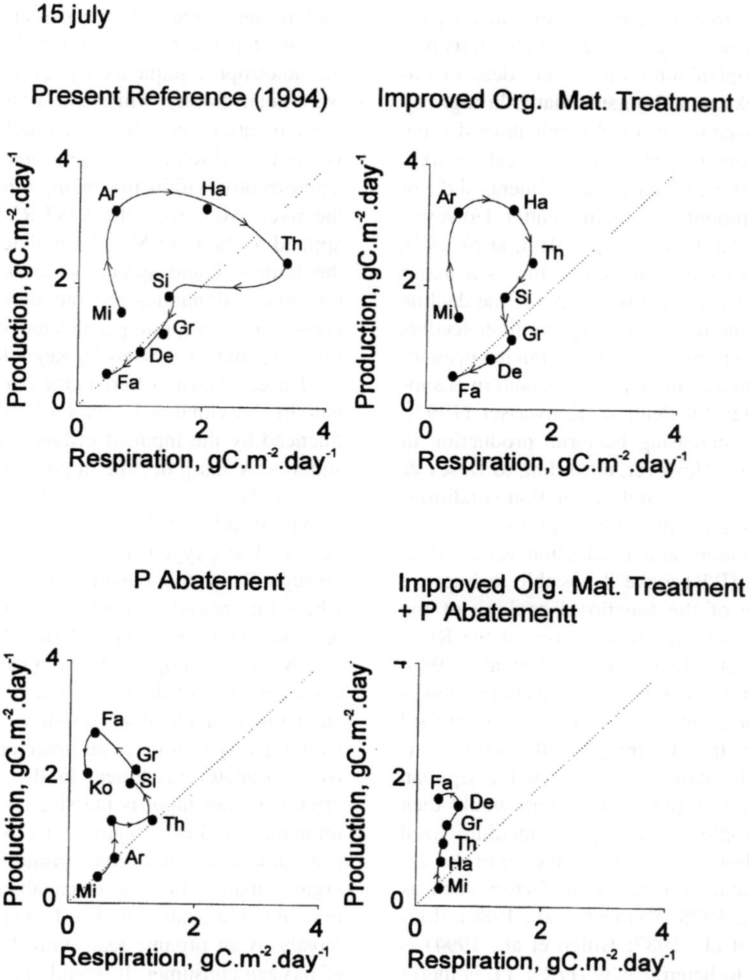

Figure 7. Production *versus* Respiration (mg C m^{-2} d^{-1}) diagrammes for the reference situation in 1994, an 80% reduction of organic matter, an 80% phosphate abatement, and a combined 80% reduction of organic matter and phosphate (Mi: Millery, pK 341; Ar: Arry, pK 314; Ha: Hauconcourt, pK 285; Th: Thionville, pK 269; Si: Sierck, pK 244; Pa: Palzem, pK 230; Gr: Grevenmacher, pK 213; De: Detzem, pK 167; Fa: Fankel, pK 59; Ko: Koblenz, pK 1).

investigations are still needed to properly take into account mussel dynamics in future models of ecological functioning.

Planktonic nitrification does not appear to be a significant factor of oxygen deficit in the Mosel. Benthic nitrification is included in the figures for benthic oxygen consumption, but constitutes only a small fraction. Investigations by Brion (1998) in large rivers showed that nitrification influences the oxygen budget in the lower part of the rivers, mainly in their estuaries where high levels of suspended matter facilitate bacterial attachment to particles.

In addition to their effect on the decline of phytoplankton biomass, indirectly influencing the oxygen input by primary production, the mussels contribute heavily to overall respiration, in a manner similar to that of planktonic bacteria. These figures are confirmed by direct *in situ* measurements by Bachmann et al. (1997, 1999) who estimate that benthic respiration amounts to 10 t O_2 d^{-1}, i.e. 25% of the oxygen budget.

Bacterial heterotrophic activity is sustained by organic matter with two origins: the autochtonous part provided by phytoplankton biomass, i.e. dead or unhealthy phytoplankton, and organic matter brought by effluents. The consequences of phytoplankton decline in the canalized sector would not be critical for oxygenation if domestic and industrial effluents did not contribute large amounts of organic matter. However, the city of Metz (226 000 inh-equivalent, at pk 297), at the beginning of the canalized sector, is a major source of organic matter that reinforces 1. the decline in phytoplankton biomass caused by the filter feeders and 2. the decrease in mean water column net primary production with increasing depth. This pattern is supported by the study by Peter & Kaltwasser (1997), showing strongly increasing bacterial production in the canalized sector. However, according to Descy & Gosselain (1994), the new light limitation conditions would not increase phytoplankton respiration.

The oxygen budget and Production *versus* Respiration diagrams (P/R) make it possible to improve the understanding of the functional ecology of the river Mosel. In developing their theory of the River Continuum Concept (RCC), Vannote et al. (1980) proposed a general scheme of an upstream downstream succession of the P/R status of unperturbed rivers. Headwater streams strongly influenced by allochtonous organic matter inputs from the riparian vegetation are heterotrophic; the system would then shift from heterotrophy to autotrophy in medium-sized rivers (stream orders 5–6), when limitation of the development of primary producers by factors such as shading (Minshall, 1978; Vannote et al., 1980), dilution rate (Descy et al., 1987; Billen et al., 1994) is smaller. A return to heterotrophy (P/R< 1), is therefore expected in higher orders (8 or 9) with increasing depth and turbidity (i.e. light limitation). The outlet of the Mosel-Meurthe sub-basin (order 6) is autotrophic, but autotrophy is followed by heterotrophy over a short distance (same order, 6), due to human perturbations (canalization, eutrophication followed by phytoplankton decline, organic pollution). As opposed to the conceptual approach (Vannote et al., 1980) and the quantitative approach based on a survey of field measurements in different river systems (Billen et al., 1995), a modelling approach allows the P/R behaviour to be explored through various scenarios of river rehabilitation. Our predictions are quite consistent with the RCC scheme for returning to 'nearly natural conditions', i.e. when both eutrophication and organic pollution are considerably lowered. The responses of each of the two tested scenarios are very informative; they show that organic pollution shortens the length of the autotrophic sequence by increasing the trend toward heterotrophy, while eutrophication (phosphorus contamination) leads to accelerated phytoplankton development which in itself reinforces the trend toward heterotrophy within the main, 6th order branch, of the river. Moreover, the RIVERSTRAHLER model applied to the river Mosel simulates '*very faithfully*', the temporal and spatial variations of phytoplankton whose dynamics are the more complex as the constraints (morphology, discharges, nutrients, etc.) fluctuate sharply (review by Reynold & Descy, 1996).

The ecological pattern in the river Mosel resembles that in the canalized sector of the river Seine, influenced by the input of effluents from Paris and its suburbs. A sharp decline in phytoplankton bloom occurs at the outlet of the canalized Marne tributary heavily invaded by *Dreissena* with a resulting oxygen deficit. The oxygen deficit grows as the river flows through Paris and beyond, until anoxia is reached where the treated effluents of 10 million inhabitants are poured into the river (Billen & Garnier, 1996).

Overall, among all the oxygen consuming processes, the model shows that bacterioplankton activity and lamellibranch impact dominate. Whereas eutrophication partly originates far upstream due to progressive phosphate enrichment in the drainage network, organic matter input is massive and its impact is felt immediately downstream of its point of discharge. Eutrophication could compensate for degradation of organic matter, but due to canalization (light limitation and colonization by bivalves) phytoplankton acts mostly as an organic load with the paradoxical role of oxygen consumer. It would appear that the targets for the rehabilitation of the river Mosel would be a decrease in phosphorus load, together with a reduction in the discharge of organic matter.

Acknowledgments

This work was undertaken in the framework of the CIPMS/IKSMS and CE-DGXI Intereg programmes. We are grateful to all members of the CIPMS/IKSMS and from the Agence de L'Eau Rhin-Meuse for helpful discussions and communication of the data base on the river Mosel. Many thanks to Xavier Philippon, assistant-ingérieur at the UMR Sisyphe, for his valuable help with the realization of the figures. Many thanks to four independant reviewers for their helpful comments on the manuscript.

References

Admiral, W., S. D. Mylius, E. D. De Ruyter van Steveninck & G. M. J. Tubbing (1993). A model of phytoplankton production in the lower River Rhine verified by observed changes in silicate concentration. J. Plankton Res. 15: 659–682.

Admiral, W., L. Breebaart, G. M. J. Tubbing, B. van Zanten, E. D. De Ruyter van Steveninck & R. Bijkerk, 1994. Seasonal variation in composition and production of planktonic communities in the lower River Rhine. Freshwat. Biol. 32: 519–531.

Bachmann, V. & P. Usseglio-Polatera & J.-C. Moreteau, 1997. Contribution of the benthic compartment. CIPMS/IKSMS report 'Oxygen budget and biological processes in the regulated rivers Moselle and Saar': 112 p.

Bachmann, V. & P. Usseglio-Polatera, 1999. Contribution of the macrobenthic compartment to the oxygen budget of a large regulated river: the Mosel. Hydrobiologia 410: 39–46.

Billen, G., P. Servais, 1989. Modélization des processus de dégradation de la matière organique en milieu aquatique. In M. Bianchi et al. (eds), Microorganismes. Dans les Écosystèmes Océaniques, Chap. 8. 219–245, Masson, Paris.

Billen, G., 1991. Protein degradation in aquatic environments. In R. Chröst (ed.), Microbial Enzymes in Aquatic Environments, Chap. 7. 122–142, Springer Verlag.

Billen, G., J. Garnier & Ph. Hanset, 1994. Modelling phytoplankton development in whole drainage networks: the 'riverstrahler' model applied to the Seine river system. Hydrobiologia 289: 119–137.

Billen, G., H. Décamps, J. Garnier, M. Meybeck, P. Servais & Ph. Boët, 1995. River and Stream ecosystems, Cushing, Cumming & Marshall (eds). Chapter 12. France, Belgium, Netherlands. Elsevier: 389–418.

Billen, G. & J. Garnier, 1996. Transfert et métabolisme de l'azote et du phosphore dans l'hydrosystème Seine. In C. Le Coz, B. Tassin & D. Thévenot (eds), 'Transfert des Polluants dans les Hydrosystèmes". Actes des Journées Sciences et Techniques de l'Environnment. Presses de l'Ecole Nationale des Ponts et Chaussées. Paris: 121–140.

Boët, Ph., M. Akopian, J. Belliard, R. Berrebi-dit-Thomas, A. Miquelis, R. Pourriot, E. Talès & P. Testard, 1998. La Faune Aquatique. In M. Meybeck et al. (eds), La Seine en son Bassin. Elsevier, Paris: 627–678.

Brion, N., 1998. Etude du Processus de nitrification à l'échelle de grands réseaux hydrographiques anthropisés. Ph.D. Thesis. Faculté des Sciences. Université Libre de Bruxelles, 85 pp.

Bultot, F. & G. Dupriez, 1976. Conceptual hydrological model for an average-sized catchment area. J. hydrol. 39: 251–292.

Capblancq, J., A. Dauta, B. Caussade & H. Décamps, 1982. Variations journalières de la production primaire du phytoplancton en rivière: modélization d'un bief du Lot. Ann. Limnol., 18: 101–132.

Caraco, N. F., J. J. Cole, P. A. Raymond, D. L.Strayer, M. L. Pace, S. E. G. Findlay & D. T. Fischer, 1997. Zebra mussel invasion in a large, turbid river: phytoplankton response in increased grazing. Ecology 78: 588–602.

Dauta, A., 1986. Modélization du développement du phytoplancton dans une rivière canalisée eutrophe: le Lot (France). Ann. Limnol., 22: 119–132.

De Ruyter van Steveninck, E. D., W. Admiraal, L. Breebaart, G. M. J. Tubbing & B. van Zanten, 1992. Plankton in the River Rhine. Structural and functional changes observed during downstream transport. J. Plankton Res. 14: 1351–1368.

Descy, J.-P., P. Servais, J. S. Smitz, G. Billen & E. Everbeck, 1987. Phytoplankton biomass and production in the river Meuse (Belgium). Wat. Res. 12: 1557–1566.

Descy, J.-P. & V. Gosselain, 1994. Development and ecological importance of phytoplankton in a large lowland river (River Meuse, Belgium). Hydrobiologia 289:139–155.

Garnier, J. & G. Billen, 1993. Ecological interactions in a shallow sand-pit lake (Créteil Lake, France). A modelling approach. In Nutrient dynamics and biological structure in shallow freshwater and brackish lakes. Hydrobiologia 275/276: 97–114.

Garnier, J., G. Billen & M. Coste, 1995. Seasonnal succession of diatoms and Chlorophyceae in the drainage network of the river Seine: Observations and modelling. Limnol. Oceanogr. 40: 750–765.

Garnier, J. & G. Billen, 1998. Développement algal et eutrophization. In Meybeck et al. (eds), La Seine en Son Bassin. Elsevier, Paris: 593–626.

Gosselain, V., J. P. Descy & E. Everbeck, 1994. The phytoplankton community of the river Meuse, Belgium: seasonal dynamics (year 1992) and the possible incidence of zooplankton grazing. Hydrobiologia 289: 179–191.

Gosselain, V., J.-P. Descy & L. Viroux, 1997. Final report of the 3 years of study on phytoplankton dynamics, primary production, *in situ* zooplankton grazing, daily cycle of oxygen content and zooplankton spatial distribution. CIPMS/IKSMS report 'Oxygen budget and biological processes in the regulated rivers Moselle and Saar': 76pp.

Gosselain, V., L. Viroux & J.-P. Descy, 1998. Can a community of small-bodied grazers control phytoplankton in rivers? Freshwat. Biol. 39: 9–24.

Hammer, M. & Mc K. A. Kichan, 1981. Hydrology and quality of water resources. J. Wiley & Sons, Inc.: 280 pp.

Kattan, Z., J.-L. Salleron & J.-L. Probst, 1986. Bilans et dynamiques du transfert de l'azote et du phosphore sur le bassin de la Moselle (Nord-Est de la France). Sc. Eau 5: 435–459.

Khalanski, M., 1997. Conséquences industrielles et écologiques de l'introduction de nouvelles espèces dans les hydrosystèmes continentaux: la moule zébrée et autres espèces invasives. Bull. fr. Piscic. 344/345: 385–404.

Kölher, J., 1995. Growth, production and losses of phytoplankton in lowland River Spree: carbon balance. Freshwat. Biol. 34: 501–512.

Lancelot, C., C. Veth & S. Mathot, 1991. Modelling ice-edge phytoplankton bloom in the Scotia-Weddell Sea sector of the Southern Ocean during spring 1988. J. Mar. system 2: 333–346.

MacIsaac, H. J., C. J. Lonnee & J. H. Leach, 1995. Suppression of microzooplankton by zebra mussels: importance of mussel size. Freswat. Biol. 34: 379–387.

Pace, M. L., E. Stuart, G. Findlay & D. Fisher, 1998. Effect of invasive bivalve on zooplankton community of the river Hudson. Freshwat. Ecol. 39: 103–116.

Peter, S. & H. Kaltwasser, 1997. Bacterioplankton activities and oxygen budget in the regulated rivers Mosel and Saar. CIPMS/IKSMS report.

Quotbi, A., 1996. Etude de la dynamique du phytoplancton et des nutriments dans une rivière aménagée le Lot (France). Modélization mathématique et simulations de scénarios. Thèse Doct. Université Paul Sabatier, Toulouse: 272pp.

Reeders, H. H., A. Bij de Vaate & F. J.Slim, 1989. The filtration rate of *Dreissena polymorpha* (Bivalvia) in three Dutch lakes with reference to biological water quality management.

Reynolds, C. S. & Descy, 1996. The production, biomass and structure of phytoplankton in large rivers. Arch. Hydrobiol. Suppl 113, Large Rivers 10: 1–4.

Schöl, A., V. Kirchesch, T. Bergfeld & D. Müller, 1997. Oxygen budget and biological processes in the regulated rivers Mosel and Saar. The BfG model. Phase 3: 80pp + annex.

Schöl, A., V. Kirchesch, T. Bergfeld & D. Müller, 1999. Model-based analysis of oxygen budget and biological processes in the regulated rivers Moselle and Saar: modelling the influence of benthic filter feeders on phytoplankton. Hydrobiologia, 410: 167–176.

Stanczykowska, A. & K. Lewandowski, 1993. Effect of filtering activity of Dreissena polymorpha (Pall.) on the nutrient budget of the littoral of the lake Mikolajskie. Hydrobiologia 251: 73–79.

Strahler, A. H., 1957. Quantitative analysis of watershed geomorphology. Trans. Am. Geophys. Union 38: 913–920.

Testard, P., 1991. Elements d'écologie de la Dreissène *Dreissena polymorpha* Pallas. Etude de la dispersion des larves et de leur fixation. Réponses à la désoxygénation de l'eau. Thèse Doct. D'Etat, Univ. P. M. Curie, Paris 6: 357pp.

Vannote, R. L., G. W. Minshall, K. W. Cummins, J. R. Sedell & C. E. Cushing, 1980. The river continuum concept. Can. J. Fish. aquat. Sci. 37: 130–137.

Hydrobiologia **410:** 167–176, 1999.
J. Garnier & J.-M. Mouchel (eds), Man and River Systems.

Model-based analysis of oxygen budget and biological processes in the regulated rivers Moselle and Saar: modelling the influence of benthic filter feeders on phytoplankton

Andreas Schöl, Volker Kirchesch, Tanja Bergfeld & Dieter Müller
Federal Institute of Hydrology (BfG), Postfach 200253, 56002 Koblenz, Germany

Key words: water quality model, regulated river, oxygen budget, phytoplankton, benthic filter feeder, *Dreissena polymorpha*

Abstract

The oxygen budget of the regulated rivers Moselle and Saar was analyzed with a water quality model for rivers (BfG model). In the river Moselle in midsummer 1994 a strong oxygen decline between Hauconcourt and Palzem was correlated with a decrease of algal biomass. Further oxygen depletion in later summer in both rivers was a result of oxygen-consuming processes, bacterial carbon oxidation and nitrification, in times of low phytoplankton biomass and photosynthesis. As computed by our model, algal biomass is reduced in summer and autumn by highly efficient grazing of the zebra mussel *Dreissena polymorpha*. From July until September 1994 in the upper part of the river Moselle, a mean reduction of chlorophyll *a* of about 50% was computed. A dynamic approach is used to compute the seasonal increase of *D. polymorpha* population density and filtering capacity as well. The longitudinal decrease of mussel population density in the Moselle towards the river mouth is a consequence of the decreasing food concentration mainly due to the filtering activity of *D. polymorpha* in upstream reaches.

Introduction

In wastewater-loaded regulated rivers, severe oxygen problems in summertime are often observed. In the rivers Moselle and Saar, oxygen depletion is strongly linked to the seasonal phytoplankton development. Both rivers are eutrophic, but nevertheless during summer low algal biomass is observed (Descy, 1993). As losses through sedimentation are assumed to be low and effective control on algal biomass by zooplankton does not happen (Gosselain et al., 1998a,b), the summer decline in phytoplankton must be explained by other processes.

Some authors emphasize the ecological impact of benthic filter feeders in the biotic processes of water bodies, earliest in estuarine systems but for lakes and rivers as well (Alpine & Cloern, 1992; Caraco et al., 1997; Cohen et al., 1984; Effler et al., 1996; Padilla et al., 1996; Stanczykowska & Lewandowski, 1993). Some of these studies were motivated by the invasion of the bivalve *Dreissena polymorpha* into the inland waters of Northern America. Since the last century *D. polymorpha* has become common in North–West Europe (Kinzelbach, 1992), and now it is a major benthic species in river inventories of the Moselle (Bachmann et al., 1997). These mussels have a high filtering capacity (Reeders et al., 1989) and are able to reduce the particle load of a river. In order to quantify the influence of *D. polymorpha* on phytoplankton and the oxygen budget of Moselle and Saar, the BfG model of the Federal Institute of Hydrology for the simulation of biotic processes in rivers was supplemented by the component 'Benthic Filter Feeder'.

Description of the BfG model

The BfG model calculates longitudinal profiles and seasonal cycles of all important water quality parameters. It assumes a turbulent flow regime and stationary discharge conditions for each longitudinal profile. The BfG model has a modular structure, main modules concerning hydraulic, physical, chemical, and biotic processes are listed in Figure 1. Additionally, a biotic

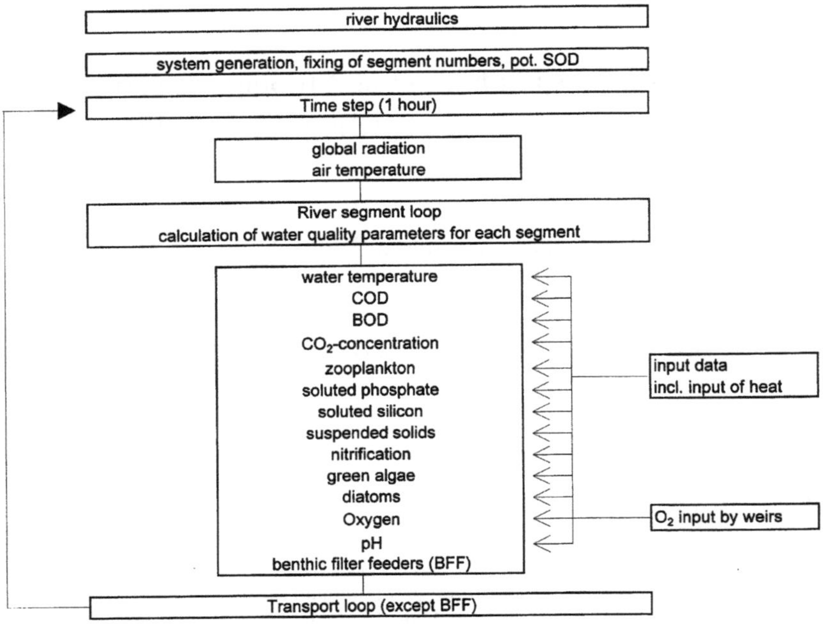

Figure 1. Scheme of the BfG model.

module simulating the growth of benthic filter feeders (BFF) was implemented here. Driving forces of the model are the discharge at the upper boundary and of main tributaries as well as meteorological conditions as global radiation, air temperature, cloudiness, and wind velocity. The hydraulic conditions are derived from gauged cross-sections every 0.1–2 km. The model requires input data of water quality parameters at the upper boundary of the modelled river stretches and of main tributaries (see Figure 2).

The model is applied on the main branches of Moselle from Millery (Moselle-km 345) to Koblenz (Moselle-km 1.5) and of Saar from Güdingen (Saar-km 92) to Konz (Saar-km 0) (Schöl et al., 1997). The transport of a distinct water mass from the upper boundary downstream is computed for every day from 1 March until 31 October 1994 with the same set of internal parameters and constants. Fixed waste water inputs by point sources were used in both rivers throughout the year 1994. For validation of the model, data of the ICPMS monitoring programme for Moselle and Saar (ICPMS, 1998) as well as data of scientific surveys (Gosselain et al., 1998b; Mosbacher & Schmitt, 1994) were used.

The mathematical description of the influence of the benthic filter feeders on the element and nutrient cycles of a river follows the modelling approaches by Schneider (1992) and Ebenhöh et al. (1995). All relevant parameters used in the BFF module refer to the zebra mussel *Dreissena polymorpha*. The assumption of an uniform population at the beginning of a year consisting of individuals with a shell length (L) of 8 mm each and a density of 3183 individuals/m^2 results in a mussel biomass of 5 g C/m^2 or 10 g dry weight/m^2, respectively.

$$D = (0.0093 * L^{2.8}) * 0.5,$$

D is soft body weight of *D. polymorpha* (mg C).

The growth of biomass in the course of a year is computed in dependence on consumed algae, diatoms (1 μg chlorophyll $a \cong 39$ μg C) and green algae (1 μg chlorophyll $a \cong 27$ μg C). The computed surface of the riverbed and the slopes is taken as colonization areas for mussels.

$$\frac{\mathrm{d}D_{seg}}{\mathrm{d}t} = (\mu - \mathrm{resp}_b - \mathrm{mort}) * D_{seg},$$

where D_{seg} is biomass of *D. polymorpha* population in one computed segment (mg C); μ is growth rate of *D. polymorpha* (day^{-1}); resp_b is basal respiration rate of *D. polymorpha* (day^{-1}); mort is mortality rate of *D. polymorpha* (day^{-1}).

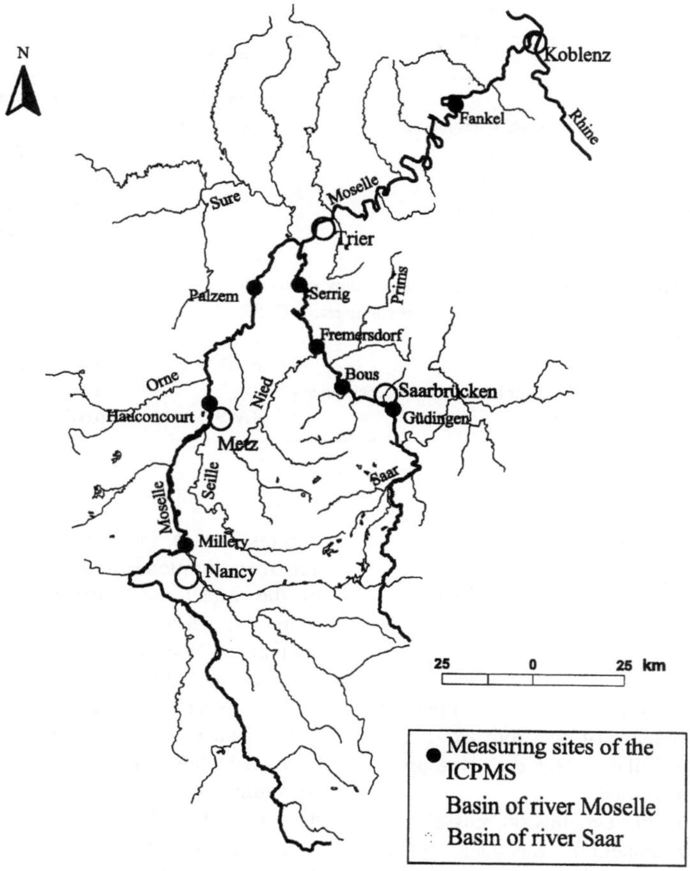

Figure 2. Basins of Moselle and Saar with main tributaries and sampling sites of the ICPMS (International Commission for the Protection of Moselle and Saar against pollution) monitoring programme.

The growth rate and the basal respiration rate are computed for one mussel. The growth rate (μ) is calculated by multiplying the ingestion rate with a yield factor (*Y*):

$$\mu = \text{ingest} * Y,$$

where

$$\text{ingest} = \varepsilon * D^{-0.213} * f(F) * f(T),$$

where ingest is ingestion rate of *D. polymorpha* (day^{-1}); ε is ingestion factor of a 1 mg C individual of *D. polymorpha* at the optimum temperature (mg C day^{-1}); *f*(*F*) is dependency of ingestion rate on food availability; with

$$\text{if } F > F_{\max} \text{ then } f(F) = 1$$

$$\text{if } F \leq F_{\max} \text{ then } f(F) = F/F_{\max},$$

where *F* is actual food concentration (mg C_{algae}/l); $F_{\max}$ is food concentration for maximal ingestion rate (mg C_{algae}/l) and *f*(*T*) is dependency of ingestion rate on water temperature with

$$f(T) = e^{-0.00605*(T_{\text{opt},1}-T)^2},$$

where *T* is actual water temperature (°C); $T_{\text{opt},1}$ is optimal water temperature for ingestion (°C).

The yield factor (*Y*) is determined by three fractions of the ingested food: egestion as pseudofaeces (FAEC), excretion and active respiration during feeding. The two latter ones are not depending on the mussel biomass. The assimilated fraction (1–FAEC)

Table 1. Parameters for the description of the growth of *D. polymorpha*

Symbol	Description	Dimension	Value	References
ε	Ingestion factor	mg C^{-1} day^{-1}	0.1105	Walz (1978)
F_{max}	Food concentration for maximal ingestion rate	mg C_{algae}/l	2.0	Walz (1978)
$T_{opt,1}$	Optimal temperature for ingestion	°C	20	Kuntzmann (1989)
α	Minimum value of egested fraction	–	0.351	Schneider (1992)
β	Coefficient for feeding level dependence	–	0.88	Schneider (1992)
q_{ex}	Excreted fraction of assimilated food	–	0.064	Schneider (1992)
q_{resp}	Respired fraction of assimilated food	–	0.15	Ebenhöh et al. (1995)
$T_{opt,2}$	Optimal temperature for basal respiration and mortality rate	°C	28	Schneider (1992)
T_{lim}	Temperature limit for basal respiration and mortality rate	°C	31	Schneider (1992)
Q_{10}	Van't Hoffscher temperature coefficient	–	3.1	Schneider (1992)
γ	Basal respiration factor	mg C^{-1} day^{-1}	0.0006	Schneider (1992)
δ	Basal respiration coefficient for biomass level dependence	–	0.25	Schneider (1992)
$mort_{max}$	Maximal mortality rate at $T_{opt,2}$	day^{-1}	0.0025	Kuntzmann (1989)

decreases with increasing food concentrations.

$$Y = (1-\mathrm{FAEC})-(1-\mathrm{FAEC})*q_{ex}-(1-\mathrm{FAEC})*q_{resp}$$

and

$$\mathrm{FAEC} = \alpha * e^{\beta^* f(F)},$$

where α is minimum value of FAEC; β is coefficient for feeding level dependence of egestion; q_{ex} is excreted fraction of the assimilated food; q_{resp} is respired fraction of the assimilated food.

Basal respiration ($resp_b$) depends on mussel biomass and a function for temperature dependence ($f(T)$) described by O'Neill et al. (1972). The function is determined by the parameters T_{lim}, $T_{opt,2}$ and Q_{10} (see Table 1) and is a multiplier ranging from 0 to 1. The maximal basal respiration ($resp_{b,max}$) is:

$$resp_{b,max} = \gamma * D^{-\delta},$$

where γ is basal respiration factor for *D. polymorpha* (mg C^{-1} day^{-1}); δ is basal respiration coefficient for biomass level dependence.

The mortality rate (mort) affects the *D. polymorpha* population of one computed segment. It is depending on a maximum value ($mort_{max}$) and the same kind of temperature function as described for basal respiration rate.

Results and discussion

The measured oxygen depletion during summer in Moselle and Saar is reproduced by the model (see Figures 3 & 4). In the river Moselle the strong oxygen oversaturation in midsummer is not well simulated by the model, particularly due to the near surface sampling, while the model calculates mean oxygen contents of the water column. During spring and early summer 1994 oxygen supersaturation caused by phytoplankton blooms occurred in the Moselle, while in the Saar only a slight spring peak was observed. Oxygen depletion in later summer is a result of oxygen consumption processes in times of low phytoplankton biomass and low biogenic oxygen input, respectively.

In the investigated stretches of Moselle and Saar the loading with organic matter and ammonium is responsible for oxygen undersaturation (Schöl et al., 1997). The oxygen demand of heterotrophic bacteria and nitrifiers is increasing with rising water temperature. The impact of these consumptions on the oxygen budget of the rivers becomes noticeable in later summer. Additionally during low discharge conditions, the oxygen input via the water surface is hindered by low flow velocities and low turbulences. Especially in the river Saar the water body of the impoundments of Mettlach, Saar-km 31, and Serrig, Saar-km 18, with a mean water depth of 4.6 m is not well aerated. Consequently the oxygen budget is stabilized by oxygen inputs at weirs, becoming important in summer.

In the river Saar, the dependence of the oxygen deficit on phytoplankton is not as strong as in the Moselle, which is particularly due to low input of algal biomass from the upper basin (see Figure 4, Saar-km 92). In contrast, the algal loading from the upper basin of Moselle and Meurthe is high (see Figure 3, Moselle-

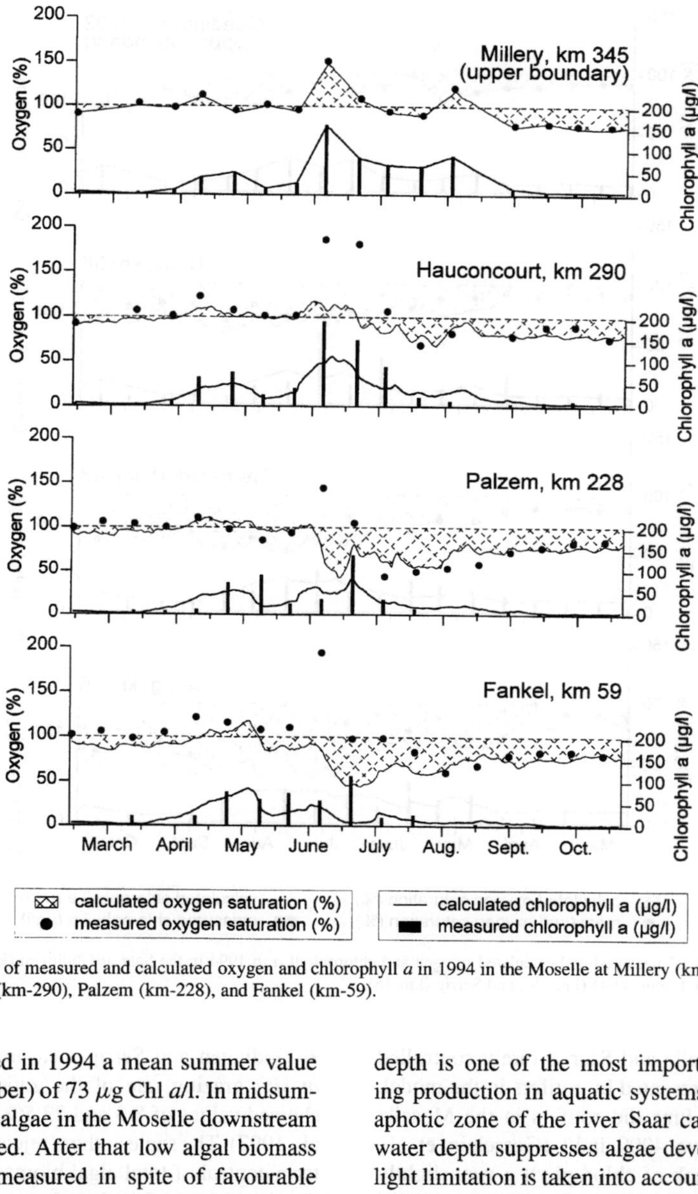

Figure 3. Seasonal cycles of measured and calculated oxygen and chlorophyll *a* in 1994 in the Moselle at Millery (km-345, upper boundary of the model), Hauconcourt (km-290), Palzem (km-228), and Fankel (km-59).

km 345), and reached in 1994 a mean summer value (June to mid September) of 73 μg Chl *a*/l. In midsummer, a decline of the algae in the Moselle downstream from Millery occurred. After that low algal biomass in both rivers was measured in spite of favourable growth conditions concerning low discharge and sufficient nutrient supply. Such seasonal phytoplankton development is documented for other temperate rivers, as well (Billen et al., 1994; Köhler, 1995).

In turbid rivers, the carrying capacity of the system is often not reached because of strong light limitation (Dokulil, 1994). Turbidity increases the ratio of mixed zone to euphotic zone, therefore the critical mixing depth is one of the most important factors regulating production in aquatic systems. As well the large aphotic zone of the river Saar caused by high mean water depth suppresses algae development. Although light limitation is taken into account in the BfG model, neither the strong algal reduction in the Moselle nor low algal biomass in summer in both rivers could be simulated sufficiently (Schöl et al., 1997).

Besides the hypothesis of algal losses due to viral or bacterial infections, it is mainly assumed that grazing losses by zooplankton are responsible for low summer algal biomass (Garnier et al., 1995). But, in 1994, according to the simulation, the phytoplank-

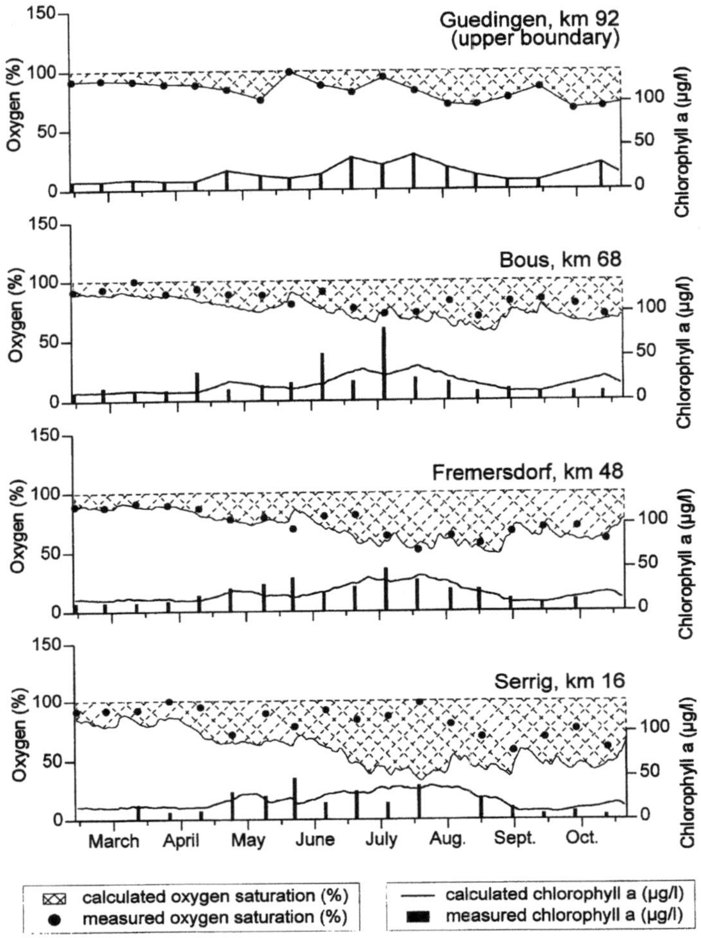

Figure 4. Seasonal cycles of measured and calculated oxygen and chlorophyll *a* in 1994 in the Saar at Güdingen (km-92, upper boundary of the model), Bous (km-68), Fremersdorf (km-48), and Serrig (km-16).

ton biomass in Moselle and Saar is rarely controlled by zooplankton, represented by rotifers in the model. Only at the end of June did rotifers in the Moselle reach densities above 4000 ind/l (Gosselain et al., 1998b), the June peak of chlorophyll *a* was slightly reduced by grazing activity of rotifers. In August 1994 measured grazing rates of the zooplankton ranged between 8 and 15% day^{-1}, while rotifer abundances were low (100–350 ind/l) (Gosselain et al., 1998b). In the Saar the computed rotifer abundances were always low, due to a low input (<500 ind/l) at the upper boundary at Güdingen. Measurements at Fremersdorf and Serrig sometimes showed high abundances of rotifers (Mosbacher & Schmitt, 1994). In August 1994 at Güdingen and Serrig measurements of the whole *in situ* grazing rate of the zooplankton community showed values of 8.3 and 23.5% day^{-1} (Gosselain et al., 1997). The same authors rejected an effective long-term control of total algal biomass by zooplankton in the river Meuse, but they demonstrate the significant role of small zooplankters in affecting phytoplankton composition due to the size-selected removal of algae and promoting large, inedible forms (Gosselain et al., 1998a).

As computed by our model, the main reason for the decrease of algal biomass in the Moselle in midsummer is the grazing by benthic filter feeders. Figure 5 shows the computed chlorophyll *a* development in

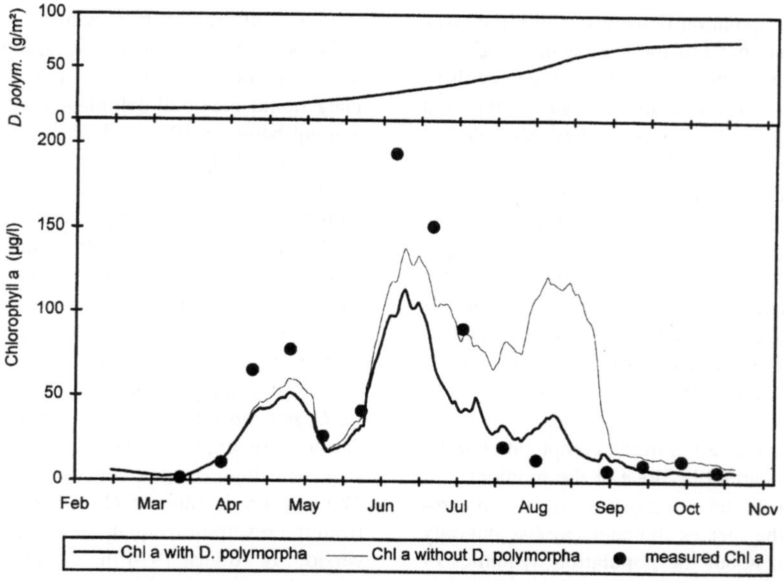

Figure 5. Measured and simulated development of chlorophyll *a* in 1994 in River Moselle at km-290, with and without consideration of *D. polymorpha*. Above, computed seasonal development of *D. polymorpha* population density (dry weight per m^2) at Moselle-km 290.

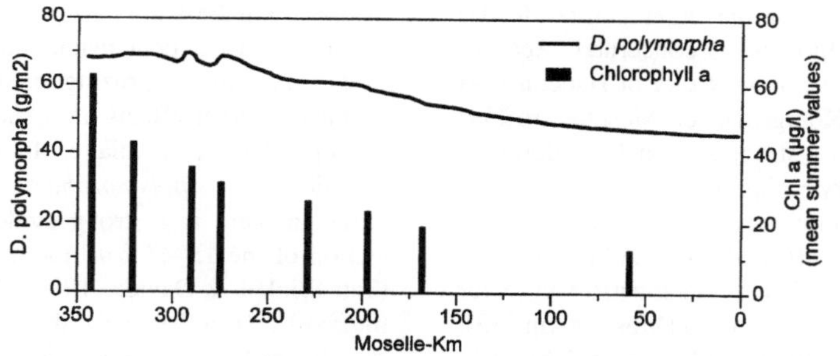

Figure 6. Simulated longitudinal profile of *D. polymorpha* biomass in river Moselle for 15 September 1994. The bars show calculated chlorophyll *a* levels as mean value of the period 1 July until 15 September 1994 at selected stations.

1994 in the Moselle at river-km 290, with and without consideration of *Dreissena polymorpha*. Additionally it presents the seasonal development of the computed biomass for *D. polymorpha* and shows the increasing influence of growing mussel population on the chlorophyll *a* level. In the beginning of July, the water volume between Millery (Moselle-km 345) and Palzem (Moselle-km 228) is filtered once during its transport downstream, and in late summer more than five times. Bachmann et al. (1997) calculated that in autumn 248.2% of the total water volume of the 30 km long sector from Hauconcourt to Königsmacker is filtered by benthic organisms per day. Comparing both simulations given in Figure 5, chlorophyll *a* losses caused by filtering activity of the mussels reach in late summer up to 85%, while for the period from July until September a mean reduction of 59% was computed. It is apparent that measured and modelled chlorophyll *a* levels agree only when the model includes effects of benthic grazing on phytoplankton biomass (see Figure 5).

Figure 6 gives a simulated longitudinal profile of *D. polymorpha* biomass in river Moselle for 15 September 1994. It also presents at selected sites the simulated chlorophyll *a* levels as mean values from July to mid September. The longitudinal decrease of

D. polymorpha population density towards the river mouth is a consequence of the decreasing food concentration due to the filtering activity of the mussels in upstream reaches. The mean summer value of the food factor (*f*(*F*)) ranges in the upper reaches (Moselle-km 345–320) between 0.89 and 0.75, while it decreases to 0.41–0.22 at the lower reaches (Moselle-km 197–km 59). In the river Rhine, the growth of the zebra mussel population was correlated to their food supply and water temperature (Smit et al., 1992a). Besides this, morphological structures and substrate availability influence mussel abundances. Both habitat factors promote the filtering capacity of *D. polymorpha* at the upper parts of the Moselle (km 345–km 277), especially in the non-channelized regions.

A satisfactory validation of the computed *D. polymorpha* biomass cannot be done at the moment. For Moselle and Saar, the composition of the macrozoobenthos and the species abundances exist, but only little information about the population dynamics of the species and their biomass along the whole river is available. Nevertheless, Bachmann et al. (1997) show that in the river Moselle *D. polymorpha* is the most important species of the benthic filter feeder group, as far as the biomass is concerned. According to their calculations, in the sector of Hauconcourt (Moselle-km 290) to Königsmacker (Moselle-km 259) *D. polymorpha* shows a strong seasonal development from spring (2.5 g dry weight/m^2) to autumn (180 g DW/m^2). A more continuous development with lower biomass level between 10 and 75 g DW/m^2 in summer and autumn is recorded for the downstream sector of Königsmacker to Grevenmacher (Moselle-km 213). Our computed data for 1994 (see Figures 5 & 6) are in the same range as the data calculated by Bachmann et al. (1997). The more sparse *D. polymorpha* population of the downstream sectors is reproduced by the BfG model as well.

In the Saar, the grazing effect of *D. polymorpha* on phytoplankton is of smaller importance than in the Moselle. This is due to the lower *D. polymorpha* biomass as a consequence of low food level and the unfavourable morphological conditions in the deep parts of the river Saar, where the ratio of bottom surface to water volume is low. Along the stretch from Bous to Serrig, which is 52 km long, in August 72% of the water volume was filtered by *D. polymorpha* during the downstream transport of water. Nevertheless, up to 59% of the computed losses of phytoplankton were caused by benthic filter feeders. The computed biomass of *D. polymorpha* reached values up to 44 g DW/m^2 in 1994. Only a rough comparison with abundances for distinct sampling sites along the river determined by Schleuter et al. (1996) is possible. They found maximal abundances of *D. polymorpha* ranging between 1000 and 2500 ind/m^2. Depending on the shell length a number of 1000 ind/m^2 corresponds to different biomasses: for individuals with a shell length of 11 mm, the biomass is 7.7 g DW/m^2, for 13 mm it is 12.2 g DW/m^2, and for 23 mm it is 60 g DW/m^2 (Bachmann et al., 1997). The computed data are in the range of the counted abundances, if it is assumed that *D. polymorpha* population is dominated by large individuals. Otherwise the computations overestimate the population by a factor of 2–5.

D. polymorpha directly affects the oxygen budget due to their high respiratory activity and, as a consequence, high oxygen demand (Effler et al., 1996). The oxygen demand of *D. polymorpha* was derived from the respiration equations by stoichiometric conversion. In August 1994 the calculated mean oxygen demand of the mussel population is 0.156 g O_2 m^{-2} day^{-1}, which makes up 16% of the total sediment oxygen demand in the sector between Hauconcourt (Moselle-km 290) and Palzem (Moselle-km 228).

Some further conclusions concerning the impact of *D. polymorpha* grazing on the foodweb should be made. Direct effects are expected concerning the phytoplankton assemblage. In our modelling food selection by *D. polymorpha* is not assumed so far, although there is a strong evidence for particle selection of the 15–45 μm fraction by *D. polymorpha* (Ten Winkel & Davids, 1982). Much more water is filtered by *D. polymorpha* than would be required to merely feed an individual and not all particles that are filtered are consumed. Excess food and particles that are too large or are of too low food quality are bound in mucus and expelled as pseudofaeces. Both small and large algae are filtered out of the water. Although these complicated feeding mechanisms are not taken into account, our simulations show that mussel grazing affects the ratio of the two algae groups, diatoms and green algae. Green algae are more successful under high grazing pressure due to their higher growth rate. Caraco et al. (1997) observed in the Hudson River a shift to more rapidly growing species after the invasion of the zebra mussels. Further the change of phytoplankton assemblage concerning the edible and unedible fraction for small-bodied zooplankton along the Moselle stretch from Hauconcourt to Trier observed by Gosselain et al. (1997) may be particularly due to the filter feeding of mussels. In the

Hudson River, Pace et al. (1998) observed a size-dependent decline in zooplankton caused by direct zebra mussel predation. Estimated consumption rates by mussels were roughly equivalent to maximum microzooplankton growth rates. More indirect effects of *D. polymorpha* grazing would be a result of nutritional competition with herbivorous zooplankton and other filter feeders. Further refinements in the assumptions of the model should allow the modelling of a more detailed, size-structured *D. polymorpha* population. Therefore the losses of *D. polymorpha* by predators should be taken into account. Smit et al. (1992b) showed high losses caused by fishes and waterfowls. Especially in winter, birds are able to reduce the mussel population (Cleven & Frenzel, 1992).

Our simulations show that *D. polymorpha* populations in rivers Moselle and Saar can strongly affect the phytoplankton biomass, especially in the river Moselle in summer and autumn. As a consequence, shown also by Effler et al. (1996) for the Seneca River, *D. polymorpha* is able to convert a phytoplankton-rich system with nearly saturated oxygen concentrations, to a system with low-phytoplankton concentrations and undersaturated oxygen concentrations. Yet the oxygen depletion itself is caused by bacterial oxygen consumption, due to heterotrophic processes as well as nitrification.

The dynamic approach in modelling *D. polymorpha* assumes a continuous density of the population along the whole river at the beginning of the season, which admittedly is a simplification. But it is still a better alternative than neglecting seasonal development by using fixed mussel densities for distinct locations. The dynamic approach affects that in the case of the Moselle the filtering capacity of the mussels is most pronounced in the upper parts and the non-channelized stretches. Here the *D. polymorpha* population grows fast mainly due to the strong coupling with optimal food supply. The computation of seasonal development of the mussel biomass leads to increased phytoplankton losses in late summer and autumn, the grazing effect is also increasing with decreasing discharge.

Acknowledgement

This study was initiated and granted by the Members of the International Commission for the Protection of Moselle and Saar against pollution and co-granted by the Interreg Programme of the European Community. Measurements were supported by the Landesamt für Wasserwirtschaft, Rheinland-Pfalz, and the crew of the research vessel MS Burgund. Skilled technical assistance was given by C. Günster and W. Krings. The authors are grateful to Prof. J. Garnier and the anonymous reviewers for their critical comments on the manuscript.

References

Alpine, A. E. & J. E. Cloern, 1992. Trophic interactions and direct physical effects control phytoplankton biomass and production in an estuary. Limnol. Oceanogr. 37: 946–955.

Bachmann, V., P. Usseglio-Polatera & J. C. Moreteau, 1997. Oxygen budget and biological processes in the regulated rivers Mosel and Saar: contribution of the benthic compartment. Final report, CREUM Metz, 110 pp.

Billen, G., J. Garnier & P. Hanset, 1994. Modelling phytoplankton development in whole drainage net-works: the RIVERSTRAHLER Model applied to the Seine river system. Hydrobiologia 289: 119–137.

Caraco, N. E., J. J. Cole, P. A. Raymond, D. L. Strayer, M. L. Pace, S. E. Findlay & D. T. Fischer, 1997. Zebra mussel invasion in a large turbid river: phytoplankton response to increased grazing. Ecology 78: 588–602.

Cleven, E. J. & P. Frenzel, 1992. Population dynamics and production of *Dreissena polymorpha* in the River Seerhein, the outlet of Lake Constance. In D. Neumann & H. A. Jenner (eds), The Zebra Mussel *Dreissena polymorpha*. – Limnologie aktuell 4: G. Fischer Verlag, Stuttgart, Jena, New York: 45–47.

Cohen, R. R. H., P. V. Dresler, E. J. P. Phillips & R. L. Cory, 1984. The effect of the Asiatic clam, *Corbicula fluminea*, on the phytoplankton of the Potomac River, Maryland. Limnol. & Oceanogr. 29: 170–180.

Descy, J.-P., 1993. Ecology of the phytoplankton of the River Moselle: effects of disturbances on community structure and diversity. Hydrobiologia 249: 111–116.

Dokulil, M. T., 1994. Environmental control of phytoplankton productivity in turbulent turbid systems. Hydrobiologia 289: 65–72.

Ebenhöh, W., C. Kohlmeier & P. J. Radford, 1995. The benthic biological submodel in the European Regional Seas Ecosystem Model. Neth. J. Sea Res. 33: 423–452.

Effler, S. W., C. M. Brooks, K. Whitehead, B. Wagner, S. M. Doerr, M. Perkins, A. C. Siegfried, L. Walrath & R. P. Canale, 1996. Impact of zebra mussel invasion on river water quality. Wat. Envir. Res. 68: 205–214.

Garnier, J., G. Billen & M. Coste, 1995. Seasonal succession of diatoms and Chlorophyceae in the drainage network of the Seine River: observations and modeling. Limnol. Oceanogr. 40: 750–765.

Gosselain, V., J.-P. Descy & L. Viroux, 1997. Final report of the 3 years of study on phytoplankton dynamics, primary production, in situ zooplankton grazing, daily cycle of oxygen content and zooplankton spatial distribution. Laboratory of Freshwater Ecology, FUNDP, Namur, Belgique: 76 pp.

Gosselain, V., L. Viroux & J.-P. Descy, 1998a. Can a community of small-bodied grazers control phytoplankton in rivers? Freshwat. Biol. 39: 9–24.

Gosselain, V., J.-P. Descy, L. Viroux, C. Joaquim-Justo, A. Hammer, A. Métens & S. Schweitzer, 1998b. Grazing by large river zoo–

plankton: a key to summer potamoplankton decline? The case of the Meuse and Moselle rivers in 1994 & 1995. Hydrobiologia 369/370: 199–216.

ICPMS (International Commission for the Protection of Moselle and Saar against pollution), 1998. Analysenergebnisse, Jahr 1994, in press.

Kinzelbach, R., 1992. The main features of the phylogeny and dispersal of the zebra mussel *Dreissena polymorpha*. In D. Neumann & H.A. Jenner (eds), The Zebra Mussel *Dreissena polymorpha*. – Limnologie aktuell 4: Gustav Fischer Verlag, Stuttgart, Jena, New York: 5–17.

Köhler, J., 1995. Growth, production and losses of phytoplankton in the lowland River Spree: carbon balance. Freshwat. Biol. 34: 501–512.

Kuntzmann, B., 1989. Aktivitätsuntersuchungen bei der einheimischen Wandermuschel *Dreissena polymorpha* PALLAS anhand von Filtration und Schalenbewegung. Diploma thesis, University of Köln: 61 pp.

Mosbacher, G. & S. Schmitt, 1994. Zooplankton composition and abundances in the River Saar, 1994. University of Saarbrücken.

O'Neill, R. V., R. A. Goldstein, H. H. Shugart & J. B. Mankin, 1972. Terrestrial ecosystem energy model. East. Decid. Forest Biome Memo Rep.: 72–91.

Pace, M. L., S. E. G. Findlay & D. Fischer, 1998. Effects of an invasive bivalve on the zooplankton community of the Hudson River. Freshwat. Biol. 39: 103–116.

Padilla, D. K., S. C. Adolph, K. L. Cottingham & D. W. Schneider, 1996. Predicting the consequences of dreissenid mussels on a pelagic food web. Ecol. Model. 85: 129–144.

Reeders, H. H., A. Bij de Vaate & F. J. Slim, 1989. The filtration rate of *Dreissena polymorpha* (Bivalvia) in three Dutch lakes with reference to biological water quality management. Freshwat. Biol. 22: 133–141.

Schleuter, M., D. Potel & S. Potel, 1996. Erhebungen der aquatischen Makrozoen an der Saar und an ihren Nebengewässern zwischen Saargemünd und Konz, Untersuchungsbericht 1994. BfG-0984, Bundesanstalt für Gewässerkunde, Koblenz: 85 pp.

Schneider, D., 1992. A bioenergetic model of zebra mussel, *Dreissena polymorpha*, growth in the great lakes. Can. J. Fish. Aquat. Sci. 49: 1406–1416.

Schöl, A., T. Bergfeld, V. Kirchesch & D. Müller, 1997. Oxygen budget and biological processes in the regulated rivers Moselle and Saar, Final report. BfG-1091, Bundesanstalt für Gewässerkunde, Koblenz: 80 pp.

Smit, H., A. Bij De Vaate & A. Fioole, 1992a. Shell growth of zebra mussel (*Dreissena polymorpha* (PALLAS)) in relation to selected physico-chemical parameters in the Lower Rhine and some associated lakes. Arch. Hydrobiol. 124: 257–280.

Smit, H., A. Bij de Vaate, H. H. Reeders, E. H. van Nes & R. Noordhuis, 1992b. Colonization, ecology, and positive aspects of zebra mussels (*Dreissena polymorpha*) in the Netherlands. In T. F. Nalepa & D. Schloesser (eds), Zebra Mussels: Biology, Impacts and Control. CRC Press, FL: 55–77.

Stanczykowska, A. & K. Lewandowski, 1993. Effect of filtering activity of *Dreissena polymorpha* (Pall.) on the nutrient budget of the littoral of Lake Mikolajskie. Hydrobiologia 251: 73–79.

Ten Winkel, E. H. & C. Davids, 1982. Food selection by *Dreissena polymorpha* Pallas (Mollusca: Bivalvia). Freshwat. Biol. 12: 553–558.

Walz, N., 1978. The energy balance of the freshwater mussel *Dreissena polymorpha* PALLAS in laboratory experiments and in Lake Constanz. Arch. Hydrobiol. (Suppl.) 55: 83–105.

Hydrobiologia **410:** 177–184, 1999.
J. Garnier & J.-M. Mouchel (eds), Man and River Systems.

Optimal design and real time control of the integrated urban run-off system

P. Harremoës & W. Rauch
Department of Environmental Science and Engineering, Technical University of Denmark, Copenhagen, Denmark

Key words: urban runoff, pollution, sewer, treatment, real time control.

Abstract

Traditional design of urban run-off systems is based on fixed rules with respect to the points of demarcation between the three systems involved: the sewer system, the treatment plant and the receiving water. An alternative to fixed rules is to model the total system. There is still uncertainty with respect to a more rational formulation of demands on the performance of combined sewer overflows and the performance of separate drainage outlets during rain. There are two approaches to management: *The prediction-design approach*: models play an essential role in the prediction of performance and evaluation of competing alternatives for design. However, the complexity of these systems is such that the parameters associated with pollution are hardly identifiable on the basis of reasonable monitoring programmes. *The empirical-iterative approach*: structures are built on simplified assumptions and their performance is evaluated. If inadequate, improvements are made by another trial on an empirical basis. When monitoring has proven the performance to be inadequate, improvements can be achieved by real time control of the whole system, which combines the two approaches. Due to the complexity of the real system, the experience is that simplicity and on-line adaptation are essential features of the real time control system.

The perspective of integration of issues in time and space

The very tradition of 'public health engineering' (GB English) or 'sanitary engineering' (US English) is associated with the depletion of oxygen in rivers. The year of 'the great stink' last century in London, the definition of BOD (GB Royal Society, 1901) and the Streeter-Phelps description of the 'oxygen sag curve' in the Ohio river (Streeter & Phelps, 1925) are historic landmarks. They are all based on the discharge of untreated urban wastewater into rivers. That is hardly the issue in Europe much longer, because the wastewater treatment standards require not only BOD removal, but also nutrient (N and P) removal. Soon, the implication is that the daily discharge of treated wastewater will no longer pose a serious threat to the oxygen regime in the rivers of Europe. However, the experience in Denmark is that the small distributed sources of wastewater in rural communities do in fact still pose a load on the rural rivers. Accordingly, standardized small treatment plants will be installed in the near future. However, this paper will be associated with the other remaining issue: the time-varying loads and effects of discharge of poorly treated wastewater with rain runoff from CSOs in urban areas.

The first distinction to be made is the identification of time scale on the pollutional effects. The oxygen regime in a river is associated with two types of discharges: degradable organic matter and nutrients.

Acute pollution

The discharge of degradable organic matter leads to consumption of oxygen from the water and from the sediments in the river. The time constant of this process is in the order of hours-days. The oxygen depletion is short term during or just after a discharge event. The fish die as a result of the discharge. The effect is associated with the individual runoff event. The formulation of the requirement is expressed as extreme statistics.

Accumulative pollution

The discharge of nutrients causes eutrophication, excess algae growth, sedimentation and oxygen depletion in the sediments and the hypolimnion of the deep lakes. The effect is accumulative and the discharge of nutrients should be evaluated as the sum of discharges, e.g. annual load.

The total system consists of the sewer system, the treatment plant and the receiving river (Harremoës et al., 1993). There are three interfaces between the three elements of the system: the transient load on the treatment plant, the transient load from combined sewer overflows to the river and the transient discharge from the treatment plant into the river. The tradition is to formulate simple rules for the permissible transient load at these three points, e.g. restriction on frequency of overflow, required dilution before discharge, percent overflow compared to annual load to the treatment plant, etc. (Harremoës & Sieker, 1993). This system has been administratively convenient; but it is not an optimal solution to the complex problem of design and operation of the total system (e.g. Harremoës & Rauch, 1996; Rauch & Harremoës, 1997).

Approaches to design, construction and control

In the official system, there is still uncertainty with respect to the formulation of demands on the performance of combined sewer overflows and the performance of separate drainage outlets during rain. Basically, there are two extreme approaches: (1) the prediction-design approach (: models play an essential role in the prediction of performance and evaluation of competing alternatives for design); (2) the empirical-iterative approach (: structures for pollution abatement are built on simplified assumptions and their performance is subsequently evaluated; when monitoring has proven the performance to be inadequate, an improved plan of action is implemented).

In scientific terms, the first approach is dominated by a deductive interpretation of the problem. It is based on the idea that if the problem is reduced to its basic components and tied together in a system of physical, chemical and biological laws of natural science, the future can be predicted with sufficient accuracy to warrant a safe design and operation. It is 'just' a question of identifying a set of laws to be used in a model structure based on prior knowledge and to calibrate the unknown parameters against data from reality.

The second approach is significantly different by advocating a purely inductive interpretation of information from experience gained from operation of the systems in question (and other systems), from which a pattern can be identified and responded to in an empirical, iterative approach to design and operation. Note that 'Reductionists' object strongly to this line of reasoning by arguing this to be "...bankrupt of understanding...' (Thomann & Mueller, 1987).

Years ago, based on the prevailing positivistic approach to natural science and engineering, hopes were high with respect to the virtue of the first approach, but frequently the second approach was applied in practise. Reductionistic modelling approaches to the description of the cause–effect relationships between nutrient load on lakes and resulting eutrophication have failed to describe the most simple parameters, like visibility. The reason is that the upper trophic level control mechanisms have not been described and do not provide unique solutions, but show chaotic attractors around pseudo-equilibria (Hosper, 1997; Scheffer, 1998). Similarly, description of the river morphology and the river aquatic habitats have not proven subject to successful modelling, but better suitable for holistic interpretations, like pattern recognition.

From the point of view of environmental management, we are still far from a clear picture of the interpretation of the relation between the urban activities and the consequences to the river environment. Further development should be a combination including elements from both approaches.

Mathematical simulation and analysis of static design

Each of the three components of the system has been modelled deterministically.

The sewer system

The deterministic modelling of the hydraulics of the sewer system has been a success, to the extent that it is the accepted experience that if the simulation does not fit the data very well, then the information about the system (the input data) may be faulty, rather than the model. In any case, no model should be used without proper calibration, if reliable model performance is demanded. The flooding issue should be modelled on the basis of extreme statistics. For that purpose, recorded extreme rain events from a time series analysis can be selected as input to the calculation – or design storms derived from IDF-curves can be used for analysis in cases of lesser importance.

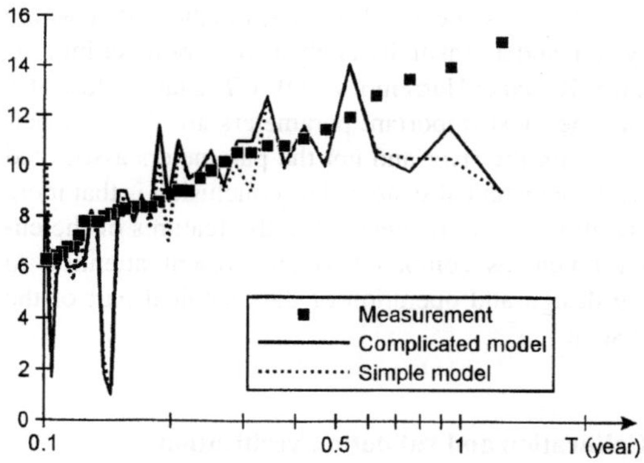

Figure 1. Simulation of small urban catchment, compared with measured data arranged as extreme statistics. The simple model gives very similar results compared to a very complicated model because the uncertainty is associated with the rain gauge (Arnbjerg-Nielsen, 1996).

Even with such an assuring performance of the models, an analysis of uncertainties reveals a significant accumulated uncertainty in the overall design. The rain input gives rise to significant uncertainty. That is why, in Denmark, significant efforts have been devoted to analysis of the properties of rain with respect to geographical distribution from gauge to gauge within a national rain gauge system (Harremoës & Mikkelsen, 1995; Mikkelsen et al., 1995, 1996).

Figure 1 illustrates one aspect of the uncertainties involved. A small catchment has been monitored in detail with respect to catchment characteristics. Rain and runoff have been monitored carefully over a period of 5 years. The modelling of the runoff has been performed for each significant event and plotted as extreme statistics. The figure shows that there is a significant discrepancy for the extreme events and that the discrepancy is the same for the simple and the very complex model. It is the uncertainty of the representability of the rain information that causes the uncertainty, not the models. That applies even to a very small catchment with the rain gauge very close to the catchment (Arnbjerg-Nielsen & Harremoës, 1996).

It has proven a lot more difficult to model the transport of pollutants with the rain runoff (Ahyerre et al., 1998). The flushing of the deposits in the sewer systems depends upon the properties of the system. It has been very difficult to identify these properties and to make adequate predictions from one event to the next – except in a few cases of significant accumulation between rain events. There is still need for a significant element of stochastic interpretation of the pollutant transport in sewer systems.

The treatment plant

The models of treatment plants have been developed successfully on the basis of an *a priori* understanding of the processes involved. The commonly accepted model structure (Henze et al., 1986) is the basis for most available computer programmes. However, that development faces the same issue of model complexity. The number of functions and parameters is so large that it is impossible to calibrate in each particular case. The question is whether treatment plants are so uniform in phenomena and function that most functions and parameters are universal and do not need calibration. The need for complexity depends on the application, design, operation or real time control.

The most difficult issue is related to the function of the clarifier and relation to the settling properties of the sludge. This cannot be modelled adequately for design and the model has to be adaptable to data on the settlability in real time control.

The receiving river

Nutrient discharge

It is well established that the accumulative contribution of the nutrients N and P is greater from the daily discharges from the treatment plant than from the combined sewer overflows and the separate drainage system. The transient discharges from the pipe system are significant only when discharging directly to a lake that does not receive discharge from the same urban catchment via the treatment plant. Even then, the greatest contribution may come from the up-stream diffuse sources. In essence, the interest in rain runoff of nutrients is limited to the accumulative load on the environment at each opening in the system. This can be calculated by use of historical rain series as input to models and accumulating the output over year or season. This will show that there is a significant difference from one year to another that has to be compared to the yearly variation of loads from diffuse sources, which show similar variation from year to year.

Oxygen depletion

There are a number of models available for simulation of the degradation of organic matter and oxygen depletion in rivers, from simple to very complicated. It is not meaningful to use design storm concepts. Rivers have diurnal oxygen fluctuations and therefore they are sensitive to transient loading during the night when the oxygen regime is low compared to transient loading during the day, when there is plenty of oxygen to draw from.

Numerous simulations of oxygen depletion in small Danish rivers and calibration against extensive data series of the oxygen dynamics have shown that it is very difficult to calibrate the river models with anything resembling a unique result. Parameters vary from one event to another in a non-systematic way (Harremoës et al., 1996).

The integrated system

These models can be combined into one model simulating the whole system. The model becomes very complex and difficult to calibrate.

These issues call for a discussion of principle. In view of the complexity of the system two different approaches have been pursued:

- Development of complex models, where every detail that can be described is incorporated into the model structure. Historically, it is based on the deductive concept of modelling. The difficulty is that such a model contains so many functions and parameters that it is impossible to identify and calibrate all. The solution is to fix the structure and a large number of parameters and to calibrate against few selected parameters, hopefully of fundamental and not marginal importance to the performance.
- Development of simple models, which incorporate only the most fundamental phenomena that may be identified and calibrated. The emphasis is more on induction by calibration against data series. The inadequacies are taken care of by statistical considerations in design and operation.

In the past, the wish to base design on analysis of historical rain series called upon simple models and fast computation. Computer development has removed that concern to a large extent. But, the increased complexity has created a demand for simplicity, simply because of the practitioners reaction to the constantly increasing complexity and difficulty in knowing how to deal in practise with such complex models and programmes. It was noteworthy at the recent conference of the triennial conferences on urban storm drainage, that for the first time there was criticism of further development in the direction of complexity, rather than the criticism of the past that models did not incorporate every known detail.

Table 1 is the result of a simulation of a whole system and a sensitivity analysis of parameter importance (Rauch & Harremoës, 1996). The table illustrates that the most important parameters are those associated with the river and not the parameters associated with the technical system. The conclusion is that more attention should be devoted to the features of the environment as compared to the present attention to the design and operation of the technical part of the system.

Calibration and validation/verification

There is reason to challenge the generally accepted practise of calibration and validation/verification. (There is confusion between disciplines with respect to the interpretation of the latter two. Here, they will be interpreted synonymously as discussed below.) The tradition is to joggle the parameters chosen for calibration such that the best fit to half the data set (or data series) is achieved. The test of the fit is carried out on the rest of the data set (or data series). The logic of this approach is questionable:

- Imagine that the two selected halves of the data are derived from the same statistical population of data, then the validation/verification must show the same result, except for statistical variation. The conclusion is that the two data sets can be merged to one and the evaluation of the model is expressed by the uncertainty associated with the fit. In that case, it is better to use statistical procedures and to talk of parameter estimation.
- Suppose that the two halves do not belong to the same statistical population of data, then the two sets include different phenomena. The poor fit of the validation/verification has falsified the model, because the model was not universal. In other words, the model has been applied outside the range in which it works. Philosophically, the fitness of calibration/validation/verification can only be expressed as statistics of uncertainty valid for the range of data available. The lack of fit may be due to a non-fitting model structure or due to statistical uncertainty. The evaluation cannot be carried to phenomena outside the phenomena observed – unless based on a universality claimed by induction from past experience from other investigations covering a wider range of phenomena (*a priori* knowledge).

The conclusion is: Extrapolation is risky!

Table 1. Linear sensitivity for 1 h mean values of dissolved oxygen in the river with a return period of 1 year; ic is the capacity of the interceptor from the CSO to the treatment plant (1 μm/s=10 (l/s)/ha) from Rauch & Harremoës, 1996).

Design	ic=0.1 μm/s	High load	ic=0.3 μm/s	Parameter	Notation	Unit
T_{Riv}	6.14	$n_{v\ Settler}$	7.27	Temperature of river	T_{Riv}	°C
Q_{Riv}	4.81	T_{Riv}	5.48	Baseflow of river	Q_{Riv}	$m^3\ s^{-1}$
wid_{Riv}	3.49	Q_{Riv}	4.98	Width of river	wid_{Riv}	m
R_{Riv}	2.99	$v_{0\ Settler}$	4.00	Respiration rate	R_{Riv}	$g\ O_2\ m^{-2}\ day^{-1}$
$K_{2\ Riv}$	2.67	wid_{Riv}	3.18	Reaeration rate	$K_{2\ Riv}$	day^{-1}
Man_{Riv}	2.50	R_{Riv}	3.05	Manning coefficient	Man_{Riv}	$m^{1/3}\ s^{-1}$
XD_{runoff}	2.40	$K_{2\ Riv}$	3.03	Sediment degradation rate	$K_{4\ Riv}$	day^{-1}
$K_{4\ Riv}$	2.09	XD_{runoff}	2.40	Sedimentation rate	w_{Riv}	$m\ day^{-1}$
w_{Riv}	1.49	T_{dry}	2.14	Maximum Photosynthesis rate	$P_{0\ Riv}$	$g\ O_2\ m^{-2}\ day^{-1}$
$P_{0\ Riv}$	1.10	Man_{Riv}	2.13	Settling parameter	$n_{v\ Settler}$	$m^3\ g^{-1}$
		$K_{4\ Riv}$	2.09	Maximum settling velocity	$v_{0\ Settler}$	$m\ day^{-1}$
		w_{Riv}	1.84	XD concentration in runoff	XD_{runoff}	$g\ COD\ m^{-3}$
		$P_{0\ Riv}$	1.50	Temperature of dry weight flow	T_{dry}	°C

Calibration/validation/verification is a rewarding exercise of the past when only few data sets were available. In recent years, extensive data sets have become available from the performance of sewer systems, treatment plants and (to a lesser extent) receiving waters. That has escalated with the introduction of reliable on-line sensors. In the future, the traditional approach will drown in data. New tools are needed!

On-line approaches to dynamic control

Real time control is a good example of the dilemma associated with choice of model complexity and unending series of data from the sensors.

The Danish experience is that simplicity is a virtue. The tradition of the trade does not make introduction of complex data interpretation and advanced control algorithms an easy task. Real time control is applied to a profession accustomed to gravity flow in fixed structures and automatic control of pumps as the most advanced experience. In Denmark, the real time control system of a sewer system that works in practice is rule-based interpretation of few selected data from strategically located sensors and commands to actuators (Andersen et al., 1997) Simple rules have been derived from analysis of chosen options on an extensively calibrated model of the system. This involves the risks described above for extreme events not covered by the data set, but hopefully taken into account by prior knowledge on the universality of hydraulic theory and hopefully adequate information on the system attributes. However, the advantage is that everybody understands the rules and can check its function. By computer runs, the rules have been compared to the most advanced approaches to real time control. The experience from use of models for this analysis is that the rules derived from experience and good engineering intuition is a good approximation to 'optimum control'.

However, the development goes in the direction of more sophisticated approaches, with practise slowly following up. The challenge is to find the best solution by interpretation of on-line data series, modelling of consequences and minimizing detrimental effects from the disturbances, under circumstances where calculations have to be done fast enough to influence the phenomena. The solution may be called 'adaptive, simplified model based predictive control'. The models have to be simplified. This can be done in two ways: (1) by model reduction, i.e. simple model structure and few parameters *versus* (2) by using complex models with most of the parameters fixed. There is no way in which a large set of model parameters can be calibrated on-line, because the data available may not contain information with which to calibrate these parameters. Because the model structure is not ideal and because only few parameters can be 'calibrated', the solution is to adapt the model parameters recursively to the most recent data. That may account for the inadequacies in the model. The goal is to combine the use of prior knowledge with data sets and statistics. Two approaches have shown promise: Grey box models, which combine simplified determinism

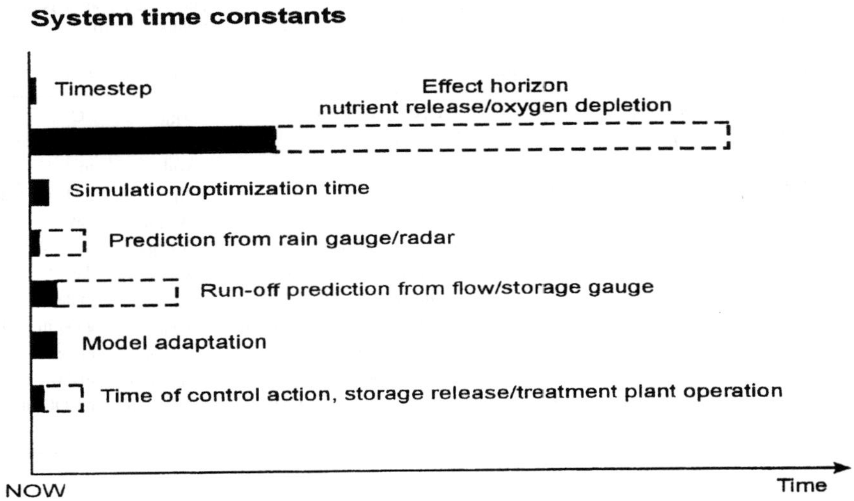

Figure 2. Characteristic time constants in real time control of integrated wastewater systems.

with Box-Jenkins stochastic time series, and Bayesian statistics. The point is that a degree of determinism is retained, the essential parameters have approximate physical interpretation and that results are expressed with an attached uncertainty.

In order to optimize the performance, the approach to real time control has to be predictive. In case of oxygen depletion in a river the prediction has to be based on calculation for each time step and into the future as far as the detrimental effects needs to be incorporated. The universe of actuator alternatives has to be analysed in order to find the optimum, from which to identify the best setting for the next time step (Rauch & Harremoës, 1998). Calculation time becomes a problem, because there is little time to find the solution if the control action is to have an effect. Time is the important element with which to obtain an overview.

These properties of a system can be illustrated by the characteristic time constants of the system (Figure 2). There are seven elements to account for:

1. the discrete time step is the computational unit, e.g. 10 s for a fully dynamic wave simulation, 2 min for a unit hydrograph;
2. the effect horizon, which may be a day in the case of oxygen depletion (travel time down the river to the critical location);
3. simulation, optimization and communication time, during which all information has to be obtained from the system, optimum identified and actuators instructed;
4. prediction from rain gauges and/or weather radars as discussed below;
5. prediction from flow and level gauges as discussed below;
6. model adaptation (the models have to have a certain amount of data in order to realise a new set of parameters suitable for the current event);
7. time of control action, e.g. storage release for a retention basin by opening a valve (minutes), delays due to the movement of the water from storage to treatment plant (minutes to hours), time to store sludge in the aeration tanks, e.g. by shifting to step feed (hour).

Figure 3 is an illustration of the significance of such time constants. The response time of the sewer system depends on the size of the system with a variation from 10 min to several hours. However, the response time of the treatment plant (the residence time) is the same whether the plant is big or small. A rain gauge can be used for prediction only a few time steps ahead and is spatially inaccurate. A weather radar can cover a large area and predict hours ahead, but only with significant uncertainty. Flow and storage gauges give much more accurate information and better predictions, but with a shorter time horizon. The irony illustrating the dilemma is that the features discussed above show that a weather radar would have a beneficial affect on treatment plant operation in

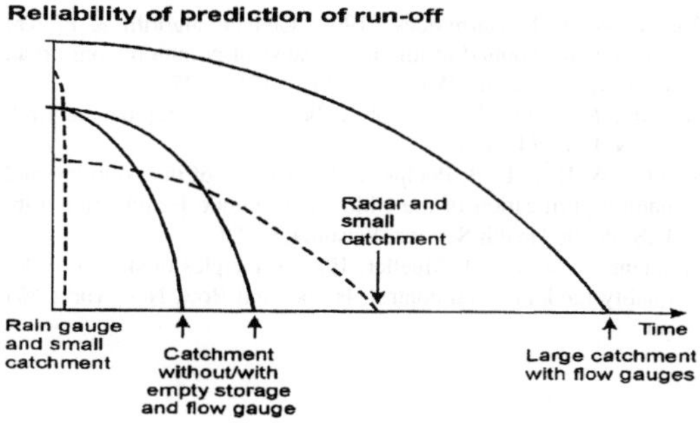

Figure 3. Relation between different measures for run-off prediction and reliability of the method.

small catchments. On the other hand, treatment plants loaded from large catchments do not need a weather radar because there is more accurate data and sufficient prediction time from gauges in a large catchment. The irony is that weather radar can be afforded by large communities, not by small.

Conclusions

There is a need for design and operation of sewer system, treatment plant and receiving water as one unit. There is a need for design and operation to be based on a more realistic set of water quality criteria to be met by the performance of the system in total. The time scale of the pollution is essential. Oxygen depletion and ammonia toxicity are acute phenomena and should be evaluated on the basis of extreme statistics, while nutrients have accumulative effects and should be evaluated on the basis of contribution to mass loading with time, e.g. a year.

Increasing data sets and increased complexity of both the problems and the systems give rise to increasing difficulty with the traditional use of calibration/validation/verification as the approach to model evaluation. In the future, more emphasis should be put on the uncertain universality of the model structure, the model parameters and the general uncertainty of model performance, especially outside the range of experience.

The dilemma is that it is very difficult to deal with still larger data sets and ever more complex models. Choices have to be made. They depend on the circumstances. Models should be dedicated to their purpose and to the reality of their uncertainty. There is a vast difference between models developed for research, planning, analysis and for real time control. The challenge is to find a suitable combination of use of prior knowledge, as a combination of theory (deduction) and experience (induction), with data from the system in question.

For real time control, simplicity is the first choice until the infrastructure and our universal and local knowledge base is ready for greater complexity. That should be developed towards 'adaptive, simplified model based predictive control', by which to optimise the performance of the system with respect to operational water quality criteria – as opposed to the simple design rules of thumb currently in use.

References

Andersen, N. K., P. Harremoës, S. Sørensen & H. S. Andersen, 1997. Monitoring and real time control in a trunk sewer, Wat. Sci. Tech. 36: 337–342.

Arnbjerg-Nielsen, K., 1996. Statistical analysis of urban hydrology with special emphasis on rainfall modelling, PhD Thesis, Department of Environmental Science and Engineering, Technical University of Denmark, ISBN 87-89220-31-5.

Arnbjerg-Nielsen, K. & P. Harremoës, 1996. The importance of inherent uncertainties in state-of-the-art urban storm drainage modelling from ungauged small catchments. J. Hydrol. 179: 305–319.

Ahyerre, M., G. Chebbo, B. Tassin & E. Gaume, 1998. Storm water quality modelling, an ambitious objective? Wat. Sci. Tech. 37: 205–213.

Harremoës, P., A. G. Capodaglio, B. G. Hellström, M. Henze, K. N. Jensen, A. Lynggaard-Jensen, R. Otterpohl & H. Søeberg, 1993. Wastewater treatment plants under transient loading: performance, modelling and control. Water Sci. Technol. 27: 71–115.

Harremoës, P. & F. Sieker, 1993. Influence of stormwater storage tanks on pollutant discharge to receiving waters. In 9th EWPCA-ISWA Symposium, Munich, 11–14 May, Documentation, Liquid Wastes Section. Abwassertechnische Vereinigung e.V. St. Augustin, Germany: 95–126.

Harremoës P. & W. Rauch, 1996. Integrated design and analysis of drainage systems, including sewers, treatment plant and receiving waters. J. Hydr. Res.

Harremoës, P. & P. S. Mikkelsen, 1995. Properties of extreme point rainfall I: results from a rain gauge system in Denmark. Atmosph. Res. 37: 277–286.

Harremoës, P., L. Napstjert, C. Rye, H. O. Larsen & A. Dahl, 1996. Impact of rain runoff on oxygen in an urban river. Water Sci. Techn. 34: 41–48.

Henze M., C. P. L. Grady, W. Gujer, G. V. R. Marais & T. Matsuo, 1987. Activated sludge model No 1. IAWPRC, London.

Hosper, H. S., 1997. Stable states, buffers and switches: An ecosystem approach to the restoration and management of shallow lakes in the Netherlands. In Proc. of Specialist Symposium – Eutrophication Research. August 97, Wageningen, The Netherlands: 221–241.

Mikkelsen, P. S., P. Harremoës & D. Rosbjerg, 1995. Properties of extreme point rainfall II: parametric data interpretation and regional uncertainty assessment. Atmosph. Res. 37: 287–304.

Mikkelsen, P. S., H. Madsen, D. Rosbjerg & P. Harremoës, 1996. Properties of extreme point rainfall III: identification of spatial inter-site correlation structure. Atmosph. Res. 40: 77–98.

Nielsen, M. K., J. Carstensen & P. Harremoës, 1996. Combined control of sewer and treatment plant during rainstorm. Water Sci. Techn. 34: 181–187.

Rauch, W. & P. Harremoës, 1996. The importance of the treatment plant performance during rain to acute water pollution. Wat. Sci. Techn. 34: 1–8.

Rauch, W. & P. Harremoës, 1997. Acute pollution of recipients in urban areas. Wat. Sci. Techn. 36: 179–184.

Rauch, W. & P. Harremoës, 1998. Genetic algorithms in real time control applied to minimize transient pollution from urban wastewater systems. Wat. Res. 33 (5): 1265–1277.

Scheffer, M., 1998. Ecology of shallow lakes, Chapman & Hall, ISBN 0 412 74920 3.

Streeter, W. H. & E. B. Phelps, 1925. A study of the pollution and natural purification of the Ohio River, Public Health Bull. 146, U.S. Public Health Service, Washington, DC.

Thomann, R. V. & J. A. Mueller, 1987. Principles of surface water quality modeling and control. Harper and Row, New York: 664 pp.

Hydrobiologia **410:** 185–193, 1999.
J. Garnier & J.-M. Mouchel (eds), Man and River Systems.

Quantification of nitrate removal by a flooded alluvial zone in the Ill floodplain (Eastern France)

José Miguel Sánchez Pérez[1], Michèle Trémolières[2], Naima Takatert[2], Philippe Ackerer[3], Arianne Eichhorn[1] & Gérard Maire[1]
[1]*Centre d'Etudes et de Recherches Eco-Géographiques (CEREG - CNRS), IFARE, Université Louis Pasteur, 3, rue de l'Argonne, F-67083 Strasbourg Cedex, France. Tel.: [+33](0)5-62-269985; Fax: [+33](0)5-62-269999; E-mail: sanchez@cesac.cemes.fr*
[2]*CEREG - CNRS, Laboratoire de Botanique et d'Ecologie Végétale, Institut Franco-Allemand de Recherche sur l'environnement (IFARE), Institut de Botanique, 28, rue Goethe, F-67083 Strasbourg Cedex, France*
[3]*Institut de Mécanique des Fluides, UMR CNRS 7507, 2 rue Boussingault, 67083 Strasbourg Cedex, France*

Key words: alluvial floodplain, nitrate removal, mass balance, groundwater modelling

Abstract

The nitrate reducing capacity of a flooded system in the Ill floodplain (Eastern France) was investigated for a period of 2 years. The methodology used consisted of a spatio-temporal monitoring of stream flow and nitrate concentrations in the groundwater and surface water, calculation of input and output fluxes and modelling of groundwater fluxes and nitrate transfer through the alluvial area. A comparison of chloride flux (used as hydrological tracer) and nitrate flux was done to determine a floodplain effect on the retention of nitrate. We show that up to 95% of the nitrate load in the groundwater is retained by the system, whereas the retention in the stream network is very low. Ammonium fluxes increased from inputs to outputs in the stream and in the groundwater. The chloride input in the groundwater is higher than the output, whereas in the surface water the output is higher than the input, the amount evacuated in streams corresponding to the losses from groundwater. The nitrate removal rate calculated for the whole modelized surface area (40 km^2) represented 559 t N yr^{-1} or 1397.7 kg N ha^{-1} yr^{-1}. The ammonium fluxes exported by the system represented 102 kg N ha^{-1} yr^{-1}. A part of nitrate is reduced and exported by the groundwater and stream network in the form of ammonium. These results can be explained by the duration of floods which controls the equilibrium between the various forms of nitrogen. Thus, long watering periods favour nitrogen removal (denitrification and plant uptake) and limit nitrate production which compensates elimination during the dry period.

Introduction

Nowadays, groundwater quality is highly disturbed by the increased inputs of nitrogen fertilizers. The riparian forests have been demonstrated to be capable of removing nitrate from agricultural drainage and thus playing an important role in the regulation of fluxes between rivers and adjacent agrosystems (Cooper, 1990; Lowrance, 1992; Pinay et al., 1993). In the Alsace Rhine floodplain the concentrations of nitrate in the groundwater in agricultural zones are higher (4.5 – 5.6 mg l^{-1} N-NO_3^-) than those measured under the alluvial forest - meadow system (less than 1 mg l^{-1}) (SEMA-DIREN, 1994). The nitrate loss through a riparian zone varies widely: the nitrate inputs were reduced by 56–100% in riparian organic soils (Cooper, 1990), by 73% in a flooded forest and by 37% in a flooded meadow in the Ill floodplain area (Takatert et al., in press) and by only 10% in the floodplain of the Adour river (southern France, Brunet et al., 1993). Moreover alluvial flooded forests greatly reduce nitrate flux of floods which infiltrate vertically through the root-soil system (Sánchez-Pérez et al., 1991a). However, this reducing capacity was modified after suppression of the natural functioning of the floodplains (Trémolières et al., 1998). The nitrate removal processes, now well-known, imply denitrification (Lowrance, 1992; Pinay et al., 1993;

Hanson et al., 1994; Lowrance et al., 1995), microbial immobilization (Groffman et al., 1992) and vegetation uptake (Sánchez-Pérez et al., 1991b; Haycock & Pinay, 1993), the importance of each of which varies according to the environmental conditions, substrate, geomorphic characteristics, the type of vegetation and hydrological regime (Pinay et al., 1995; Sánchez-Pérez & Trémolières, 1997). Catchment hydrology interacts with the biological processes responsible for nitrate depletion. Thus streams which flow through the riparian zone can have a strong influence on nitrate reduction by promoting plant uptake, as has been shown by Cooper (1990) and Fustec et al. (1991). This demonstrates that in evaluating the filtering effect of riparian forests we have to take into account the whole zone, including terrestrial and aquatic systems in the same area.

This study attempts to quantify the fluxes of water and nitrate which flow through an alluvial flooded system in the ground and surface water, and thus to evaluate the part of nitrate removal of the alluvial system. Secondly, we try to determine indicators which control the nitrate efflux from the system.

Study area

The study site is an area of 4000 ha which is covered by forests and meadows, flooded by the river Ill, the main tributary of the Rhine in the East France. A large network of phreatic streams and diffluents of the Ill flow in this sector over a length of 51 km and cover a surface of 52 ha (Figure 1). The study sector is surrounded by heavily fertilized crops (especially maize) which provide high concentrations of nitrate, around 5.5 mg l^{-1} N-NO_3^-in the groundwater under the cultivated fields. The Ill hydrological regime and consequently the groundwater regime are characterized by high water in late winter and the beginning of spring (December – March) and low water in summer. Interannual mean precipitations for the years 1994 – 1996 were 546 mm (S.E. Mean = 15 mm) and rainfall inputs during low water periods (May – September) represent almost 50% of the total annual inputs. Soils are gleysoils characterized by the presence of an hydromorphic horizon (gley or pseudogley) at a depth of 0.7m. These soils are subjected to permanent reducing conditions. Two sectors are to be distinguished, a sector located in the east with soils enriched in organic matter (about 20% in the topsoil) and in the west a sector with clayey-silty organic matter-poor soils (less than 7%). During low water period, groundwater level is at 1 m depth in the western sector where overflowing by river occurs every year and at 0.7 m depth in the eastern sector where there is only groundwater rising without overflowing.

Material and methods

The study involved spatio-temporal monitoring of stream flow and nitrate concentrations in the ground and surface waters, calculation of input and output fluxes and modelling of nitrate transfer through the area.

Sampling of ground and surface water

Groundwater samples were collected in a network of 31 piezometers installed in the gravel layer at 4.5 m depth and perforated over the forest border, 6 upstream (4 in the west sector and 2 in the east sector), and 7 downstream (5 in the west sector and 2 in the east sector) (Figure 1). Monthly surveys of concentrations of nitrate and chloride used as a hydrological tracer were made in the groundwater for the two years 1996 and 1997. The periodicity used for sampling is highly representative of the variations occurring during a water year (Sánchez-Pérez, 1992). Surface waters were sampled and the discharge measured upstream and downstream of all the streams which flow through the forest (Figure 1). Sampling was done three times in 1996 and five times in 1997. In order to evaluate the balance between inputs and outputs, ground and surface waters were sampled on the same dates. The data concerning daily discharge of the Ill River were provided by the "Service des Eaux et des Milieux Aquatiques" (SEMA), French office of waters and the aquatic environment.

Groundwater samples were collected after removal of ten times the water volume of the piezometer by a motorized pumping. Collection was made using an electrical pump. The depth of the water table was measured in each well prior to pumping. Water was pumped from each well for approximately 1 min prior to collecting the sample in order to clear the tubing of the sample from the previous well. The water samples were collected in polyethylene bottles after pumping, filtered in the laboratory through a 0.45 μm filter and stored in the dark at 4 °C until chemical analyses were performed.

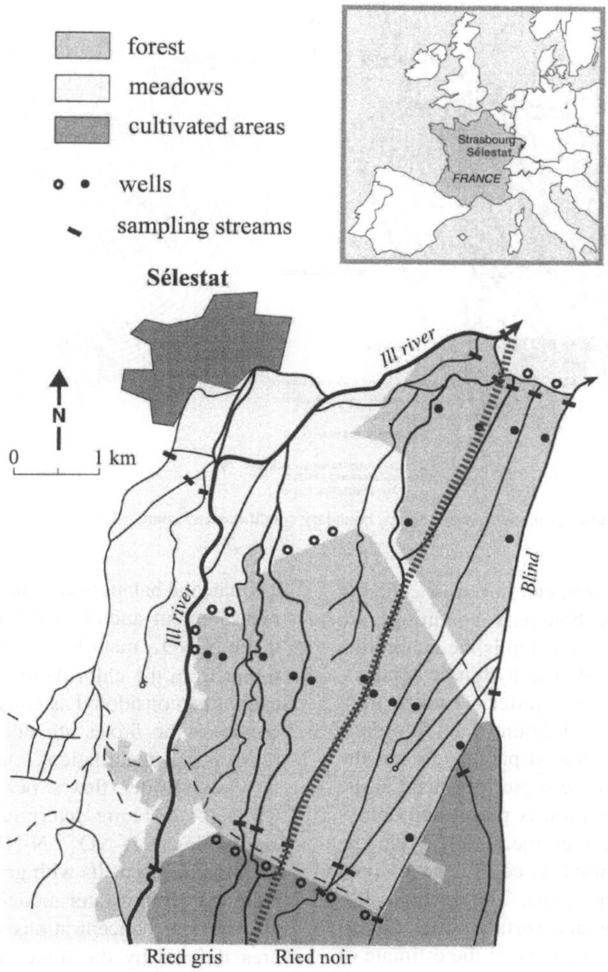

Figure 1. Study area, location of sampling sites in ground (piezometers) and surface water.

Water chemistry

We measured N-NO_3^- and N-NH_4^+ by colorimetry with a microflux automated analyzer (Alliance instruments, Integral 4). Cl^- was analyzed by ionometer with a specific Ag electrode. The procedures used are specified in APHA (1985): ammonium was determined by the indophenol blue method, nitrate by the cadmium reduction method and phosphate by the ascorbic acid method.

Standard chemical analysis procedures were followed including use of replicates and blanks. Each analysis is controlled through a series of standards, which are selected according to the measured concentrations. Data were checked and replicates with a difference of 5% were re-analyzed.

Groundwater modelling

A groundwater flow model based on finite elements is used to quantify the water fluxes in the groundwater and in the rivers. Groundwater modelling is made for a surface of 40 km^2 (Figure 2). The mesh size is 200 m for the area outside the stream network and 60 m in the stream network. The average thickness of the aquifer in this sector is about 100 m. The transmissivity varied from 1.10^4 m^2 d^{-1} on the west side to 6.10^4 m^2 d^{-1} on the east side of the site.

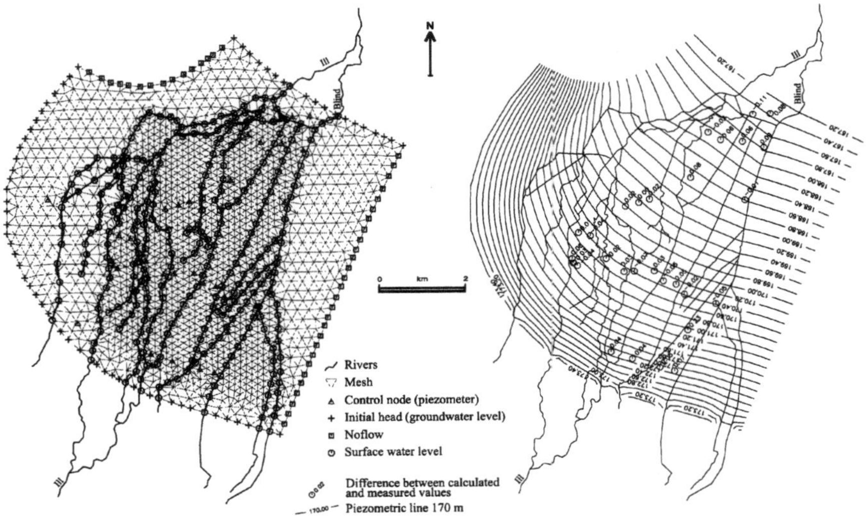

Figure 2. Groundwater model: control nodes, boundary conditions and contour lines of the groundwater flux.

The control nodes (piezometers used to calibrate the model) and the flow boundary conditions are presented in Figure 2. The first step is the calibration of a steady state flow model. The hydraulic conductivity and the thickness of the aquifer are taken from numerous modelling studies of groundwater flow done in the same area. The calibrated parameters are the river bed conductance and the prescribed head at the upstream boundary. Calibration is performed on average piezometric head values measured within the domain (29 piezometers used as control nodes) and average water fluxes in the rivers. The double calibration (groundwater heads and surface water fluxes) is expected to ensure the reliability of the estimate of water fluxes.

Determination of ground and surface water fluxes

The effects of the vegetation and the soil on nitrate transport in the groundwater is estimated by the difference between the input fluxes at the upstream boundary and the output fluxes at the downstream boundary. These fluxes depend on the water fluxes and nitrate concentrations in the river and streams, and in the groundwater. The different data required are obtained by *in situ* measurements (N-NO_3^- concentrations in the surface and ground water, flow rate of surface water) or by modelling (which provides the water input and output flow rates in the aquifer). The mass balance of chloride, used as a conservative tracer, is performed to estimate the balance error due to insufficient sampling, measurement and/or model errors. We assume that, if the N-NO_3^- mass balance error is significantly different from the chloride mass balance error, N-NO_3^- cannot be considered as conservative and that the decrease in the fluxes are due to additional processes (absorption, degradation,....).

Groundwater fluxes per surface unit (kg N or Cl^- ha^{-1} d^{-1}) are determined by coupling the concentrations of N-NO_3^-, N-$NH_4^+$and Cl^- measured in the monitoring wells with groundwater flux estimated from the groundwater modelling. The concentrations are average concentrations over the cross-sectional area defined by the dimensions of the model. The volume of the groundwater leaving the control volume per day (Q = daily flow rate m^3 d^{-1}) was determined by the hydrogeological model for the section area.

The flux of (N-NO_3^-, N-NH_4^+ and Cl^-) (M) leaving the system was calculated by:

$$M = M_{\text{input}} - M_{\text{output}} = (([C_{\text{input}}]\text{x } Q_{\text{input}}) - ([C_{\text{output}}]\text{x } Q_{\text{output}})),$$

where M = mass flux of element leaving (kg d^{-1}), $[C_{\text{input}}]$ = average concentration of the element in the input monitoring well (g cm^{-3}), $[C_{\text{output}}]$ = average concentration of the element in the output monitoring well (g cm^{-3}), Q_{input} = daily flow rate in the input section (m^3 d^{-1}), Q_{output} = daily flow rate in the output section (m^3 d^{-1}).

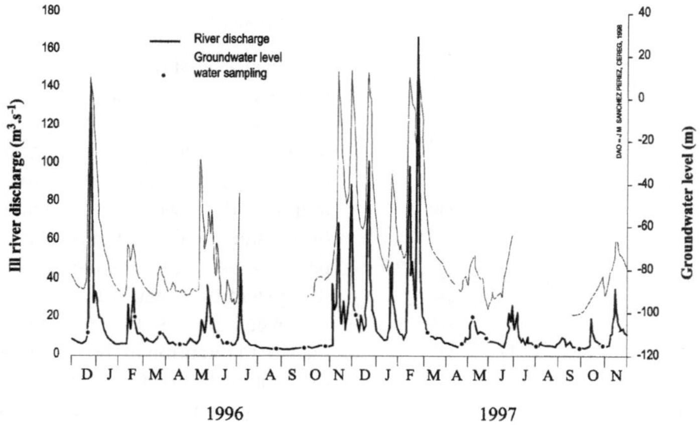

Figure 3. Discharge of the river Ill over the study period (January 1996–December 1997) and corresponding dates of ground and surface water sampling.

The groundwater removal rate per surface unit was then obtained by dividing the mass removed per day (M) by the surface area (40 km^2).

Results

River discharge

During the sampling low water period, the total river flow range was 8.9 – 22.3 m^3 s^{-1} for the inputs and 12.8 – 27.6 m^3 s^{-1} for the outputs, which corresponds to a river discharge of around 20 m^3 s^{-1}. We do not measure the discharge during the overflowing period. According to the daily discharge monitoring of the river Ill, discharge ranges from 160 m^3 s^{-1} during the flood events to 3.8 m^3 s^{-1} during low water (Figure 3). When we compare the occasional measurements in the river and diffluents to the daily discharge of the river Ill in the study site, the river discharge measured in the network during low water period represented 50% of the total surface water fluxes including the flood water fluxes for the study period (December 1995 – November 1997).

Groundwater modelling

The water table computation presented in Figure 2 by the contour lines at intervals of 0.2, indicated a groundwater flux SW – NE; differences in control nodes were less to 0.1 m. Groundwater flow calculated by the modelling represented 3.98.10^5 m^3 d^{-1} for the inputs and 1.06.10^5 m^3 d^{-1} for the outputs.

Table 1. Concentrations of N-NO_3^-, N-NH_4^+ and Cl^-in the groundwater, expressed in mg l^{-1}

Date	N-NO_3^- (mg l^{-1})		N-NH_4^+ (mg l^{-1})		Cl^- (mg l^{-1})	
	Input	Output	Input	Output	Input	Output
23/12/95	2.71	0.31	0.03	0.57	72.1	64.5
24/02/96	5.87	0.85	0.03	0.41	77.7	61.3
22/03/96	3.63	0.52	0.18	0.60	79.7	71.5
22/04/96	4.02	0.91	0.07	0.08	79.3	80.1
30/05/96	6.03	0.80	0.02	0.06	83.2	78.1
24/06/96	2.66	0.59	0.08	0.07	77.8	68.7
20/07/96	5.87	0.63	0.06	0.44	86.0	67.0
27/08/96	4.46	0.46	0.02	0.22	82.2	68.7
6/10/96	5.81	1.15	0.01	0.68	82.1	77.4
1/11/96	5.36	0.91	0.04	0.10	81.3	84.3
8/12/96	1.37	1.38	0.05	0.07	76.8	89.7
23/03/97	2.68	0.69	0.03	0.03	68.7	67.4
26/04/97	2.48	0.43	0.01	0.08	69.0	74.7
10/06/97	3.30	0.53	0.06	0.57	80.5	69.7
19/07/97	4.80	0.65	0.03	0.66	68.1	68.3
14/09/97	4.63	1.34	0.03	0.35	79.5	74.8
19/10/97	2.29	0.34	0.08	0.50	81.5	80.0
23/11/97	4.79	0.58	0.01	0.33	80.3	81.6
Mean	4.04	0.73	0.05	0.32	78.1	73.8

Drainage represented 2.92.10^5 m^3 d^{-1} (or 3.4 m^3 s^{-1}) and corresponded to the differences of surface flow measurements between upstream and downstream in the river network (3.2.10^5 m^3 d^{-1} or 3.7 m^3 s^{-1}, Table 2).

Table 2. Mean daily fluxes in streams and groundwater over the study period

	Inputs (I)	Outputs (O)	O - I	Δ(O-I/I)
Streams				
Flow ($m^3.d^{-1}$)	1 347 840	1 667 520	319 680	24%
N-NO_3^- ($kg.d^{-1}$)	3986.9	4209.3	222.4	6%
N-NH_4^+ ($kg.d^{-1}$)	102.9	198.5	95.6	93%
Cl- ($kg.d^{-1}$)	80 730	104 663	23 933	30%
Groundwater				
Flow ($m^3.d^{-1}$)	398 000	106 000	-292 000	-73%
N-NO_3^- ($kg.d^{-1}$)	1608.7	77.0	-1531.8	-95%
N-NH_4^+ ($kg.d^{-1}$)	18.2	34.4	16.2	89%
Cl- ($kg.d^{-1}$)	31 085	7819	-23 267	-75%
Groundwater + Streams				
Flow ($m^3.d^{-1}$)	1 745 840	1 773 520	27 680	2%
N-NO_3^- ($kg.d^{-1}$)	5595.6	4286.3	-1309.4	-23%
N-NH_4^+ ($kg.d^{-1}$)	121.1	232.9	111.8	92%
Cl- ($kg.d^{-1}$)	111 815	112 482	666	1%

Inputs and outputs of nitrate and ammonium nitrogen and chloride in ground and surface water

Concentrations of nitrate nitrogen in the groundwater input collected through the study period can be relatively high and vary widely from 1.37 to 6.03 mg l^{-1} whereas the output concentrations are very low, always less than 1.38 mg l^{-1} (Table 1). The groundwater in flux into the system was 546.9 – 2399.9 kg d^{-1} and the efflux was 32.9 – 146.3 kg d^{-1}. The nitrate nitrogen fluxes in the stream network are evaluated at 3986.9 kg d^{-1} input and 4209.3 kg d^{-1} output (Table 2). Ammonium nitrogen inputs fluctuate over the study period between 8.2 and 440.6 kg d^{-1} (mean 102.9 kg d^{-1}) in the surface water and between 2.9 and 70.6 kg d^{-1} in groundwater. Outputs are always higher than inputs, between 19.5 and 922.9 kg d^{-1} (mean 198.5 kg d^{-1}) in the surface water and 2.8 – 71.8 kg d^{-1} in groundwater flow. Outputs by streams are close to the inputs except in April and June 1996. The chloride fluxes represented an input of 80 730 kg d^{-1} and 104 663 kg d^{-1} output in the streams. In the groundwater, the concentrations, about 75 mg l^{-1}, do not vary between inputs and outputs. Expressed in terms of the fluxes, 27 110 – 34 228 kg d^{-1} enter the system and 6498 – 9503 kg d^{-1} flow out the system (Table 2).

Variations of inputs and outputs

The discharge of the river Ill is well correlated with the groundwater level and shows several overflow peaks in winter (Figure 3). The lowest drainage effect is observed for the highwater period (December survey) when the difference of discharge between input and output is the lowest for the whole period. In the surface water, the change of nitrate fluxes over the whole period shows a trend to the elimination of nitrate from the system by streams. In winter, nitrate nitrogen output over input rises to 36%, whereas in summer we measured retention in the system of the nitrate brought by streams, (around 10–20%) (Figure 4). In contrast to nitrate, variations of ammonium fluxes show periods of elimination in summer and periods of retention in the system in winter. The mean nitrate evacuation in the stream system is negligible (Table 2). About 30% (15 – 43%) of chlorides are evacuated from the stream system which is close to the increase of output discharge (Figure 4). Chloride flux evacuated by stream system represented 23 933 kg d^{-1}.

In groundwater, nitrate fluxes are reduced by 95%. The ammonium output fluxes are 89% higher than the input flux. Chloride concentrations show little differences between input and output from the area over the study period whereas fluxes of the inputs were reduced by 75%, i.e. by 23 267 kg d^{-1} (Table 2).

It is assumed that the groundwater fluxes remain constant over time. In this domain, groundwater level fluctuations are small due to the numerous rivers which stabilize the water level by infiltration/drainage. Therefore, this assumption may be considered as valid. For chloride, it is assumed that the efflux of chloride from surface waters were 30% higher than the influx and was compensated by the reduction of chloride in groundwater fluxes. The loss of nitrate in the groundwater is significantly higher, about 95% on average. If we consider that the loss due to the rivers is negligible, then nitrogen influx is largely absorbed or degraded by the vegetation and/or the soil.

Estimate of nitrate removal

To evaluate the nitrate removal by the system we calculate balance of inputs and outputs of nitrate and chloride by including fluxes in the streams and in groundwater. The concentrations used were those measured at 5 m depth, but we assume that these measurements are representative of the concentrations measured at a depth of less than 50 m (our measurements correspond to the measurements of the monit-

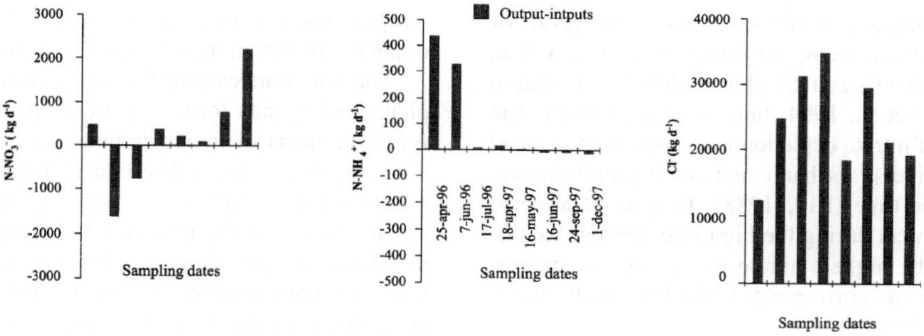

Figure 4. Nitrate, ammonium and chloride fluxes in the stream network of the study area.

oring network done by French office of waters and the aquatic environment – APRONA at around 50 m depth). Under these conditions, we can calculate the nitrate removal at a depth of 100 m (depth used in the model) and the surface area of the model (40 km^2); the estimation of nitrate removal by the alluvial zone studied was 559 t N yr^{-1}, which represented a global NO_3^--N removal capacity of about 1397.7 kg N ha^{-1} yr^{-1}.

Discussion

Nitrate depletion in groundwater

Nitrate removal seems to be particularly efficient with up to 95% reduction when the groundwater flux flows through the system, even in winter when there is no vegetation effect. The chloride used as hydrological tracer varies between 60 and 90 mg l^{-1} in the zone and is preserved globally during the transfer in groundwater through the system. Comparisons of chloride and nitrate fluxes showed that the alluvial floodplain effectively retained nitrate entering the system, findings which were in good agreement with studies on the fate of nitrate in groundwater using Br^- as a tracer (Hubbard & Lowrance 1996; Starr et al., 1996).

By contrast, ammonium fluxes increase from upstream to downstream. This result demonstrates that a part of nitrate is reduced (Schipper et al., 1994), but this cannot explain the overall removal. Denitrification (measurements in progress) might be involved in nitrate removal by a permanent reducing horizon of soil in our study area as has commonly been shown by numerous studies (e.g. Lowrance, 1992; Pinay et al., 1995). This process was demonstrated as the main process in the elimination of nitrate in flooded zones (Leonardson et al., 1994).

However, a large network of streams in this sector drains the groundwater and thus can have an effect on nitrate removal from groundwater. In this study we try to combine the relative roles of the soil-forest system and the drainage by streams.

Effect of drainage

Overall, the discharge of the streams was high and thus streams transport high quantities of chloride and nitrate, about 10 and 50 times more, respectively, than in the groundwater. Moreover these streams drain the groundwater with a regular (or constant) increase of about 24% of flow entering the system. Globally over the whole study period, nitrate input corresponds to nitrate output. The influence of the high waters is low. In fact, during the flood period the inputs of chloride and nitrate are diluted by the large water volumes brought by the flood (see the episode of December 1996) The retention of nitrate by streams seems to be zero, as has been shown by Jansson et al. (1994a) who measured total retention of less than 3% of the total nitrogen transported in the stream. By contrast, chloride output values from streams are closely correlated with those of discharge increase and, consequently, the overall balance is always positive and indicates an elimination of the pollution by drainage. So the differences observed between the behaviour of these two elements show that the nitrate balance could express a retention of nitrate in streams, albeit low. In fact, in summer the trend is to retention in the streams and could reflect a stream plant uptake (Cooper, 1990) and/or of terrestrial plant uptake in groundwater. Svendsen & Kronvang (1993) show that in-stream retention can effectively reduce nutrient export to downstream aquatic systems in low water periods and that areas with macrophytes have the highest potential for storage nutrients. It is also suspected that the decrease of nitrate concentration in surface water is the result of upwelling of nitrogen-poor groundwater rather than nitrate removal by aquatic plant and/or denitrification (Trémolières et al., 1994; Jansson et al., 1994b). The efficiency of nitrate depletion in streams is influenced by nitrate inputs, discharge and consequently by water residence time (Hill, 1988). In spite of absence of measurement during the highwater period, we observe that the highest nitrate evacuation by streams of our study site corresponds to the highest discharge (December, 1997).

Nitrate removal rates from the groundwater

The differences in nitrate input and output fluxes may be due to drainage by the surface water, mass balance error, nitrate reduction or plant uptake. The first assumption can be rejected, the effect of drainage is significant for choride fluxes and negligible for nitrate fluxes. However, the differences found in the nitrate balance in the groundwater are significant and a forest – meadow effect (plant uptake and denitrification) could be proposed for nitrate removal. Thus we can conclude *a priori* that the drainage of groundwater has a reduced effect on nitrate removal from groundwater. Moreover the groundwater dilutes nitrate concentrations in streams.

The estimated nitrate nitrogen removal rate is 559 t N yr^{-1} in the study area. This calculation represents a global elimination from the study area (40 km^2), and thus corresponds to 1397.7 kg N ha^{-1} yr^{-1} within the area. This rate seemed high when we compared it with a riparian poorly drained zone which retains 120 kg ha^{-1} yr^{-1} (Nelson et al., 1995). Leonardson et al. (1994) report retention varying from 390 to 490 kg N ha^{-1} yr^{-1} in wet meadows whereas in artificially flooded meadows retention was 0 – 100 kg ha^{-1} yr^{-1} in spite of a plant uptake amounting to 50–70 kg N ha^{-1} yr^{-1} and high denitrification losses between 190 and 220 kg N ha^{-1} yr^{-1}.

In flooded zones, the nitrate removal depends more particularly on the flood conditions (duration, alternating periods of wet and dry) which control the equilibrium between the various forms of nitrogen. Thus long watering periods favour nitrogen retention or elimination and limit nitrate production which compensates elimination during the dry period. In fact, it is highly probable that the nitrate loss is compensated during the flow by nitrification which increases the amount of nitrate in the groundwater, in spite of favourable conditions for nitrate elimination (permanent hydromorphic horizon and, in some parts of the system, organic soils which increase denitrification potential).

The data collected in this study need to be supplemented by data for the highwater periods in order for the influence of stream drainage on nitrate removal efficiency to be clearly determined. The modelling of nitrate transfer could solve some uncertainties concerning the part of drainage and that of nitrate removal by alluvial system. Moreover, denitrification measurement and monitoring of groundwater in wells installed at various distances from the entrance to the alluvial system will allow us to confirm and improve our initial data. Moreover we have to take into account the role of an unsaturated zone which could provide nitrate in the alluvial soils.

Acknowledgements

This research was supported by IFARE (Institut Franco-Allemand de Recherche sur l'Environnement), PNRH (French National Programme of Research in Hydrology) and PIR Environment SEAH (Systèmes Ecologiques et Actions de l'Homme).

References

APHA, 1985. Standard methods for the examination of water and wastewater, 16th edn., American Public Health Association: 1268 pp.

Brunet, R. C., G. Pinay, F. Gazelle & L. Roques, 1993. Role of the floodplain and riparian zone in suspended matter and nitrogen retention in the Adour river, south-west France. Reg. Riv. Res. Mgmt. 8: 55–63.

Brüsch, W. & B. Nilsson, 1993. Nitrate transformation and water movement in a wetland area. Hydrobiologia 251: 103–111.

Cooper, A. B., 1990. Nitrate depletion in the riparian zone and stream channel of a small headwater catchment. Hydrobiologia 202: 13–26.

Fustec, E., A. Mariotti, X. Grillot & J. Sajus, 1991. Nitrate removal by denitrification in alluvial groundwater. Role of former channels. J. Hydrol. 123: 337–354.

Groffman, P. M., A. J. Gold & R. C. Simmons, 1992. Nitrate dynamics in riparian forests: Microbial studies. J. envir. Qual. 21: 666–671.

Hanson, G. C., P. M. Groffman & A. J. Gold, 1994. Denitrification in riparian wetlands receiving high and low groundwater inputs. J. envir. Qual. 23: 917–922.

Haycock, N. E. & G. Pinay, 1993. Nitrate reduction in grass and poplar vegetated riparian buffer strips during the winter. J. envir. Qual. 22: 273–278.

Hill, A. R., 1988. Factors influencing nitrate depletion in rural stream. Hydrobiologia 160: 111–122.

Hubbard, R. K. & R. R. Lowrance, 1996. Solute transport and filtering through riparian forest. Am. Soc. Agricult. Eng. 39: 477–488.

Jansson, M., R. Andersson, H. Berggren & L. Leonardson, 1994b. Wetlands and lakes as nitrogen traps. Ambio 23: 320–325.

Jansson, M., L. Leonardson & J. Fejes, 1994a. Denitrification and nitrogen retention in a farmland stream in Southern Sweden. Ambio 23: 326–331.

Leonardson, L., L. Bengtsson, T. Davidsson, T. Persson & U. Emmanuelsson, 1994. Nitrogen retention in artificially flooded meadows. Ambio 23: 332–341.

Lowrance, R., 1992. Groundwater nitrate and denitrification in a coastal plain riparian forest. J. envir. Qual. 21: 401–405.

Lowrance, R, G., Vellidis & R. K. Hubbard, 1995. Denitrification in a restored riparian forest wetland. J. envir. Qual. 24: 808–815.

Nelson, W. M., A. J. Gold & P. M. Groffman, 1995. Spatial and temporal variation in groundwater nitrate removal in a riparian forest. J. envir. Qual. 24: 691–699.

Pinay, G., L. Roques & A. Fabre, 1993. Spatial and temporal patterns of denitrification in a riparian forest. J. appl. Ecol. 30: 581–591.

Pinay, G., C. Ruffinoni & A. Fabre, 1995. Nitrogen cycling in two riparian forest soils under different geomorphic conditions. Biogeochemistry 30: 9–29.

Sánchez-Pérez, J. M. 1992. Fonctionnement hydrochimique d'un écosystème forestier inondable de la plaine du Rhin. La forêt alluviale du secteur de l'île de Rhinau en Alsace (France). PhD, Université Louis Pasteur, Strasbourg: 176pp.

Sánchez-Pérez, J. M., M. Trémolières & R. Carbiener, 1991a. Une station naturelle d'épuration des phosphates et des nitrates apportés par les eaux de débordement du Rhin: la forêt alluviale à frêne et orme. C. r. hebd. Séanc. Acad. Sci., Paris 112: 395–402.

Sánchez-Pérez, J. M., M. Trémolières, A. Schnitzler & R. Carbiener, 1991b. Evolution de la qualité physico-chimique des eaux de la frange superficielle de la nappe phréatique en fonction du cycle saisonnier et des stades de succession des forêts alluviales rhénanes. Acta Oecologica 12: 581–601.

Sánchez-Pérez, J. M. & M. Trémolières, 1997. Groundwater nutrient levels in the Rhine alluvial forest ecosystems depending on the hydrological regime and the soil texture. Global Ecol. Biogeograph. Letters 6: 211–217.

Schipper, L. A., A. Z. B. Cooper, C. G. Harfoot & W. J. Dyck, 1994. An inverse relationship between nitrate and ammonium in an organic riparian soil. Soil Biol. Biochem. 26: 799–800.

Schnabel, R. R., L. F. Cornish, W. L. Stout & J. A. Shaffer, 1996. Denitrification in a grassed and a wooded, valley and ridge, riparian ecotone. J. envir. Qual. 25: 1230–1235.

SEMA-DIREN, 1994. Qualité de la nappe phréatique de la plaine d'Alsace. Teneurs en nitrates 1991.

Svendsen, L. M. & B. Kronvang, 1993. Retention of nitrogen and phosphorus in a Danish lowland river system: implications for the export from the watershed. Hydrobiologia 251: 123–135.

Starr, J. L., A. M. Sadeghi, T. B. Parkin & J. J. Meisinger, 1996. A tracer test to determine the fate of nitrate in shallow groundwater. J. envir. Qual. 25: 917–923.

Takatert, N., J. M. Sánchez-Pérez, M. Trémolières, 1999. Spatial and temporal variations of nutrients concentration in the groundwater of a floodplain: effect of hydrology, vegetation and substrate. Hydrol. Process. 13: 1511–1526.

Trémolières, M., U. Roeck, J. P. Klein & R. Carbiener, 1994. The exchange process between river and groundwater on the central Alsace floodplain (Eastern France): II. The case of a river with functional floodplain. Hydrobiologia 273: 19–36.

Trémolières, M., J. M. Sánchez-Pérez, A. Schnitzler & D. Schmitt, 1998. Impact of river management history on the community structure, species composition and nutrient status in the Rhine alluvial hardwood forest. Plant Ecol. 135: 59–78.

Hydrobiologia **410:** 195–206, 1999.
J. Garnier & J.-M. Mouchel (eds), Man and River Systems.

The impact of hydrological fluctuations on shallow groundwater hydrochemistry under two alluvial meadows

L. Krebs[1], J. Corbonnois[2] & S. Muller[1]
[1]*Université de Metz, UPRES 'Ecotoxicité, Biodiversité et Santé Environnementale', Laboratoire de Phytoécologie, Ile du Saulcy, 57045 Metz Cedex 01, France*
[2]*Université de Metz, Laboratoire de Géographie physique, CEGUM, Faculté des Lettres et Sciences Humaines, Ile du Saulcy, 57045 Metz Cedex, France*

Key words: alluvial grassland, tall forbs formation, groundwater hydrochemistry, nutrient availability, water-table level

Abstract

Groundwater ionic profiles, with particular regard to bioavailable nutrient (N, P, and K) levels, were studied in a flooding alluvial grassland of the river Meuse. The observed variations were related to different groundwater levels and hydrological events (rises in groundwater), the vegetation types and the human practices (cutting, ploughing). A network of 20 piezometers was installed with regard to the vegetation type, the groundwater table level and agricultural practices and was located in the fairly wet meadows. Comparison of water ionic levels between Meuse surface water and groundwater table shows the efficiency of nutrient retention of the different compartments of the plant–soil ecosystem. Grassland ecosystem, by its absorption and transfer functions, modifies the chemical characteristics of shallow groundwater. A comparative study was carried out during the growing season on groundwater hydrochemistry in two meadow ecosystems (a tall forbs formation with *Filipendula ulmaria* and its homologous grassland type, the *Senecioni-Oenanthetum mediae*) and a ploughed soil. At the same time sampling and nutrient content analyses of above and underground biomass were carried out in the two meadows which differ not only by their floristic composition but also by their groundwater table level. The lowest groundwater N, P and K levels were found under the wettest meadow with the tall forbs formation, but total nutrient amounts in vegetation (above- and below-ground biomass) did not differ consistently between the two meadows. Plant pools of N, P and K were mainly located in below-ground organs in both meadows and the annual outputs through hay removal were quantitatively the most important in the *Senecioni-Oenanthetum mediae* association (the driest meadow). It appeared from these results that groundwater N, P and K availability was lower in the wettest meadow. The lower N, P and K concentrations in the shallow groundwater under the tall forbs formation can not be closely related to the vegetation but rather to a higher water table level.

Introduction

Flood plains are terrestrial–aquatic interface zones and are closely dependent on hydrogeomorphological conditions which are generated by the river. These riverine wetlands intervene decisively in water-flow regulation and the regular supply of sediments and nutrients by floods. Wetlands carry out hydrological functions such as groundwater recharge and they can also amplify the effects of any potential pollution. Flood plain hydrology is poorly understood at present but is widely recognised as a primary driving force influencing wetland ecology (Carter, 1986; Hunt et al., 1996).

In the last ten years several studies have examined the hydrochemical functioning of alluvial forest (Carbiener & Trémolières, 1990; Pinay, 1986; Sanchez Perez, 1992; Trémolières et al., 1991). The relationship between grassland vegetation and water level fluctuations has been studied in many French alluvial plains (Carbiener, 1969; Magnanon, 1991; Pautou, 1975) and in other countries, such as England (Denny, 1993) and The Netherlands (Barendregt et al., 1986), but the impact of hydrological fluctuations on the purification of water nutrient content of a flooded meadow ecosystem has not been studied in detail. But studies have indicated that water-table control was a promising tool for managing soil water to reduce agrochemical loads in surface and ground waters on high-water-table soils (Thomas et al., 1992; Willis et al., 1997). Investigations of nitrogen retention in artificially flooded meadows were performed in southern Sweden and showed that denitrification and plant uptake of nitrogen were stimulated by the artificial flooding (Leonardson et al., 1994).

Some more targeted studies have examined the influence of agricultural practices on groundwater N and P levels in non-alluvial meadows (Dorioz & Ferhi, 1994; Gaury, 1992) or the effect of grassland management and changes in groundwater level on the yield (Oomes,1991; Oomes & Van der Werf, 1997; Oomes et al., 1996), but only rarely groundwater hydrochemistry in alluvial flooded meadows (Ackerer et al., 1990; Kugler, 1988). On the other hand, several papers have documented the location and transfer of nitrogen and phosphorus in soil/plant compartments of flooding grassland (Chaneton et al., 1996; Lavado et al., 1993; Rubio et al., 1995) and nitrogen and phosphorus tissue concentrations in wetland plants (McJanet et al., 1995).

In the flooding alluvial grassland of the river Meuse, two plant communities which derive from each other through changes in cutting date were the subject of different studies to assess their capacity to purify shallow groundwater. The content of macronutrients (N, P, K) is studied in plants, roots, soils and groundwater table according to the vegetation type (tall forbs formation and its homologous grassland type) and the water table level. This tall forbs formation is a very impoverished plant community where the vegetation dynamic favours the development of one species (*Filipendula ulmaria*[1]) to the detriment of others. The vegetation in this tall forbs formation is higher and the root system is more developed.

This paper reports on the interactive effects of water table level and botanical composition on nutrient availability in shallow groundwater.

Description of the study area

The river Meuse rises in the Plateau de Langres and flows into the North Sea 900 km downstream. This basin, like the whole of the Lorraine, is subject to an oceanic climate with a continental influence (Harmand, 1992). Precipitation is abundant and regular throughout the year with a mean annual rainfall value of 904 mm (in the 1969/1990 period). The mean annual air temperature is 9.1°C and the temperature range between the coldest and the warmest month is 16.3°C (Faber, 1994). However, these rainfall and temperature means conceal striking interannual and seasonal differences. The quantity of precipitated water can exceed 1000 mm in wet years (1981, 1983, 1995), and can be less than 600 mm during dry years (1971, 1976).

The discharge of the Meuse at the Stenay station is 46.5 m^3/s (watershed area, 3904 km^2) and varies greatly according to the season. We can distinguish two main hydrological periods: a low-water period during the summer months when the discharge is barely supported by the hydrogeological contribution of the substratum, and a high-water period from December to March. During this high-water period an average of 63 days of significant flooding (discharge exceeding 100 m^3/s) has been calculated.

The study was conducted in the Mouzay flood plain, located 40 km downstream of the town of Verdun just beyond the Ardennes (Figure 1a). In this area the river Meuse follows a meandering course with many abandoned meanders. The valley floor, which is 3 km wide and flat, consists of 18-m thick quaternary calcareous gravel alluvium covered by about 60 cm of silt. It contains a circulating water table contained by impervious Callovien clays (Figure 1b). This area is frequently flooded in winter and spring.

A phytosociological study (Grévilliot, 1996) allowed us to distinguish three successive associations along the hydric gradient. The experiment was carried out in parcels which are situated at the intermediate topographical level. At this level we could distinguish the *Senecioni-Oenanthetum-mediae*, or SOM, association. It is a meadow community dominated by grasses, but mesohygrophilic and mesophilic species particularly characterise this intermediate grassland type. Three sub-units have been distinguished, *Colchicetosum pratensis*, or SOM1, in the upper level and *typicum*, or SOM2, in the medium level and *Myosotetosum palustris*, or SOM3, in the lower level. More extensive management practices (reduction of the regulation by cutting) leads this association to evolve into a tall forbs formation dominated by *Filipendula ulmaria* (Grévilliot, 1996).

[1] The nomenclature follows Lambinon et al. (1992).

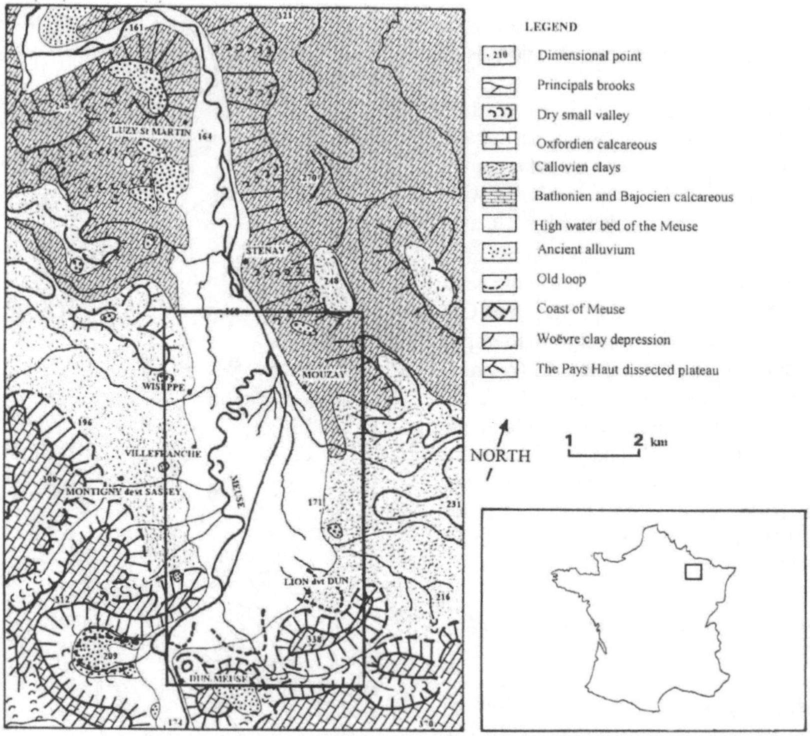

Figure 1a. (a) The Meuse valley in Mouzay area; geomorphologic sketch. (b) Transverse section of the Meuse flood plain.

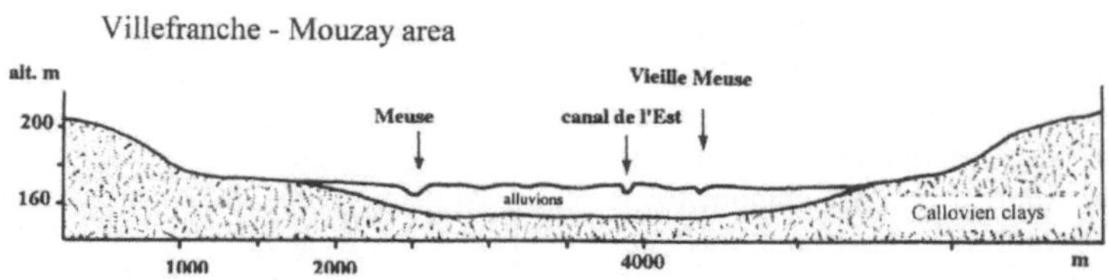

Figure 1b. (b) Transverse section of the Meuse flood plain.

Materials and methods

Sampling and analysis of shallow groundwater

A network of 20 piezometers (Figure 2) was set up in the Mouzay flood plain a site enclosed between the Meuse and the northern branch of the East Canal. The piezometers, 1 and 3 m deep, were installed in the gravel at the input and output of 100-m^2 permanent plots in order to study the groundwater table level changes during the seasonal cycle and to sample groundwater. The three permanent plots are situated in non-fertilised sites, the cutting dates were variable and the groundwater level in the tall forbs formation is 15–20 cm higher than in the grassland type.

The piezometers were closed to their extremities and perforated either at the last metre (for the 3-m deep piezometer) or at the last 80 cm (for the 1-m deep piezometer). Sixteen piezometers were installed with regard to the vegetation type (eight in the wet sub-unit of the *Senecioni-Oenanthetum mediae* SOM3 and eight in the tall forbs formation with *Filipendula ulmaria*). These two sites were compared with a ploughed soil without vegetation (four piezometers).

Sampling was performed after pumping at the level of the groundwater table saturating the subsoils. The chemical composition of these samples can only be representative of the aquifer after several renewals of the piezometer volume.

Groundwater samples were taken every 15 days (occasionally weekly) during the growing period and the Meuse and canal water samples were taken at the same time. The variables which were analysed were the following: NO_3^-, NH_4^+, PO_4^{3-}, K^+, Mg^{2+}, Ca^{2+}, Na^+, SO_4^{2-}, Cl^-, pH, temperature and conductivity. The variables were measured according to methods detailed in AFNOR (1986). NO_3^-, SO_4^{2-} and Cl^- concentrations were analyzed by ion chromatography using a Dionex 4500i system using an AS4A anion column. Ammonium concentrations have been determined by the colorimetric method using indophenol blue (AFNOR T90-015). Mg^{2+} and Ca^{2+} contents were determined by atomic absorption (AFNOR T90-005), K^+ and Na^+ contents by flame emission (AFNOR T90-019), and P content colorimetrically (AFNOR T90-023).

Sampling and analysis of above and below-ground phytomass

The aerial productivity was calculated in 1997 by sampling of plant biomass. The sampling for the two types of vegetation occurred at the flowering peak (06/06/97 for the SOM3 association and 08/07/97 for the tall forbs formation), because the maximum of nutrients in the plant is reached at this stage of development. The sampling area for above-ground phytomass (all aerial parts) recommended in numerous studies (Dutoit & Alard, 1995; Magnanon, 1991; Vertes, 1993) was about 1 m^2. This area was subdivided into four elementary units of 0.25 m^2. The vegetation was cut in June and July, according to the plant communities, to a stubble height of 4 cm.

Root productivity was also measured by sampling of underground biomass. This sampling was carried out after the growing season (01/10/97), with the same treatments and analysis as for the aerial biomass. The sampling area for underground biomass was limited to about 0.25 m^2 (four elementary units of 6.25. 10^{-2} m^2), because the method is very destructive. The roots samples were taken off with the top soil (0–20 cm layer) and were separated from the soil by washing using water spray jets. The root comparment encompassed all below-ground organs, including rhizomes and bulbs of various species.

Roots and above-ground biomass were dried and ground. K, Mg, Ca, Na and P were quantified by being ashed at 450°C, dissolved in nitric acid and the mineral content was determined by atomic absorption for the cations and by colorimetric method for P. To determine total N, a subsample of the dried and ground sample was digested with sulfuric and phosphoric acids (95% H_2SO_4 – 5% H_3PO_4) in a 420°C block digester (TECATOR 1015 model) according to the Kjeldahl method. All these elements were measured according to methods (Lambert, 1994) usually employed at the Laboratoire d'Ecologie des Prairies of Michamps (Belgium).

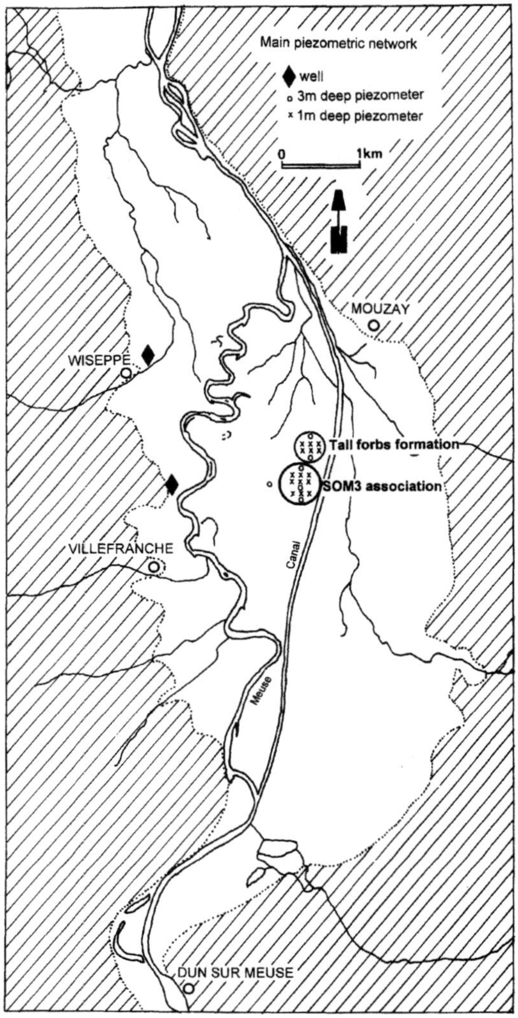

Figure 2. Localization of the piezometric network in the Mouzay area.

Table 1. Mean concentrations and extreme values of bioavailable nutrients and physical variables in surface water and groundwater (3 m deep) of the study area during the 1997 period. The averages were calculated from 240 groundwater samples and 32 surface water samples

	Surface water	Groundwater table	Difference (%)
$N\text{-}NO_3^-$ ($mg\ l^{-1}$)	2.600	0.032	–99%
	(1.360 – 3.700)	(0 – 0.433)	
$N\text{-}NH_4^+$ ($\mu g\ l^{-1}$)	58.0	38.9	–33%
	(9.3 – 179.0)	(9.0 – 134.0)	
$P\text{-}PO_4^{3-}$ ($\mu g\ l^{-1}$)	42.5	16.0	–62%
	(0 – 135.7)	(0 – 35.0)	
K^+ ($mg\ l^{-1}$)	2.28	0.57	–75%
	(1.86 – 3.15)	(0.26 – 1.04)	
Mg^{2+} ($mg\ l^{-1}$)	7.12	5.69	–20%
	(6.13 – 8.65)	(4.48 – 6.52)	
pH	8.18	7.50	–
	(7.96 – 8.33)	(7.36 – 7.83)	
Conductivity	502	556	–
(μS/cm)	(437 – 564)	(435 – 644)	

Table 2. Comparison of water table level (mean and extreme values for year 1997), soil texture and mean nutrient concentrations in soils (0–20 cm depth) between the two meadows. The averages (a) of soil nutrient concentrations were calculated from spring (b) and autumn (c) values

		SOM3 association	Tall forbs formation
Water table level (cm)		–41	–28
		(–64; +7)	(–50; +5)
Clay content (%) (0–20 cm depth)		64	71
pH (H_2O)	(a)	7.3	7.2
	(b)	7.7	7.7
	(c)	6.8	6.6
Organic matter (%)	(a)	16	21
	(b)	15	22
	(c)	17	20
C org. ($g\ kg^{-1}$)	(a)	98.0	117.3
	(b)	92.5	122.6
	(c)	103.5	111.9
N org. ($g\ kg^{-1}$)	(a)	10.8	12.6
	(b)	10.4	12.8
	(c)	11.1	12.3
CEC (meq 100 g^{-1} dry soil)	(a)	43.6	50.4
	(b)	40.5	51.5
	(c)	46.6	49.3
Ca^{2+} (meq 100 g^{-1} dry soil)	(a)	78.5	85.9
	(b)	76.3	82.7
	(c)	80.6	89.0
Na^+ (meq 100 g^{-1} dry soil)	(a)	0.19	0.23
	(b)	0.15	0.19
	(c)	0.22	0.27
Mg^{2+} (meq 100 g^{-1} dry soil)	(a)	2.33	2.32
	(b)	2.41	2.29
	(c)	2.25	2.34
K^+ (meq 100 g^{-1} dry soil)	(a)	0.51	0.43
	(b)	0.51	0.45
	(c)	0.50	0.41
available P ($mg\ kg^{-1}$)	(a)	170	191
	(b)	114	133
	(c)	225	248

Sampling and analysis of soil

Soil samples of the top soil (0–20 cm layer) were taken in the two meadows before (2/05/1997) and after (1/10/1997) the growing season. At each grassland plot, six soils cores of 8-cm diameter were extracted with a stainless steel corer. Nutrient determinations were based on a composite sample of each plot obtained from the six soils cores.

Soil texture was determined for each meadow. Organic carbon and organic nitrogen were measured according to the Anne method (oxidation by sulfochromic acid at 135°C) and the Kjeldahl method (after digestion by sulphuric acid at 350°C). Exchangeable cations (Ca^{2+}, Mg^{2+}, Na^+ and K^+) concentrations in the soil were determined by flame emission after extraction with 1 N ammonium acetate. Exchange capacity was determined by extraction with 1 N ammonium acetate according to the Metson method. Available phosphorus was determined by extraction with 0.1 N ammonium oxalate according to the Joret–Hebert method.

Results

Comparison of nutrient concentrations between surface water and groundwater table

When hydrological stability prevails, the concentrations of nutrients in groundwater are much lower than those found in surface water (Table 1). More particularly for nitrate-N, ammonia-N, orthophosphate and potassium which show a decrease respectively of 99, 33, 62 and 75% in comparison with surface water.

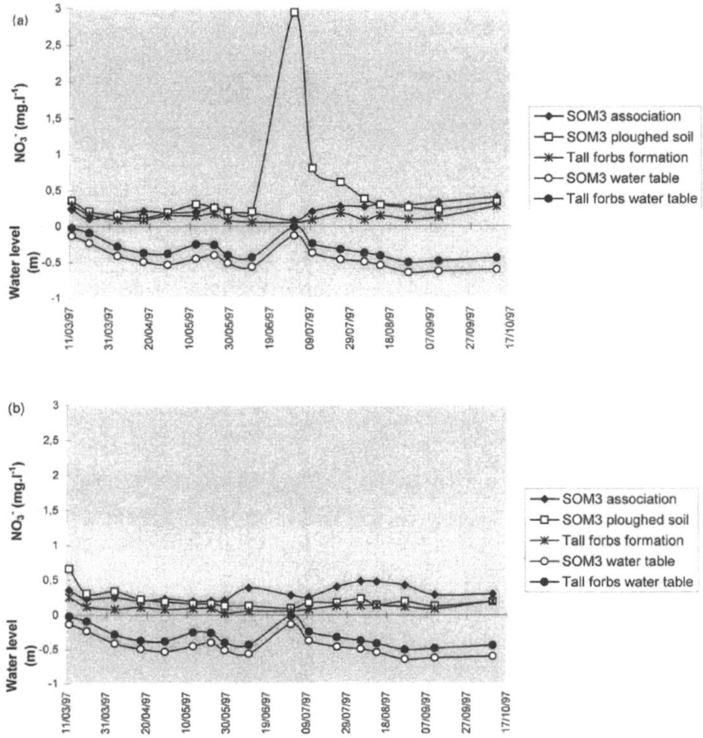

Figure 3. Mean (a) 1 m and (b) 3 m deep groundwater content of nitrate according to vegetation type, ploughing and water table level.

Soil nutrient content

The soils of the two grassland plots have a very fine texture with a high clay content (Table 2). At the 0–20 cm depth, the soils showed different clay content with a mean of 71% under the tall forbs formation and 64% for the SOM3 grassland type. The rate of organic matter was also higher in the wettest meadow than in the SOM3 meadow (21% and 16%, respectively). In the wettest site, the organic matter decomposition rate would be restricted due to more anaerobic conditions in the topsoil. The cation exchange capacity seemed to be lower in the SOM3 grassland and concentrations in nutrients were higher in the tall forbs wettest meadow except for potassium and magnesium.

Changes in groundwater nutrient content

Comparison of groundwater hydrochemistry between the SOM3 association, the tall forbs formation and a ploughed soil showed differences in groundwater nitrate, N-ammonia, K and P levels. But during the growing season, differences were really found only for N-NH_4^+ and K^+. N-NO_3^- and P-PO_4^{3-} showed values which were at the limit of the detection point (between 0 and 15 μg l^{-1} for ortho-phosphates and between 0 and 150 μg l^{-1} for N-nitrate).

However, the lowest mean groundwater nitrate level (at 1 and 3 m deep) was found under the tall forbs formation (Figure 3). The rate of nitrogen mineralization seemed to be reduced in this formation because the proportion of soil organic matter is higher (21% as against 16% in the grassland type).

The 1 m deep nitrate content at the output of the ploughed plot showed negligibly higher concentrations in comparison with the SOM3 grassland plot, but these concentrations were strongly affected by hydrological events such as an important rainfall and a rise in groundwater table (there was no flood in 1997). A peak in nitrate of about 3 mg/l occurred after important precipitation in the second fortnight of June (mean rainfall value of 112 mm during this period and a maximum water table level in the order of –13.5 cm).

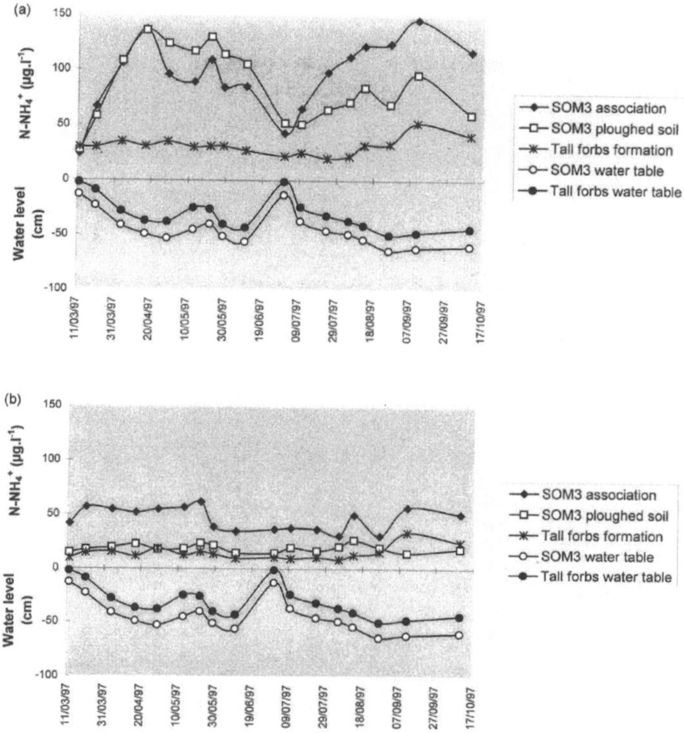

Figure 4. Mean (a) 1 m and (b) 3 m deep groundwater content of N-ammonia according to vegetation type, ploughing and water table level.

The highest 3 m deep nitrate content was found under SOM3 but the peak of nitrate found at 1 m in the ploughed plot did not appear. A great deal of organic matter disappeared under ploughed soils by nitrate leaching but an increase in nitrate concentrations could only be detected in groundwater in the vegetation root zone. This phenomenon resulted certainly from a higher dilution effect in the gravel level. From these results it could be seen that there was a stratification of groundwater nutrient content under SOM3 plots (ploughed or not) which was not found under the tall forbs meadow.

The 1 m deep groundwater mean N-ammonia level (Figure 4) under the three plots showed the lowest concentrations under the tall forbs meadow in comparison with the SOM3 and the ploughed plots. This difference of concentrations diminished in the 3-m deep groundwater with a high decrease of N-ammonia content under the SOM3 and the ploughed plots. The groundwater N-ammonia content in the root zone of the two meadows seemed to be affected by the level of N-mineralisation in the soil.

Comparison of the K content (Figure 5) of the 1- and 3-m deep groundwater between the three plots showed clear differences. The lowest concentrations emerged under the tall forbs formation and the highest under the SOM3 association.

The change of groundwater K levels under each plot during the growing season was almost similar, particularly at 3 m deep. The concentrations under the tall forbs meadow were relatively identical at 1 and 3 m deep. The stratification of groundwater nutrient content was also confirmed for K under the SOM3 plots, with the highest concentrations in the shallow groundwater (the same phenomenon was observed for groundwater N-ammonia concentrations).

No P was detected in groundwater during the growing season under the tall forbs meadow and the concentrations were very low under the SOM3 and the ploughed plots (about 15 μg/l).

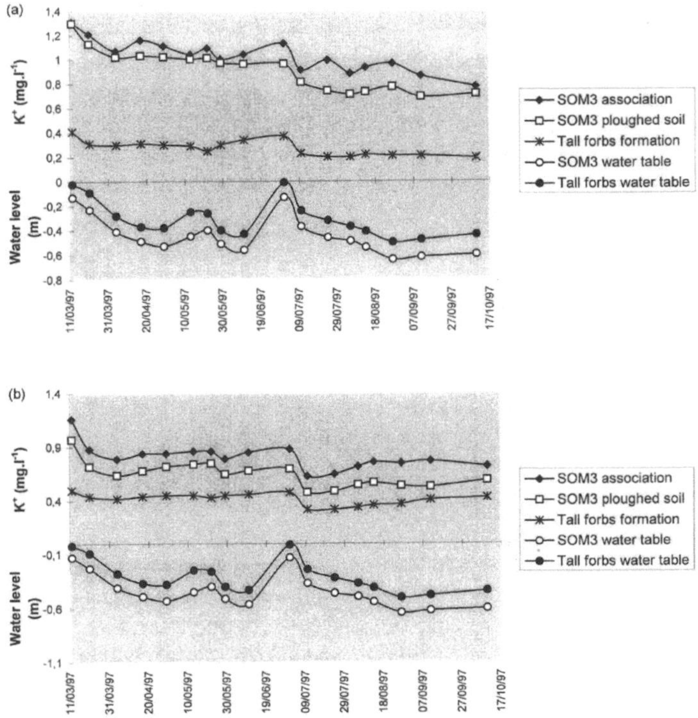

Figure 5. Mean (a) 1 m and (b) 3 m deep groundwater content of potassium according to vegetation type, ploughing and water table level.

Table 3. Mean (n=4, ±SE) grassland above and below-ground yields according to vegetation type. Sampling occurred at the flowering peak (06/06/97 for the SOM3 and 08/07/97 for the tall forbs formation)

	SOM3 association ($t\ ha^{-1}$)	Tall forbs formation ($t\ ha^{-1}$)	Difference ($t\ ha^{-1}$)
Above-ground biomass	4.2 ± 0.3	4.6 ± 0.5	0.4
Root biomass	10.6 ± 0.8	13.4 ± 0.3	2.8

Changes in dry matter yields and tissue nutrient content

The dry matter yield of above ground biomass was not strongly affected by management (late cutting). The difference in dry matter yield during the 1997 growing season (Table 3) was not very high between the two meadows (+ 0.4 ton/ha for the tall forbs formation).

However, the underground dry matter yield was higher in the tall forbs formation in comparison with the homologous grassland type (+ 2.8 ton/ha). Root tissue nutrient content, particularly N, P and K, was affected by the meadow type (Table 4). Higher amounts of these nutrients were found in the roots of the tall forbs formation.

On the other hand, the highest amounts of N, P and K in the above ground biomass were found in the SOM3 association. The tissue Mg and Ca contents were always higher in the tall forbs formation (both for plants and roots).

These latter results were at variance with the findings concerning hydrochemistry for the two meadows, especially for nitrogen and potassium.

Discussion

These initial results demonstrated the efficiency of alluvial meadows ecosystems in purifying shallow groundwater nutrients. In comparison with the Meuse surface water, the concentrations of N-nitrate, N-ammonia, ortho-phosphate and potassium in the groundwater table were lower.

The groundwater chemistry, with regard to the two factors nitrate and orthophosphate during the growing season, does not distinguish the different meadows very well because the concentrations were very low everywhere. It is the immediate absorption by vegetation and soil micro-organisms which is responsible for these low groundwater nutrient concentrations (Buttler, 1992).

On the other hand, nitrate leaching under ploughed soil was rapidly observed in shallow groundwater after important precipitation because nitrate is a very water-soluble ion. The high concentrations in shallow groundwater were submitted to a dilution effect in the deeper layer.

The groundwater hydrochemistry of both meadows seemed to show a different nutrient uptake (particularly N-ammonia and K) by vegetation according to the stage of development. The groundwater nutrient content was always higher in the SOM3 meadow in comparison with the tall forbs meadow. But these results did not coincide with the tissue nutrient content in above ground and underground biomass.

Root nutrient content analysis showed a higher storage capacity for the tall forbs species, but this increase resulted only from a higher yield of underground biomass. Only tissue concentrations of Mg were affected by the species itself and not by the vegetation biomass. This last observation could also be related to an increase in the activity of bivalent cations during the summer period (Duchaufour, 1997) which favours higher concentrations in the soil solution. On the contrary, the monovalent cations (K and NH_4) increase in the soil solution during the wet periods and nutrient leaching occurs more intensively because it coincides with the drainage period. Potassium is mostly present in easy decomposable organic matter and is known as a mobile ion which can easily be leached. Moreover, the rise in groundwater which occured early july affected the groundwater K level in each plot.

The comparison of the nutrient amounts in above ground vegetation between the two meadows revealed a higher K and N exportation in the SOM3. Potassium intervenes in many physiological functions and represents, with nitrogen, the nutrient most exported by cutting and removing the dry matter. But potassium concentrations also vary strongly with the species type, Readman (1993) showed that *Symphytum officinale*, which characterizes the SOM3 grassland type, is an excellent potassium source thanks to its tap root which comes down to the saturated soil zone. This observation can be closely related with the findings of Gerloff (1966) who proved that some species well provided for K extraction could enrich the whole plant community.

The tissue Mg content in above ground and underground biomass was higher in the tall forbs formation with *Filipendula ulmaria*. This species can accumulate magnesium thanks to its root system which is better at exploring the soil (Denudt, 1975).

At this stage of the study we can affirm that the nutrient uptake in the tall forbs meadow is not higher than in the SOM3 association, in spite of a higher underground biomass. On the other hand, the uptake of Mg is higher for the tall forbs species such as *Filipendula ulmaria*.

So, the lower nutrient concentrations in the shallow groundwater under the tall forbs formation can not be closely related to the vegetation, but a water table which is higher by about 15–20 cm in this meadow could explain these differences. In fact nitrogen mineralization rates can be expected to be strongly affected by different groundwater levels (Oomes, 1991). Under wet and anaerobic conditions in the topsoil, the decomposition and mineralization rate of the organic matter is reduced (Grootjans et al., 1985) and a greater part of the inorganic nitrogen is lost by denitrification (Berendse et al., 1994). The decrease in N availability in the tall forbs formation is indicated by the decrease in N concentrations in shallow groundwater and by the decrease of the aerial biomass–underground biomass ratio. Van Duren et al. (1997), in a comparative study between undrained and drained *Calthion* meadows (characterized by *Caltha palustris* and *Filipendula ulmaria*) supported the hypothesis that under undrained conditions N-availability to the vegetation is relatively low. This is thought to be caused by a reduced mineralization and increased denitrification.

Koerselman et al. (1993) concluded that nutrient ratios were suitable parameters to diagnose nutrient deficiencies of soils. According to Koerselman & Verhoeven (1995), an N:P ratio of less than 14 is supposed to indicate nitrogen deficiency while a ratio of more than 16 should point to phosphorus deficiency. The N:P ratios found in the two experimental sites (10 for SOM3 and 10.4 for tall forbs) clearly point to nitrogen rather than to P deficiency. On the other hand, Pegtel et al. (1996) calculated mean N:K ratios of 1.2 and K:P ratios of 8.6 as being threshold values. The

Table 4. Comparison of nutrient amounts (Kg ha^{-1} (%)) in roots and above-ground vegetation between the two types of meadow. Sampling of the underground biomass occurred after the growing season (01/10/97)

		Total N	K	P	Na	Mg	Ca
SOM3	Above-ground vegetation	67.2 (1.6)	75.0 (1.79)	6.7 (0.16)	4.3 (0.10)	8.8 (0.21)	29.9 (0.72)
association	Roots	96.7 (0.9)	51.0 (0.48)	14.2 (0.13)	35.5 (0.36)	25.7 (0.24)	182.2 (1.72)
	Total	163.9 (2.5)	126.0 (2.27)	20.9 (0.29)	39.8 (0.46)	34.5 (0.45)	212.1 (2.44)
Tall forbs	Above-ground vegetation	49.1 (1.1)	51.2 (1.12)	4.7 (0.10)	1.2 (0.03)	15.5 (0.34)	37.2 (0.82)
formation	Roots	97.3 (0.7)	70.2 (0.52)	15.2 (0.11)	21.2 (0.16)	48.8 (0.36)	224.2 (1.67)
	Total	146.4 (1.8)	121.4 (1.64)	19.9 (0.21)	22.4 (0.19)	64.3 (0.70)	261.4 (2.49)
Difference	Above-ground vegetation	−18.1 (−0.5)	−23.8 (−0.67)	−2.0 (−0.06)	−3.1 (−0.07)	+6.7 (+0.13)	+7.3 (+0.10)
	Roots	+0.6 (−0.2)	+19.2 (+0.04)	+1.0 (−0.02)	−14.3 (−0.20)	+23.1 (+0.12)	+42.0 (−0.05)
	Total	−17.5 (−0.7)	−4.6 (−0.63)	−1.0 (−0.08)	−17.4 (−0.27)	+29.8 (+0.25)	+49.3 (+0.05)

The sign – indicates gain for SOM3; the sign + indicates gain for tall forbs and nutrients values expressed in% are between brackets.

N:K and K:P ratios found in our study (0.89 – 11.1 for SOM3 and 0.95 – 10.9 for tall forbs, respectively) indicated that K not appeared to be a limiting factor for the above-ground phytomass.

Available P contents in soils were relatively high in the two sites. These results can be attributed to clay minerals which have the greatest capacity to physically and chemically bind organic P in soils (Tiessen et al., 1983). The relatively sharp decline in nutrient content in these wet fields might be due to P being fixed in the soil by the Ca-rich groundwater (Koerselman et al., 1993). The highest soil contents in organic carbon and available P were found in the wettest meadow (117.3 and 191 mg kg^{-1}, respectively). These results can be attributed to a main factor: in the lower site, the organic matter decomposition rate would be more restricted due to waterlogging, favoring organic C and P accumulations (Rubio et al., 1995).

A high groundwater table also resulted in less K being available (Oomes et al., 1996) and the decrease in K availability is indicated by the decrease in the nutrient and dry matter yields (Oomes & Altena, 1994). These observations were confirmed by the low groundwater K concentrations and the decrease of K yield (–23.8 kg ha^{-1}) in the tall forbs meadow.

Conclusion

At the start of this research we expected to find that tall species were more adapted to nutrient uptake in shallow groundwater. But the first results showed us that lower nutrient content in groundwater resulted from higher water table levels and not from vegetation.

It appears that the groundwater table level determines the vegetation composition but also the availability of nutrients. Humid conditions, which result from a high water table, and late cutting influence the vegetation by stimulating tall-species growth.

The process of shallow groundwater impoverishment causes a decline in nutrient yield. A high groundwater level caused the yield of N and K to fall markedly in above ground biomass and the yield of P to fall slightly. The decrease in nutrient content in the tall species is caused by a decrease in N mineralization, a stronger P fixation on soil colloïds and a leaching of K in wet conditions.

However, data on nutrient content of the different species present in both meadows are necessary to ascertain the causes of this decline of shallow groundwater N, P and K concentrations. So, it would be interesting to compare the nutrient content in roots and plants of *Filipendula ulmaria* with samples taken in each meadow.

Acknowledgements

This investigation was financially supported by the Agence de l'Eau Rhin Meuse. The authors would like to thank the Laboratoire d'Ecologie des Prairies of Michamps for the vegetation analysis, Roger Poinsaint for his technical assistance and Craig Powel for correcting the English.

References

Ackerer, P., R. Carbiener, D. Laigle, P. Muntzer, C. Schenck, M. Trémolières & L. Zilliox, 1990. Contamination des eaux souterraines par les nitrates dans la plaine d'Alsace. Incidences de l'agriculture. In R. Calvet (ed.), Nitrates-Agriculture-Eau. INRA, Paris-Grignon: 245–249.

Afnor, 1986. Eaux, méthodes d'essai. Afnor ed., Paris.

Barendregt, A., J.T. De Smidt & M.J. Wassen, 1986. The impact of groundwater flow on wetland communities. In Colloque Phytosociologique, XIII. Végétation et Géomorphologie. Bailleul: 603–612.

Berendse, F., M.J.M. Oomes, H.J. Altena & W. De Visser, 1994. A comparative study of nitrogen flows in two similar meadows affected by different groundwater levels. J. appl. Ecol. 31: 40–48.

Buttler, A., 1992. Hydrochimie des nappes des prairies humides de la rive Sud du lac de Neuchâtel. Bull. Ecol. 23: 35–47.

Carbiener, R., 1969. Le grand Ried d'Alsace, Ecologie d'un paysage. Bull. Soc. Ind. de Mulhouse 734: 15–44.

Carbiener, R. & M. Trémolières, 1990. The Rhine rift valley groundwater-river interactions: evolution of their susceptibility to pollution. Regul. Rivers Res. Mgmt 5: 375–389.

Carter, V., 1986. An Overview of the hydrologic concerns related to wetlands in the United States. Can. J. Bot. 64: 364–374.

Chaneton, E.J., J.H. Lemcoff & R.S. Lavado, 1996. Nitrogen and phosphorus cycling in grazed and ungrazed plots in a temperate suhumid grassland in Argentina. J. appl. Ecol. 33: 291–302.

Denny, P., 1993. Water Management Strategies for the Conservation of Wetlands. J. IWEM 7: 387–394.

Denudt, G., 1975. Essai de caractérisation de la flore et de la végétation prairiale à l'aide des taneurs minérales. Thesis, Univ. Cat. de Louvain: 118 pp.

Dorioz, J.M. & A. Fehri, 1994. Pollution diffuse et gestion du milieu agricole: transferts comparés de phosphore et d'azote dans un petit bassin versant agricole. Wat. Res. 28: 395–410.

Duchaufour, P., 1997. Abrégé de pédologie, 5. Masson, Paris: 291 pp.

Dutoit, T. & D. Allard, 1995. Mineral contents and plant diversity in chalk grassland under different management. A case study in Normandy, France. Vth International Rangeland Congress, Salt Lake City.

Faber, M., 1994. Contribution à l'étude hydroclimatique de la Meuse de sa source à la station de Stenay. Mémoire de maîtrise, Nancy: 76 pp.

Gaury, F., 1992. Systèmes de culture et teneurs en nitrates des eaux souterraines. Thesis, INRA, Département SAD, Mirecourt: 229 pp.

Gerloff, G.C., D.G. Moore & J.L. Curtis, 1996. Selective absorption of mineral elements by native plants of Wisconsin. Plant Soil XXV: 393–405.

Grévilliot, F., 1996. Les écosystèmes prairiaux de la plaine alluviale de la Meuse lorraine: phytosociologie, dynamique et fonctionnement, en relation avec les gradients hydriques et les modifications des pratiques agricoles. Thesis, Univ. Metz: 217 pp.

Grootjans, A.P., P.C. Schipper & H.J. Van der Windt, 1985. Influence of drainage on N-mineralization and vegetation response in wet meadows. In *Calthion Palustris* stands, Acta oecol., oecol. Plant. 6: 403–417.

Harmand, D., 1992. Histoire de la vallée de la Meuse Lorraine. Presse universitaire de Nancy: 146 pp.

Hunt, R.J., D.P. Krabbenhoft & M.P. Anderson, 1996. Groundwater inflow in wetlands systems. Wat. Res. 32: 495–507.

Koerselman, W. & J.T.A. Verhoeven, 1995. Eutrophication of fen ecosystems; external and internal nutrient sources and restoration strategies. In Wheeler, B.D., S. Shaw, W. Fojt & R.A. Robertson (eds), Restoration of temperate wetlands. Wiley, Chichester: 91–112.

Koerselman, W., M.B. Van Kerkhoven & J.T.A. Verhoeven, 1993. Release of inorganic N, P and K in peat soils: effect of temperature, water chemistry and water level. Biogeochemistry 20: 63–81.

Kugler, J., 1988. Approche méthodologique pour la modélisation du transfert des nitrates vers la nappe phréatique d'Alsace: cas de la zone inondable du Ried central de l'Ill. Thesis, Strasbourg: 186 pp.

Lambert, J., 1994. Modes opératoires, analyses des végétaux. Fac. des Sc. Agro. de Louvain, UCL: 49 pp.

Lambinon, J., J.E. De Langhe, L. Delvosalle & J. Duvigneaud, 1992. Nouvelle flore de Belgique, du G.D. du Luxembourg, du Nord de la France et des régions voisines, 4. Patrimoine du Jardin Botanique Naturel de Belgique (ed): 1092 pp.

Lavado, R.S., G. Rubio, G. Casasola & C.R. Alvarez, 1993. Effect of waterlogging and application of nitrogen and phosphorus fertilizer on the production of two species of an Argentine natural grassland. In N.J. Barrow (ed.), Plant Nutrition from Genetic Engineering to Field Practice. Kluwer, Dordrecht: 341–344.

Leonardson, L., L. Bengtsson, T. Davidsson, T. Persson & U. Emanuelsson, 1994. Nitrogen retention in artificially flooded meadows. Ambio 23: 332–341.

Magnanon, S., 1991. Contribution à l'étude des prairies naturelles inondables des marais de Donges et de l'estuaire de la Loire. Phyto-écologie, Phytosociologie, valeur agronomique. Thesis, Univ. de Nantes: 269 pp.

McJanet, C.L., P.A. Keddy & F.R. Pick, 1995. Nitrogen and phosphorous tissue concentrations in 41 wetland plants: a comparison across habitats and functional groups. Funct. Ecol. 9: 231–238.

Oomes, M.J.M., 1991. Effects of groundwater level and the removal of nutrients on the yield of non fertilized grassland. Acta oecol. 12: 461–469.

Oomes, M.J.M. & H.J. Altena, 1994. Decrease in grassland production by management of vegetation and groundwater. In Mannetje, Lt and J. Frame (eds), Grassland and Society. Proceedings of the 15th General Meetings of the European Grassland Federation: 328–331.

Oomes, M.J.M. & A. Van Der Werf, 1997. Restoration of species diversity in grasslands: the effect of grassland management and changes in ground water level. Acta bot. Gal. 143: 451–461.

Pautou, G., 1975. Contribution à l'étude écologique de la plaine alluviale du Rhône entre Seyssel et Lyon. Thesis, Univ. de Médecine, Grenoble: 375 pp.

Pegtel, D.M., J.P. Bakker, G.L. Verweij & L.F.M. Fresco, 1996. N, K and P deficiency in chronosequential cut summer-dry grasslands on gley podzol after cessation of fertilizer application. Plant Soil 178: 121–131.

Pinay, G., 1986. Relation sol-nappe dans les bois riverains de la Garonne. Etude de la dénitrification. Thesis, Univ. Lyon I: 200 pp.

Readman, J., 1993. Ces herbes qu'on dit mauvaises. Terre vivante (ed), Paris: 64 pp.

Rubio, G., R.S. Lavado, A. Rendina, M. Bargiela, C. Porcelli & A.F. de Lorio, 1995. Wetlands 15: 386–391.

Sánchez Pérez, J.M., 1992. Fonctionnement hydrochimique d'un écosystème forestier inondable de la plaine du Rhin. Thesis, CEREG, Univ. Louis Pasteur, Strasbourg: 176 pp.

Sánchez Pérez, J.M., M. Trémolières, A. Schnitzler & R. Carbiener, 1991. Evolution de la qualité physico-chimique des eaux de la frange superficielle de la nappe phréatique en fonction du cycle saisonnier et des stades de succession des forêts alluviales rhénanes (*Querco-Ulmetum minoris*). Acta oecol. 12: 581–601.

Thomas, D.L., C.F. Douglas, M.C. Smith, M.B. Parker & M.G. Stephenson, 1992. Controlled-drainage and nitrogen rate effects on nitrate-nitrogen losses. ASAE Paper No 92-2646. Am. Soc. Agric. Eng. St. Joseph, MI.

Tiessen, H., J.W.B. Stewart & J.O. Moir, 1983. Changes in organic and inorganic phosphorus composition of two grassland soils and their particle size fractions during 60–90 years of cultivation. J. Soil Sci. 34: 815–823.

Trémolières, M., D. Carbiener, R. Carbiener, I. Eglin, F. Robach, J.M. Sanchez Perez, A. Schnitzler & D. Weiss, 1991. Zones inondables, végétation et qualité de l'eau en milieu alluvial rhénan: l'île de Rhinau, un site de recherches intégrées. Bull. Ecol. 22: 317–336.

Van Duren, I.C., D.M. Pegtel, B.A. Aerts & J.A. Inberg, 1997. Nutrient supply in undrained and drained *Calthion* meadows. J. Veg. Sci. 8: 829–838.

Vertes, F., 1983. Contribution à l'étude phytosociologique et écologique des prairies et alpages de moyenne Tarentaise. Application à l 'évaluation des potentialités fourragères de la vallée de Peusey-Nancroix. Thesis, INAPG: 163 pp.

Willis, G.H., L.M. Southwick, J.L. Fouss, C.E. Carter & J.S. Rogers, 1997. Nitrate losses in runoff and subsurface drain effluent from controlled-water table plots. Bull. Envir. Contam. Toxico. 58: 566–573.

Hydrobiologia **410:** 207–212, 1999.
J. Garnier & J.-M. Mouchel (eds), Man and River Systems.

Scenario analysis of nutrient management at the river basin scale

B. Kronvang[1], L. M. Svendsen[1], J. P. Jensen[2] & J. Dørge[3]
[1]*National Environmental Research Institute, Department of Streams and Riparian Areas, Vejlsøvej 25, DK-8600 Silkeborg, Denmark*
[2]*National Environmental Research Institute, Department of Lake and Estuarine Ecology, Vejlsøvej 25, DK-8600 Silkeborg, Denmark*
[3]*VKI Water Quality Institute, Agern Alle 11, DK-2970 Hørsholm, Denmark*

Key words: river basin model, source management, transport management

Abstract

A new river basin model (TRANS) for studying the transport, removal and accumulation of nutrients in rivers, lakes and riparian areas has been developed and tested on data from a 115 km^2 river basin in Denmark (river Gjern). The model combines catchment information on soil type and land use with a physical hydrodynamic modelling system and several semi-dynamic empirical models on diffuse nutrient loading and nutrient retention in rivers, lakes and riparian areas. Three main river basin management scenarios were defined for this basin and the environmental effects analysed using the model: 1. Improved sewage treatment at sewage works; 2. 20% set-aside of arable land; 3. river system rehabilitation. Combining the three scenarios in an overall river basin management strategy reduced annual total nitrogen export from the river basin by 93 tonnes, corresponding to a 53% reduction in nitrogen export during the hydrological year 1994–95. Similarly, annual total phosphorus export was reduced by 2.6 tonnes corresponding to a 46% reduction in phosphorus export.

Introduction

Pollution of the aquatic environment with nitrogen (N) and phosphorus (P) is still one of the main environmental problems facing both Denmark and other European countries (European Environment Agency, 1995; Kronvang et al., 1996). Average annual diffuse losses of total N and total P thus amounted to 18.4 kg N ha^{-1} and 0.33 kg P ha^{-1} from approx. 80 small Danish catchment areas during the period 1989–1996 (Kronvang et al., 1996). The result is toxic algae and oxygen deficit in our lakes, estuarine fjords and inner marine waters and nitrate pollution of our groundwater (Kronvang et al., 1993). The Danish Action Plan on the Aquatic Environment has effectively reduced nutrient discharges from industry and urban sewage (Kronvang et al., 1996).

In contrast, efforts to reduce nutrient discharges from agriculture have been less successful. Recent data from the Danish Nationwide Monitoring Programme thus show that nitrogen leaching from the root zone of farmland has only been fallen by approx. 17% during the 1990s, while riverine nitrogen from diffuse sources has only fallen by approx. 9% (Windolf et al., 1997). Moreover, diffuse phosphorus loss from farmland is greater than previously believed (Kronvang et al., 1997).

The most important task in Denmark and most other European countries is therefore to reduce nutrient losses from fields as much as possible. This can be accomplished in to main ways: (i) source management, e.g. improved utilization of livestock manure and general reductions in fertilization at the field level; (ii) transport management, e.g. restoration of streams, wetlands and lakes in order to utilize their potential to remove nitrate by denitrification and store phosphorus by sedimentation and adsorption processes (Fleischer et al., 1991; Haycock & Burt, 1993; He & Walling, 1996; Hoffmann et al., 1998; Kronvang et al., 1999).

The main step taken in Denmark to combat diffuse nutrient pollution is to reduce nutrient losses from arable land, including requirements limiting livestock density at the farm level, nine-month storage capacity for liquid manure, regulations on fertilizer application practice and input per hectare arable land, requirements for fertilization plans and budgets at the farm level, green cover on fields in winter and set-aside of agricultural land.

Another major step is measures aimed at restoring the natural self-purification potential of streams, lakes and wetlands, a step which has recently been adopted in a new Danish Action Plan on the Aquatic Environment (Hansen, 1996; Iversen et al., 1998).

This paper aims to demonstrate the benefits of combating diffuse nutrient pollution through source and transport management at the river basin scale. A recently developed Danish river basin model was used to examine two different scenarios in the Gjern river basin in Denmark.

Description of the river basin

Scenarios were established for the Gjern river basin in central Jutland, Denmark, which is a typical example of the Danish moraine landscape. The river basin has a catchment area of 115 km^2. Long-term average precipitation and runoff in the river basin is 770 mm (1961–1990) and 333 mm (1974–1994), respectively. Soil type varies from sandy to loamy, with 68% of the river basin being arable land. The river system comprises approx. 85 km of stream channel, a 39 ha upstream lake and 895 ha of riparian areas.

Description of the monitoring methods and the model concept

Water sampling and discharge measurements were undertaken at 35 monitoring stations within the river basin on a monthly basis. Water level was recorded continuously at nine of the 35 monitoring stations and mean daily discharge was calculated using normal stage-discharge relationships. Water samples were analyzed for total nitrogen and total phosphorus according to the methods described in Svendsen et al. (1995).

TRANS, the river basin model developed for studying the transport, removal and accumulation of nutrients in rivers, lakes and riparian areas, combines catchment information on soil type and land use with a physical hydrodynamic modelling system, a model for macrophyte growth and several semi-dynamic empirical models on diffuse nutrient loading and nutrient retention in rivers, lakes and riparian areas (Figure 1 & Table 1).

Table 1. The processes modelled in the river basin model (TRANS) together with the reference for the corresponding submodel

Submodels in the river basin model (TRANS)	Reference
Hydrodynamics	Havnø et al., 1995
Diffuse nutrient loading	Kronvang et al., 1995
Retention of N in shallow lakes	Windolf et al., 1996
Retention of P in shallow lakes	Jensen et al., 1994
Release of P from lake sediment	Jensen et al., 1994
Sedimentation of P in lowland streams	Svendsen, 1992
Resuspension of P in lowland streams	Svendsen, 1992
Macrophyte production and respiration in lowland streams	Thyssen et al., 1990
Dentrification in streams	Thyssen et al., 1990
Dentrification in flooded riparian areas	Kronvang (Unpubl.)
Sedimentation during flooding of riparian areas	Kronvang (Unpubl.)

The river basin model (TRANS) was set-up and calibrated to describe daily water and nutrient fluxes in the entire river basin during the hydrological year June 1994–May 1995. Modelled and measured concentrations of nitrogen and phosphorus at the 35 monitoring stations and in the lake were in reasonably good agreement (Figure 2).

Denitrification in wet meadows is not modelled directly. Instead, the effect of this process in the river basin was estimated from GIS maps of wet meadows in the riparian areas in 1876 and in 1995. The difference in wet meadow acreage between these two years was approx. 300 ha. In the river system rehabilitation scenario we assumed a denitrification rate of 100 kg N ha^{-1} following re-establishment of these wetlands (Hoffmann et al., 1998; Kronvang et al., 1999).

Results

Nutrient sources and export from the river basin

Nitrogen and phosphorus export from the river basin amounted to 15.0 kg N ha^{-1} and 0.57 kg P ha^{-1} during 1995. Diffuse nutrient sources accounted for 85% of the total nitrogen export and 45% of the total phosphorus export (Figure 3).

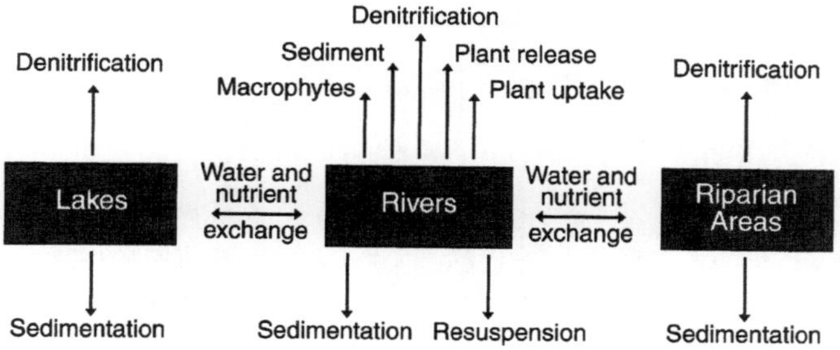

Figure 1. Model structure and the physical, chemical and biological processes described in the river basin model (TRANS).

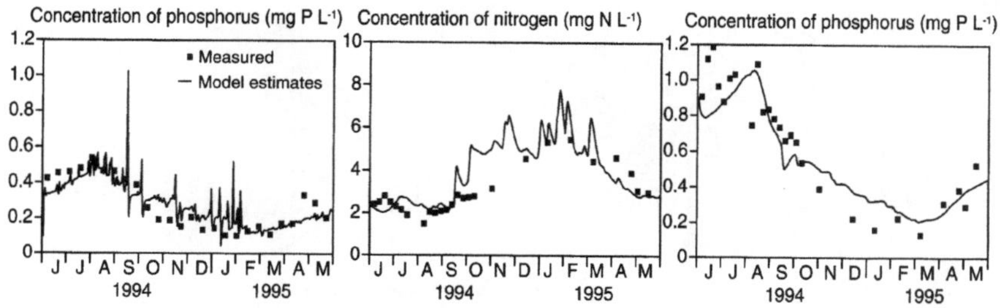

Figure 2. Comparison of modelled and measured total P concentration at the monitoring station at the river basin outlet, and of daily mean concentrations of total N and total P, respectively, in the lake during the hydrological year June 1994–May 1995.

Monitoring of nutrient losses from 35 small sub-catchments of the Gjern river basin revealed high spatial variation attributable to differences in runoff, soil type and land use (Figure 4).

Scenario for source management in the river basin

A scenario with improved sewage treatment (denitrification and chemical treatment) at the five sewage works located in the river basin reduced nitrogen and phosphorus loading of the surface water by 28.5 tonnes N (13.2%) and 0.6 tonnes P (12.5%) (Table 2). Export of nitrogen and phosphorus from the river basin was only reduced by 11.4 tonnes N (6.4%) and 0.3 tonnes P (5.3%), however (Table 2). This apparent inconsistency can be ascribed to a decrease in nutrient removal by denitrification processes and nutrient retention by sedimentation in the river basin especially in the shallow lake.

In comparison, annual export of total nitrogen and total phosphorus from the river basin was reduced by 29.6% and 7.1%, respectively, when 20% of the arable land was set aside (Table 2).

Table 2. Predicted effects of improved sewage treatment and 20% set-aside of arable land within the Gjernriver basin on nutrient loading of surface waters, nutrient retention in the watercourses, lake and riparian areas and nutrient export from the river basins as compared to model estimates for the current situation (hydrological year June 1994–May 1995)

Source management scenario	Total nitrogen (tonnes N)	Total phosphorus (tonnes P)
Current situation		
Surface water loading	215.8	4.8
Retention in the river system	38.8	−0.8
Export from the river basin	177.0	5.6
Improved sewage treatment		
Surface water loading	187.3	4.2
Retention in the river system	21.7	−1.1
Export from the river basin	165.6	5.3
20% set-aside		
Surface water loading	150.0	4.4
Retention in the river system	24.4	−0.8
Export from the river basin	124.6	5.2

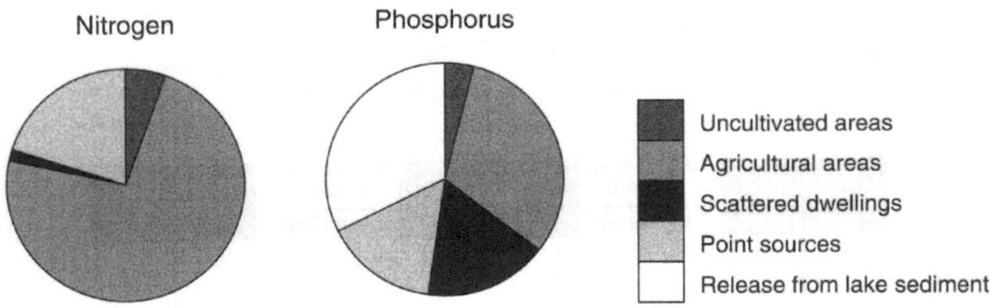

Figure 3. Source apportionment of total N and total P export from the Gjern river basin during 1995.

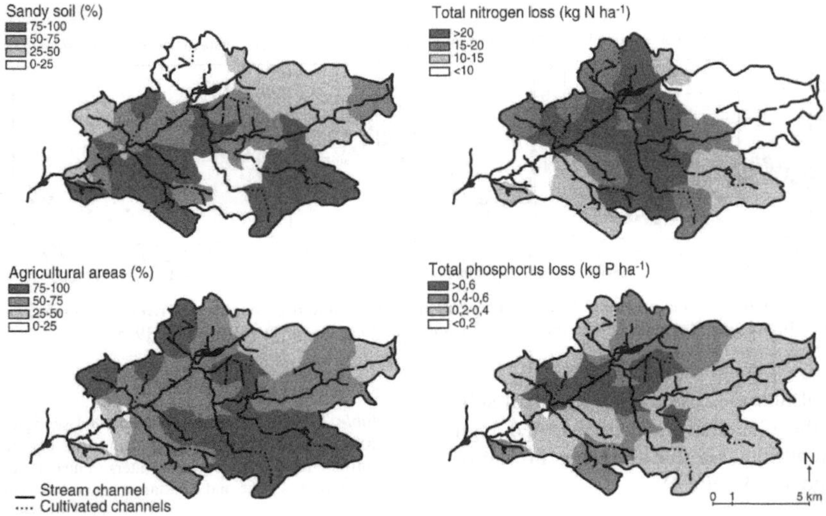

Figure 4. Soil type, proportion of agricultural land, total N loss and total P loss in the Gjern river basin in 1995.

Scenario for transport management

The transport management scenario encompasses re-establishing the natural sinuosity and dimensions of all stream channels in the river basin, dredging 50 cm of the lake sediment which is highly enriched with phosphorus due to former discharge of untreated sewage water from an upstream town, and re-establishing 300 ha of riparian wet meadows.

Rehabilitation of the entire river system increased annual removal of nitrogen from 38.8 tonnes N to 81.4 tonnes N (110%) (Table 3). Moreover, annual retention of total phosphorus was increased from net loss of 0.8 tonnes P from the lake sediment to net retention of 0.9 tonnes P (Table 3). The major part of the increase in nitrogen removal following river system rehabilitation was attributable to denitrification of nitrate in wet meadows. In the case of phosphorus, the major part

Table 3. Predicted effects of river system rehabilitation within the Gjern river basin on nutrient loading of surface waters, nutrient retention in the watercourses, lake and riparian areas and nutrient export from the river basin as compared to model estimates for the current situation (hydrological year June 1994–May 1995)

Transport management scenario	Total nitrogen (tonnes N)	Total phosphorus (tonnes P)
Current situation		
Surface water loading	215.8	4.8
Retention in river systems	38.8	−0.8
Export from the river basin	177.0	5.6
River system rehabilitation		
Surface water loading	215.8	4.8
Retention in river systems	81.4	0.9
Export from the river basin	134.4	3.9

Table 4. Predicted effects of an overall river basin management strategy combining nutrient source and transport management within the Gjern river basin on nutrient loading of surface waters, nutrient retention in watercourses, lake and riparian areas and nutrient export from the river basin as compared to the model estimates for the current situation (hydrological year June 1994–May 1995)

Combined source and transport management scenario	Total nitrogen (tonnes N)	Total phosphorus (tonnes P)
Current situation		
Surface water loading	215.8	4.8
Retention in the river system	38.8	−0.8
Export from the river basin	177.0	5.6
Overall river basin management strategy		
Surface water loading	121.6	3.8
Retention in the river system	38.2	0.8
Export from the river basin	83.4	3.0

of the increase in retention was attributable to sedimentation of P in the shallow lake and on temporarily flooded riparian areas.

Overall river basin management strategy

Combining the source and transport management scenarios into an overall river basin management strategy reduced annual total N export from the river basin by 93.6 tonnes N, corresponding to a 53% reduction in nitrogen export during the hydrological year 1994–95 (Table 4). Similarly, the annual total P export from the river basin was reduced by 2.6 tonnes P, corresponding to a 46% reduction in phosphorus export during the hydrological year 1994–95 (Table 4).

Discussion and conclusions

During the past decade, it has become more and more obvious that environmental research has to consider how to upscale results from the field scale to a larger scale in order to enforce local environmental planning. Recent attention has therefore focused on how to combine simple empirical and more process-oriented models for hydrology, diffuse nutrient losses and nutrient turnover in integrated models (Styczen & Storm, 1993; Lunn et al., 1996; Witgren & Arheimer, 1996).

Operating within a GIS environment, such models can be important tools for environmental planning, monitoring and research since they provide insight into the coupling between the location and development of driving forces and pressures that determine nutrient loss within a catchment and the resulting environmental impact.

The new Danish river basin model (TRANS) links several submodels for diffuse nutrient loss to freshwater and nutrient turnover in stream channels, lakes and riparian areas with a one-dimensional hydrodynamic model and a GIS platform thereby forming a unified model for use in scenario analysis of the environmental consequences of changes in land use, river system rehabilitation, etc. at the river basin scale.

TRANS requires further development, especially with regard to the very important processes that govern subsurface nitrogen retention along the nitrogen transport pathways from the root zone to surface waters.

The scenarios examined clearly reveal that nutrient management in this specific river basin should be directed towards control of diffuse sources, primarily agricultural land. A combination of improved sewage treatment, 20% set-aside of agricultural land and river system rehabilitation at the basin scale significantly reduced both nitrogen (53%) and phosphorus (46%) export from this river basin.

In Denmark, such measures have until now been implemented along approx. 50 km of river reach as river rehabilitation demonstration projects. Large-scale river system rehabilitation projects are an important future step in Denmark to be used in conjunction with regulation of the agricultural use of fertilizers in the newly adopted Action Plan for the Aquatic Environment II (Iversen et al., 1998).

Acknowledgement

This study was conducted as part of the Danish Environmental Research Programme and European Community, Environment and Climate Programme (Contract No. ENV4-CT97-0435).

References

European Environment Agency, 1995. Europe's Environment – The Dobřis Assessment. D. Stanners & P. Bourdeau (eds), European Environment Agency. Copenhagen, Denmark: 676 pp.

Fleischer, S., L. Stibe & L. Leonardson, 1991. Restoration of wetlands as a means of reducing nitrogen transport to coastal waters, AMBIO 20: 271–272.

Hansen, H. O., 1996. River Restoration – Danish experience and examples, National Environmental Research Institute, Denmark: 99 pp.

Havnø, K., M. N. Madsen & J. Dørge, 1995. MIKE 11 – A generalized river modelling package, In V. P. Singh (ed.), Computer models of watershed hydrology. Wat. Res. Publication, Colorado, ISBN 0-918334-91-8: 733-782.

Haycock, N. E. & T. P. Burt, 1993. Role of flood plain sediments in reducing the nitrate concentration of subsurface runoff: A case study in the Cotswolds, U.K. Hydrol. Processes 7: 287–295.

He, Q. & D. E. Walling, 1996. Use of fallout Pb-210 measurements to investigate longer-term rates and patterns of overbank sediment deposition on the flood plains of lowland rivers. Earth Surface Processes and Landforms 21: 141–154.

Hoffmann, C. C., M. L. Pedersen, B. Kronvang & L. Øvig, 1998. Restoration of the rivers Brede, Cole and Skerne: A joint Danish and British EU-LIFE project, IV – implications for nitrate and iron transformation. Aquat. Conserv. 8: 223–240.

Iversen, T. M., R. Grant, G. Blicher-Mathiesen, H. E. Andersen, E. Skop, J. J. Jensen, B. Hasler, J. Andersen, C. C. Hoffmann, B. Kronvang, H. E. Mikkelsen, J. Waage-Petersen, A. Kyllingsbæk, H. D. Poulsen & V. F. Kristensen, 1998. Danish Action Plan on the Aquatic Environment II – a scientific evaluation. Danish Ministry of Environment and Energy, National Environmental Research Institute, Silkeborg, Denmark: 44 pp. (in Danish).

Jensen, J. P., E. Jeppesen, J. Bøgestrand, A. R. Petersen, M. Søndergaard, J. Windolf & L. Sortkjær, 1994. Aquatic Environment Monitoring programme 1993. Freshwater-lakes. National Environmental Research Institute, Silkeborg, Denmark, Technical report 121: 115 pp.

Kronvang, B., G. Ærtebjerg, R. Grant, P. Kristensen, M. Hovmand & J. Kirkegaard, 1993. Nationwide Monitoring of Nutrients and Their Ecological Effects: State of the Danish Aquatic Environment. AMBIO 22: 176–187.

Kronvang, B., R. Grant, S. E. Larsen, L. M. Svendsen & P. Kristensen, 1995. Non-point-source nutrient losses to the aquatic environment in Denmark: Impact of agriculture. Mar. Freshwat. Res. 46: 167–177.

Kronvang, B., P. Græsbøll, S. E. Larsen, L. M. Svendsen & H. E. Andersen, 1996. Diffuse nutrient losses in Denmark. Wat. Sci. Tech. 33: 81–88.

Kronvang, B., A. Laubel & R. Grant, 1997. Suspended sediment and particulate phosphorus transport and delivery pathways in an arable catchment, Gelbæk stream, Denmark. Hydrol. Processes 11: 627–642.

Kronvang, B., C. C. Hoffmann, L. M. Svendsen, J. Windolf, J. P. Jensen & J. Dørge, 1999. Retention of nutrients in river basins. Aquat. Ecol. 33, 29–40.

Lunn, R. J., R. Adams, R. Mackay & S. M. Dunn, 1996. Development and application of a nitrogen modelling system for large catchments. J. Hydrol. 174: 285–304.

Styczen, M. & B. Storm, 1993. Modelling of N-movements on catchment scale – a tool for analysis and decision making. Fertilizer Res. 36: 1–6.

Svendsen, L. M., 1992. Dynamics of phosphorus, organic nitrogen and organic matter in watercourses – method, retention, transport and model. Ph.D. Thesis. National Environmental Research Institute, Ministry of the Environment, Silkeborg, Denmark: 260 pp.

Svendsen, L. M., B. Kronvang, P. Kristensen & P. Græsbøll, 1995. Dynamics of phosphorus compounds in a lowland river system: importance of retention and non-point sources. Hydrol. Processes 9: 119–142.

Thyssen, N., M. Erlandsen, B. Kronvang & L. M. Svendsen, 1990. Dynamic and empirical modelling of biological structure and physical factors in lowland streams. In: Nitrogen and phosphorus in fresh and marine waters. Project abstracts of the Danish NPo Research Programme, C-abstracts. Danish EPA, Copenhagen: 147–171.

Windolf, J., E. Jeppesen, J. P. Jensen & P. Kristensen, 1996. Modelling the seasonal variation in in-lake concentration and retention of nitrogen in 16 shallow lakes: A four-year mass balance study. Biogeochemistry 33: 25–44.

Windolf, J., L. M. Svendsen, B. Kronvang, J. Skriver, N. B. Ovesen, S. E. Larsen, A. Baattrup-Pedersen, H. L. Iversen, J. Erfurt, D. Müller-Wohlfeil & J. P. Jensen, 1997. Monitoring Programme 1996 – Freshwater – streams and springs. Technical report 214, National Environmental Research Institute, Silkeborg, Denmark: 109 pp.

Witgren, H. B. & B. Arheimer, 1996. Source apportionment of riverine nitrogen transport based on catchment modelling. Wat. Sci. Tech. 33: 109–115.

Hydrobiologia **410:** 213–221, 1999.
J. Garnier & J.-M. Mouchel (eds), Man and River Systems.

Methodology for multifunctional assessment of riverine wetlands in the Seine river basin

Eliane Fustec[1], Philippe Boët[2], Aïcha Amezal[3] & Nils Fauchon[4]
[1]*UMR 7619-Sisyphe, Laboratoire de Géologie Appliquée, Université P. et M. Curie, 4 place Jussieu, F-75252 Paris Cedex 05, France*
[2]*Cemagref, Division QHAN, parc de Tourvoie, BP 44, F-92163 Antony Cedex, France*
[3]*Agence de l'Eau Seine-Normandie, 51 rue Salvador Allende, F-92027 Nanterre Cedex, France*
[4]*VIVENDI et Syndicat des Eaux d'Ile-de-France, Quartier valmy, 32 Place Ronde, F-92982 Paris-la-Défense Cedex 82, France*

Key words: riverine wetlands, functional assessment, flood control, water supply, habitat, nitrate removal

Abstract

A new policy for wetlands conservation in France induced the development of research on the main functions of these ecosystems. This paper summarizes the results of four studies initiated in the Seine River basin by scientists, the regional water agency and a water supply syndicate to assess some of the functions of wetlands, mainly of riverine wetlands. One was directly conducted at the scale of the whole basin, focusing on the role of wetlands in water supply, flood control and nutrient retention. A second one was centred on the importance of waterbodies more or less connected to the channel for fish communities and on the role of flooded grasslands for the reproduction of pike. From the results obtained in large sectors of the fluvial corridor, a method is proposed for a regionalization of the hydrosystem. Two other approaches were tested in large areas of fluvial corridors to classify nitrate removal capacities of riverine wetlands through the identification of physical and biological indicators or modelling. What emerges from these studies is the necessity to undertake a common classification of wetlands on a hydrogeomorphic basis and then to develop multifunctional assessment in reference zones from the identified indicators or tested models.

Introduction

All wetlands have to be protected in France in the same way as water resource and aquatic ecosystems since a new law was promulgated in 1992. However, lack of informations regarding inventories, classifications and functions of wetlands as well as the impact of numerous human activities promoting their degradation, caught resource planners and managers somewhat unaware. This problem is highly topical in many countries where the protection of wetlands is lawfully imposed and has led to serious debates, partly due to the high economic and political stakes.

Discussions centre everywhere on the legal definition of wetlands, their delineation and their value. Recent economic studies emphasized that wetlands, especially floodplains, are among the ecosystems which reach the highest values (Barbier et al., 1997; Costanza et al., 1997). Socio-economic values of wetlands depend on their ability to perform a series of functions as identified both by scientists and managers (Adamus & Stockwell, 1983; Mitsch & Gosselink, 1993). The main wetlands functions may generate services or products useful for society: inundation of upstream floodplains can prevent damage to downstream urban or cultivated areas, sediments and associated pollutants trapping or nitrate removal can improve water quality, submersion or waterlogged conditions in soils, specific landforms and waterbodies in wetlands lead to the development of unique plant communities and to numerous habitats that contribute to the maintenance of biodiversity. They offer also recreational and educational services.

The main problem is to determine at what level a specific function or a grouping of functions is realized for different types of wetlands. These functions depend on the regional location of the ecosystems, on their dynamics and degree of human disturbance. Recently, researchers from state or regional agencies or private companies have undertaken efforts in many countries to classify wetlands and to assess their main functions. Several approaches are proposed, the majority of them emphasizing the importance of relationships between hydrology and geomorphology, and wetlands functions (Adamus, 1988; Brinson, 1993b; Larson & Mazzarese, 1994; Novitzki, 1994; Maltby et al., 1994; Ferren et al., 1996b; Maltby et al., 1996). The historical dimension may be also considered to assess the extent of natural evolution and disturbances as human impacts on the functioning of wetlands (Hook, 1993; Ferren et al., 1996a). In the Seine river basin, several methods were simultaneously developed by scientists, by the regional water agency and a water supply company to initiate functional and value assessment of its wetlands, especially the riparian wetlands.

Approaches at the scale of the whole basin

Among two studies developed at the scale of the whole Seine river basin, the first one focused on the role of wetlands in water resource control and the second one looked at the factors controlling fish communities in the hydrographic network.

Evaluation of functional importance of wetlands as applied to water resources

In order to make all actors in the Seine basin aware of the importance of wetlands and to rapidly improve innovative planning for better protection and management of these ecosystems, the water agency initiated a classification and a map of functional potential of the main wetlands according to their influence on water resources. This study was developed by research consultants with scientific oversight (BURGEAP & Bravard, 1995). It focused on the role of wetlands in water supply, flood control and nutrient retention. The great majority of identified wetlands were of riverine type but some remarkable lacustrine, palustrine and estuarine wetlands taking part in water regulation were also considered.

The procedure for this assessment used several indicators of 'positive effects' of wetlands for each of these functions. For example, pumping flow in wells established in wetlands gave information on their water supply capacity which depends on water inputs in the wetlands, especially on the connection between these ecosystems and deep aquifers. Three classes of pumping flow were considered: more than 100 m^3 h^{-1} per well, between 25 and 100 m^3 h^{-1} and less than 25 m^3 h^{-1} (legend of Figure 1). To identify areas liable to be flooded, geological maps were used to determine the width of channel banks grouped in three classes: less than 1 km, between 1 and 3 km and more than 3 km. A second indicator was derived from a preliminary map of flooded areas giving informations on levels reached by overbank flows during historical inundations. Nutrient retention capacities of wetlands were estimated from the percentage of areas covered by riparian forests, wet grasslands, swamps, oxbow lakes etc... in fluvial corridors, assuming that these systems act as sinks for nutrients. Three classes were determined: more than 20% of effective systems areas, between 20 and 2% and less than 2%.

On the other hand, the procedure used two indicators of 'negative effects' induced by human activities which can alter these functions. The first one was the part of the surface of fluvial corridors occupied by sand-pit lakes and three classes were determined: less than 2% of the total area of the banks, between 2 and 15% and between 15 and 50%. The second one was the quality of running waters in accordance with the classification previously adopted by water agencies. The different classes of water quality (excellent, good, passable, mediocre and bad) are supposed to integrate the influence of urban areas on river banks able to alter water quality downstream and to reduce flooded areas through embankments protecting constructed areas.

Numerical indexes were assigned to each class of the indicators of 'positive' and 'negative' effects and their synthesis results in seven classes assessing the potential affects of the main wetlands of the Seine river basin on water resource (Figure 1).

Assessment of habitat function for fish communities

Several studies on the distribution of fish communities in the hydrographic network of the Seine river, developed since 1990 in the Programme PIREN-Seine emphasized the importance of spatial and temporal scales in the system and the great influence of human impacts (channelization, reservoirs, waste and storm urban waters) on the distribution and biology of different species (Belliard, 1994; Boët et al., 1994; Belliard et al., 1995). Furthermore, they suggested that in the

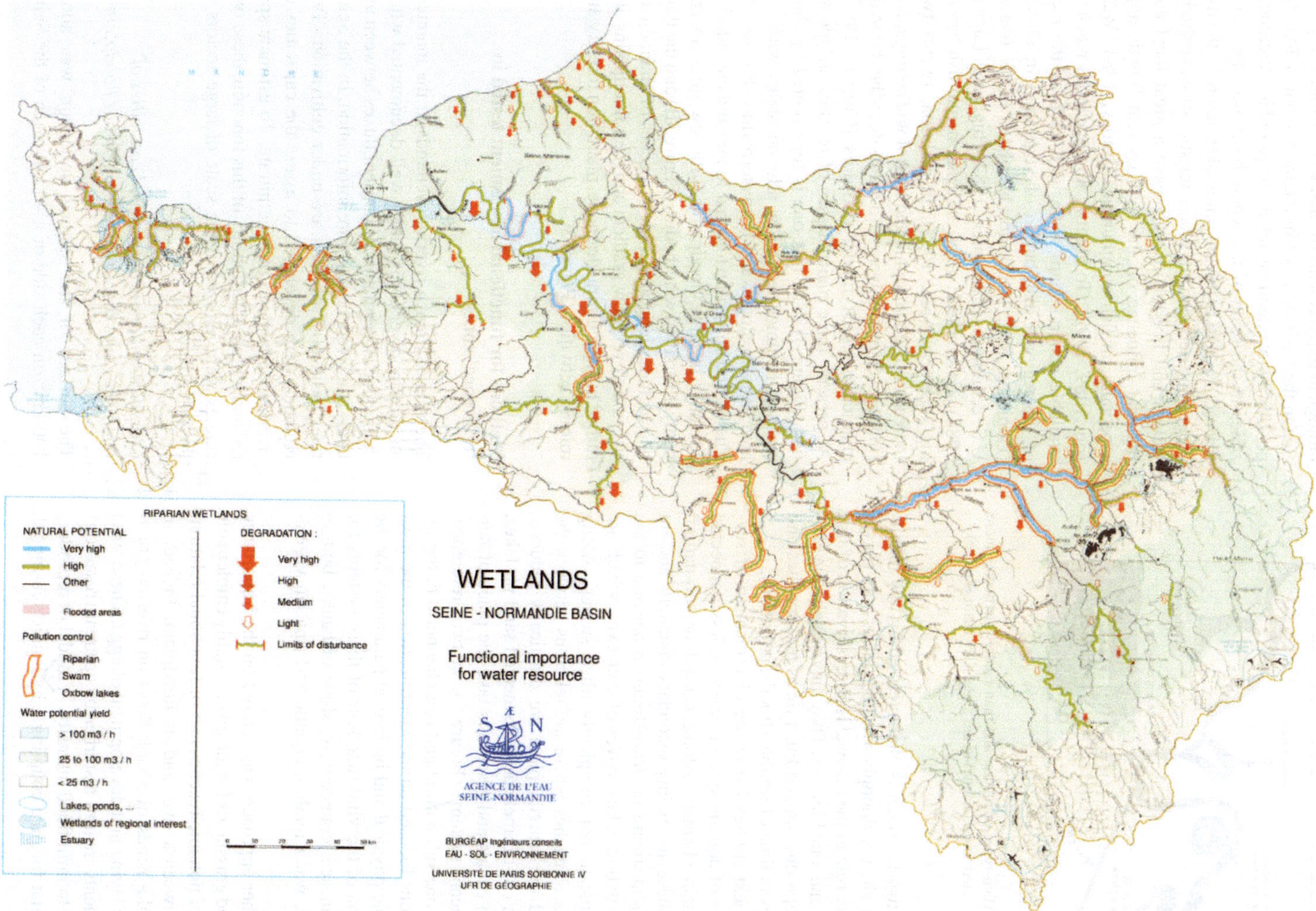

Figure 1. Synthetic map of functional potential of wetlands for water resources in the Seine river basin and Normandy (France).

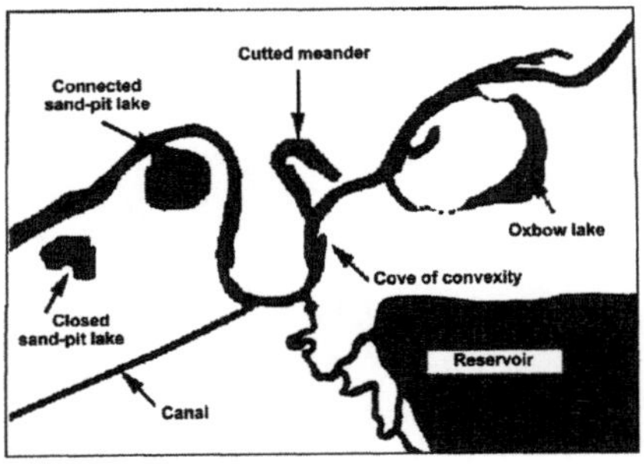

Figure 2. Main types of natural and man-made waterbodies more or less connected to the channel in the Seine river basin (France).

Seine network, which is characterized by low longitudinal gradients, floodplain waterbodies, may play a significant role in the renewal and preservation of fish diversity and total biomass. They may be valuable habitat for reproduction and function as nursery sites and as refuges in time of trouble (floods, pollution).

The next studies led to the identification of several types of aquatic systems more or less connected to the main channel. Along natural or sub-natural reaches these are mainly secondary channels, ancient meanders upstream or downstream connected to the channel, oxbow-lakes, coves of convexity which are comparable to not completely filled swales. Along channelized reaches where the relations between the river and the alluvial plain are drastically reduced, upstream cut-off meanders, connected sand-pit lakes, diversion canals and reservoirs take the place of natural adjacent systems (Figure 2). In two 'reference' reaches, one sub-natural and one channelized, the investigations focused on the identification of the most favourable physical and biological parameters for the realization of the vital functions of these systems for fishes (surface, connectivity, slope of banks, bottom structure, water quality, aquatic and riparian vegetation).

Another approach was centred on the importance of flooded grasslands for the pike, a highly emblematic species of potamic reaches and of functional relationships between a river and its floodplain. Indeed, in spring pike needs overbank flows on riverine grasslands for laying and development of eggs. These strict requirements are now disturbed by upstream reservoirs intended to limit flood intensity. A model of reproduction habitat for pike coupling a hydraulic sub-model, topographic and land use maps of riparian zones and biological requirements of pike was developed.

On the basis of a hydrosystem regionalization, the possibility to extend the results acquired in 'reference' zones to the scale of the whole basin was then considered. Parallel to the longitudinal structure of the hydrographic network, the characteristics of watersheds, valleys and floodplains may have a great local and regional influence on the hydrosystem functioning, especially on fish communities (Bailey, 1984; Wasson, 1989, 1992). To identify homogeneous reaches, available general maps were used such as the map of the regional distribution of vegetation in France (Dupias & Rey, 1985) closely related to the one of U.S. ecoregions (Omernick, 1987; Hughes & Larsen, 1988) and synthetic maps of departmental fish breeding schemes (Area, 1992). In addition, twenty-two indicators of geomorphic and hydrologic settings, and riparian vegetation were utilized to describe the main characteristics of fluvial corridors (Gorges, 1996). At present, three main classes of reaches are identified (Figure 3). The first one is represented by short reaches, hemmed in narrow alluvial plains with relatively steep slopes and no waterbodies (Figure 3A) The second one group together longer and meandering stretches in large floodplains with low gradient and many diversified waterbodies (Figure 3B) and the third one includes long and straight reaches of channelized rivers with very low gradient and their floodplains disturbed by agricultural drainage, plantation forestry and mining activities (Figure 3C).

Methods for functional assessment tested in 'reference' zones

The role of French riverine wetlands in the maintenance of biodiversity is rather well documented while their value as regulators of nutrient fluxes between uplands and rivers needs more information. In the Seine river basin 60% of soils are under cultivation. Two methods were developed to assess the capacities of riverine wetlands to remove nitrate. To estimate spatial and temporal variability of this function, these two studies were realized at the scale of large sectors of fluvial corridors.

Classification of nitrate removal capacities of reference zones in a large heterogeneous floodplain

The surface of the studied floodplain was about 100 km^2 on each side of a 40 km reach of the Seine

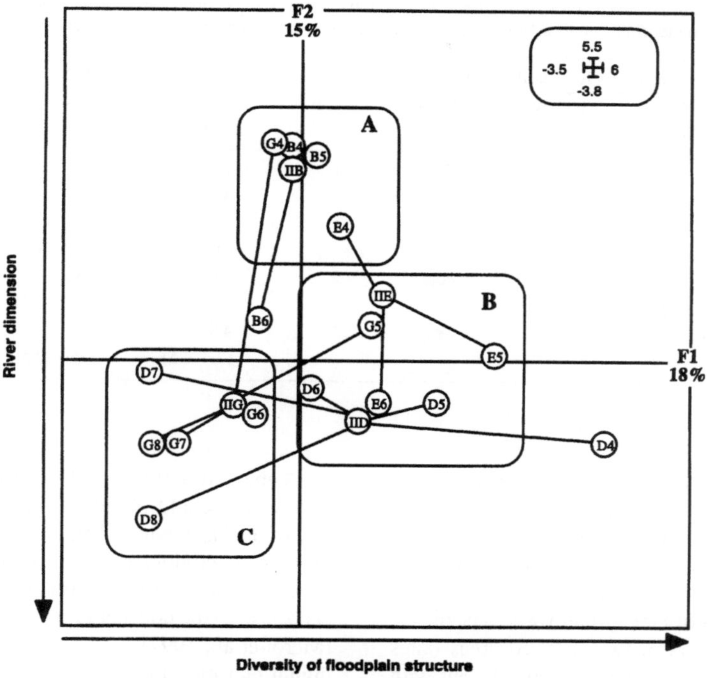

Figure 3. Principal component analysis (PCA) of 21 morphological characteristics describing the major and minor riverbeds of 119 reaches at least or greater fourth-order (2–52 km) from the Seine river and its major tributaries, Aube, Yonne, Marne and Oise. F1F2 factorial map representation of the centroids stream-order (X4 to X8) within the four largest natural regions from the sedimentary part of the Seine river basin (IIB Jurassic formations, IIE lower Cretaceous formations, IID upper Cretaceous formations, IIG Tertiary formations).

river upstream and downstream of its confluence with the Aube river. The width of the floodplain varies from 2 to 5 km. The control of this alluvial valley on floods is preserved for the protection of the downstream urban district of Paris. The shallow alluvial aquifer is more or less connected to a deep aquifer in chalk and the floodplain is more or less inundated every year mainly by overbank flows and partly by swollen aquifers. Hundred piezometers were established at regular distance in the alluvial plain to follow the variations of groundwater level and quality. Nitrate concentrations in groundwater show significant variations in space which may be related to the heterogeneity of the floodplain. The mosaïc structure depends on sediment deposition, intensity and duration of inundation by overbank flow and/or groundwater rising, and land use (30% of the study area is under cultivation). Nitrate concentrations remain very low during the year in many sites, while significant variations are observed in others. In a few sites nitrate content remains very high (Greiner, 1997).

Eight reference zones were selected in the study area to initiate a classification of nitrate removal capacity of the different types of patches. Nitrate removal in such a complex system may be realized through denitrification in anoxic layers of soils and/or absorption by plants. Anoxic conditions are influenced by water level and soil texture which determine oxygen diffusion and residence time of water and nutrients. Absorption by plants depends on roots depth and differ under crops and alluvial forests in the areas where the water level is rather deep during low water periods.

To classify nitrate removal capacity of the different systems in the floodplain, three types of hydrologic conditions were considered (water table level never under 1 m depth, decreasing to 2 m depth and decreasing under 2 m), three clay content in soils (clay content ranging from 40 to 60%, from 20 to 40% and less than 20%) and two types of land use (forested and cultivated areas). Three classes of nitrate removal efficiency were determined: 1. total, 2. partial and 3. null. This classification was validated with oxygen concen-

Table 1. Classification of four reference zones of riparian wetlands in the Seine river floodplain according to their nitrate removal capacity. Annual mean and extreme values (in parenthesis) for water level, dissolved oxygen, redox potential and nitrate (NO_3) concentration

Reference zone	Indicators for classification			Class	Indicators for validation		
	Water level (m)	Texture (clay %)	Land use		Oxygen (mg^{-1})	Redox (mV)	Nitrate ($mg^{-1}\ l^{-1}$)
A	0.7 (0–1.3)	38	Forest	1	1.4 (0.9–2.6)	250 (166–359)	0.2 (0–0.5)
B	0.7 (0.2–1.0)	45	Crops	1	1.7 (0.9–2.9)	240 (174–302)	0.6 (0.1–2.0)
C	1.7 (0.4–2.6)	45	Forest	2	2.9 (0.9–5.2)	338 (203–434)	17.4 (9.1–33.5)
D	0.5 (0–1.3)	20	Crops	3	4.9 (2.4–6.5)	422 (354–486)	64.1 (38–94)

tration, redox potential and nitrate concentration in groundwater (Reddy & d'Angelo, 1994).

Results are given in Table 1 for four of the reference zones selected in the floodplain. They show the great influence of water level variations and of superficial clayey deposits above the groundwater. The nitrate removal capacity differs in two forested areas (sites A & C) with relatively similar clay content but with different hydroperiods. For cultivated areas (sites B & D) the pattern is similar for water level but a low clay content strongly limits nitrate retention in site D.

Modelling of nitrate removal potential in riparian wetlands

Another method for determining nitrate removal potential was initiated by a regional water supply syndicate in a sub-basin (Grand-Morin river) of the Marne river (Figure 4a). The whole basin is intensively cultivated and up to 85% of arable lands have artificial drainage by-passing riparian buffer zones. The quality of surface water is strongly degraded: nitrate concentration in the Grand-Morin river is generally between 20 and 40 $mgNO_3\ l^{-1}$ and increases to 100 $mgNO_3\ l^{-1}$ during flood events in winter (Penven et al., 1998). This pollution is causing increasing problems for drinking water pumping stations downstream. In partnership with agricultural managers and farmers, the regional water supply syndicate takes a very active role in the development of actions intended to reduce non point source pollution. Regulation procedures to reduce nitrogen inputs in upland cultivated areas are tested and a restoration of efficient buffer zones appears to be a complementary way to reduce nitrate loads in surface water (Novotny & Olem, 1994).

A method aiming to identify the sites where denitrification could be active was tested in the lower part of the basin where riparian zones are less affected by human activities (Figure 4a). It was based on the development of a model similar to hydrological models based on the concept of variable contributive areas (Mérot et al., 1997). The procedure used an available digital elevation model and two discriminant factors, soil moisture and residence time of water, to determine the areas where denitrification may occur. The different steps of this approach are shown in Figure 5 (Conesa-Alcolea, 1997). A map of relative denitrification potentials was established for the study area (Figure 4b).

Discussion and conclusions

New policies have led scientists, planners and managers to develop studies for a better understanding of the multiple functions of wetlands acting as regulators of water and matter fluxes and in the maintenance of biodiversity. The urgent implementation of the new policy in France when knowledge of the different types of wetlands and of their functioning was insufficient induced simultaneaous research at several scales on wetlands classification, quantification of the main functions, identification of indicators and methods for functional assessment (Fustec & Frochot, 1996). The first four attempts in the Seine river basin brought new information and also showed several deficiencies.

At the scale of the whole basin, the first study focussed on the role of wetlands towards water resources, used available data on some hydrological characteristics of wetlands while geomorphic settings

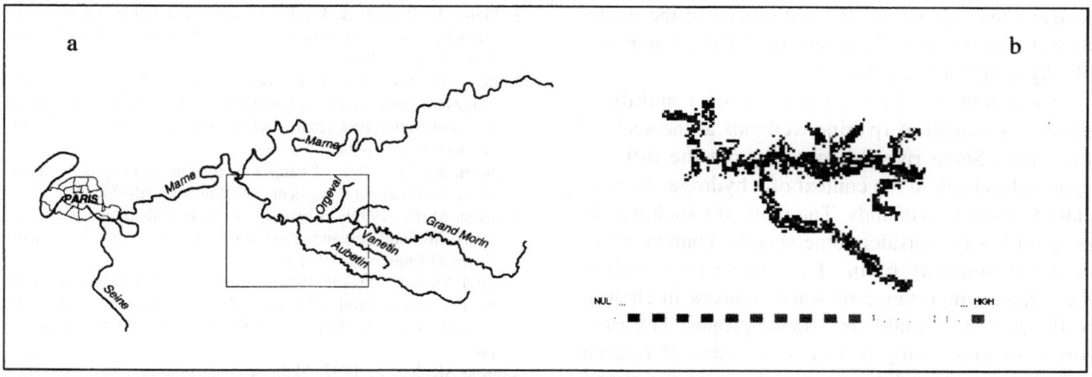

Figure 4. Map of the study site (a) and of denitrification potential of riparian zones (b) determined from transfer time of water in the Grand-Morin basin (France).

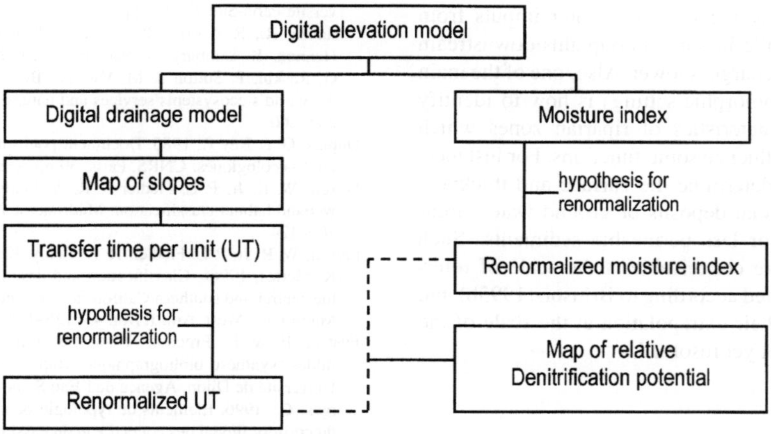

Figure 5. Method for the assessment of nitrate removal capacitiy of riparian wetlands applied in the Grand-Morin basin (France).

were not considered. Because of a lack of informations, some 'negative effects' were not taken into account (embankments, artificial drainage of floodplains, pumping of water, cultivation...), and some have to be qualified. For instance, sand-pit lakes strongly modify ground-water flows but act as sinks for nutrients, especially for nitrogen (Helmer & Labroue, 1993; Schanen, 1998). The second study was directed towards the importance of connected waterbodies and flooded meadows for fishes communities. It included more information on the geomorphic characteristics of floodplains and the method for a regionalization of the hydrosystem correctly underlines the importance of longitudinal gradients. However, the transverse dimension of floodplains is limited to rather narrow strips along the channels and maps used in this procedure (scale of 1/25 000) gave useful informations on riverine wetlands for high order streams (equal and greater than 4) but not for low order streams. Much emphasis is now attached to the role of riparian buffer zones associated with low order streams because of their total length (1–3 order streams represent 85% of the total length of the Seine hydrographic network), their place as interfaces directly connected to cultivated uplands and because they are more affected by human impacts than large floodplains (Brinson, 1993a).

The two procedures which aimed at the assessment of nitrate removal capacities of riparian wetlands used first geomorphic and hydrological characteristics of floodplains but the results need to be validated by *in situ* measurement of denitrification in more or less effective sites. Above all, the real impact of these sites on water quality must be determined in the framework of a hydrodynamic approach.

So, it appears that the challenge of a multifunctional assessment of riverine wetlands at the scale of the whole Seine river basin, must put the different approaches back in the context of a hydrogeomorphic classification of wetlands. The method which will be adopted has to consider some specific characteristics of the sedimentary basin of the Seine river such as the relative importance of water sources in riparian wetlands. For example, the scheme proposed by Brinson (1993a) showing a decreasing ratio of riparian transport to overbank transport with increasing stream order has to be adapted according to the importance of deep ground-water discharge. Streamsides along first order streams may receive great water imputs from deep aquifers while in some floodplains downstream ground-water discharge is lower. Also, one of the main questions for geomorphic settings is how to identify the vertical characteristics of riparian zones which could strongly influence some functions. For instance, it is difficult to determine the surface and thickness of clayey superficial deposits or ground-water circulations in more or less permeable sediments. Such determinations are conceivable at the scale of reference zones, selected according to Brinson (1993b), but the question of their extrapolation at the scale of the whole basin is not yet resolved.

References

Adamus, P. R. & L. T. Stockwell, 1983. A method for wetland functional assessment. Vol. 1: critical review and evaluation concepts. Report FHWA IP.82.24, U.S. Department of Transportation: 171 pp.

Adamus, P. R., 1988. The FHWA/Adamus (WET) Method for Wetland Functional Assessment. In Hook, D.D. (ed.), The Ecology and Management of Wetlands, vol. 2, Timber press, Portland: 128–136.

AREA, Commission de bassin Seine-Normandie, 1992. Cartographie de synthèse des Schémas départementaux de vocation piscicole et des ZNIEFF (Zones Naturelles d'Intérêt Ecologique Floristique et Faunistique) humides du bassin Seine-Normandie. Définition des secteurs d'intérêt écologique majeur préalablement à des interventions d'aménagement: 64 pp.

Bailey, R. G., 1984. Testing an Ecosystem Regionalization. Envir. Mgmt 19: 239–248.

Barbier, E. B., M. Acreman & D. Knowler, 1997. Economic Valuation of Wetlands: a guide for Policy Makers and Planners. In Pack, D. & V. Pearson Higgins (eds), UICN Public., Cambridge, UK: 127 pp.

Belliard, J., 1994. Le peuplement ichtyologique du bassin de la Seine: rôle et signification des échelles temporelles et spatiales. Thèse Doct. Université Paris VI: 97 pp.

Belliard, J., P. Boët & J. Allardi, 1995. Evolution à long terme du peuplement piscicole du bassin de la Seine. Bull. Fr. Pis. 337/338/339: 83–91.

Boët, P., B. Duvoux, J. Allardi & J. Belliard, 1994. Incidence des orages estivaux sur le peuplement piscicole de la Seine à l'aval de l'agglomération parisienne (bief Andrésy-Méricourt), La Houille Blanche 1/2: 141–147.

Brinson, M. M., 1993a. Changes in the functioning of wetlands along environmental gradients. Wetlands 13: 65–74.

Brinson, M. M., 1993b. A hydrogeomorphic classification for wetlands. Wetland research program technical report, U.S. Army Corps of Engineers: 79 pp.

BURGEAP & P. Bravard, 1995. Cartographie de l'intérêt fonctionnel des zones humides du bassin Seine-Normandie *vis-à-vis* des ressources en eau. Rapport AESN R.1677-A.4035/294218: 32 pp.

Conesa-Alcolea, I., 1997. Maîtrise de la pollution par les nitrates dans le bassin versant du Grand Morin. Proposition d'indicateurs de suivi et d'une méthodologie pour l'identification des zones naturelles d'épuration. Mém. DESS Poll.Chim.et Envir., Université Paris-Sud XI: 137 pp.

Costanza, R., R. d'Arge, R. de Groot, S. Farber, M. Grasso, B. Hannon, K. Limburg, S. Naeem, R.V. O'Neill, J. Paruelo, R. G. Raskin, P. Sutton & M. Van den Belt, 1997. The value of the world's ecosystems services and natural capital. Nature 387: 253–260.

Dupias, G. & Rey P., 1985. Document pour un zonage des régions phyto-écologiques. CNRS, Paris: 39 pp +carte.

Ferren, W. R. Jr, P. L. Fiedler & R. A. Leidy, 1996a. History of wetland habitat classification. Madrono: a West. Am. J. Bot. 43: 105–124.

Ferren, W. R. Jr, P. L. Fiedler, R. A. Leidy, K. D. Lafferty & L. A. K. Mertes, 1996b. Classification and description of wetlands of the central and southern California coast and coastal watersheds. Madrono: a West. Am. J. Bot. 43: 125–182.

Fustec, E. & B. Frochot, 1996. Les fonctions des zones humides, synthèse bibliographique. Rapport Université Paris VI, Université de Dijon, Agence de l'Eau Seine-Normandie: 144 pp.

Gorges, G., 1996. Eléments de typologie pour une régionalisation du corridor fluvial des cours d'eau du bassin de la Seine. Rapport MST-GSE, Université Paris XII: 34 pp.

Greiner, I., 1997. Analyse de la variabilité fonctionnelle de la plaine alluviale de la Seine et modélisation des transferts de nitrates. Thèse Doct. Université Paris VI: 222 pp.

Helmer, C. & L. Labroue, 1993. Denitrification in gravel-pit lakes. Hydrobiologia 252: 35–44.

Hook, D. D., 1993. Wetlands: history, current status and future. Envir. Toxicol. Chem. 12: 2157–2166.

Hugues, R. M. & D. P. Larsen, 1988. Ecoregions: an approach to surface water protection. J. Wat. Pollut. Cont. Fed. 60: 486–493.

Larson, J. S. & D. B. Mazzarese, 1994. Rapid assessment of wetlands: history and application to management. In Mitsch, W. J. (ed.), Global Wetlands: Old World and New. Elsevier Sc., Amsterdam, NL: 625–636.

Maltby, E., D. V. Hogan, C. P. Immirzi, J. H. Tellam & M. J. Van der Peijl, 1994. Building a new approach to the investigation and assessment of wetland ecosystem functioning. In Mitsch, W. J. (ed.), Global Wetlands: Old World and New Elsevier Sc., Amsterdam, NL: 637–658.

Maltby, E., D. V. Hogan & R. J. Mc Innes, 1996. Functional analysis of european wetland ecosystems. Phase 1 (FAEWE), the function of river marginal wetland ecosystems: improving the science base for the development of procedures of functional analysis. Final report EC DG XII CT90-0084: 408 pp.

Mérot, Ph. & P. Durand, 1997. Modelling interactions between buffer zones and the catchment. In Haycock, N., T. Burt, K. Goulding & G. Pinay (eds), Buffer Zones: Their Processes and Potential in Water Protection. Quest Envir., Harpenden, GB: 208–217.

Mitsch, W. J. & J. G. Gosselink, 1993 (2nd ed). Wetlands, Van Nostrand Reinhold, NY: 722 pp.

Novitzki, R. P., 1994. EMAP-Wetlands: a program for assessing wetland condition, functioning. In Mitsch, W.J. (ed.), Global Wetlands: Old World and New. Elsevier Sc., Amsterdam, NL: 691–709.

Novotny, V. & H. Olem, 1994. Water Quality: Prevention, Identification and management of Diffuse pollution. Van Nostrand Reinhold, NY: 387–438.

Omernick, J. M., 1987. Ecoregions of the conterminous United States. Annls Ass. Am. Geogr. 77: 118–125.

Penven, M. J., T. Muxart, F. Bartoli, P. Bonté, D. Brunstein, C. Cosandey, V. Gouy, S. Irace, T. Leviandier & S. Sogon, 1998. Petits bassins ruraux et pollutions diffuses. In Meybeck, M., G. de Marsily & E. Fustec (eds), La Seine en son Bassin: Fonctionnement Écologique d'un Système Fluvial Anthropisé. Elsevier, Paris, France: 159–210.

Reddy, K. R. & E. M. d'Angelo, 1994. Soil processes regulating water quality in wetlands. In Mitsch, W. J. (ed.), Global Wetlands: Old World and New. Elsevier Sc., Amsterdam, NL: 309–324.

Schanen, O., 1998. Analyse et modélisation de l'impact hydrodynamique et biogéochimique des lacs de gravières sur la nappe alluviale du Val de Seine. Thèse Doct. Université Paris VI: 272 pp.

Wasson, J. G., 1989. Elements pour une typologie fonctionnelle des eaux courantes: 1. Revue critique de quelques approches existantes. Bull. Ecol. 20: 109–127.

Wasson, J. G., 1992. La rivière et l'Homme: vers une gestion de bassin intégrant la dimension écologique. Rev. Géogr. Lyon 67: 333–343.

Hydrobiologia **410:** 223–240, 1999.
J. Garnier & J.-M. Mouchel (eds), Man and River Systems.

Linking land use, erosion and sediment yields in river basins

D. E. Walling
Department of Geography, University of Exeter, Exeter, EX4 4RJ, U.K.

Key words: erosion, sediment yield, sediment storage, sediment budget, floodplain sedimentation.

Abstract

Results obtained from erosion plots and catchment experiments provide clear evidence of the sensitivity of erosion rates to land use change and related human activity. Evidence for the impact of land use on the sediment yields of world rivers is less clear, although examples of rivers where sediment yields have both increased and decreased in recent decades can be identified. The apparent lack of sensitivity of river sediment loads to land use change reflects, at least in part, the buffering capacity associated with many river basins. This buffering capacity is closely related to the sediment delivery ratio of a river basin, in that basins with high sediment delivery ratios are likely to exhibit a reduced buffering capacity. Investigations of the impact of land use and related human activity on sediment yields should consider the overall sediment budget of a catchment rather than simply the sediment output. Information on the sediment budget of a drainage basin is difficult to assemble using traditional techniques, but recent developments in the application of fingerprinting techniques to establish sediment sources and in the use of environmental radionuclides, such as caesium-137 and lead-210, to document sediment storage offer considerable potential for providing such information. Sediment storage within a river basin can give rise to environmental problems where sediment-associated pollutants accumulate in sediment sinks. The accumulation of phosphorus on river floodplains as a result of overbank sedimentation can, for example, represent an important phosphorus sink.

The context

The results obtained from erosion plot experiments and experimental catchment studies in many different areas of the world provide clear evidence of the sensitivity of erosion rates to land use and related human activity. Table 1, for example, shows that the change from natural vegetation to cultivation can increase soil erosion rates by an order of magnitude or more. Increased rates of soil erosion can be expected to give rise to increased sediment transport by local streams and the literature contains results from many experimental catchment studies which show that vegetation clearance or land use change can again cause increases in sediment yield of an order of magnitude or more (cf. Table 2). Abernethy (1990) has similarly reported results from longer-term investigations of several small reservoir catchments in Southeast Asia which showed that during the present century annual sediment yields have progressively increased by between 2.5 and 6% per year as a result of land clearance and subsequent intensification of land use (Figure 1). Abernethy linked these increases to the rate of population growth in the areas concerned and found that the ratio of the increase in sediment yield to that for population was greater than unity and averaged ca. 1.6:1. Based on these findings, he suggested that annual suspended sediment yields from river basins in many developing countries could be expected to double in about 20 years. If the above generalisations are coupled with information on the global increase in population and the global extent of land clearance and intensification of land use, which indicates, for example, that the area of the earth's surface given over to crop production and livestock grazing has increased by more than five-fold over the past 200 years (cf. Buringh & Dudal, 1987), major increases in the sediment loads of the world's rivers could be inferred (cf. Douglas, 1967).

The available evidence regarding the impact of human activity and land use change on the sediment loads of the world's rivers, is however, less clearcut. Existing estimates of the global transfer of sediment from the land surface of the globe to the oceans sug-

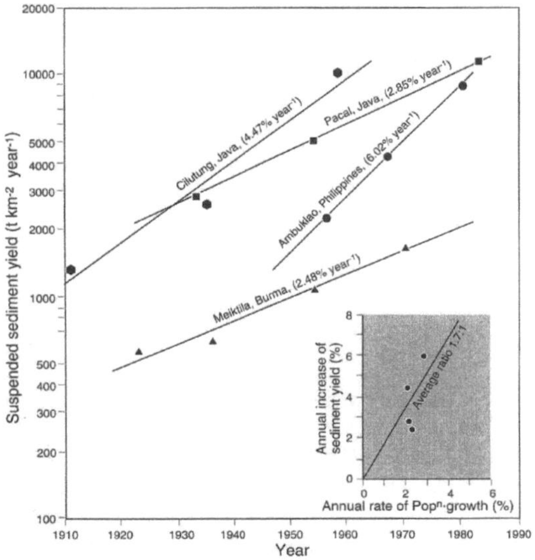

Figure 1. Trends of increasing sediment yield in several reservoir catchments in Southeast Asia (based on Abernethy, 1990).

gest that this flux is currently about 20×10^9 tonnes per year (cf. Milliman & Syvitski, 1992; Walling & Webb, 1996), but the extent to which this value has increased over and above the 'natural value' remains uncertain. In the absence of long-term records of sediment transport for rivers in most areas of the world, it is necessary to look for other sources of evidence of the magnitude of this increase. In their classic study *Erosion and Sediment Yield on the Earth*, which was based on sediment yield data assembled from more than 3600 measuring stations on the world's rivers, Dedkov & Mozzherin (1984) used space–time substitution to estimate the extent to which the sediment yields of the world's rivers had increased due to catchment disturbance by land use activity. They grouped the available data from both the mountainous and non-mountainous (plains) regions of the world into large (>5000 km^2) and small (<5000 km^2) basins and further classified these according to the degree of disturbance by agricultural activity. Category I basins were largely undisturbed and characterized by either a forest cover >70% or a cultivated area <30%. Category II basins exhibited an average degree of disturbance (i.e. area under forest or cultivation of between 30 and 70%), whereas category III basins were characterized by the greatest degree of disturbance, with an area under forest of <30% or a cultivated area >70%. By comparing the average sediment yields of category

Table 1. A comparison of soil erosion rates under natural undisturbed conditions and under cultivation in selected areas of the world

Country	Natural (kg m^{-2} year^{-1})	Cultivated (kg m^{-2} year^{-1})
China	<0.20	15.00–20.00
U.S.A.	0.003–0.30	0.50–17.00
Ivory Coast	0.003–0.02	0.01–9.00
Nigeria	0.05–0.10	0.01–3.50
India	0.05–0.10	0.03–2.00
Belgium	0.01–0.05	0.30–3.00
U.K.	0.01–0.05	0.01–0.30

Based on Morgan (1986).

I basins, with the equivalent values for category III basins, it was possible to derive an approximate measure of the magnitude of the increase in sediment yield associated with land disturbance by human activity (cf. Table 3). Although the results of this analysis are necessarily limited by the fact that they represent a simple comparison of mean values of sediment yield for drainage basins in different categories, and are therefore heavily dependent on the representativeness of the sample of river basins involved, they suggest that the sediment yields of many of the world's rivers may have increased ca. 5-fold as a result of disturbance by human activity.

Milliman et al. (1987) provide provide a similar, although somewhat greater, estimate for the magnitude of the change in the sediment load of the Yellow River in China since the expansion of agriculture into the loess plateau, which began about 200 BC. In this case, an analysis of sediment deposits in the Yellow Sea was used to estimate that the sediment load of the Yellow river during the early and middle Holocene was an order of magnitude less than that associated with the more recent post-agriculture period.

Other lines of evidence are, however, less conclusive in terms of confirming a substantial increase in the sediment loads of the world's rivers as a result of human activity and land use change. The work of Tardy et al. (1989), which has attempted to reconstruct global sediment yields over the geological past (cf. Walling, 1995), indicates that present-day sediment loads are ca. 1.6 times greater than those in the recent geological past (i.e. the Cenozoic, ca. 33 million years ago) (Figure 2A). However, although the impact of human activity is invoked as one explanation of this increase, these authors indicate that it may

Table 2. Some results from experimental basin studies of the impact of land use change on sediment yield

Region	Land use change	Increase in sediment yield	Reference
Westland, New Zealand	Clearfelling	×8	O'Loughlin et al. (1980)
Oregon, U.S.A.	Clearfelling	×39	Fredriksen (1970)
Northern England	Afforestation (ditching and ploughing)	×100	Painter et al. (1974)
Texas, U.S.A.	Forest clearance and cultivation	×310	Chang et al. (1982)
Maryland, U.S.A.	Building construction	×126–375	Wolman & Schick (1967)

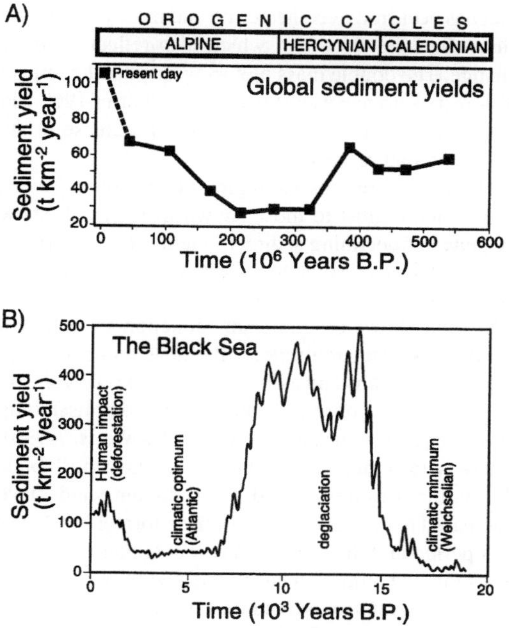

Figure 2. Variation of sediment yields over geological time. Based (A) on Tardy et al. (1989) and (B) on Degens et al. (1991).

also partly reflect the progressive increase in tectonic activity since Jurassic times. Studies of long-term sedimentation in water bodies receiving drainage from large areas, such as the Black Sea (cf. Degens et al., 1976, 1991), further suggest that recent increases in sediment input, although clearly evident, are of limited magnitude when compared with climatically induced changes occurring during the late Pleistocene and the Holocene (Figure 2B).

The apparent contradictions in the information presented above and the lack of clear evidence of substantial increases in sediment transport by the world's rivers as a result of human activity and land use change could reflect the averaging associated with larger river basins, whereby increased erosion rates may be limited to only parts of the basin. Equally, global variability in the magnitude of land use impacts will inevitably mean that significant increases are only found in some areas of the world. There is, however, also evidence to suggest that the marked increases in rates of soil loss cited above may not be reflected by increases of an equivalent magnitude in the sediment loads of world rivers. Alford (1992), for example, reports a study of the 14 028 km^2 Chao Phraya river basin draining the highlands of northern Thailand which showed no evidence of a significant increase in sediment yield during the period extending from the late 1950s to the mid 1980s, despite substantial deforestation and extensive swidden agriculture within the basin. This apparent discrepancy may reflect the importance of sediment delivery processes in attenuating upstream increases in sediment mobilisation (cf. Walling, 1983), such that much of the additional sediment mobilised within a river basin may be stored within the system, either on the lower slopes and in the small tributary streams, or on the floodplains bordering the main channel system, and will not reach the basin outlet. There are also situations where human activity has caused a reduction, rather than an increase, in sediment yield. More particularly, the construction of reservoirs will commonly trap the majority of the sediment load formerly transported by the river. The closure of the Aswan Dam on the River Nile has, for example, reduced the annual sediment yield of that river from ca. 100×10^6 t $year^{-1}$ to almost zero, and Meade & Parker (1985) cite the case of the Mississippi River, U.S.A., where the construction of five major dams on the Missouri River, one of its major tributaries, reduced the sediment yield at its mouth by more than 50% between the 1950s and the 1980s. In an attempt to scale up such findings to the global scale, Vorosmarty et al. (1997) estimated that more than 40% of the global river discharge is currently intercepted by large impoundments and that the global sediment retention by reservoirs may exceed

Table 3. Increases in the sediment yields of world rivers resulting from catchment disturbance by land use activities, (based on data presented by Dedkov & Mozzherin (1984))

Group	Small basins	Large basins	All basins
Lowland Rivers (n=1854)	×13.0	×8.1	×10.0
Mountain Rivers (n=1811)	×2.2	×3.8	×2.8

25% of the global flux from the land to the oceans. It therefore seems likely that in some areas of the globe increases in sediment loads caused by land clearance and land use change may be balanced by reductions associated with reservoir development.

Changing sediment yields

The problem of assessing the magnitude of the impact of land use change and related human activity on river sediment loads is exacerbated by the general lack of long-term records of sediment transport by the world's rivers. Only a few records extend back to the 1920s and 1930s, and most cover only a few decades. Furthermore, some sediment load records may be of limited reliability (cf. Walling & Webb, 1981). Analysis of the available records can, however, provide more specific evidence regarding the sensitivity of river sediment loads to land use change, although in many cases there will be a need to disentangle the effects of both land use change and climate change, which may be closely linked. Figure 3 presents information concerning long-term trends in the sediment load of two rivers in the former Soviet Union. The record for the 12 500 km^2 basin of the Dema River at Bochkarevo in Russia presented in Figure 3A shows a clear trend of increasing suspended sediment yield during the period of record which extends from 1949 to 1985. This upward trend of sediment yield is statistically significant (>95% level), but the minor increase in annual runoff over the period of record is not statistically significant. The double mass plot of sediment yield versus runoff suggests that the main increase in sediment yields dates from ca. 1963 and that sediment loads have increased by about 1.4 times since that time. Bobrovitskaya (pers. com.) has indicated that the increased sediment loads evidenced by this river reflect the expansion of cultivation within the drainage basin. The second example, which is presented in Figure 3B, relates to the Dnestr River at Sambur, which drains an 850-km^2 catchment in the Ukraine. Here the trend line fitted to the annual sediment yields is again highly significant statistically and suggests that these have increased by as much as 5-fold since the early 1950s. This increase undoubtedly reflects the influence of forest clearance within the headwaters of the basin (Bobrovitskaya, pers. com.), but it is also a response to the general increase in runoff amounts that has occurred over the period and more particularly since the late 1960s and which also reflects climate change. The double mass plot suggests that the impact of forest clearance was particularly felt after 1968 and that this itself accounts for a 1.8-fold increase in the sediment load of the river.

Figure 4 presents examples of two other rivers, which, in contrast to those shown in Figure 3, show evidence of declining sediment yields. The record for the River Isar at Munchen, Germany, which drains a catchment area of 2855 km^2, spans the period 1930–1990 and shows a significant reduction in sediment yield. This reduction reflects the development of hydropower stations and associated storage reservoirs on this river, and, more particularly, the commissioning of the Sylvenstein Dam in 1959 (cf. Weiss, 1996). There is no significant trend in the annual runoff totals during the period of record, but sediment yields have decreased to only about 20% of their former level over this period and the trend of the double mass plot suggests that the reduction has intensified in recent years. The other river represented on Figure 4 is the Yellow River at Longmen, China, which drains a catchment area of 497 561 km^2. This river is well known for its very high sediment load and for the very high erosion rates associated with the extensive area of thick loess deposits which occupies much of its middle reaches. The available record extends over 55 years and in this case shows evidence of a significant decline in both runoff and sediment yield during this period. Sediment yields can be seen to reduce by about 50% during the period of record and much of this reduction can be attributed to the extensive soil and water conservation works undertaken in the middle reaches of the Yellow River basin since the 1970s, which have reduced both runoff and sediment mobilisation. The reduction in runoff over this period also reflects a shift to a drier climate and reduced annual rainfall totals. Attempts by Chinese workers to assess the relative importance of soil and water conservation works and climate change

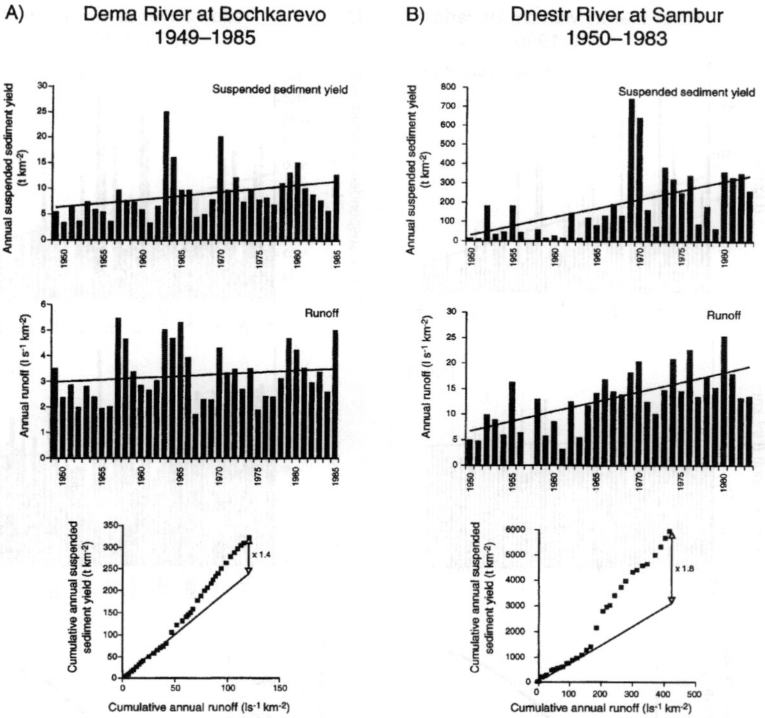

Figure 3. Trends in suspended sediment yield and runoff for the Dema River at Bochkarevo, Russia, 1949–1985 (A) and the Dnestr River at Sambur, Ukraine, 1950–1983 (B). (Based on data supplied by Professor Bobrovitskaya, State Hydrological Institute, St Petersburg, Russia.)

in accounting for the reduced runoff and sediment loads in the main Yellow River have suggested that the former accounts for ca. 60% of the reduction in runoff and ca. 50% of the reduction in sediment yield (cf. Gu, 1994; Mou, 1996). The double mass plot further emphasises that the reduced sediment load is not simply the result of reduced runoff, since, if the relationship between annual sediment load and annual runoff can be assumed to be linear, no change in the slope of the double mass plot would be expected if the reduction in sediment yield was solely a reflection of reduced runoff. In this case, the double mass plot suggests that a reduction in sediment yield of approximately 40% has occurred since the late 1970s, which cannot be directly linked to the reduction in runoff during this period. The potential significance of soil and water conservation measures in reducing the sediment loads of large river basins is further emphasized by the records for the tributaries of the Yellow River which drain solely areas within the loess region. In the case of the Wuding River (29 662 km^2), sediment yields have declined by almost 90% between the 1950s and the early 1990s (Figure 5), although part of this decline is again due to climate change and the associated reduction in precipitation. The double mass plot presented in Figure 5 suggests that a reduction in sediment yield of ca. 60% has occurred since 1971, which cannot be directly linked to the reduced runoff.

As a final example of the sensitivity of river sediment loads to land use change, Figure 6 presents data for two rivers which show little evidence of changing sediment loads over a periods of more than 40 years. The record of annual sediment loads for the River Lech above Fussen in Bavaria (1422 km^2) extends over more than 60 years and, since this river has been essentially unaffected by impoundment and regulation, it affords a valuable example of the response of an Alpine river to changing land use. The annual runoff record shows evidence of a small reduction over the period of record and this trend is statistically significant at the 95% level, but the sediment record shows little evidence of change, despite the likelihood of significant changes in both land use and land management practices within the catchment over the past

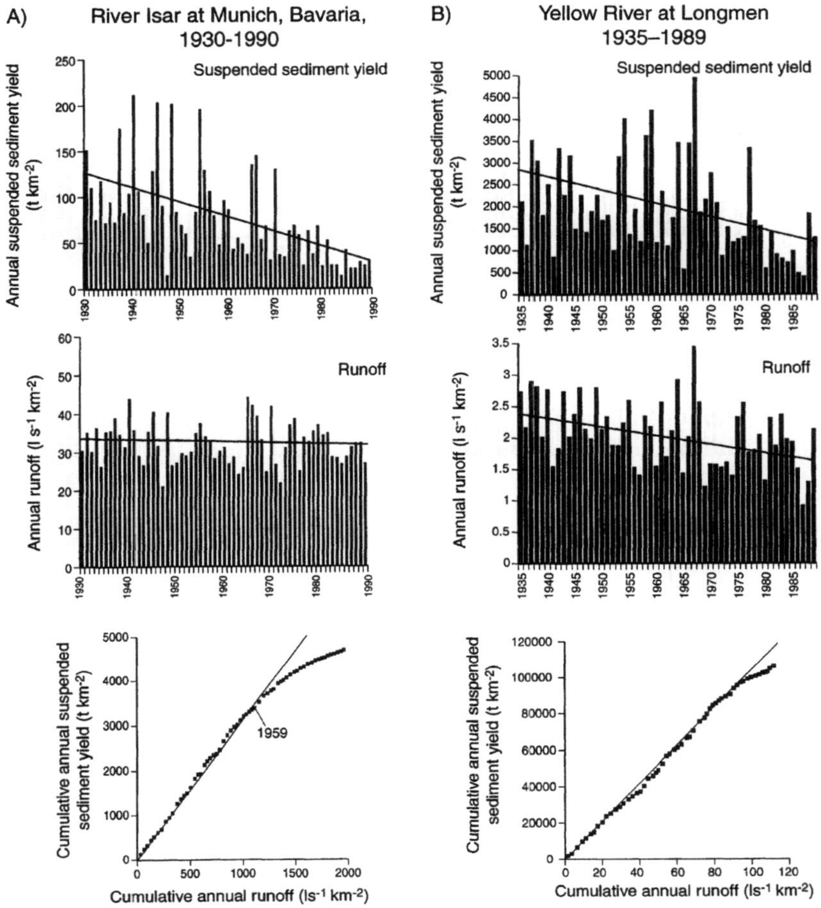

Figure 4. Trends in suspended sediment yield and runoff for the River Isar at Munich, Bavaria, 1930–1990 (A) and the Yellow River at Longmen, China, 1935–1989 (B). (Based on data supplied by Dr F.H. Weiss, Bayer Landesamt fur Wasserwirtschaft, Munich, and Dr Fang Xuemin, IWHR, Beijing.)

60 years (cf. Summer et al., 1996). A similar picture is provided by the annual runoff and sediment load data for the Upper Yangtze River at Yichang, China (Figure 5B). This river drains an area of 1 005 501 km^2, which supports a population of ca. 140 million. The population has increased rapidly from ca. 60 million in 1953 and there are many reports of increased soil loss (cf. Dai & Tan, 1996). Despite such evidence of increased population pressure and intensification of land use, the time series of annual sediment loads is essentially stationary and there is no evidence of decreasing runoff, which might offset the impact of increased erosion. Some evidence of increased sediment loads is provided by the shorter records for some of the tributary rivers within the Upper Yangtze basin (cf. Lu & Higgitt, 1998), but again this suggests that sediment loads have exhibited only limited change.

The results presented above afford coverage of only a small number of rivers and a limited range of physiographic conditions. Nevertheless, they provide evidence of the complexity of the causal link between land use change and associated increases in rates of erosion and river sediment loads. The records from the two rivers in the former Soviet Union depicted in Figure 3 show significant increases in sediment yields consequent upon land use change and land disturbance, and the records from the Yellow River presented in Figure 4B and Figure 5 similarly demonstrate the

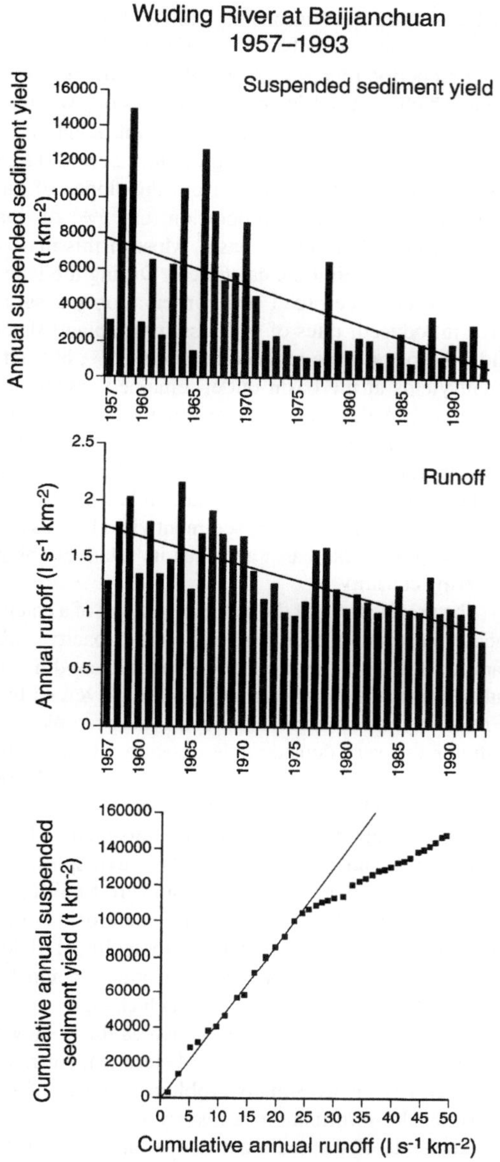

Figure 5. Trends in suspended sediment yield and runoff for the Wuding River, China, 1957–1993. (Based on data supplied by Mr Mou Jinze, YRCC, Zhengzhou.)

sensitivity of the sediment yields from this river basin to reduction by means of soil and water conservation measures. Conversely, however, the records from the River Lech and the Upper Yangtze presented in Figure 6 provide examples of situations where sediment loads appear to be essentially insensitive to changes occurring within the catchment. These apparently conflicting findings reflect, at least in part, variations in what might be termed the 'buffering capacity' of the catchments involved. The work of Meade & Trimble (1974) and Phillips (1992) in assessing the impact of post-colonial agricultural activity on the sediment loads of rivers draining to the coastal seaboard of the eastern United States has usefully demonstrated how much of the sediment eroded from their watersheds was deposited or stored within the river basin and failed to reach the basin outlet. Subsequent reductions in soil erosion associated with the decline of plantation agriculture and the introduction of soil conservation measures were also not reflected by downstream sediment loads, since reductions in sediment transport were balanced by remobilisation of stored sediment. Major changes in land use which would have been accompanied by changing erosion rates were therefore not reflected by major changes in sediment transport within the lower reaches of these rivers. The effectiveness of such buffering can be related to the classic sediment delivery ratio concept (cf. Walling, 1983). Rivers with a low sediment delivery ratio will be characterized by a high degree of buffering, since much of the sediment mobilized within the catchment will be deposited and these deposits can in turn be remobilized if sediment supply to the river system declines. Where, however, sediment delivery ratios are high, little sediment will be deposited and there will be only limited sediment available to be remobilized during times of reduced supply. Such river basins can therefore be expected to be poorly buffered and the sediment output will be sensitive to changing erosion rates associated with changing land use. Estimates of sediment delivery ratios available for the Middle Yellow River basin (cf. Gong & Xiong, 1980; Mou & Meng, 1980) indicate that much of this region is characterised by values close to 1.0 or 100%, due in part to the high drainage density and the frequent occurrence of hyperconcentrated flows, which reduce the potential for sediment deposition. These values are consistent with the low degree of buffering found in this river basin. Values of the sediment delivery ratio cited for the Upper Yangtze River basin by Dai & Tan (1996) and Liu & Zhang (1996) are much lower than those for the Middle Yellow River basin and averaged ca. 0.34 or 34%. Such values are likewise consistent with the much higher degree of buffering suggested by the record of sediment loads for the Upper Yangtze (cf. Figure 6B). No corresponding information on sedi-

Table 4. The suite of sediment properties used to fingerprint suspended sediment sources and the results of the Multivariate Discriminant Function analysis used to select this composite fingerprint

Tracer property	Cumulative % samples classified correctly
N	51.9
Total P	74.3
Sr	77.9
Ni	82.4
Zn	83.8
^{226}Ra	83.8
^{137}Cs	88.2
unsupported ^{210}Pb	88.2
Fe	92.7
Al	94.1

Based on Walling et al. (1999a).

ment delivery ratios are available for the other rivers documented in Figures 3–6, although existing evidence would suggest that SDR values for the River Lech would be relatively low and thus indicative of a well-buffered system, as suggested by the stability of the long-term record of sediment loads.

Sediment budgets

The above discussion emphasizes that any attempt to consider the impact of land use on erosion and sediment transfer through a river basin should direct attention not only to changes in sediment load at the catchment outlet, but also to the entire delivery or conveyance system and to the sinks or stores involved. This broader perspective will permit improved understanding of the link between land use, erosion and sediment yield as well as focusing attention on the fate of mobilised sediment. Although the detrimental effects of increased sediment loads in rivers are well documented (cf. Clark et al., 1985), accumulation of sediment in sinks and stores within a drainage basin is known to give rise to other environmental problems. The sediment budget concept (cf. Dietrich & Dunne, 1978; Walling, 1988) affords a convenient means of representing and interpreting this link and Figure 7, based on the classic work of Trimble (1981), provides a useful example of its application. In this case, data from a range of field and documentary sources were used to reconstruct the sediment budget of Coon Creek, a 360-km^2 basin in the Driftless Area of Wisconsin, U.S.A., for two periods in the past, namely 1853–1938 and 1938–1975. The first period represented one of poor land management which resulted in severe soil erosion, whereas the second was characterised by the introduction of conservation measures. During the first period, large volumes of soil were eroded from the slopes of the basin, but only a small proportion (ca. 5%) of this was transported out of the basin. Most of this material was stored within the catchment. During the latter period when widespread soil conservation measures were introduced, rates of soil loss from upland sheet and rill erosion were reduced by about 25%, but sediment yields at the basin outlet remained essentially the same, because sediment stored in the tributary valleys and upper main valley was remobilized. In this case, therefore, substantial changes in land use and land management within the upstream basin were not reflected in the downstream sediment yield because of the low SDR of the basin and thus its relatively high buffering capacity.

The precise form of the sediment budget of a catchment, and thus its SDR and buffering capacity, will vary according to the local physiographic conditions, but considerable variation may also occur even within a relatively small area. Figure 8, for example, illustrates the very considerable diversity of sediment budget structure reported by Golosov et al. (1992) for small- and medium-sized drainage basins on the Russian Plain, a region which is heavily impacted by land use activities and soil erosion. In these examples, the sediment delivery ratios ranged from zero to 89%. In the former case, changes in land use and erosion rates would not be reflected in changes in sediment yield, whereas in the latter case sediment yields could be expected to be highly sensitive to such changes.

Scaling up sediment budgets such as those shown in Figures 7 and 8 to much larger heterogeneous river basins clearly involves many problems in terms of assembling the information necessary to establish the key elements of the sediment budget and such studies have hitherto been constrained by the limitations of traditional monitoring techniques for quantifying sediment budgets. As a result much of our existing understanding of the sediment budgets of larger river basins is based on theory and inference rather than direct empirical evidence. However, recent work has demonstrated the potential of newly developed techniques for establishing the relative importance of different sediment sources within a drainage basin and for assessing the importance of overbank flood-

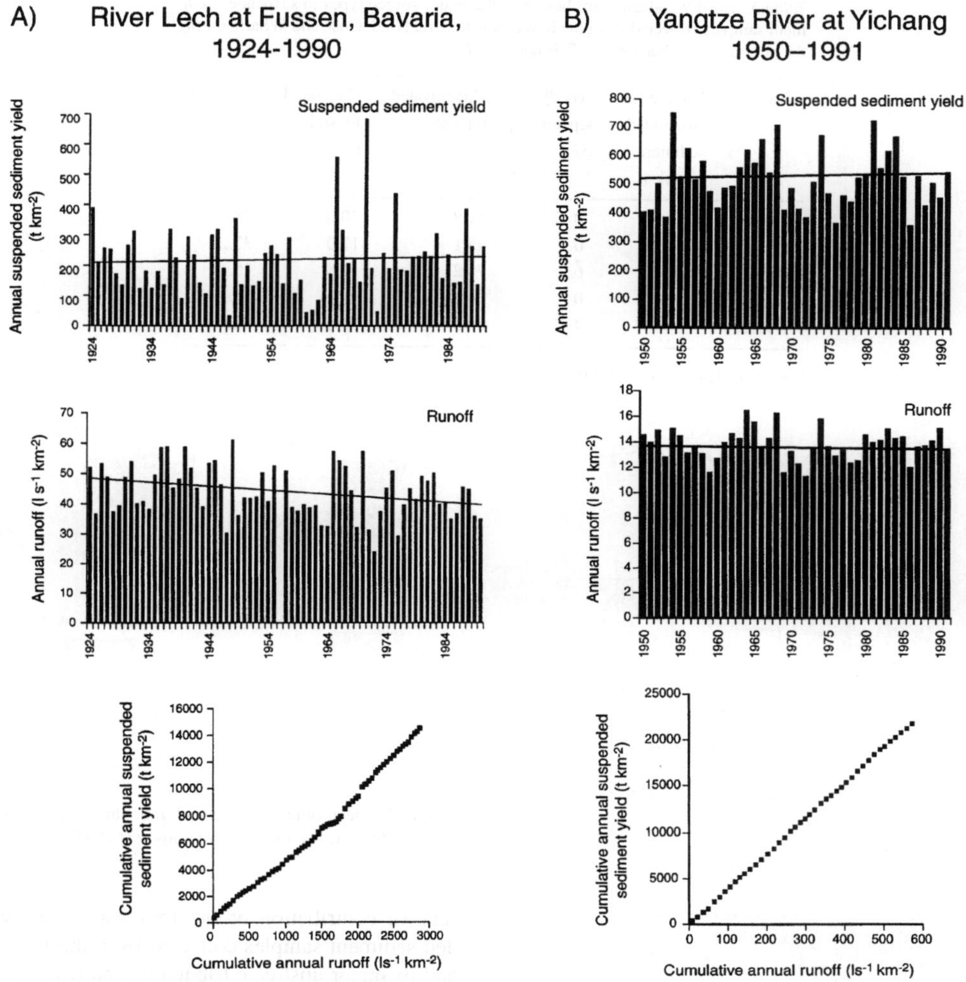

Figure 6. Trends in suspended sediment yield and ruonff for the River Lech at Fussen, Bavaria, 1924–1990 (A) and the Yangtze River at Yichang, China, 1950–1991 (B). (Based on data supplied by Dr F. H. Weiss, Bayer Landesamt fur Wasserwirtschaft, Munich and Dr Fang Xuemin, IWHR, Beijing.)

plain deposition and associated floodplain storage as a component of the sediment budget. This potential can usefully be demonstrated by referring to the results of some recent work undertaken by the author and his co-workers in the catchment of the River Ouse in Yorkshire within the framework of the U.K. Land–Ocean Interaction Study (LOIS). This work was aimed at elucidating the key features of the sediment budget of this 4000-km^2 catchment and involved establishing the relative importance of the primary sediment sources and quantifying the role of floodplain sedimentation as a sediment sink (cf. Walling et al., 1998, 1999a,b).

The fingerprinting approach, involving multicomponent fingerprints (cf. Walling et al., 1993; Walling & Woodward, 1995), was used in the Ouse catchment to determine the relative importance of four potential sediment sources within the catchment, namely, the surface of cultivated areas, areas of permanent pasture and upland moorland, and areas of woodland, and channel banks and other sources of subsurface material. In essence, the fingerprinting approach in-

Table 5. Load-weighted contribution of the major source types to suspended sediment samples collected from the Rivers Swale, Ure, Nidd, Ouse and Wharfe during the period November 1994 to February 1997

River	Number of suspended sediment samples	Woodland topsoil (%)	Uncultivated topsoil (%)	Cultivated topsoil (%)	Channel bank material (%)
Swale	19	0	41.8	30.0	28.2
Ure	14	0.7	45.1	17.0	37.2
Nidd	14	6.9	75.2	2.8	15.1
Ouse	30	0	24.6	38.1	37.3
Wharfe	7	4.4	69.5	3.6	22.5

Based on Walling et al. (1999a).

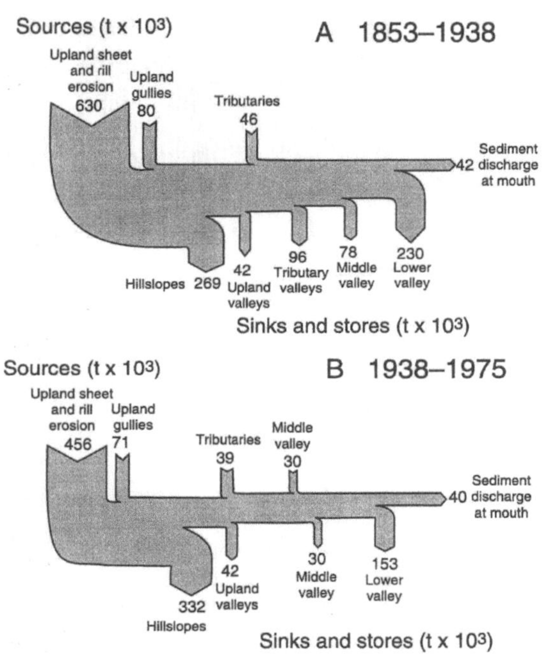

Figure 7. The sediment budgets of Coon Creek, Wisconsin, U.S.A., reconstructed by Trimble (1983) for the periods 1853–1938 and 1938–1975.

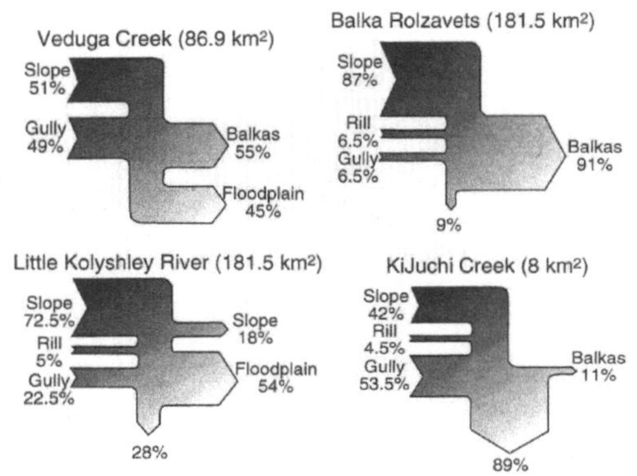

Figure 8. The sediment budgets of four drainage basins on the Russian Plain, as documented by Golosov et al. (1992).

volves comparison of the geochemical properties of suspended sediment transported by the river with those of potential sources, in order to establish their relative importance. Use of statistically verified composite fingerprinting signatures involving several different sediment properties (cf. Table 4), enabled the four potential sources to be clearly discriminated and a multivariate mixing model was used to determine the relative contribution of the four sources to suspended sediment samples collected from the River Ouse and its major upstream tributaries and from the River Wharfe, which joins the Ouse below its tidal limit (cf. Figure 9). The load-weighted estimate values for the relative contributions from the four different sediment sources are listed in Table 5. These results confirm that a major proportion of the suspended sediment load of the River Ouse and its tributaries is derived from surface sources and is thus likely to be directly influenced by land use activities, including cultivation and grazing pressure.

In upscaling from smaller to larger catchments, it is important to recognize that the river floodplains which border the main channel systems in larger river basins can represent an increasingly important sediment sink, which will attenuate changes in upstream sediment inputs. Walling & Quine (1993), for example, estimated that floodplain storage accounted for ca. 23% of

Table 6. A comparison of the estimates of total storage of sediment on the floodplains bordering the main channel system of the River Ouse and its primary tributaries, with the estimated suspended sediment loads for the study rivers

River	Floodplain storage (t year^{-1})	Mean annual suspended sediment load (t year^{-1})	Total sediment derived to channel (t year^{-1})	Floodplain storage as % of sediment input
Swale	19214	42352	61566	31.2
Nidd	7573	7719	15292	49.5
Ure[a]	15125	28887	44012	34.3
Ouse[a]	18733			
Total to Ouse gauging station	49041	75111	124152	39.5
Total to tidal limit	60645			
Wharfe	10325	10816	21141	48.8

[a] Ure is the River Ure to its confluence with the River Swale, while Ouse refers to River Ure/Ouse from below this point to the tidal limit. Based on Walling et al. (1998).

the sediment delivered to the main channel system of the 6850-km^2 catchment of the River Severn in the U.K. Attempts to document rates of overbank sedimentation on river floodplains, in order to establish the importance of such conveyance losses, face many practical problems. However, recent developments in the application of the environmental radionuclide Cs-137 to estimate rates of overbank sedimentation offer considerable potential in this context (cf. Walling & He, 1997a,b). This approach was used by Walling et al. (1998) to establish the significance of floodplain storage in the suspended sediment budget of the main channel systems of the Rivers Ouse (3315 km^2) and Wharfe (818 km^2). In this case, more than 250 sediment cores were collected from 26 representative transects located along the main channel systems of the Yorkshire Ouse and its major tributaries and the River Wharfe (cf. Figure 9) for Cs-137 analysis. The estimates of mean sedimentation rates for the individual transects obtained from these cores, which averaged ca. 0.2 g cm^{-2} year^{-1}, were extrapolated to the individual reaches between adjacent transects and the mean annual conveyance loss associated with overbank sedimentation on the floodplain bordering the main channel system was calculated. By relating these losses to the mean annual sediment loads of the rivers (Table 6), it was possible to establish the relative importance of floodplain storage in the sediment budget of the main channel system (Table 6, Figure 10). In the case of the main River Ouse system, floodplain deposition accounted for ca. 40% of the total amount of suspended sediment delivered to the main channel system over the past 40 years and for the River Wharfe the equivalent value was ca. 49%. These conveyance loss values of 40% or more, associated with the floodplains bordering the main channel system of a 4000-km^2 river basin, serve to emphasize further the potential importance of storage in attenuating the link between upstream erosion and downstream sediment loads and thus the impact of land use change on sediment yield.

The application of environmental radionuclides in quantifying the role of river floodplains as sediment sinks also offers the possibility of investigating changes in sedimentation rates through time. In this way, studies of changing sediment output from a drainage basin could be paralleled by studies of changing sediment storage. By measuring both the Cs-137 and unsupported Pb-210, content of floodplain cores, it is possible to derive estimates of sedimentation rates over the past 35–45 years (Cs-137) and 100–120 years (unsupported Pb-210) which can be compared (cf. Walling & He, 1994, 1999). The basis of these estimates precludes separation of the record into consecutive periods, but, by comparing the longer-term with the shorter-term estimate, it is possible to establish whether sedimentation rates have increased or decreased in recent years. Table 7, which is based on the work of Walling & He (1999), identifies the trends evidenced by single cores collected from the floodplains of a representative selection of 21 rivers

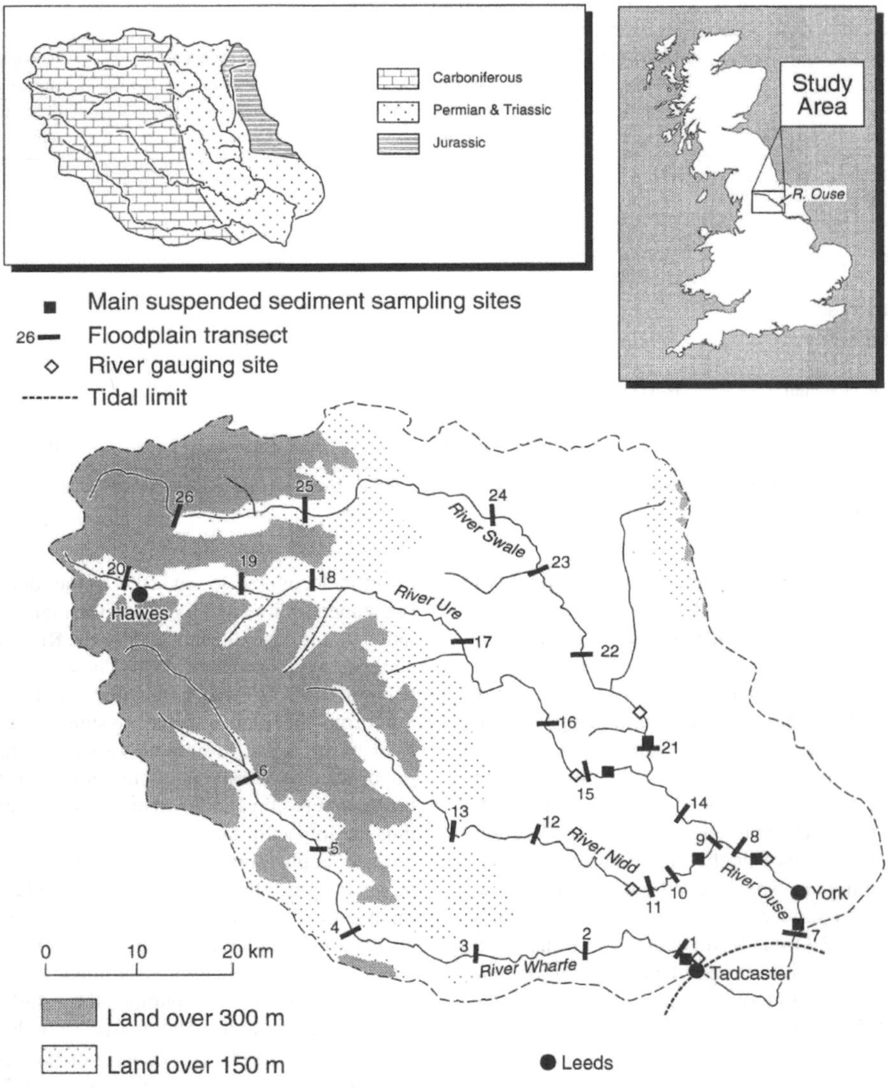

Figure 9. The catchments of the River Ouse and Wharfe, Yorkshire, U.K.

in the U.K. The results show no evidence of major changes in sedimentation rates over the past 100 years, but the individual rivers provide examples of situations where sedimentation rates have both increased and decreased and remained effectively stable. Increases and decreases have been defined as instances where the sedimentation rate for the past 33 years has increased or decreased by more than 10% relative to the sedimentation rate for the past 100 years. This period was marked by significant changes in land use and land management practices, but it appears that such changes had relatively little impact on rates of floodplain sedimentation and storage. In these circumstances, it is likely that sediment outputs from the river basins represented also changed relatively little. Thus, although floodplain storage could represent an important buffer for changing sediment fluxes, such effects would appear to have been of limited import-

Table 7. A comparison of mean annual sedimentation rates for the past 33 and 100 years estimated for a selection of sites on the floodplains of British rivers

River/Location	Sedimentation rate (g cm^{-2} $year^{-1}$)		Trend[a]
	past 33 years	past 100 years	
1. River Ouse near York	0.95	1.04	Stable
2. River Vyrnwy near Llanymynech	0.21	0.46	Decrease
3. River Severn near Atcham	1.22	1.42	Decrease
4. River Wye near Preston on Wye	0.15	0.28	Decrease
5. River Severn near Tewkesbury	0.86	0.95	Stable
6. Warwickshire Avon near Pershore	0.46	0.66	Decrease
7. River Usk near Usk	0.88	1.01	Decrease
8. Bristol Avon near Langley Burrell	0.39	0.33	Increase
9. River Thames near Dorchester	0.51	0.64	Decrease
10. River Torridge near Great Torrington	0.70	0.93	Decrease
11. River Taw near Barnstaple	0.60	0.65	Stable
12. River Tone near Bradford on Tone	0.56	0.43	Increase
13. River Exe near Stoke Canon	0.45	0.42	Stable
14. River Culm near Silverton	0.35	0.32	Stable
15. River Axe near Colyton	0.51	0.40	Increase
16. Dorset Stour near Spetisbury	0.04	0.04	Stable
17. River Rother near Fittleworth	0.11	0.14	Decrease
18. River Arun near Billingshurst	0.39	0.48	Decrease
19. River Adur near Partridge Green	0.51	0.71	Decrease
20. River Medway near Penshurst	0.15	0.23	Decrease
21. River Start near Slapton	0.51	0.45	Increase

[a]The trend indicates the change in sedimentation rate when comparing the past 33 years with the past 100 years. A change of $>\pm 10\%$ is taken to represent a significant increase or decrease.
Based on Walling & He (1999).

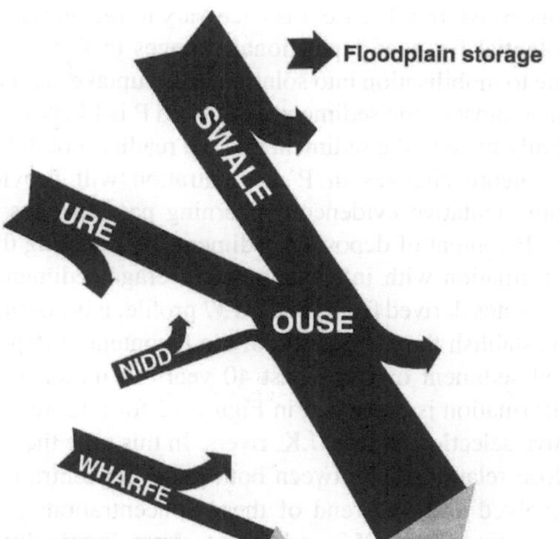

Figure 10. The role of overbank floodplain sedimentation in the sediment budget of the main channel systems of the River Ouse and its major tributaries and the River Wharfe, Yorkshire, U.K.

ance in these river basins. It is, nevertheless, possible that significant changes in erosion rates were buffered by loose coupling of the slope and channel systems and sediment storage in headwater areas, since such sinks are likely to be of similar importance to floodplain storage within the lower reaches of the main river system. Further work is clearly required to elucidate the behaviour of this other potentially important component of the catchment sediment budget.

Sediment storage

Recognition of the potential importance of sediment storage within a drainage basin and its role in attenuating the impact of land use change on downstream sediment loads also directs attention to the wider environmental significance of such storage. Deposition of sediment and, more importantly, sediment-associated nutrients and contaminants in sediment sinks could constitute a significant environmental problem in terms of both their accumulation and storage and the potential for subsequent remobilisation, such that the storage response of a river basin to land use change could be of equal importance to changes in sediment output. Marron (1987), for example, cites the case of the Belle Fourche River in South Dakota, U.S.A., where gold mining within the upstream catchment during the period extending from the 1880s to 1977 caused the accumulation of large quantities of arsenic-contaminated sediment within the river floodplain and it was estimated that, even with no further inputs from mining activity, arsenic contamination would remain a problem within this river system for many centuries to come due to remobilisation of contaminated sediment.

Recent concern for phosphorus cycling within terrestrial and freshwater ecosystems has also directed attention to the potential role of river floodplains in storing sediment associated-P mobilized from the upstream catchment. Since particulate-P can account for a major proportion of the total-P load of a river, overbank sedimentation on floodplains can result in substantial conveyance losses of P and appreciable quantities of P can accumulate in such floodplain sinks. Figure 11 provides information on the P content of a sediment core collected from the floodplain of the River Culm in Devon U.K. In this case, total-P concentrations in the upper horizons of the floodplain sediment, which average ca. 2600 mg kg^{-1} are more than twice those associated with soils in adjacent areas above the level of the floodplain.. These higher concentrations reflect the deposition of sediment-associated P mobilised from the upstream watershed P in association with fine sediment. Selective erosion, involving preferential mobilisation of fines and organic matter, causes enrichment of the P content of suspended sediment relative to the bulk soil, and this enrichment is in turn reflected in the high concentrations found in the floodplain sediments. In addition to being characterized by higher total-P concentrations, the floodplain sediments also possess a much greater P stock and this stock will increase through time as deposition continues. Figure 11 also presents information on the downcore variation of Cs-137 concentrations in the floodplain core and this information can be used to estimate the rate of sediment deposition, because the level associated with the peak Cs-137 concentration can be used to establish the floodplain surface at the time of peak fallout rates which occurred in 1963 (cf. He & Walling, 1996). In this case the deposition rate is estimated to be 0.31 g cm^{-2} $year^{-1}$ and, if this value is combined with the average total P content of sediment in the upper part of the core (i.e. post 1963), an annual input of sediment-associated P to the floodplain surface of 8.2 g m^{-2} $year^{-1}$ can be estimated. Equivalent values for the total P concentration of post 1963 floodplain sediment deposits and the P accumulation rate have been obtained for sediment cores collected from 19 further sites on the floodplains of U.K. rivers and these values are presented in Table 8. The range of P concentrations associated with floodplain sediments listed in Table 8 closely reflects the intensity of agricultural activity in the upstream catchment, although point-source inputs of P to the river system from sewage works will clearly also exert a significant influence. The values of P accumulation presented in Table 8 are of a similar order of magnitude to that of 9 g m^{-2} $year^{-1}$ reported by Fustec et al. (1995) for a floodplain site on the River Seine at Maizières, France, and further underscore the potential importance of river floodplains as a phosphorus sink. Remobilisation of these sediment stores could release large amounts of P back into the rivers, although rates of bank erosion associated with U.K. rivers are such that this sediment is likely to have a residence time of 10^2 to 10^4 years and can be viewed as being in long-term storage. Major changes in channel mobility and floodplain erosion, such as might accompany climate change, must, however, be seen as a potential cause of more rapid remobilisation.

Consideration of downcore changes in the P content of floodplain sediment also affords a means of deriving some information on past changes in the P content of suspended sediment transported by the river concerned. In this case it is necessary to recognize the potential for post-depositional changes in P content, due to mobilisation into solution, plant uptake etc. but, since most of the sediment-associated P is likely to be firmly fixed to the sediment and not readily mobilised, downcore changes in P concentration will provide some tentative evidence concerning past changes in the P content of deposited sediment. By coupling this information with information on average sedimentation rates derived from the Cs-137 profile, it is possible to establish the likely trend of the P content of deposited sediment over the past 40 years or more. Such information is presented in Figure 12 for a representative selection of five U.K. rivers. In this case there is close relationship between both the P concentrations involved and the trend of these concentrations over the period since 1950, and the intensity of agricultural activity in the upstream catchment. The catchments of both the Usk and the Upper Severn are characterized by low intensity agriculture with substantial areas of upland sheep pasture and the sediment collected from

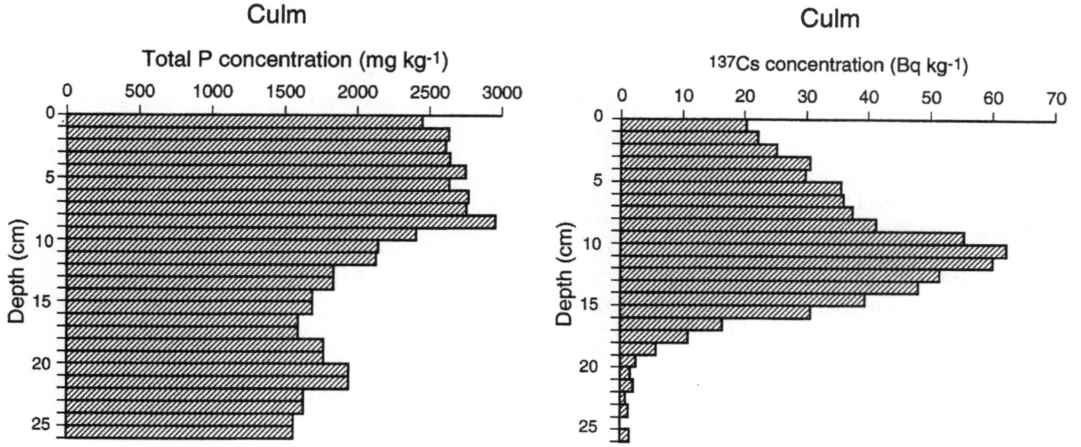

Figure 11. Down-core variations in total phosphorus (A) and caesium-137 (B) for a sediment core collected from the floodplain of the River Culm near Columbjohn, Devon, U.K.

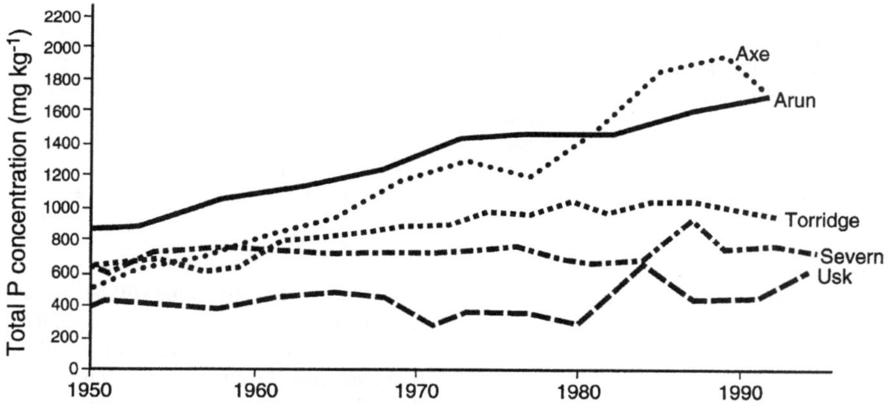

Figure 12. Variations in the total phosphorus content of sediment deposited on the floodplains of several British rivers during the period 1950–1990.

the floodplains of these two rivers exhibits relatively low P concentrations with only limited evidence of an upward trend. This is consistent with the relatively low levels of fertiliser application in these two catchments. In contrast, the data obtained for the Rivers Axe, Arun and Torridge show both higher P concentrations and a more well-defined upward trend, which are consistent with the more intensive agriculture in their catchments. The catchment of the River Torridge is occupied primarily by permanent pasture, whereas the catchments of the Axe and Arun are characterized by areas of both both pasture and arable cultivation. In the case of the River Torridge, the P content of sediment deposited on the floodplain has increased by nearly 50% over the past 40 years, whereas the increases for the Arun and Axe are ca. 94 and 170%, respectively. Equivalent values for 15 other sites, for which information on the P content of floodplain sediment and P accumulation rates were provided in Table 8, are also are presented in Table 8. This larger data set shows further evidence of substantial increases in the P content of sediment deposited in the floodplain store of British rivers over the past 40 years, which are likely to reflect land use change and increased fertilizer application within the upstream catchments. Recognition of the significance of sediment as a carrier for nutrients and contaminants clearly introduces a need to take account of both the amount and the properties of sediment

Table 8. Phosphorus storage on the floodplains of British Rivers

River/Location	Mean Total P concentration of post 1963 sediment (mg kg^{-1})	Total P storage since 1963 (g m^{-2} year^{-1})	Increase in Total P content of deposited sediment 1950–1992 (%)
1. River Vyrnwy near Llanymynech	760	1.79	
2. River Severn near Atcham	717	9.0	23
3. River Wye near Preston on Wye	1031	1.6	17
4. River Usk near Usk	417	3.8	25
5. River Teme near Broadwas	507	2.0	45
6. River Torridge near Great Torrington	955	6.5	45
7. River Taw near Barnstaple	830	4.8	75
8. River Exe near Stoke Canon	1090	4.8	65
9. River Culm near Silverton	2660	8.2	33
10. River Start near Slapton	1931	8.4	40
11. River Tone near Bradford on Tone	1047	5.8	55
12. Bristol Avon near Langley Burrell	1266	5.0	28
13. River Thames near Dorchester	1532	7.8	30
14. Warwickshire Avon near Pershore	2374	10.6	23
15. River Axe near Colyton	1466	7.6	170
16. Dorset Stour near Shillingstone	1321	11.6	53
17. River Rother near Fittleworth	1895	2.0	33
18. River Arun near Billingshurst	1436	5.7	94
19. River Adur near Partridge Green	795	4.0	147
20. River Medway near Penshurst	913	1.3	10

stored within the delivery system and the influence of land use on both aspects.

Perspective

This contribution has endeavoured to review existing understanding of the relationship between land use change, erosion and sediment yield. Although the impact of land use change on rates of soil loss, and more particularly the impact of land clearance and cultivation in increasing erosion rates, has been extensively documented, the evidence for major changes in the sediment loads of larger rivers is less clear. Existing long-term records provide evidence of cases where sediment loads have both increased and decreased as a result of land use change and other human activity, but it is clear that river basins and more particularly river systems possess considerable capacity to buffer such changes. The extent of this buffering capacity can be expected to be closely related to the sediment delivery ratio of the basin, such that river basins with low sediment delivery ratios will be characterized by a high buffering capacity and vice versa. Recognition of the importance of sediment storage within a drainage basin necessarily means that attempts to understand the linkages between land use, erosion and sediment yield should consider the overall sediment budget and the associated sources and sinks, rather than only the sediment outputs. Changes in the functioning of these stores could be of greater significance than changes in sediment output from a drainage basin. Recent work has demonstrated an improved capacity for assembling the information needed to construct a sediment budget and investigations of the role of floodplain sedimentation as a sediment sink have increasingly emphasised the potential importance of this store. By directing attention to the overall sediment budget of a river basin and to the associated sediment stores and sinks, as well as the sediment output, the wider environmental significance of these components of the sediment budget becomes increasingly apparent. They can represent important sinks for sediment-associated contaminants and any attempt to fully understand the cycling and transfer of phosphorus within a river basin must, for example, take account of the potential importance of river floodplains as P sinks.

Acknowledgements

The work reported in this paper draws both on data generously provided by others and on the results of several investigations undertaken by the author and his co-workers in recent years. These inputs are gratefully acknowledged. Particular thanks are extended to Professor N. Bobrovitskaya, Dr F.H. Weiss, Mr Mou Jinze and Dr Fang Xuemin for their help in providing the long-term sediment yield data presented in Figures 3–6, to Dr Qingping He and Dr Phil Owens for their valuable collaboration over the past 5 years, to Art Ames and Will Blake for their help with P analysis and to Mr Terry Bacon for producing the diagrams.

References

Abernethy, C., 1990. The use of river and reservoir sediment data for the study of regional soil erosion rates and trends. Paper presented at the International Symposium on Water Erosion, Sedimentation and Resource Conservation (Dehradun, India, October, 1990).

Alford, D., 1992. Streamflow and sediment transport from mountain watersheds of the Chao Phraya basin, southern Thailand: a reconnaissance study. Mountain Res. Devel. 12: 257–268.

Buringh, P. & R. Dudal, 1987. Agricultural land use in space and time. In Wolman, M. G. & F.G.A. Fournier (eds), Land transformation in agriculture, SCOPE Report no, 32, Wiley, Chichester: 9–43.

Chang, M., F.A. Roth & E.V. Hunt, 1982. Sediment production under various forest site conditions. In Recent developments in the explanation and prediction of erosion and sediment yield, IAHS Publication no. 137: 13–22.

Clark, E.H., J.A. Haverkamp & W. Chapman, 1985. Eroding soils: the off-farm impacts. The Conservation Foundation, New York.

Dai, D. & Y. Tan, 1996. Soil erosion and sediment yield in the Upper Yangtze basin. In Walling, D. E. & B. W. Webb (eds) Erosion and sediment yield: global and regional perspectives, IAHS Publication no. 236: 191–203.

Dedkov, A. P. & V. T. Mozzherin, 1984. Eroziya i stock nanosov na zemle, Izdatelstvo Kazanskogo Universiteta.

Degens, E. T., A. Paluska & E. Eriksson, 1976. Rates of soil erosion In Svensson, B. H. & R. Soderlund (eds), Nitrogen phosphorus and sulphur: global cycles, SCOPE Report no. 7, Stockholm: 185–191.

Degens, E. T., S. Kempe & J. E. Richey, 1991. Summary: Biogeochemistry of major world rivers. In Degens, E. T., S. Kempe & J. E. Richey (eds), Biogeochemistry of Major World Rivers. Wiley, Chichester: 323–347.

Dietrich, W. B. & T. Dunne, 1978. Sediment budget for a small catchment in mountainous terrain. Z. Geomorph. Supplbd. 29: 191–206.

Douglas, I., 1967. Man, vegetation and the sediment yields of rivers. Nature 215: 925–928.

Fredriksen, R. L., 1970. Erosion and sedimentation following road construction and timber harvest on unstable soils in three small western Oregon watersheds. U.S. Forest Service Research Paper, PNW 104.

Fustec, E., P. Bonte, J. C. Fardeau, L. Khebibeche, A. Chesterifoff & A. M. Carru, 1996. La rétention des MES et des polluants associés dans les zones inondables. In Piren-Seine Rapport 1996/II, Thème 'Corridor Fluvial', Laboratoire de Géologie Appliquée, Université Pierre et Marie Curie, Paris: 3-3–3-16.

Golosov, V. N., N. N. Ivanova, L. F. Litvin & A. Yu. Sidorchuk, 1992. Sediment budgets of river catchments and river channel aggradation on the Russian plain. Geomorphology (Moscow) 4: 62–71 [In Russian].

Gong, S. & G. Xiong, 1980. The origin and transport of sediment of the Yellow River. In Proceedings of the first international symposium on river sedimentation, Beijing, China: 43–52.

Gu, W., 1994. On the reduction of water and sediment yield of the Yellow River in later years. Int. J. Sediment. Res. 9: 1–12.

He, Q. & D. E. Walling, 1996. Rates of overbank sedimentation on the floodplains of British lowland rivers documented using fallout caesium-137. Geogr. Ann. 78A: 223–234.

Liu, Y. & P. Zhang, 1996. Erosion of material in ground surface and sediment yield in the Upper Yangtze River. Int. J. Sediment. Res. 11: 43–53.

Lu, X & D. L. Higgitt, 1998. Recent changes of sediment yield in the Upper Yangtze, China. Env. Mmt. 22: 697–709.

Marron, D. C., 1987. Floodplain storage of metal contaminated sediments downstream of a goldmine at Lead, South Dakota. In Averett, R. C. & D. M. McKnight (eds), Chemical Quality of Water and the Hydrologic Cycle. Lewis, Chelsea, MI: 193–209.

Meade, R. H. & R. S. Parker, 1985. Sediment in rivers of the United States. In National waters Summary 1984, U.S. Geological Survey Water Supply Paper no. 2275: 49–60.

Meade, R. H. & S. Trimble, 1974. Changes in sediment loads in rivers of the Atlantic Drainage of the United States since 1900. In Effects of man on the interface of the hydrological cycle with the physical environment, IAHS Publication no. 113: 99–104.

Milliman, J. D. & J. P. M. Syvitski, 1992. Geomorphic/tectonic control of sediment discharge to the ocean: the importance of small mountainous rivers. J. Geol. 100: 325–344.

Milliman, J. D., Y.-S. Qin, M-E. Ren & Y. Saito, 1987. Man's influence on the erosion and transport of sediment by Asian rivers: The Yellow River (Huanghe) example. J. Geol. 95: 751–762.

Morgan, R.P.C., 1986. Soil erosion and Conservation. Longman, Harlow.

Mou, J., 1996. Recent studies of the role of soil conservation in reducing erosion and sediment yield in the loess plateau area of the Yellow River basin. In Walling, D. E. & B.W. Webb (eds), Erosion and Sediment Yield: Global and Regional Perspectives, IAHS Publication no. 236: 541–548.

Mou, J. & Q. Meng, 1980. Sediment delivery ratio as used in the computation of the watershed sediment yield. Beijing, China.

O'Loughlin, C. L., L. K. Rowe & A. J. Pearce, 1980. Sediment yield and water quality responses to cleafelling of evergreen mixed forests in Western New Zealand. In The influence of man on the hydrological regime,with special reference to representative and experimental basins, IAHS Publication no. 130: 285–292.

Painter, R. B., K. Blyth, J. C. Mosedale & M. Kelly, 1974. The effect of afforestation on erosion processes and sediment yield. In Effects of man on the interface of the hydrological cycle with the physical environment, IAHS Publication no. 113: 150–157.

Phillips, J. D., 1992. Pre- and post-colonial sediment sources and storage in the Lower Neuse basin, North Carolina. Phys. Geogr. 14: 272–284.

Summer, W., E. Klaghofer & K. Hintersteiner, 1996. Trends in soil erosion and sediment yield in the alpine basin of the Austrian Danube. In Walling, D. E. & B. W. Webb (eds), Erosion and Sediment Yield: Global and Regional Perspectives, IAHS Publication no. 236: 473–479.

Tardy, Y., R. N'Kounkou & J.-L. Probst, 1989. The global water cycle and continental erosion during Phanerozoic time. Am. J. Sci. 289: 455–483.

Trimble, S. W., 1981 Changes in sediment storage in Coon Creek Basin, Driftless Area, Wisconsin, 1853–1975. Science 214: 181–183.

Vorosmarty, C. J., M. Meybeck, B. Fekete & K. Sharma, 1997. The potential impact of neo-Castorization on sediment transport by the global network of rivers. In Walling, D. E. & J.-L. Probst (eds), Human Impact on Erosion and Sedimentation, IAHS Publication no. 245: 261–273.

Walling, D. E., 1983. The sediment delivery problem. J. Hydrol. 65: 209–237.

Walling, D. E., 1988. Measuring sediment yield from river basins. In Lal, R. (ed.), Soil Erosion Research Methods. Soil Wat. Conserv. Soc., Ankeny, Iowa: 39–73.

Walling, D. E., 1995. Suspended sediment yields in a changing environment In Gurneu, A. & G. Petts (eds), Changing River Channels, Wiley, Chichester: 149–176.

Walling, D. E. & Q. He, 1994. Rates of overbank sedimentation on the flood plains of several British rivers during the past 100 years. In Variability in stream erosion and sediment transport, IAHS Publication no. 224: 203–210.

Walling, D. E. & Q. He, 1997a. Investigating spatial patterns of overbank sedimentation on river floodplains. Wat. Air Soil Pollut. 99: 9–20.

Walling, D. E. & Q. He, 1997b. Use of fallout caesium-137 in investigations of overbank sedimentation on river floodplains. Catena 29: 263–282.

Walling, D. E. & Q. He, 1999. Changing rates of overbank sedimentation on the floodplains of British rivers during the past 100 years. In Brown, A. G. & T. A. Quine (eds), Fluvial Processes and Environmental Change. Wiley, Chichester (in press).

Walling, D. E. & T. A. Quine, 1993. Using Chernobyl-derived fallout radionuclides to investigate the role of downstream conveyance losses in the suspended sediment budget of the River Severn, UK. Phys. Geogr. 14: 239–253.

Walling, D. E. & B. W. Webb, 1981. The reliability of suspended sediment load data. In Erosion and sediment transport measurement, IAHS Publication no. 133: 79–88.

Walling, D. E. & B. W. Webb, 1996. Erosion and sediment yield: a global overview. In Walling, D. E. & B. W. Webb (eds), Erosion and Sediment Yield: Global and Regional Perspectives, IAHS Publication no. 236: 3–19.

Walling, D. E. & J. C. Woodward, 1995. Tracing suspended sediment sources in river basins: a case study of the River Culm, Devon, UK. J. mar. Freshwat. Res. 46: 327–336.

Walling, D. E., P. N. Owens & G. J. L. Leeks, 1998. The role of channel and floodplain storage in the suspended sediment budget of the River Ouse Yorkshire, UK. Geomorphology 22: 225–242.

Walling, D. E., P. N. Owens & G. J. L. Leeks, 1999a. Fingerprinting suspended sediment sources in the catchment of the River Ouse, Yorkshire, UK, Hydrol. Proc. 13: 955–975.

Walling, D. E., P. N. Owens & G. J. L. Leeks, 1999b. Rates of contemporary overbank sedimentation and sediment storage on floodplains of the main channel systems of the Ouse and Tweed catchments, UK. Hydrol. Proc. 13: 993–1009.

Walling, D. E., J. C. Woodward & A. P. Nicholas 1993. A multi-parameter approach to fingerprinting suspended-sediment sources. In Tracers in hydrology, IAHS Publication no. 215: 329–338.

Weiss, F. H., 1996. Sediment monitoring, long-term loads, balances and management strategies in southern Bavaria. In Walling, D. E. & B. W. Webb (eds), Erosion and Sediment Yield: Global and Regional Perspectives, IAHS Publication no. 236: 575–582.

Wolman, M. G. & A. P. Schick, 1967. Effects of construction on fluvial sediment: urban and suburban areas of Maryland. Wat. Res. 6: 1312–1326.

Hydrobiologia **410:** 241–250, 1999.
J. Garnier & J.-M. Mouchel (eds), Man and River Systems.

High frequency variations of water flux and sediment discharge during the Little Ice Age (1586–1725 AD) in the Rhône Delta (Mediterranean France). Relationship to the catchment basin

G. Arnaud-Fassetta[1] & M. Provansal[2]
[1]*Université de Paris VII-Denis Diderot, Département de Géographie, U.M.R. 183 C.N.R.S.-P.R.O.D.I.G. Tour 54–64, P.O. Box 7001. 2, Place Jussieu, F-75251 Paris Cedex 05, France. Tel./fax: [+33](0)1-44-277669; Fax: [+33](0)1-44-27-6083; E-mail: fassetta@paris7.jussieu.fr*
[2]*Université d'Aix-Marseille I, Institut de Géographie, U.M.R. 6635 C.N.R.S. (C.E.R.E.G.E.). 29, Avenue Robert Schuman. F-13621 Aix-en-Provence Cedex 01, France. Tel.: [+33](0)4-42-64015; E-mail: provansal@cerege.fr*

Key words: Rhône delta, fluvial dynamics, sediment discharge, source zones, fluvial metamorphosis, Little Ice Age

Abstract

During the 17th and 18th centuries, the Rhône delta recorded high frequency fluctuations in water and sediment influx. These variations resulted from the drastic climatic changes that took place during the Little Ice Age, which were intensified by dense land settlement in the catchment basin. The use of complementary types of information (iconographic and textual archives, photo-interpretation of traces of fluvial metamorphosis, grain size distribution and mineralogy of alluvial infillings) allows a precise study of the major change that affected the Rhône delta in the second part of the 17th and the beginning of the 18th centuries. This change corresponds to a rapid response by the fluvial system and the occurrence of floods that were higher and more frequent on the lower Rhône. It resulted in a rapid change in fluvial environments, i.e. channel infilling, raising of river banks, appearance of crevasses and secondary channels, and accelerated delta progradation. It is associated with the influx of an abundant and exceptionally coarse-grained bottom load. The provenance of these sediments is discussed, using the heavy mineral assemblages they contain, with reference to reworking of previous alluvial deposits, probably due to changes in the active band and episodic loads in the whole catchment basin. The short response time to hydrologic impulse and the importance of the metamorphosis are related to the sediment influx from the Massif Central, whose steeply inclined rivers are close to the delta. The contribution of the northern Alps and the durancian basin (southern Alps) is not as important because of their distance from the delta and the time lag affecting the bottom load transfer downstream. The aims of this study are to investigate the response time, transfer velocity of sandy deposits, and the contribution of different catchment sub-basins to the supply of the Rhône delta. The role of climatic and anthropic factors are also discussed.

Introduction

During the 17th and 18th centuries, the Rhône delta underwent a major change, characterized by the metamorphosis of fluvial environments and considerable coastal progradation. This phase, which lasted several hundred years, included many fluctuations. The first and the most important fluctuation took place during the end of the 17th and the beginning of the 18th century, concluding with the avulsion of the river that in 1712, displaced the river to its present-day location. Sedimentological and historical data corresponding to these decades are detailed enough to allow chronological analysis of its development and determine the relationship with solid influx from the catchment basin.

The hydrology and the sediment load of the Rhone's drainage basin (95 500 km^2) depend on topographical characteristics and the type of surface, the nature of the bedrock and surficial formations, and on the regional climates of each catchment sub-basin (Pardé, 1925; Bourdier, 1961). The climatic deterioration to the cool and humid Little Ice Age modified the formation of solid and liquid discharge and coincided with a dense settlement of the countryside.

There is a lot of evidence for this change in the Rhône basin, i.e. the advance of glaciers in the northern Alps (Leroy-Ladurie, 1967) and the southern Alps (Evin, 1983; Jorda, 1985; Lagier & Masson, 1997), the lowering of the upper limit of forests (Tessier et al., 1993), the increase in lacustrine levels (Magny, 1993) and the torrential nature of flooding (Brochier, 1983; Bravard, 1989; Salvador, 1991; Pfister, 1992; Miramont, 1998; Pichard, in press), the phenologic data, the archives and first instrumental measurements (Alexandre, 1987; Pfister, 1992; Roditis, 1992; Serre-Bachet et al., 1992). At this stage in the reseach, it appears that not all the regions in the catchment basin behave the same way and that the climatic parameters (temperature, precipitation) are not covariant. Jorda & Roditis (1993) and Pichard (1995) show that the hydrologic peak of the lower Rhône between 1680 and 1710 does not coincide with the most drastic periods of fluvial ice at Arles. In the durancian basin, these decades correspond to a glacial retreat (Lagier & Masson, 1997) and a less dynamic hydrology (Guilbert, 1994; Pichard, in press).

Schumm (1977), Starkel (1983) and Brown (1997) describe the modes of adjustments of rivers in terms of variations in liquid and solid discharge. Evidence of the climatic deterioration of the Little Ice Age can be found in the basin upstream of the Rhône (Bravard, 1989) and on the Durance (Guilbert, 1994; Pichard, 1995) as early as the 14th century. However, the metamorphosis of the upper Rhône and the Isère is not recorded until the 18th century (Bravard, 1989; Salvador, 1991), as is that of the Buëch and the Durance (Gautier, 1992; Miramont, 1998). The map of Cassini (1788) shows the transformation of fluvial environments at the end of the 18th century. The river is braided along its whole length, occupying a wide river bed made of coarse material. The alluvial fans of tributaries, multiple oxbow lakes, and swampy plains developed behind bank deposits, constitute major zones of sediment storage.

Downstream, the delta records the solid transfers originating in the entire catchment basin. Although the suspended sediment arrives almost instantaneously during floods, the transport of the bottom load is slower and discontinous in space and time (Roditis & Pont, 1993). It depends on the climatic geography of each drainage basin, the gradient of tributaries and their position in the drainage network, structural dams and/or intermediate sediment traps. Features of detrital influx can be misinterpreted due to chronological shifting, which makes it difficult to differentiate between the effect of climatic and anthropic forcing. Salvador (1991) estimates the displacement velocity of the bed load in Isère at 10 km/century during the 18th century. It is probable that the sandy load, which was the major component of the bottom load in the delta, was transported more rapidly.

In the delta, the major hydrosedimentary change occurs many decades earlier than those occurring upstream of the basin. Therefore, the characteristics (deformation of fluvial sediment beds) and the chronology are studied in this paper. The relationship between the upstream basin and an estimation of transfer velocities is investigated using heavy mineral assemblages to determine sediment origin.

Paleohydrology and fluvial metamorphosis of the Rhône delta from the end of the 16th century to the beginning of the 18th century

The Rhône delta is situated downstream of a catchment basin characterized by a large hydrologic and climatic diversity (Pardé, 1925; Figure 1). The alpine and mediterranean tributaries play a decisive role in the regime, abundance, and sediment load of the river. Over the last 100 km, the river receives sediment influx on its left bank from the Durance flowing from the southern Alps and on its right bank from the Ardèche, the Cèze, and the Gardon, flowing from the Massif Central. This 'torrential' regime and the sedimentary load of its tributaries appears decisive for delta morphogenesis. However, the transport of bottom load is probably more rapid for right bank tributaries with steep gradients (8–10%) than for the lower Durance (gradient: 2–3%).

Many paleochannels resulted from delta construction during the Holocene (L'Homer et al., 1981). To the east, the Bras de Fer is the most recent of these channels (Figure 2). Its period of activity is accurately determined by the iconography of texts between 1586 and 1712. Its course makes a wide loop, that cuts across many beach ridge systems. Seismic probes and offshore drill cores confirm the remnants of a large detrital cone (PNOC, 1993).

Several remarkable historical documents exist that aid in understanding the evolution of the Bras de Fer: the hydrologic chronicals of Arles, whose raw data are discussed by Pichard (1995), environmental historian,

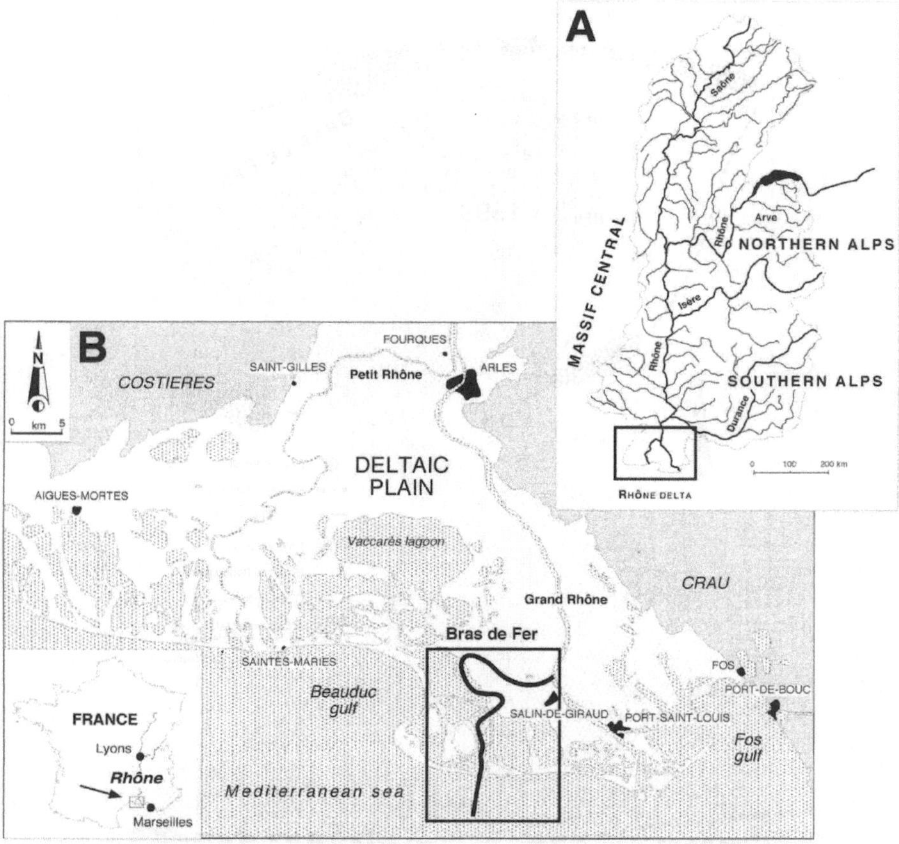

Figure 1. Location map showing the Rhône catchment basin (A) and the Bras de Fer paleostream in the Rhône delta (B).

and a continuous chronological series of very precise old maps (Caritey, 1995). The final morphology of fluvial environments were determined by topographic surveys (L'Homer, 1975a, b) and photo-interpretation. Deep cores indicate their vertical extent (Figure 3).

The periodicity of floods > 5.25 m NGF[1] is established by Pichard (1995) at Arles (Figure 4). The general increase in hydrologic flux between the middle of the 16th and the end of the 19th centuries is characterized by two distinct peak periods (1700–1710 and 1750–1810), during which high floods varied from once a year to once every decade. The first peak is preceded by a long period of instability, among which the most important floods occurred during the years 1560–1600 and 1670–1700. The period 1700–1710 corresponds to a phase of major hydrologic degradation that ends in the deflection of the Rhône into the delta.

The iconography of river mouths confirm this chronology and shows that the hydrologic change corresponds to a considerable increase in solid influx. Between 1660 and 1730, measurements made at 1- to 10-y intervals were mapped for the Royal Navy. These maps were digitalized to allow their superposition and the evaluation of the volume of sediments deposited in the channel and at its mouth (Caritey, 1995). They show the change in the fluvial landscape with the increasing emergence of islands, followed by the separation of the main channel in two subchannels between 1680 and 1710. The progradation of the Rhone's mouth at 80 m/y from 1660 to 1700 accelerates to reach 180 m/y from 1700 to 1710.

Using photo-interpretation, fluvial environments

[1] NGF: 'Nivellement Général de la France', general levelling of France.

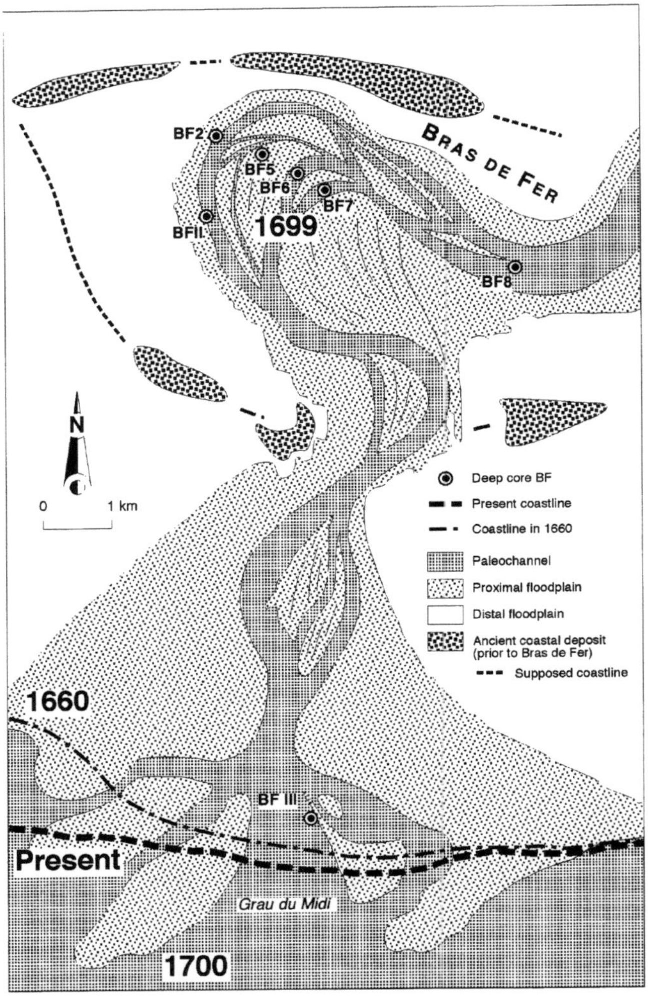

Figure 2. Paleogeography of the Bras de Fer and its mouth obtained by photo-interpretation. Location of the core sites.

before the deflection of 1712 could be reconstructed. The environments were particularly unstable: involving the multiplication of islands and crevasses and division of the river bed along the axis of meander that appears to be in the process of oxbow-lake formation. The overall structure of the beds (estimated by cores made in alluvial deposits, Figure 3), confirm the following hypothesis: the channel, a dozen meters deep and 700 m wide upstream, becomes progressively shallower in the meander and to the mouth (4 – 7 m), whereas the bank deposits multiply and thicken.

The analysis of fluvial dynamics is based on the sedimentology of seven cores (4–12 m in length) taken along the axis of the paleochannels and in bank deposits. All cores reached the marine deposits (Figure 3). Grain size analysis was performed on 200 sediment samples taken every 30 cm in the sandy layer and every 10 cm in the more silty layer. The granulometry and the C/M diagram were used to reconstruct the depositional environments (Folk & Ward, 1957; Passega, 1957; Bravard, 1983). Sediment structure, revealed by X-ray photography, was interpreted in terms of dynamics (Reineck & Singh, 1980). Based on these criteria, three types of hydrological regime (river flow) were identified (Arnaud-Fassetta, 1998): type A, characterized by a homogeneous and coarse

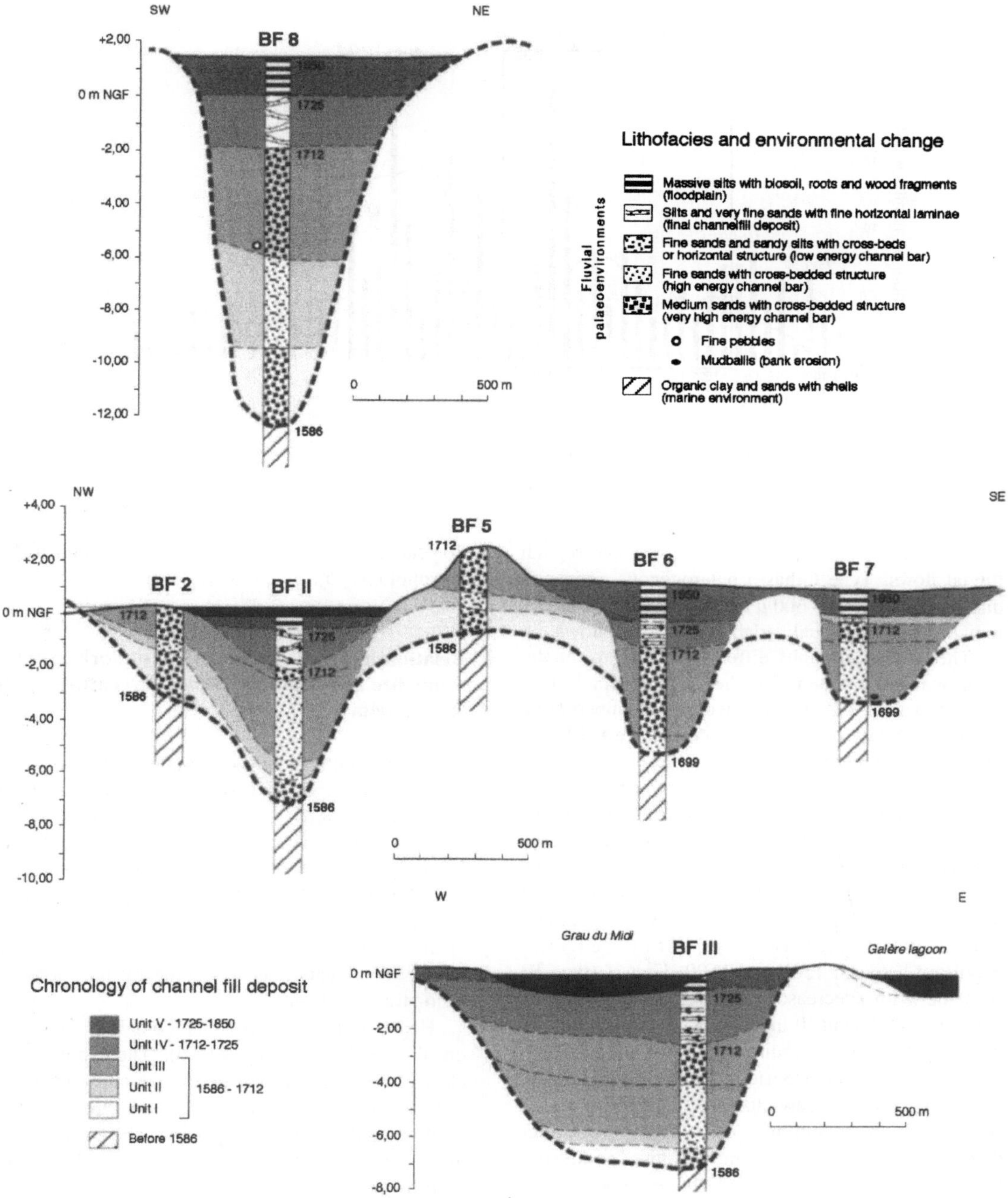

Figure 3. Bras de Fer paleostream structure and channel fill deposits.

grain size distribution (C_s: 700–800 μ), with cross-beds or oblique structures in the channel and the part of the bank close to the river, corresponds to a high and constant fluvial flows. Type B is homogeneous and has a medium grain size distribution (C_s: 400–500 μ), with a subhorizontal structure coarsely bedded in

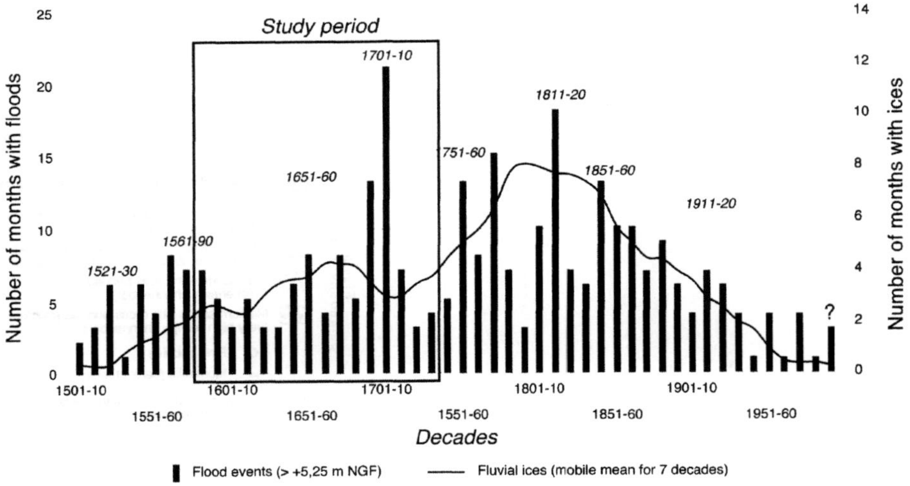

Figure 4. Flood frequency of the Rhône at Arles, since 1500 AD (after Pichard, 1995).

the channel, corresponds to a decreasing and regular fluvial flows. Type C has a heterogenous grain size distribution (C_S: 400–600 μ) with a variable structure, indicates a pluri-annual variability of fluvial flows.

The core stratigraphy differs according to the depositional environment. In the main channel, four units (types C–B–A–B) are successively superposed. In the secondary channels and on the external banks only the two upper units (A–B) are present. The chronological variablity of the dynamics, characterized by a maximum, can be compared to the hydrologic periodicity proposed by Pichard (1995). The spatial distribution of sedimentary units can be explained by the changes in fluvial landscapes and the widening of the active band, coinciding with the hydrologic maximum (Caritey, 1995). The lower unit I (type C) was deposited between 1580 and 1600. Unit II (type B) coincides with a decrease in the river's energy between 1600 and 1640. Unit III indicates a gradual increase in hydrologic activity, culminating in the peak of 1700–1710 and the lateral extension of the active band, with the creation of two new channels. Unit IV (type B) corresponds to the infilling of the Bras de Fer after deflection (1712–1725). At the top, unit V indicates floodplain deposits of the Grand Rhône, which cap the underlying units.

A rapid morpho-sedimentary response by the Rhône to hydrologic variations takes place over several years (Figure 5). The highest peak at the end of the 17th to the beginning of the 18th century is of short duration (10 years), but it is preceded by 30 – 40 years of medium to high floods, which prepares the morphologic adjustment of the river bed.

Relationship to the catchment basin: origin of the sandy bed load from the 16th–18th centuries in the Rhône delta

Numerous studies analyze the contribution of different catchment basins to the solid load transported in the hydrographic network. Some of these studies use heavy mineral as markers to determine the paleogeographic origin of solid fluxes (Hassan, 1976; Tourenq, 1986; Hamrouch & Stanley, 1990). In the Rhône catchment basin, Van Andel (1955) and Petit et al. (1996) have isolated significant assemblages whose identification is facilitated by the variety of litho-structural facies.

Heavy mineral analysis of core BF8 was made on 37 samples (Figures 3 & 6). The samples were chosen in relation to the grain size analysis. Heavy mineral analysis was only made on the fine sand fraction (160–50 μ) because it is the only fraction that is transported directly from its source, coarse sand-being frequently transported in steps (Arnaud-Fassetta, 1997). Sediment influx from the Alps are characterized by pistachite, common hornblende and glaucophane, originating in the durancian basin upstream (Dubar, 1982). Sediment influx from the Massif Central (Cévennes, Vivarais, Velay) contains augite, aegyritic augite, basaltic hornblende, hyper-

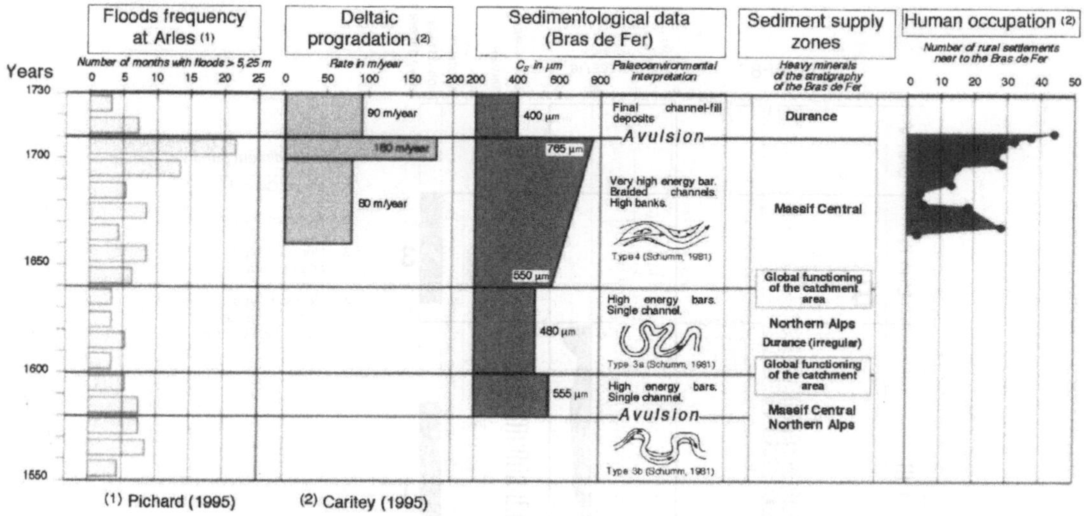

Figure 5. Table showing a compilation of water flux and sediment discharge in the Rhône delta from the 16th to 18th centuries.

stene, andalusite and titanite. In addition, two types of minerals indicate modifications in the active band of the river: the abundance of minerals resistant to weathering (zircon, tourmaline, staurolite) reflect the reworking of alluvial sediments and older detrital deposits (Gandolfi & Paganelli, 1977). Heavy minerals of high density ($d > 3.7$) include garnet, zircon and staurolite. Their transport is related to high energy currents and their deposition depends on local hydrodynamic fluctuations.

At the top of unit I and the boundary of units II and III, two short phases of sediment influx originate from the entire catchment basin. The long distance and the heterogeneity of source zones explain why these phases do not correspond to the coarser sequences. Unit II and in part, unit I, correspond to sediment influx mainly originating in the northern Alps and to a renewel of erosion of river banks, with an irregular participation of the Durance and an accessory load originating in the Massif Central. These deposits coincide with a reduction in hydrodynamics, confirmed by the presence of high density minerals, which result either from the effect of the distance of source zones or a decrease in flood energy. Unit III is unique in that its sediments originate exclusively from the Massif Central. They indicate the effect of extensive flooding due to tributaries located close to the delta and with a steep gradient. The relative absence of northern alpine and durancian sediment influx is indicative of the problem of the transfer time taken by the sandy load originating from upstream sub-basins in the Rhône catchment basin. Unit IV is associated with solid influx originating in the southern Alps (Durance). Their late arrival in the delta, compared to the beginning of change in Durance (1660–1670), indicates that the sediment was transported over a period of 40–50 years in the durancian basin before reaching the delta plain.

Discussion and conclusions

This study reveals a major change in the Rhône hydrosystem at the turn of the 17th century, which leads to a metamorphosis of the lower Rhône river bed. The river changes from meandering to 'deltaic braiding', characterized by the multiplication of channels, the deposition of numerous sandy banks, and rapid channel infilling.

The 'deltaic braiding' is a local morphosedimentary response to the major hydrologic change from 1700–1710. The rapid alteration of fluvial environments, which occurring to the peak of the change, is the result of the magnitude of fluvial activity during this period (with a frequency of many high floods occurring every year to every decade). It is also a consequence of the river bed development and the reworking of sediment load from the catchment basin during the preceding 30–40 years of hydrologic transition.

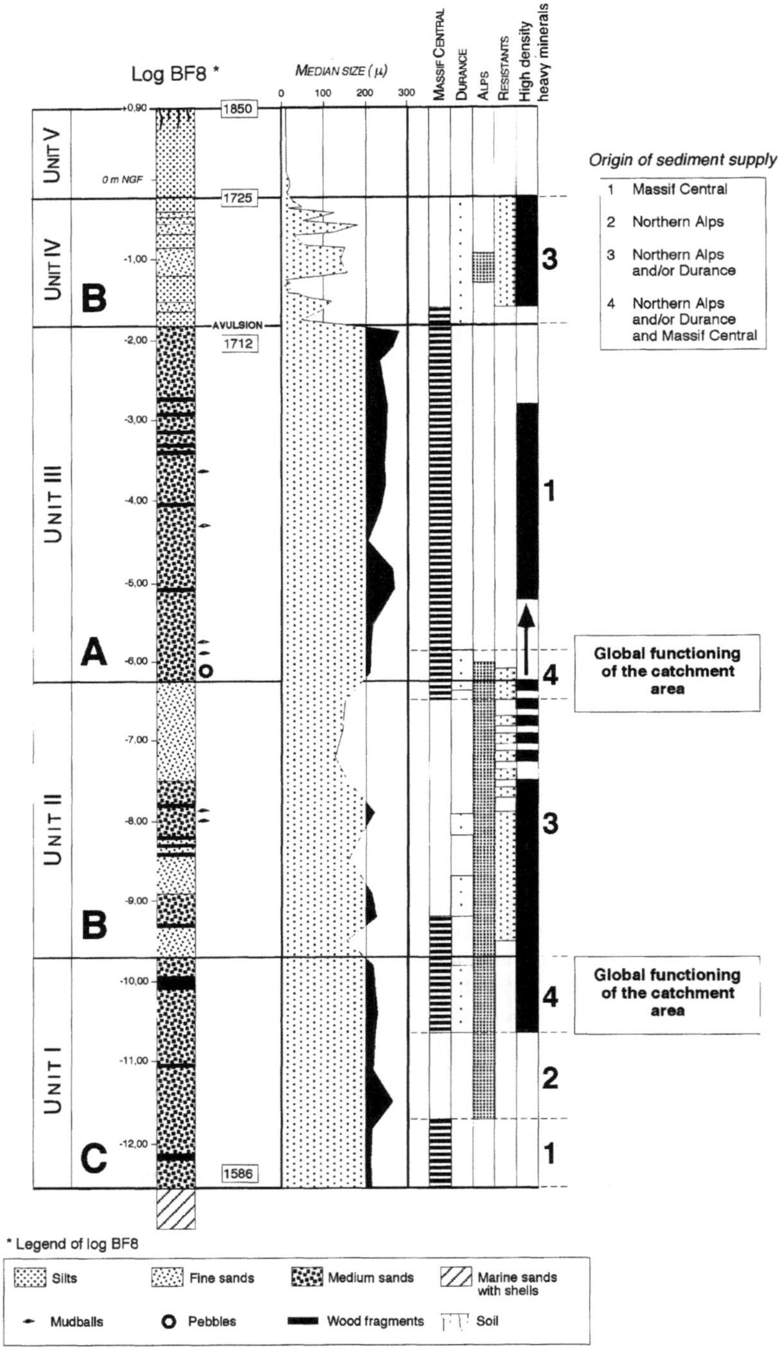

Figure 6. Heavy mineral and grain size analyses of core BF8 (Bras de Fer).

The rapid adjustment of fluvial environments also assumes the availability of abundant sediment. This is assured at the end of the 16th century by the destabilization of the hydrosystem (Little Ice Age), which starts to infill the channel during the 17th century. The hydrologic change and the metamorphosis are due to the decisive influence of the source zones in the Cévennes and the Vivarais, which are closest and have the highest hydraulic energy. Tributaries from the Cévennes and Ardèche, at a distance of only 100–150 km and with a steep gradient, have the capacity to rapidly supply the Rhône delta during hydrologic peaks. The alpine tributaries and the upstream Durance basin, because of their relatively great distance away (> 250–300 km), supply a smaller amount of sediment, whose transfer is chronologically shifted. In the durancian basin, the hydroclimatic change begins around 1660–1670, but the solid influx does not arrive in abundance at the delta until 40 – 50 years later, i.e. not before 1710. They contribute moderately to the metamorphosis of the lower Rhône at the beginning of the 18th century. Nevertheless, the effect of the possible paleoclimatic fluctuations between the Massif Central and the Alps cannot be excluded.

In light of the data, it appears that it is difficult to distinguish the effects of the climatic history and land occupation in the catchment basin. The abundance of sediment is a function of both the river's capacity of transportation and the sediment supply coming from drainage basins, in other words the effects of erosion, accelerated by settlement of the countryside beginning in the 17th century. However, the transfer of detrital sediment downstream of the fluvial system depends essentially on the hydroclimatic data (volume, intensity and seasonal distribution of precipitation, frequency and levels of floods). In the case of the lower Rhône, the exact anthropic impact and the development of fragile eastern slopes of the Massif Central, compared to those of the northern and southern Alps remains in question.

This study compares the transport velocity of sediment load in the catchment basin and the metamorphosis of the lower Rhône. Durancian sands arrive at the delta with a time lag estimated at 40–50 years. Their velocity over the distance from upstream of the basin to its mouth (250–300 km) can be evaluated at 5–8 km/y. The contribution of durancian sediment load from upstream to the metamorphosis of the lower Rhône depends on the duration of the change. In the case of a short lived change (one to many decades) the durancian deposits only contribute moderately to lower Rhône metamorphosis. The changes in fluvial landscapes is mainly due to the 'nearby' hydrogenesis, in which the Massif Central plays a decisive role. During a longer period of change (lasting one or many centuries), e.g. the First Iron Age or Late Antiquity, the overall catchment basin, including the upper Durance, contributed to the metamorphosis of the lower Rhône. Thus, the Rhône delta became the catchment basin that recorded diachronous disequilibria affecting the different catchment sub-basins, even while undergoing specific climatic constraints (Arnaud-Fassetta, 1998).

In conclusion, this metamorphosis of the delta occurs before and is chronologically disconnected from the one described by Bravard (1989) and Salvador (1991) in the northern Alps. The Isère and the Drac change from meandering to braided rivers at the end of the 18th century, whereas the Bras de Fer changes its fluvial regime at the end of the 17th century. The metamorphosis is even more rapid when the bottom load is fine-grained, and thus, more mobile. This undoubtedly explains the time lag (50–100 years) between delta metamorphosis, related to mainly sandy bottom loads that are transported rapidly (5–8 km/y) and those related to the northern Alps, associated with a slow-moving coarse bottom load (100 m/y; Salvador, 1991).

Acknowledgements

Our thanks go to M. Guillemard of the 'Centre d'Etudes Techniques de l'Equipement' of Aix-en-Provence for the carrying out of the core sampling, M. Dubar and M. Ricq de Bouard of the 'Centre des Recherches Archéologiques' of Valbonne-Sophia-Antipolis for mineralogical advice and some students of the 'Institut de Géographie' of Aix-en-Provence for their participation in the measurement campaigns. We are also grateful to the landowners of the Camargue for allowing us access to the paleostream. This study has been supported by grants of the European Economic Community (Environment Research Programm, Directorate General for Science, Research and Development under Contracts N°EV5V-CT94-0465).

References

Alexandre, P., 1987. Le climat en Europe au Moyen Age. Paris, EHESS, Recherches d'Histoire et de Sciences Sociales 24: 828.

Arnaud-Fassetta, G., 1997. Evolution du plancher alluvial du Petit Rhône à l'échelle pluriannuelle (delta du Rhône, France du sud). Géomorphologie: relief, Processus, Environnement 3: 237–256.

Arnaud-Fassetta, G., 1998. Dynamiques fluviales holocènes dans le delta du Rhône, thèse Université d'Aix-Marseille I.

Bourdier, F., 1961 Le bassin du Rhône au Quaternaire, Géologie et Préhistoire, CNRS, Paris.

Bravard, J. P., 1983. Les sédiments fins des plaines d'inondation dans la vallée du Haut-Rhône (approche qualitative et spatiale), Rev. Géographie Alpine 71: 363–379.

Bravard, J. P., 1989. La métamorphose des rivières des Alpes françaises à la fin du Moyen Age et à l'Epoque Moderne, Rev. Géographie de Liège 25: 145–157.

Brochier, J. E., 1983. Deux mille ans d'histoire du climat dans le midi de la France: étude sédimentologique, Anna. Economies, Sociétés, Civilisations 2: 425–438.

Brown, A. G., 1997. Alluvial geoarchaeology. Floodplain archaeology and environmental change, Cambridge, Cambridge University Press.

Caritey, C., 1995 L'évolution de l'embouchure du Rhône du milieu du XVIIe à la fin du XIXe s. Relations avec le régime du fleuve, Maîtrise Université Aix-Marseille I.

Dubar, M., 1982. Stratigraphie des dépôts du Néogène supérieur et du Pléistocène du bassin de la moyenne Durance; interprétation géodynamique et paléogéographique, thèse d'Etat, Marseille.

Evin, 1983. Structure et mouvements des glaciers rocheux des Alpes du sud, thèse Université de Grenoble.

Folk, R. L. & W. C. Ward, 1957. Brazos river bar, a study in the significance of grain size parameters, J. Sedimentary Petrol. 29: 3–27.

Gandolfi, G. & L. Paganelli, 1977. Le province petrografiche del litorale toscano, Boll. Soc. Geol. 96: 653–663.

Gautier, E., 1992. Recherches sur la morphologie et la dynamique fluviales dans le bassin du Buëch (Alpes du sud), thèse Université de Paris X-Nanterre.

Guilbert, X., 1994. Les crues de la Durance depuis le XIVe s. Fréquence - Periodicité - Essai d'interprétation paléoclimatique, Maîtrise Université d'Aix-Marseille I.

Hamroush, H. A. & D. J. Stanley, 1990. Paleoclimatic oscillations in East Africa interpreted by analysis of trace elements in Nile delta sediments. Episodes 13: 264–269.

Hassan, F. A., 1976. Heavy minerals and the evolution of the modern Nile. Quat. Res. 6: 425–444.

Jorda, M., 1985. La torrentialité holocène des Alpes françaises du sud, facteurs anthropiques et paramètres naturels de son évolution. Cahiers ligures de Préhistoire et Protohistoire 2: 49–70.

Jorda, M. & J. C. Roditis, 1993. Les épisodes du gel du Rhône depuis l'An mil. Périodisation, fréquence, interprétation paléoclimatique. Méditerranée 3: 19–30.

L'Homer, A., 1975 a. Notice explicative de la carte géologique d'Istres au 1/50000e, BRGM.

L'Homer, A., 1975 b. Notice explicative de la carte géologique des Saintes-Maries au 1/50000e, BRGM.

L'Homer, A., A. Bazile, J. Thommeret & Y. Thommeret, 1981. Principales étapes de l'édification du delta du Rhône de 7000 BP à nos jours; variations du niveau marin, Oceanis 7: 389–408.

Lagier, T. & S. Masson, 1997. Les glaciers de la Vallouise (glacier blanc-glacier noir): héritages holocènes et évolution climatique, Maîtrise Université de Provence.

Le Roy Ladurie E., 1967 Histoire du climat depuis l'an mil, Flammarion. Paris.

Magny, M., 1993. Les fluctuations des lacs jurassiens et subalpins et l'histoire du climat au Moyen Age. Histoire et Mesure 8: 5–17.

Miramont, 1998. Morphogénèse, activité érosive et détritisme alluvial holocènes dans le bassin de la Moyenne Durance (Alpes françaises du sud), thèse Université d'Aix-Marseille I.

Pardé, M., 1925. Le régime du Rhône. Etude hydrologique, Etude et Travaux de l'Institut de Géographie Alpine de l'Université de Grenoble, Lyon, thèse d'Etat Université de Grenoble.

Passega, R., 1957. Texture as characteristics of clastic deposition, American Association Petrol. Geol. Bull. 41: 1952–1964.

Petit, C., M. Campy, J. Chaline & J. Bonvalot, 1996. Major palaeohydrographic changes in Alpine foreland during the Pliocene-Pleistocene. Boreas 25: 131–143.

Pfister, C., 1992 Monthly temperature and precipitation in central Europe 1527-1979: quantifying documentary evidence on weather and its effects. In Bradley & Jones (eds), Climate since 1500 AD: 118–142.

Pichard, G., 1995. Les crues sur le bas Rhône de 1500 à nos jours. Pour une histoire hydro-climatique, Méditerranée 3–4: 105–116.

Pichard, G., Les eaux du ciel. Sur l'histoire des pluies, des neiges et des débordements. Revue Marseille (in press).

PNOC, 1993. Bilan et évolution à long terme des écosystèmes côtiers, rapport d'activité 1992-1993 du Programme National d'Océanographie Cotière.

Reineck, H. E. & I. B. Singh, 1980. Depositional sedimentary environments, Springer-Verlag, Berlin, Heidelberg, New York.

Roditis, J. C, 1992. Le gel du Rhône depuis l'an Mil, Maîtrise Université d'Aix-Marseille II.

Roditis, J. C. & D. Pont, 1993. Dynamiques fluviales et milieux de sédimentation du Rhône à l'amont immédiat de son delta. Méditerranée 3: 5–18.

Salvador, P. G., 1991. Le thème de la métamorphose fluviale dans les plaines alluviales du Rhône et de l'Isère - Bassin de Malville et ombilic de Moirans - Bas Dauphiné, thèse Université Jean Moulin-Lyon III.

Schumm, S. A., 1977. The fluvial system. J. Wiley and sons, New-York.

Serre-Bachet, F., J. Guiot & L. Tessier, 1992. Dendroclimatic evidence from southwestern Europe and northwestern Africa. In Bradley & Jones (eds), Climate since 1500 AD: 349–365.

Starkel, L., 1983. Climatic change and fluvial response. In R. Gardner & H. Scoging (eds), Mega-Geomorphology: 195–211.

Tessier L., J. L. De Beaulieu, M. Coutaux, J. L. Edouard, P. Ponel, C. Rolando, M. Thinon, A. Thomas & K. Tobolski, 1993. Holocene paleoenvironments at the timberline in the Alps (Taillefer massif, french Alps), a multidisciplinary approach. Boreas 22: 43–254.

Tourenq, J., 1986. Etude sédimentologique des alluvions de la Loire et de l'Allier, des sources au confluent. Les minéraux lourds des roches des bassins versants. Documents du BRGM 108.

Van Andel, T. H., 1955. Sediments of the Rhone delta; sources and deposition of heavy minerals. Verhandelingen van het Koninklijk Mijnbouwkundig Genootschap, Geologische serie deel 15: 502–555.

Hydrobiologia **410**: 251–261, 1999.
J. Garnier & J.-M. Mouchel (eds), Man and River Systems.

Estimation of sediment yield and soil loss using suspended sediment load and ^{137}Cs measurements on agricultural land, Brie Plateau, France

S. Sogon[1,2], M-J. Penven[2,3], P. Bonte[1] & T. Muxart[2]
[1]*Laboratoire des Sciences du Climat et de l'Environnement, UMR 1572, CEA-CNRS, Gif/Yvette, France*
[2]*Laboratoire de Géographie Physique, UMR 8591, CNRS et Université Paris 1, Meudon, France*
[3]*Université de Rennes 2, Rennes, France*

Key words: agricultural land, hillslope and watershed scales, underground drainage network, suspended sediment, soil erosion, caesium-137

Abstract

Recent land use changes have resulted in environmental disturbances on agricultural land in NW Europe. The development of underground drainage networks in regions characterized by temporary hydromorphic soils has altered the delivery of water discharge and sediment loads in rivers. To implement soil and water conservation strategies, space-time movements of fine sediment particles must be investigated at watershed scale. A cultivated watershed in the Brie Plateau (upstream of Paris) was chosen for this study. In the short-term, sediment fluxes were measured at three monitoring stations within nested watersheds (drainage pipe under a cultivated field of 6.4 ha, Vannetin River 4.6 km^2 and 30.6 km^2 areas). Suspended sediment loads, estimated over a four year period (1991–1995), revealed wide seasonal and interannual variations reflecting the seasonal rainfall distribution. Sediment yields at the outlet of the drainage system were large (0.24 t ha^{-1} yr^{-1}) during a year with high rainfall. This incurs a risk of impoverishment of fine particles in soil over the medium-term. Sediment yields in the river could reach 0.30–0.47 t ha^{-1} yr^{-1} for a rainy year. Large sediment deliveries only accompanied a widespread overland flows on slopes. In the medium-term, the time-integrated ^{137}Cs technique is ideal for the Brie Plateau, where overland flow does not produce significant rill erosion features in the landscape. The ^{137}Cs sampling grid was relatively dense because soil erosion and soil redistribution varied greatly on the gentle slopes (4–5% mean slope angle). The ^{137}Cs tracer was used to integrate soil redistribution for a medium term-period (33 years). The study of a 7 ha field (downstream part of a 24 ha crop catchment) revealed a complex soil redistribution pattern and enabled the construction of a sediment budget. Five representative topographic units were investigated. Two of them exhibited net soil loss (sediment delivery ratio 74% on the hillslope and 83% at the outlet unit of the drainage basin). In the other spatial units, deposition exceeded erosion because of upslope and/or lateral sediment transfers. The latter units represent the main buffer areas of the slope.

Introduction

Recent changes in land use and management in NW European temperate plains have created strong environmental disturbances in agricultural ecosystems (on-farm impacts) (Auzet, 1987) and, because of slope-stream linkages, in downstream rivers (off-farm impacts). Among these disturbances, recent soil erosion (Boardman et al., 1990; Poesen et al., 1995; Morgan, 1996) enhances the risk of land degradation and raises the problem of the conservation of a natural resource that is essential for food production. Furthermore, large sediment delivery to rivers, with associated contaminants and pollutants such as fertilisers and pesticides, represent a real problem for river ecosystems and for drinking water treatment plants, which draw water from the rivers. Soil erosion and water pollution problems are therefore social problems because they present a threat to society.

On the Brie plateau (East Central part of the Paris Basin, Figure 1), agricultural intensification has resulted in land regrouping, mechanization, chemical

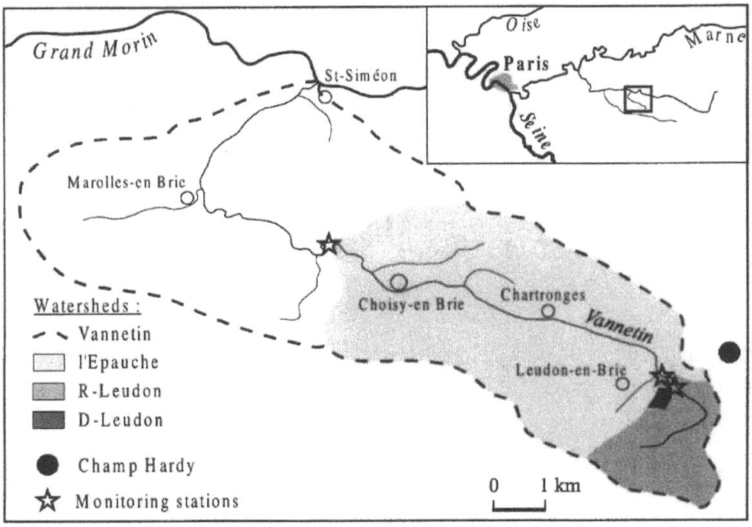

Figure 1. The Brie Plateau and the location of the study sites.

component supplies (Mazoyer & Roudart, 1998) and the widespread use of underground drainage networks beneath cultivated, temporarily waterlogged, clayey silty soils (Riedel & Franc de Ferrière, 1951; Jamagne, 1973; Bartoli et al., 1995; Penven et al., 1998).

The region produces large crop yields. Each year, winter-sown cereals (wheat and barley) and rapeseed cover slightly more than 50% of the cultivated area of the plateau. Bare soils are, therefore, exposed to rain drop detachment and runoff at critical periods in the year (late autumn, winter and early spring). This aggravates the risk of soil erosion even though slope angles on the plateau are rather gentle (ca. 4–5%; Penven et al., 1998).

The central part of the Brie Plateau is drained by the Grand Morin river and its tributaries. Information on variations of water quality in these rivers is required for the supply of water to part of the Paris urban area, as a drinking water treatment plant is located on the Marne river, downstream of its confluence with the Grand Morin.

The aims of this paper are

1. to estimate the effect of the underground drainage network on processes of soil loss and sediment delivery to the Vannetin river (a tributary of the Grand Morin);
2. to compare these data with sediment yields measured in the Vannetin river at different spatial scales (monitoring of nested catchments during four years);
3. to map and estimate soil erosion and soil redistribution on cultivated fields in a small upstream catchment and to assess soil loss into the river since agricultural intensification, using a time integrated method based on ^{137}Cs; and
4. to compare and discuss the results of slope process monitoring with those of stream-based sediment yield estimation.

Assessment of sediment delivery to the river using a monitoring station network

A monitoring station network has been installed in the headwaters of the Vannetin river watershed (Penven & Muxart, 1993). It includes three nested monitoring stations, their location based on stream ordering and land use (Figure 1). The D-Leudon station is located at the outlet pipe of an underground drainage network which collects water and sediment from a cultivated field of 6.43 ha area. The other two stations, R-Leudon and Epauche, are located on the river, downstream of an agricultural watershed (size 4.6 km^2, including the D-Leudon cultural plot) and downstream of a rural area (30.6 km^2), including the above two areas, respectively. Rainfall, water discharge and suspended sediment loads have been measured at these three stations over a period of four hydrological years (September 1991–August 1995) in order to determine suspended sediment yields related to individual storms and to yearly outputs. Since water discharge increase sharply during flood events at all the stations (i.e. small watersheds), suspended sediment concentrations (filtration of particles at a 0.45 μm grid mesh) have been measured every 45 min at the D-Leudon and R-Leudon stations and every hour at the Epauche station. The first two years were relatively dry (625 and 651 mm) whereas the next two were wet (815 and 886 mm) compared with the estimated 678 mm average annual rainfall.

Variations in suspended sediment concentrations during flood events

An example of variations in suspended sediment concentration is given for a series of flood events in January 1995, which are among the main events recorded during four years (Table 1). During this long wet winter period (low rainfall intensity, but high rainfall total), overland flow was widespread on bare soils. Suspended sediment concentrations rose to as high as 414 mg l^{-1} at the outlet (D-Leudon) of the drainage system, 1382 mg l^{-1} at R-Leudon station and 1467 mg l^{-1} at Epauche station in the Vannetin river. Temporal variations in suspended sediment concentration appear to be very similar in the two stations, but particulate matter concentrations did not vary with water discharge in a simple relationship. For example, during a succession of flood events, maximum suspended sediment concentrations at D-Leudon and R-Leudon stations were higher for the smallest last flood event than for previous high magnitude events.

Seasonal variations in sediment deliveries

Sediments were calculated for all flood events at the various stations. Figure 2 shows the results obtained at D-Leudon and R-Leudon stations during the wet September 1994–August 1995 hydrological year. Sediment deliveries of the river were particularly high during the series of flood events in late January 1995, which were preceded and accompanied by overland flow as well as soil erosion on slopes. Sediment deliveries at R-Leudon station reached 66.1 t during the first flood event (21–24 January) and 89.0 t during the three events (28–29), approximately 47% and 64% of the annual sediment load at this station respectively. The behaviour of the drainage system was rather similar to that of the river, although it was not directly affected by surficial soil erosion. Sediment deliveries at D-Leudon station were as high as 0.5 t during the first flood event and 0.7 t during the three flood events, accounting for 55% and 69% of the total annual suspended sediment load respectively.

Annual variations in sediment deliveries

Annual suspended sediment yields were determined at the three stations of the Vannetin river basin (Table 2). At the outlet of the drainage system (D-Leudon) suspended sediment yields ranged from 0.001 t ha^{-1} yr^{-1} in the dry 1991–1992 year (625 mm rainfall) to 0.15 t ha^{-1} yr^{-1} and 0.24 t ha^{-1} yr^{-1} in the wet 1994–1995 period (886 mm) and 1993–1994 (815 mm) hydrological years respectively.

Suspended sediment yield in the river at R-Leudon was particularly low for the dry 1991–1992 year (0.0004 t ha^{-1} yr^{-1}), equal in the next two years (0.08 t ha^{-1} yr^{-1}), despite different total annual rainfall (651 mm in 1992–1993 and 815 mm in 1993–1994), and high for the very wet 1994–1995 year (0.30 t ha^{-1} yr^{-1}). Suspended sediment yields at Epauche station were slightly higher than those calculated at R-Leudon station (0.09 t ha^{-1} yr^{-1} in 1993–1994 and 0.47 t ha^{-1} yr^{-1} in 1994–1995). The difference in suspended sediment yields are mainly related to the seasonal rainfall distribution.

Measurements of sediment loads at watershed outlets should be compared with estimates of the scale of cropland erosion and sediment transfers on hillslopes in order to link the field to the river and to estimate the sediment delivery ratio (Walling 1988, 1990), namely the proportion of eroded slope sediment reaching the river.

Assessment of soil redistribution and soil loss over the medium-term period using ^{137}Cs

Soil erosion processes are characterized by marked spatial and temporal heterogeneities (de Roo, 1991; Sutherland, 1994; Chappell, 1996). The use of the ^{137}Cs technique allows an integrated assessment of soil erosion over a long time interval (ca. 33 years) which smoothes these heterogeneities.

^{137}Cs Technique and its application to the study site

The ^{137}Cs technique, first used in the 1970s (Ritchie & McHenry, 1975), and largely developed in the 1980s

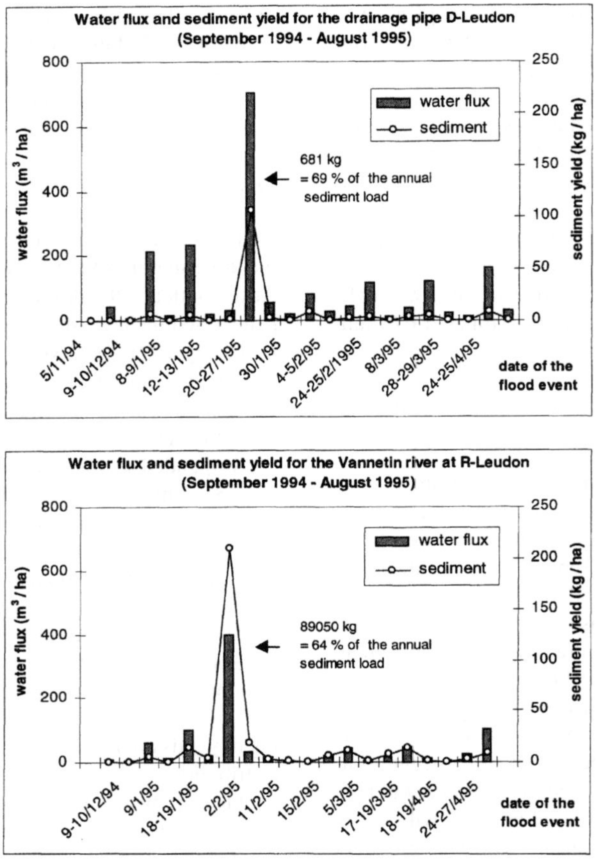

Figure 2. The sediment deliveries at D-Leudon and R-Leudon stations during the very wet September 1994–August 1995 hydrological year.

(Elliott et al., 1984; Walling & Bradley, 1988; Bernard & Laverdière, 1990), is a powerful tool which can be used to determine the spatial pattern of soil redistribution over medium-term periods on agricultural land. The technique is based on the labelling of clay particles in the topsoil at the soil surface by ^{137}Cs fallout during the atmospheric bomb test period (1954–1963) in the northern hemisphere. The subsequent ^{137}Cs redistribution over the slopes represents particle movements generated by all bio-physical processes. Comparison of ^{137}Cs loadings in different cultivated field sites with an average reference value, representing the local fallout input established on undisturbed areas (i.e. mainly flat pastures), gives qualitative information on soil redistribution (see for example Ritchie & McHenry, 1990). Quantitative estimates of soil redistribution rates can be calculated using an appropriate model to convert ^{137}Cs measurements into values of soil loss or gain (Kachanoski & De Jong, 1984; Walling & Quine, 1990; Kachanoski, 1993; Walling & He, 1997).

The technique was applied to the Champ Hardy small zero order watershed (24 ha area, located 1 km east of the Vannetin river basin). This watershed was selected as a representative area of the cultivated Brie plateau (same gentle slopes, soil units and land use) after a study of the geological and the topographical maps and of an aerial IR photograph series taken in 1992 for the participants in the PIREN Seine programme. A detailed topographic survey (2000 points) was conducted on the whole watershed using a SOKKISHA theodolite, and the map, charted by a triangulation, was overlaid on the aerial IR photograph (Figure 3). The watershed is blocked downstream by a dirt track that is perpendicular to the valley axis. A tile drainage system was installed in the 1950s under

Table 1. Maximum concentrations of suspended sediment, maximum water discharges and sediment loads during the main flood events in January 1995 at the three monitoring stations in the Vannetin watershed

January 1995	D-Leudon			R-Leudon			Epauche		
	Max. conc. (mg l^{-1})	Max. water discharge (l s^{-1})	Sediment load (t)	Max. conc. (mg l^{-1})	Max. water discharge (l s^{-1})	Sediment load (t)	Max. conc. (mg l^{-1})	Max. water discharge (l s^{-1})	Sediment load (t)
21–24	414	18.3	0.55	1009	912	66.1	1467	7572	620.3
25–26	182	12.8	0.08	690	417	11.8	846	4229	124.0
26–27	256	10.6	0.05	1382	335	11.1	1318	3408	94.3
Total			0.68			89.0			837.6

the cultivated fields of the watershed. The outlet of this drainage network is located in the thalweg, under the track, and now represents the source of the river.

The study field (7 ha), located upstream of the track, includes opposite slopes (slope angle 1%–9%) of the watershed and the thalweg. Since a preliminary study in the region revealed wide lateral variations in ^{137}Cs budgets (Sogon et al., 1994), the field was sampled (Figure 3) on a relatively dense grid framework (20×20 m) in 21–22nd September 1995, (90 cores) and 26–28th September 1996 (70 cores). This difference in sampling time is considered to have no effect on the ^{137}Cs budgets because the rainfall during this period was extremely low (442 mm). Each core (double U corer: 20 cm^2 square cross sectional area; Figure 4) was sectioned according to: 0–35 cm (tilth) and every 5 cm below the ploughed horizon down to 60 cm depth (tile drainage level). Each sample was dried, weighed, crushed, sieved (2 mm) and prepared for analysis by gamma spectrometry.

Reference ^{137}Cs inventory and Chernobyl fallout in the study area

^{137}Cs inventories of 17 cores collected with the double U corer in adjoining undisturbed pastures gave an average value of 3170±500 Bq m^{-2} (corrected to 1st January 1996). In 1986, a new ^{137}Cs input was introduced in Europe by the Chernobyl nuclear plant accident which tended to complicate the bomb fallout signal (De Roo, 1991) and was characterized by a large spatial fallout heterogeneity. In this paper, the input in the Champ Hardy area was estimated to be approximately 1000±200 Bq m^{-2} (in 1986) from independent approaches:

Table 2. Annual suspended sediment yields at the three monitoring stations in the Vannetin river basin (t ha^{-1} yr^{-1})

Year	D-Leudon	R-Leudon	Epauche	Rainfall (mm)
1991–92	0.001	0.0004	–	625
1992–93	–	0.08	–	651
1993–94	0.24	0.08	0.09	815
1994–95	0.15	0.30	0.47	886

1. ^{137}Cs activities of soil samples collected before and after the accident (25–30% ^{137}Cs activity excess);
2. direct ^{137}Cs fallout determination south of Paris related to local rainfall and recalculated to account for Brie plateau rainfall; and
3. difference between the local reference inventory (3170 Bq m^{-2}) and the Chernobyl free reference inventory, estimated at 2400 Bq m^{-2} in 1996 (Sogon et al., 1994).

This fallout, assumed to be uniform at field scale, can modify the estimates of erosion and aggradation rates, but has little effect on the soil redistribution pattern.

Mapping of soil redistribution over the slope

Two calibration models were tested to assess soil redistribution rates, considering a plough layer of 0.30 m, the simple mass balance model of Kachanoski (1993) and the mass balance model 2 of Walling & He (1998), using a proportional factor γ of 0.1 and fallout input data reconstructed from bomb fallout of 2400 Bq m^{-2} (Cambray et al., 1989) and Chernobyl fallout, to obtain 3170 Bq m^{-2} in 1996. Regardless of the model, the soil redistribution rates are approximately the same. For example, the most eroded location

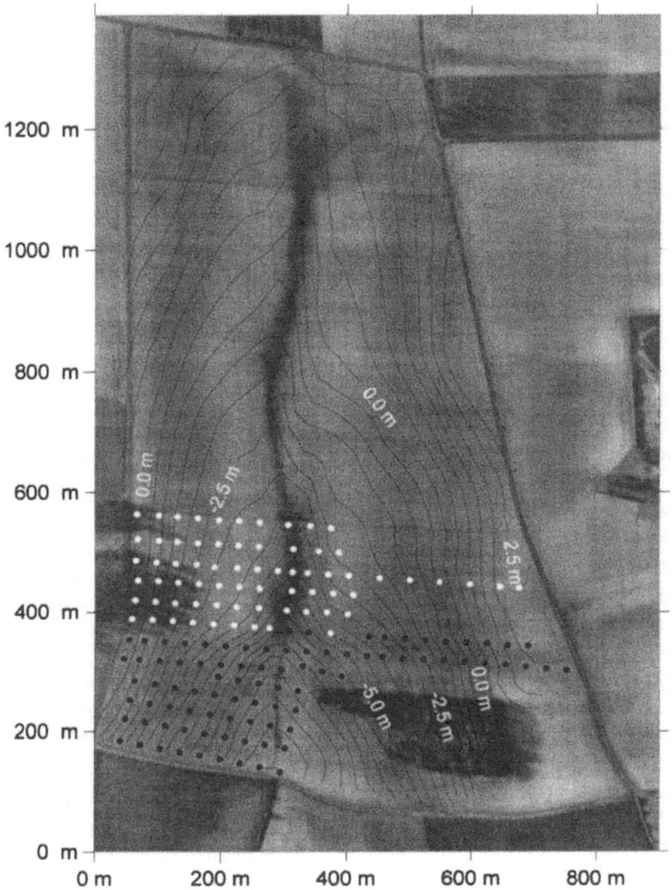

Figure 3. Aerial photography, topography and sampling network of the 'Champ Hardy' watershed (black dots: sampled in 1995, white dots: sampled in 1996).

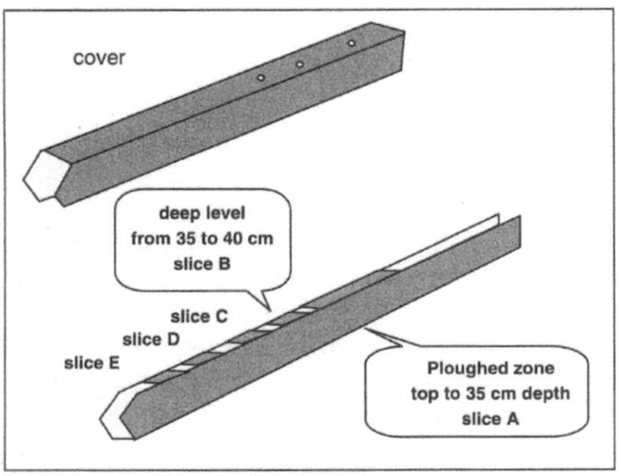

Figure 4. Double U corer.

(1700 Bq m^{-2}) gives 78 and 85 t ha^{-1} yr^{-1} with the Kachanoski model and model 2, respectively, while the location with the highest sedimentation (4587 Bq m^{-2}) gives 47 and 59 t ha^{-1} yr^{-1}, respectively. On the other hand, in regions such as our study area, where accumulation and erosion are of the same order of magnitude, the reference value is very critical. For example, the use of a reference value of 3100 Bq m^{-2} instead of 3170, causes a shift in soil redistribution rates (about 3 t ha^{-1} yr^{-1} on the whole gamut).

The iso-erosion patterns (Figure 5) show no clear evidence of alternation of erosion and aggradation areas. The results suggest that iso-erosion and iso-aggradation patterns are associated with variations in

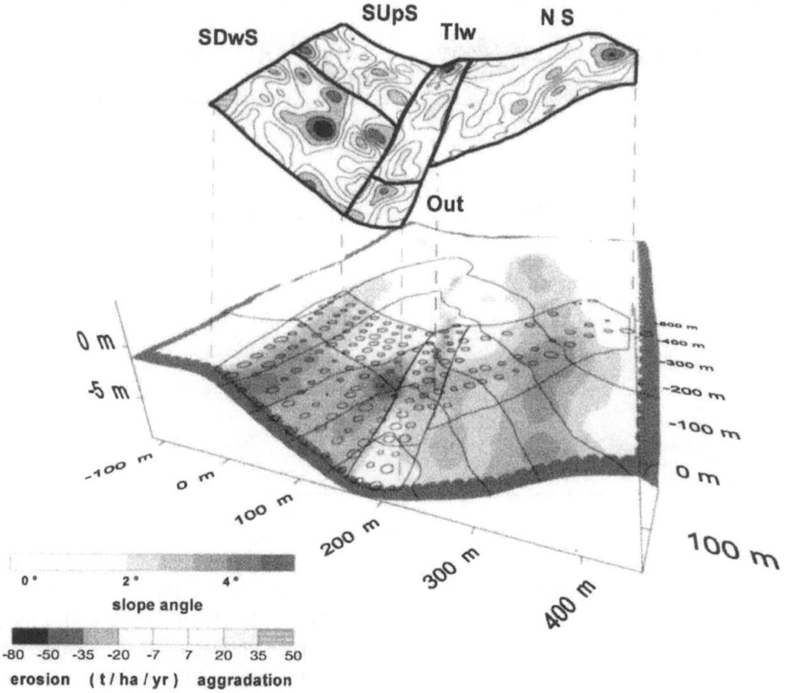

Figure 5. Estimated soil redistribution rates for each core, iso-erosion curves and landform units overlaid on the slope map of the 'Champ Hardy' watershed.

slope steepness, which is one factor of slope soil redistribution (Martz & De Jong, 1987, 1991).

The range of soil redistribution rates is wide, with very high local aggradation and erosion rates (70 t ha^{-1} yr^{-1}), and large areas with positive residual rates possibly representing short term soil particle storage. These values are similar to those documented for N. W. European loamy cropland (see for example Bollinne, 1982; Walling, 1990; Morgan, 1996).

Assessment of sediment budgets

It is not yet possible to construct a sediment budget for the Champ Hardy watershed because data are lacking for the upper part of the watershed. However, ^{137}Cs inventories can be used to assess soil redistribution and soil loss for different homogeneous microscale landform units of the study area. A first attempt was made (Figure 5) using geomorphological characteristics (slope angle, upstream/downstream location) to select five spatial units. Sediment budgets were obtained from soil loss/deposition values by calculating the kriging of the different areas corresponding to the variation range of soil redistribution (steps of 5 t ha^{-1} yr^{-1}). A more rigourous geostatistical study will be carried out in the future.

The assessment of soil redistribution on slopes for these spatial units gives contrasting results (Table 3). In three units, South Upstream Slope (SUpS), Thalweg (Tlw) and North Slope (NS), deposition exceeds erosion because of upslope and/or lateral sediment transfers. In fact, the upstream part of the study area has a radial concentrated morphology and rather low slope angles at the foot of the slope which allow sedimentation. This observation reveals the importance of storage effects with reduced slope angle. The fate of a soil particle in this situation is to advance downslope step by step rather than advance directly to the river.

The South Downstream Slope (SDwS) is mainly eroded for slopes higher than 5%, locally 9%. This area exhibits net soil loss of 7.6 t ha^{-1} yr^{-1}, corresponding to a Sediment Delivery Ratio (SDR) of 74%. The thalweg Outlet (Out) sector is the connection zone with the river. Despite the very low slope angles (2%),

Table 3. Assessment of soil redistribution, soil loss sediment delivery ratio and sediment budget in different areas of Champ Hardy watershed

Hillslope unit	Soil erosion ($t\ ha^{-1}\ yr^{-1}$)	Soil deposition ($t\ ha^{-1}\ yr^{-1}$)	Soil loss ($t\ ha^{-1}\ yr^{-1}$)	SDR	Area (ha)	Sediment budget ($t\ yr^{-1}$)
South upstream slope: **S Up S**	4.8	7.0	–	–	2.10	+4.7
Thalweg: **Tlw**	4.2	8.3	–	–	0.49	+2.0
North slope: **N S**	4.3	4.9	–	–	2.60	+1.6
South downstream slope: **S Dw S**	10.2	2.6	7.6	74%	1.75	−13.3
Outlet: **Out**	12.3	2.0	10.3	83%	0.09	−1.0

the sediment budget is negative. Net soil loss reaches 10.3 t ha^{-1} yr^{-1} for this very small area.

Discussion

Role of the drainage network

Suspended sediment yield estimates at the outlet of the underground pipe (D-Leudon) highlight the role played by modern drainage networks (perforated plastic pipes which replaced the old tile drainage) in the supply of particulate matter to the Vannetin river. This network represents an artificial extension of the hydrologic system under cultivated fields. When the drainage system functions continuously (from late autumn to early spring in 'wet' years), regular and – in some wet years – rather high sediment delivery was provided to the rivers of the Brie Plateau at every flood event. The annual sediment load depended on both total rainfall input and seasonal variations. Sediment yields were as high as 0.24 t ha^{-1} in 1993–1994 (total annual rainfall 815 mm; non-uniform monthly rainfall distribution, with 2 months (September & December) with over 100 mm) and 0.15 t ha^{-1} in 1994–1995 (annual rainfall 886 mm; more uniform monthly rainfall distribution; 4 successive months (January–April) with over 100 mm). The high sediment yield for the 1993–1994 year (0.24 t ha^{-1}) was probably caused by emptying the underground pipe network of the particulate deposits accumulated in the previous two dry years (625 mm in 1991–1992 and 651 mm in 1993–1994). At the decade time-scale, the sediment yields of the Brie Plateau drainage system imply a risk of soil impoverishment by washing away the finest particles, which are the most active and important in terms of agronomic potential. In addition, leached particles, naturally stopped in the illuvial horizon, are now removed, probably giving rise to progressive changes in the hydraulic properties of the soils.

^{137}Cs was used as a sediment tracer to identify the source of the fine particles delivered by the drainage system. ^{137}Cs activities were measured along the soil profile at different depths and compared with the ^{137}Cs activities associated with fine particles at the outlet of the D-Leudon drain pipe during flood events. ^{137}Cs activities were high in the horizon mixed by tillage (6 Bq kg^{-1}) and were negligible under this layer (less than 0.3 Bq kg^{-1}). On the other hand, ^{137}Cs activities of suspended sediments delivered by the pipe were rather high (25–30 Bq kg^{-1}), showing that they were essentially fine particles leached from the ploughed horizon. Combined with the results demonstrating soil erosion by surficial runoff, these data illustrate the vulnerability of the ploughed horizon, incurring some risk of land degradation.

Short-term space-time variations in sediment yield in the Vannetin river

A comparison of annual suspended sediment yields at the two Vannetin river stations reveals great temporal variability when compared with the amount and seasonal distribution of rainfall and the vegetation cover in the fields. High sediment yields occurred in two main situations. First, in the wet autumn or winter period without any high rainfall intensity when the soils are bare (low cover by winter cereals, ploughed fields for spring sowing). Second, in spring when a short but heavy rainfall event occurs on bare fields prepared for spring barley, maize and sunflower seeds.

As to annual sediment yields, it is interesting to note that in some years the rate of soil loss was higher under cultivated soils (D-Leudon: 6.43 ha) due to the drainage system, than in the Vannetin river at R-Leudon (4.6 km^2) and Epauche (30.6 km^2) stations (Table 2; dry year 1991–1992 and first wet year 1993–1994, after a period of relative dryness). Conversely, sediment yields in the river were high (0.30 and 0.47 t ha^{-1} yr^{-1}, respectively, at R-Leudon and Epauche stations) in a wet year (886 mm in 1994–1995), as widespread overland flow caused soil erosion and sediment delivery to rivers.

On the other hand, no variation in suspended sediment yields with space scale occurred for the two small nested Vannetin river catchments, which have the same topographic conditions (gentle slopes) and the same farming practices. In fact, the erosion rates in the Brie Plateau rivers were similar to those found in the same bioclimatic zone of north-western Europe, where suspended sediment yields in rivers are usually less than 0.5 t ha^{-1} yr^{-1} and occasionally as high as 5 t ha^{-1} yr^{-1} (Walling & Webb, 1987; Morgan, 1996).

Medium-term soil loss and sediment yield assessment

The spatial distribution of ^{137}Cs loadings converted to net soil flux in the Champ Hardy watershed showed that soil loss and sediment delivery were non-uniform. However, the partition of the study area into several spatial units, according to their geomorphological and topographic characteristics, led to more uniform patterns. There is a clear evidence that the three upstream units are dominated by aggradation, since the suspended matter inputs (both upslope and lateral) are mostly higher than the sediment outputs. At the opposite end, the other two units, south downstream slope and downstream dry thalweg, are characterized by rather high soil loss and SDR (7.6 and 10.3 t ha^{-1} yr^{-1} and 74% and 83%, respectively) and negative sediment budgets (−13.3 and −1.0 t yr^{-1}, respectively) indicating a net loss of sediment from these two landscape units.

The soil loss rates in these two units of the Brie Plateau are comparable to estimated annual erosion rates in north-western Europe (Bollinne, 1977, 1978; Morgan, 1985; Boardman, 1990) which vary, for example on cultivated fields, from 0.1 to 20 t ha^{-1} yr^{-1} in the U.K. and 3 to 30 t ha^{-1} yr^{-1} in Belgium.

Annual net soil loss from the downstream thalweg area is approximately 1.0 t (10.3 t ha^{-1} $yr^{-1}\times 0.09$ ha). This figure roughly represents the lower limit of soil loss for the whole watershed area (24 ha). On this basis, sediment exported from the fields of the watershed to the river is 0.05 t ha^{-1} yr^{-1}.

Note that this soil loss figure corresponds to the mean sediment yield over 33 years moving off a small watershed (24 ha). It lies within the range of sediment yields measured over a short period (four years) at the two Vannetin river monitoring stations (e.g. 0.0004–0.30 t ha^{-1} yr^{-1} for a 4.6 km^2 watershed at R-Leudon and up to 0.47 t ha^{-1} yr^{-1} for a 30.6 km^2 basin at Epauche).

Conclusion

In the context of Brie Plateau, which is characterized by gentle slopes, clayey-loamy soils, intensive farming and a widespread underground drainage system, the most common erosive processes in the fields are

1. leaching of the finest soil particles from the ploughed horizon to the drainage network; and
2. detachment and transport of topsoil particles by overland flow and subsequent deposition (temporary storage) on the slope as soon as the slope angle decreases.

Rill and gully erosion are rare and take place on bare soils only in extreme rainfall events. Soil redistribution and soil loss vary considerably, some landscape units being mostly sediment sinks for the particles, while others are eroded to a greater or lesser extent. On the steepest slopes, the results (representative of 33 years) show that annual soil loss can be 25–100 times higher than sediment yields in the small rivers of the Brie Plateau.

Using two suitable (but offering different time-scale information) techniques (stream monitoring and ^{137}Cs as a tracer of particle transfers), an attempt was made to relate slope processes to sediment transport in rivers, and to determine the relative importance of soil erosion, sediment delivery and sediment yield at various space-time scales. Comparisons of sediment yield estimates, using these two techniques, show reasonable agreement.

Further work is needed to study the fate of the suspended sediment load in downstream rivers, and particularly to estimate sedimentation rates in the channel and the valley floor.

Acknowledgements

The authors thank Mr Crépin, the 'Champ Hardy' farmer, for authorizing large samplings in his field and Mr Corbedanne and Mr Beaufort, for their cooperation in permitting the installation of monitoring systems. We are grateful to Prof. Y. Dewolf, and E. Joannin and M. Prat of Université Paris 7 for their help in sample collection, and A. Andrieu and S. Chambolle of LGP for soil particle analyses. Samples from experimental plots were provided by M. Tercé (INRA at Versailles) and S. Bourgeois (INA at Grignon), and data on ^{137}Cs fallout was provided by P. Bouisset of LMRE (CEA-IPSN, Orsay). The authors are grateful for comments from Phil Owens and the other two referees, which helped improved the original manuscript. This study was funded by GDR 'Analyse et Modélisation des Systèmes Fluviaux Anthropisés', commonly referred to as 'PIREN Seine'. LSCE contribution No. 180.

References

Auzet, A.-V., 1987. L'érosion des sols par l'eau dans les régions de grandes cultures: aspects agronomiques, C.E.R.E.G., Ministère de l'agriculture, Paris: 60 pp.

Bartoli, F., G. Burtin, J.-J. Royer, M. Gury, V. Gomendy, R. Philippy, T. Leviandier & R. Grafrej, 1995. Particle transport from soils to streams: spatial variability of topsoil characteristics within one soil type. Geoderma 68: 279–300.

Bernard, C. & M. R. Laverdière, 1990. Variation spatiale de l'activité en Césium-137 et ses répercussions sur l'estimation de l'érosion hydrique. Pédologie 40: 299–309.

Boardman, J., 1990. Soil erosion on the South Downs: a review. In J. Boardman, I. D. L. Foster & J. A. Dearing (eds), Soil Erosion on Agricultural Land. J. Wiley & Sons, Chichester: 87–105.

Boardman, J., I. D. L. Foster & J. A. Dearing, 1990. Soil Erosion on Agricultural Land. J. Willey & Sons, Chichester: 687 pp.

Bollinne, A., 1977. La vitesse de l'érosion sous culture en région limoneuse. Pédologie 27: 191–206.

Bollinne, A., 1978. Study of the importance of splash and wash on cultivated loamy soils of Hesbaye (Belgium). Earth Surface Processes 3: 71–84.

Bollinne, A., 1982. Etude et prévision de l'érosion des sols limoneux cultivés en moyenne Belgique. Thèse de Doctorat, Université de Liège, Belgique: 356 pp.

Chappell, A., 1996. Modelling the variation of processes in the redistribution of soil: digital terrain models and ^{137}Cs in Southwest Niger. Geomorphology 17: 249–261.

De Roo, A. P. J., 1991. The use of ^{137}Cs as a tracer in an erosion study in South Limburg (The Netherlands) and the influence of Chernobyl fallout. Hydrol. Processes 5: 215–227.

Elliott, G. L., B. L. Campbell & R. J. Loughran, 1984. Correlation of erosion and erodibility assessments using ^{137}Cs. J. Soil Conserv. 40: 24–29.

Kachanoski, R. G., 1993. Estimating soil loss from changes in soil cesium-137. Can. J. Soil Sci. 73: 629–632.

Kachanoski, R. G. & E. De Jong, 1984. Predicting the temporal relationship between soil cesium-137 and erosion rate. J. envir. Qual. 13: 301–304.

Jamagne, M., 1973. Contribution à l'étude pédogénétique des formations loessiques du Nord de la France. Thèse Doctorat, Gembloux (Belgique): 445 pp.

Martz, L. W. & E. De Jong, 1987. Using Cesium-137 to assess the variability of net soil erosion and its association with topography in a Canadian prairie landscape. Catena 14: 439–451.

Martz, L. W. & E. De Jong, 1991. Using Cesium-137 and landform classification to develop net soil erosion budget for small Canadian prairie watershed. Catena 18: 289–308.

Mazoyer, M. & L. Roudart, 1998. Histoire des agricultures du monde. Du Néolithique à la crise contemporaine, Le Seuil, Paris: 534 pp.

Morgan, R. P. C., 1985. Assessment of soil erosion risk in England and Wales. Soil Use and Management 1: 127–131.

Morgan, R. P. C., 1996. Soil Erosion and Conservation, 2nd Edn, Longman, Harlow (Essex), U.K: 198 pp.

Penven, M.-J. & T. Muxart, 1993. Assessment of soil losses in Brie (France): measuring suspended loads in rivers with a graduated monitoring network. In S. Wicherek (ed.), Farm Land Erosion in Temperate Plains Environment and Hills. Elsevier, Amsterdam: 207–220.

Penven, M.-J., C. Montier, D. Brunstein, T. Muxart & H. Sebbab, 1998. Carte de sensibilité des sols à l'érosion hydrique dans le bassin versant du Grand Morin. Géochronique 65: 18 & 37.

Poesen, J., G. Govers & D. Goossens (eds), 1995. Experimental Geomorphology and Landscape Ecosystem Changes. Catena, special issue, vol. 25: 1–4, Elsevier.

Riedel, C. E. & J. Franc de Ferrière, 1951. Les sols et les climats de la Brie. Chez les auteurs, Melun, France: 241 pp.

Ritchie, J. C. & J. R. McHenry, 1975. Fallout ^{137}Cs: a tool in conservation research. J. Soil Conserv. Res. 30: 283–286.

Ritchie, J. C. & J. R. McHenry, 1990. Application of radioactive fallout Cesium-137 for measuring soil erosion and sediment accumulation rates and patterns: a review. J. envir. Qual. 19: 215–233.

Sogon, S., T. Muxart, M.-J. Penven, D. Brunstein, V. De La Pouge & P. Bonté, 1994. Is Cs-137 a reliable tool for soil redistribution study in western Europe since Chernobyl fallout? Eur. Union Geosci. 9–13 April, Strasbourg, France (8).

Sutherland, R. A., 1994. Spatial variability of ^{137}Cs and the influence of sampling on estimates of sediment redistribution. Catena 21: 57–71.

Walling, D. E., 1988. Measuring sediment yield from river basins. In R. Lal (ed.), Soil Erosion Research Methods. Soil and Water Conservation Society, Ankeny, Iowa: 39–73.

Walling, D. E., 1990. Linking the field to the river: sediment delivery from agricultural land. In J. Boardman, I. D. L. Foster & J. A. Dearing (eds), Soil Erosion on Agricultural Land. John Wiley and Sons, Chichester, Sussex, England: 129–152.

Walling, D. E. & B. W. Webb, 1987. Material transport by the world's rivers: evolving perspectives. In J. C. Rodda & N. C. Matalas (eds), Water For The Future: Hydrology in Perspective. I.A.H.S. Publ. no. 64: 313–329.

Walling, D. E. & S. B. Bradley, 1988. The use of Caesium-137 measurements to investigate sediment delivery from cultivated areas in Devon, U.K. In Sediment Budgets, Proceedings of the Porto Alegre Symposium, December 1988, I.A.H.S. Publication no. 174: 325–335.

Walling, D. E. & T. A. Quine, 1990. Calibration of Caesium-137 measurements to provide quantitative erosion rate data. Land Degradation and Rehabilitation 2: 161–175.

Walling, D. E. & Q. He, 1997. Models for Converting ^{137}Cs Measurements to Estimates of Soil Redistribution Rates on Cultivated and Uncultivated Soils. A Contribution to the IAEA Coordinated Research Programmes on Soil Erosion (D1.50.05) and Sedimentation (F3.10.01): 29 pp.

Hydrobiologia **410:** 263–268, 1999.
J. Garnier & J.-M. Mouchel (eds), Man and River Systems.

Assessment of soil erosion in a small agricultural basin of the St. Lawrence River watershed

L. Mabit[1], C. Bernard[2], M. R. Laverdière[3] & S. Wicherek[1]
[1]*Centre de Biogéographie-Écologie, UMR-CNRS 8505, École Normale Supérieure de Fontenay-Saint-Cloud, Le Parc, 92211 Saint-Cloud Cédex, France*
[2]*Institut de recherche et de développement en agroenvironnement, 2700 rue Einstein, Sainte-Foy (Québec), Canada G1P 3W8*
[3]*Université Laval, Département des sols et de génie agroalimentaire, Sainte-Foy (Québec), Canada G1K 7P4*

Key words: erosion, ^{137}Cs, nonpoint pollution, ecosystems, mapping, St. Lawrence watershed

Abstract

In the last decades, the conservation of soil and water resources has become a major environmental concern worldwide. The authors used an anthropogenic radiotracer (^{137}Cs) to assess the magnitude and the spatial distribution of soil movement in a 80 ha agricultural watershed located in Southeastern Québec (Canada). This project is part of a larger one aimed at increasing our understanding of erosive processes, in order to minimize their impacts on soil resources and downstream hydrosystems. Using this technique, it was possible to establish a budget and a map of soil redistribution. The results could help farm operators to implement a resource conservation strategy. It was also possible to assess the environmental impact of this soil movement, by estimating the magnitude of the net output of sediment from the studied watershed. This was estimated at an average of 237 tons per year, over a period of some 33 years. The output represents 93% of the total soil displacement within the watershed boundaries. Although slightly higher, these results are consistent with those obtained, over a time period limited to 2 years, by water quality monitoring at the watershed outlet and by model simulation. This intensive study, involving more than 500 sampling points for ^{137}Cs, has yielded valuable information on the assessment and the spatial representation of erosive processes. Such information is needed to improve the management of watersheds and can help identify the areas most at risk and establish priorities for corrective actions. In this way, nonpoint pollution of downstream hydrosystems may be controlled more efficiently.

Introduction

Soil erosion and water pollution are two closely interrelated problems. Intensified agricultural practices frequently result in increased and accelerated pollutant (nutrients, pesticides, heavy metals) inputs to water bodies. Sheet and rill erosion play a major role in the spatial redistribution and the transfer of these pollutants within drainage basins, leading to environmental pollution problems. To reduce the severity of nonpoint pollution, the use of preventive measures, aimed at reducing problems at the source, has to be considered. For pollutants related to soil erosion, this means identifying the areas most vulnerable to soil losses, so soil conservation measures can be implemented in those priority areas. This way it is possible to prevent, or at least to reduce, the pollution of downstream water bodies and to minimize the need for costly remedial measures.

Direct measurement techniques have a relatively low potential for quickly identifying the sources of water erosion and quantifying their magnitude. Soil loss measurements on runoff plots or at the outlet of small watersheds have to be carried out over many years, in order to integrate the interannual variability of climate, soil and crop conditions. Besides, even if these measurements can produce reliable soil loss or sediment export figures, they cannot, however, identify the spatial origin of these sediments. Soil redistribution within the boundaries of the plots or watersheds,

Figure 1. Location map of the experimental site.

which may have as negative impacts on yields as net losses of soil, cannot be estimated either.

The Cesium-137 (^{137}Cs) technique overcomes some of these limitations. It has been used worldwide in the last decades, as reviews of the literature indicate (Ritchie & McHenry, 1990; Walling & Quine, 1992; Bernard et al., 1998a). This paper presents an example of the use of this method on a small Canadian agricultural watershed.

The experimental site

The study watershed is located in the Eastern Township region, on the Appalachian Piedmont, in Southeastern Québec (Canada). The watershed covers an area of 0.8 km^2 (80 ha). It is totally included within the boundaries of the Agriculture and Agri-Food Canada experimental farm, in Lennoxville (45° 22′ N, 71° 50′ W), some 100 km east of Montréal (Figure 1).

The climate is typically continental. The annual mean temperature is around 5.5°, with a historic high and low of +37° and −44° respectively. On average, the growing season extends from May 10–15 to October 10–15. The frost-free period covers 100–120 days. The average annual precipitation is 1004±132 mm. Around 20% of it falls as snow and covers the soil surface for 4–5 months. The topography is rolling and slope gradients locally reach 15%, especially in the lower half of the watershed. The total relief is 80 m. All the surface runoff from the cultivated fields is channelized through a 650 m long permanent stream, flowing towards an tributary of the Saint-François River. The latter discharges into the St. Lawrence River.

The soils of the watershed originate from sandy to clay lacustrine materials deposited at the end of the Wisconsín glacial era. The surface textures are dominated by sandy loams in the upper half of the watershed and by clay loams in the lower half.

The average size of the cultivated fields is 3.4 ha (Figure 2, Table 1). The land use is varied, and includes forages, small grains and corn. Generally, annual crops cover no more than 20% of the watershed area. The balance is under forage cover, either cultivated or pastured (Table 1). Soil tillage operations are not very intensive, given the land use and the plow depth seldom exceeds 20 cm.

Materials and methods

In the early 1960s, the use of radioactive tracers to measure soil erosion was proposed. Of all the possible isotopes, Cesium-137 proved to be the most successful, due to its characteristics and its environmental behavior (Ritchie & McHenry, 1990). Starting in the 1970s, it has been used by many researchers, almost worldwide, under varied agro-environmental conditions, to assess long term soil erosion rates and patterns (Walling & Quine, 1992). Soil movements are estimated from ^{137}Cs measurements by comparing the inventory of cultivated soils to that of a noneroded control site. The losses or gains of soil, over the last 30–40 years, are assumed to be more or less proportional to the loss or gain of the ^{137}Cs areal activity of the cultivated sites (Ritchie & McHenry, 1990).

Due to its many advantages, this technique was used to investigate soil movement in the Lennoxville watershed. All the cultivated fields included in the watershed were sampled on a regular 25*30 m grid, for a total of 539 points. Nine control points located within 500 m all around the watershed, under forest or permanent prairie cover, were also sampled to establish the base level of the ^{137}Cs activity of the soils of the area.

At each sampled point, undisturbed soil samples were taken to a minimum depth of 50 cm with a motorized soil corer. Three cores were taken within 1 m from each other and bulked by depth increment. In the watershed, the 0–25, 25–35 and 35–45 cm depth incre-

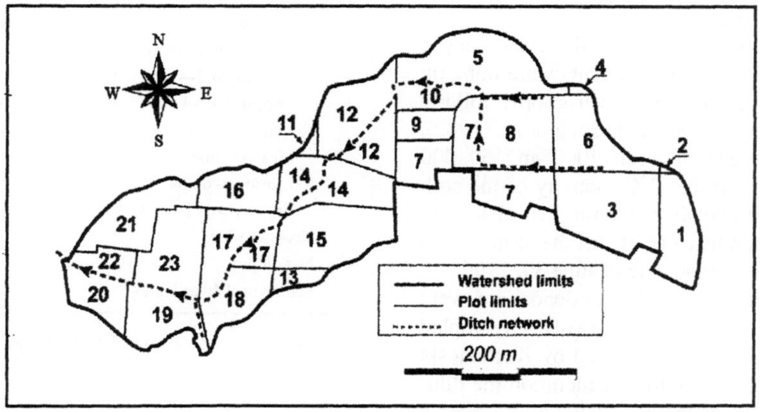

Figure 2. Map of the fields in the Lennoxville watershed.

Table 1. Land use of the fields of the experimental watershed for the 1991–1996 period

Field	Area (ha)	Land use					
		1991	1992	1993	1994	1995	1996
1	4.8	Hay	Hay	Hay	Hay	Hay	Corn
2	0.4	Hay	Hay	Barley	Barley	Corn	Barley
3	6.9	Hay	Hay	Hay	Hay	Barley	Corn
4	0.5	Hay	Hay	Hay	Hay	Hay	Barley
5	4.8	Hay	Hay	Hay	Hay	Corn	Hay
6	6.5	Barley	Barley	Hay	Hay	Hay	Hay
7	2.9	Corn	Corn	Barley	Barley	Barley	Corn
8	5.1	Corn	Barley	Barley	Barley	Barley	Corn
9	1.5	Fallow	Hay	Hay	Hay	Hay	Hay
10	1.0	Corn	Corn	Hay	Hay	Corn	Barley
11	0.4	Pasture	Hay	Hay	Corn	Barley	Hay
12	5.2	Hay	Hay	Corn	Corn	Barley	Hay
13	4.7	Hay	Hay	Hay	Hay	Hay	Pasture
14	2.8	Pasture	Pasture	Pasture	Pasture	Pasture	Pasture
15	4.8	Corn	Barley	Corn	Corn	Hay	Barley
16	2.4	Hay	Hay	Corn	Corn	Corn	Barley
17	3.6	Pasture	Pasture	Hay	Hay	Hay	Pasture
18	3.0	Pasture	Pasture	Hay	Hay	Hay	Pasture
19	3.2	Barley	Hay	Hay	Hay	Hay	Corn
20	4.7	Hay	Hay	Barley	Wheat	Barley	Hay
21	4.6	Building	Building	Building	Building	Building	Building
22	1.6	Fallow	Fallow	Fallow	Fallow	Fallow	Fallow
23	2.6	Pasture	Hay	Hay	Hay	Hay	Pasture

Source: Agriculture and Agri-Food Canada (Lennoxville research station.)

ments were kept for cesium determinations. In control sites, the 0–10, 10–20 and 20–30 cm depth increments were counted for ^{137}Cs activity.

The soil samples were dried and sieved using a 2 mm mesh. The ^{137}Cs measurements were done on the fine fractions, by gamma spectroscopy, following the procedure described by de Jong et al. (1982). Counting times ranged between 7000 and 50 000 seconds, depending on the ^{137}Cs activity of the soil. This was sufficient to obtain a counting error less than 10% at the 95% confidence level. All the cesium data were adjusted to a common date of June 30, 1996.

Net soil movements, either loss or deposition, were estimated from the total ^{137}Cs activity of each sampled point using the relationship proposed by Kachanoski (1993). This model takes into consideration the dilution of cesium fallouts in the plow layer by successive tillage operations and the enrichment of eroded sediments in ^{137}Cs, resulting from the preferential export of fine and cesium rich, particles by erosive processes. On the basis of past experiments (Bernard et al., 1992), an enrichment ratio of 2.0 was used. The resulting soil movement estimates were spatialized, by direct interpolation, using the Surfer 6.0 software package (Golden Software Inc, 1995).

The impact of the Chernobyl accident, which introduced significant new radio-cesium amounts in most European countries, could be ignored, given the low level of fallout in North America (Anspaugh et al., 1988).

Results and discussion

The base ^{137}Cs level, i.e. the residual amount left from the historic fallout in the absence of erosion or deposition and allowing for radioactive decay, was estimated at 2650 Bq m^{-2}, with a 95% confidence interval of $\pm$250 Bq m^{-2}. All soil movements in the Lennoxville basin were thus estimated by comparing the soil ^{137}Cs activity at any given point to that value. Generally, all the detected cesium was found in the first two sampled depths and none was detected in the 35–45 layer. Once transformed to soil movement, the point data were interpolated to the whole watershed. A map of the soil redistribution in the watershed, covering a 33 year period (1963–1996), was thus produced (Figure 3).

The soil redistribution estimates varied from a net loss of 20 t ha^{-1} y^{-1} to a net gain of 12 t ha^{-1} y^{-1}. The location of the areas of net soil erosion and deposition reflects the morphology and the topography of the watershed (Figure 3). Globally, most of the watershed has suffered moderate erosion and only a few hectares lost more than 6 t ha^{-1} y^{-1} of soil. The stable and depositional areas are limited in space and are generally located in low spots, in areas of convergent flow, or along the main drainage ditch (Figure 3). There is no evident relationship between estimated soil losses and land use. These suggest that most of the eroded sediments are exported out of the watershed.

Table 2. Sediment budget in the Lennoxville watershed

Stable area	15 ha
Net deposition area	4 ha
Mean deposition rate	3.2 t ha^{-1} y^{-1}
Deposition 2–4 t ha^{-1} y^{-1}	2.7 ha
Deposition >4 t ha^{-1} y^{-1}	0.5 ha
Net erosion area	61 ha
Mean erosion rate	4.1 t ha^{-1} y^{-1}
Erosion 2–6 t ha^{-1} y^{-1}	54 ha
Erosion >6 t ha^{-1} y^{-1}	7 ha
Gross erosion[a]	3.1 t ha^{-1} y^{-1}
Net output	2.9 t ha^{-1} y^{-1}
Sediment delivery ratio	93%

[a]Mean erosion averaged over the whole watershed area.

Similarly to what was done in a previous study (Bernard et al., 1998b), a sediment budget for the watershed was established from the spatialized data. The main results are presented in Table 2. The stable area, the one for which the ^{137}Cs soil activity is within the confidence interval around the mean activity of the control sites, covers 15 ha. Net deposition occured on only four ha. Some 61 ha have experienced a net soil loss, with an average of four t ha^{-1} y^{-1}. Net soil losses were between two and six t ha^{-1} y^{-1} for almost 90% of the eroded area (54 ha out of 61). Only seven ha exhibit an estimated net erosion rate greater than six t ha^{-1} y^{-1}. When averaged to the whole watershed area, the net soil loss, called gross erosion, was 3.1 t ha^{-1} y^{-1}. From the spatial integration of all soil movements, it is estimated that 237 t y^{-1} (2.9 t ha^{-1} y^{-1}) of sediment reach the outlet of the watershed, which represents 93% of the gross erosion. This ratio is frequently referred to as the sediment delivery ratio.

For an uncultivated watershed, under permanent vegetative cover, the natural erosion rate should be around 0.01 mm y^{-1} (Roose, 1993). The estimated net loss of sediment in the Lennoxville watershed represents an average loss of 0.34 mm of soil per year (on the basis of a mean bulk density of 1.1 g cm^{-3}).

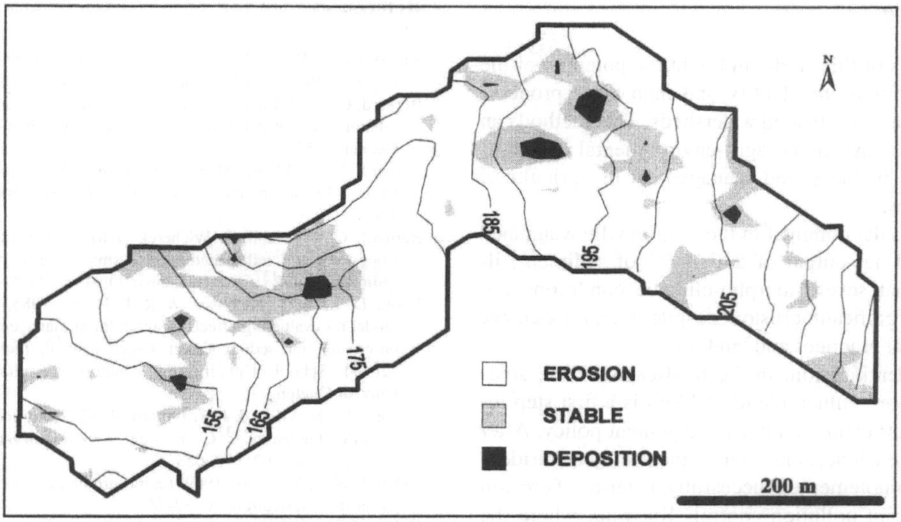

Figure 3. Soil movements map for the Lennoxville watershed.

Consequently, it seems that the erosive processes on the studied watershed are some 30 times higher than under a permanent vegetative cover. Over the last 33 years, it can be estimated that an average 1.1 cm of soil was lost from the whole watershed area.

At first glance, less erosion and net sediment production might have been expected on a watershed that is not intensively cultivated. However, the nature of the climate has to be considered. Over 20% of the annual precipitation falls as snow. In the spring time, this snow melts rapidly, under conditions of frozen or partly thawed soils and runoff and soil erosion result from this situation. Kirby & Mehuys (1987) have shown that between 10 and 88% of the annual soil loss, under bare soil conditions, may take place during this intensive runoff period. Dubé & Mailloux (1969) reported that the snowmelt period may account for a significant proportion of annual soil loss: 23% for potatoes, 40% for bare soil and 69% for cultivated hay. Rudra et al. (1989) reached similar conclusions for Ontario. The relatively pronounced topography of the watershed probably also has something to do with the estimated erosion rates and high net sediment output.

The sediment loads leaving the experimental watershed were monitored on a flow proportional basis for a two year period, stretching from June 1991 to May 1993 (Salehi, 1996). The author reported a net output of sediment of 0.6 t ha^{-1} y^{-1}, 94% of which were suspended sediments and the balance, bed load. Some 70% of the total exports occured in the spring time. The author also used the 'GAMES' model (Cook et al., 1985) to simulate the erosive behaviour of the watershed. For the two years of the simulation, the model estimated a gross erosion of 0.8 and 2.6 t ha^{-1} y^{-1} and a net sediment output of 0.5 and 0.6 t ha^{-1} y^{-1}, for a sediment delivery ratio of 63 and 23%, respectively. It was also estimated that 44% of the watershed area experienced a net soil loss.

One must be cautious when comparing erosion estimates from ^{137}Cs and other sources. Cesium data integrate the soil movements that took place since 1963, regardless of the source of these movements (natural, anthropogenic..) or of the erosive process (rainfall, snowmelt runoff, tillage operations). Also, the estimated soil movements are net figures, i.e. the result of erosion plus deposition. A 0 t ha^{-1} y^{-1} estimate does not preclude any soil movement, but rather means that erosion and deposition were of the same order of magnitude. It should also be remembered that despite the fact that estimates from ^{137}Cs measurements are reported as annual rates, most of the soil movement may have occurred in a limited period of time. So, comparing erosion rates estimated from ^{137}Cs measurements to direct monitoring data that may be limited in time, may be misleading.

Conclusion

The results of this study highlight the potential of the ^{137}Cs technique to identify and map areas prone to soil erosion in cultivated watersheds. This method can be used to investigate agro-environmental problems and assist in the sound management of agricultural ecosystems.

The results obtained in the Lennoxville watershed (estimated net output of 237 t y^{-1} of sediment) illustrate that severe morpho-climatic conditions can lead to significant erosion despite rather extensive agricultural practices and land use.

The identification, in the landscape, of the areas that are more vulnerable to soil loss is a first step towards an efficient resource management policy. After this has been done, conservation measures can be identified and implemented succesfully, in terms of erosion and nonpoint pollution control. Knowing where the zones at risk are, makes it possible to restrict the corrective actions to those areas that really need them. This way, conservation policies may be optimized, i.e. they will produce the best results at the least cost.

Finally, although this study provides valuable data on soil redistribution at the small watershed scale, its time and resource intensive character makes it difficult to duplicate the approach on many watersheds. However, important data sets like the one produced in this study could be used to develop less intensive sampling strategies and to check the accuracy of the soil movement predictions that can be obtained from them.

Acknowledgement

This study was supported by a research grant from the Conseil des Recherches en Pêche et en Agroalimentaire du Québec (CORPAQ), project No 3511.

References

Anspaugh, L. R., R. J. Catlin & M. Goldman, 1988. The global impact of the Chernobyl reactor accident. Science: 1513–1519.

Bernard, C., M. R. Laverdière & A. R. Pesant, 1992. Variabilité de la relation entre les pertes de césium et de sol par érosion hydrique. Geoderma 52: 265–277.

Bernard, C., L. Mabit, M. R. Laverdière & S. Wicherek, 1998a. Césium-137 et érosion des sols. Cahiers Agricultures 7: 179–186.

Bernard, C., L. Mabit, S. Wicherek & M. R. Laverdière, 1998b. Long-term soil redistribution in a small French watershed as estimated from ^{137}Cs data. J. envir. Qual. 27: 1178–1183.

Cook, D. J., W. T. Dickinson & R. P. Rudra, 1985. The Guelph model for evaluating effects of agricultural management systems on erosion and sedimentation. User's manual. Technical report 126–71, School of engineering, Ontario Agricultural College, Univ. of Guelph.

de Jong, E., H. Villar & J. R. Bettany, 1982. Preliminary investigations on the use of ^{137}Cs to estimate erosion in Saskatchewan. Can. J. Soil Sci. 62: 673–683.

Dubé, A. & A. Mailloux, 1969. La mesure de l'érosion à Cap-aux-Corbeaux. Agriculture 26: 32–36.

Golden Software Inc., 1995. Surfer for Windows, version 6. User's Guide. Golden, CO.

Kachanoski, R. G., 1993. Estimating soil loss from changes in soil cesium-137. Can. J. Soil Sci. 73: 629–632.

Kirby, P. C. & G. R. Mehuys, 1987. The seasonal variation of soil erosion by water in Southwestern Quebec. Can. J. Soil Sci. 67: 55–63.

Ritchie, J. C. & J. R. McHenry, 1990. Application of radioactive fallout cesium-137 for measuring soil erosion and sediment accumulation rates and patterns: a review. J. envir. Qual. 19: 215–233.

Roose, E., 1993. Erosion, a current environmental problem? The GCES, a new strategy for fighting erosion to resolve this dilemna of a growing society. In S. Wicherek (ed.), Farm Land Erosion in Temperate Plains Environment and Hills. Elsevier, Amsterdam, The Netherlands: 571–585.

Rudra, R. P., W. T. Dickinson & G. J. Wall, 1989. The role of hydrometeorological and soil conditions in soil erosion and fluvial sedimentation, Can. Agr. Eng. 31: 107–116.

Salehi, F., 1996. Mesure et modélisation des sédiments d'un petit bassin versant agricole. Thèse de doctorat, Université Laval, Québec, 121 pp.

Walling, D. E. & T. A. Quine, 1992. The use of caesium-137 measurements in soil erosion surveys. In J. Bogen, D. E. Walling & T. J. Day (eds.), Erosion and Sediment Transport Monitoring Programmes in River Basins. Publ. 210. International Association of Hydrological Sciences, Wallingford, England: 143–152.

Hydrobiologia **410:** 269–277, 1999.
J. Garnier & J.-M. Mouchel (eds), Man and River Systems.

Challenges in relating land use to sediment yield in the Upper Yangtze

D. L. Higgitt & X. X. Lu
Department of Geography, University of Durham, Science Laboratories, South Road, Durham, DH1 3LE, U.K.

Key words: land use, sediment yield, soil erosion, Yangtze River, Three Gorges Project

Abstract

Observations of the spatial pattern of sediment yields within the Upper Yangtze basin have suggested the importance of agricultural land as a major source area. The paper attempts to examine the relationship between land use and sediment yield by incorporating land cover types, derived from AVHRR data, into a multivariate analysis of factors controlling sediment yields. The extraction and analysis of catchment variables is undertaken in a GIS framework. Relationships between indices of agricultural land and sediment yields are highly scattered and the precise influence of land use on regional sediment yields remains rather inconclusive. There is a tendency for catchment areas with high proportions of arable land to exhibit high sediment yields but the individual catchments generating the highest sediment yields within the Upper Yangtze appear to relate to specific conditions such as landsliding or loess cover.

Introduction

Numerous attempts have been made to explain global and regional patterns of sediment yield in terms of climate and topography (Langbein & Schumm, 1958; Wilson, 1973; Jansen & Painter, 1974; Jansson, 1988; Milliman & Syvitski, 1992; Summerfield and Hulton, 1994). By comparison, limited attention has been paid to the effect of land use on sediment yield at a regional scale. This has mainly been due to the difficulty of obtaining detailed land use information for large areas and of isolating land use effects from other controls.

The development of global environmental databases offer the opportunity to extract spatially referenced information about catchment characteristics. Employing a Geographical Information System (GIS) framework, data can be clipped from watershed boundaries in order to examine the interaction between sediment yields and potential controlling factors. Until recently, most studies employing such techniques have concentrated at the global scale (Summerfield & Hulton, 1994; Ludwig & Probst, 1996) and have been confined to natural (i.e. non-anthropogenic) controls. Improving resolution of databases has permitted investigation of regional and intra-basin variability and many of the data sets are available commercially or can be downloaded from the Internet.

This paper describes an attempt to incorporate land cover information into a multivariate analysis of controls over sediment yields in the Upper Yangtze.

The Upper Yangtze and the Three Gorges Project

The initiation of construction of the Three Gorges Project (TGP) on the River Yangtze has drawn wide interest from environmental commentators. Dwarfing the largest existing Hydro-Electric Power (HEP) schemes, the proposed benefits of flood control, power generation and improved navigation have been scrutinised in relation to the environmental consequences of regulation. One of several forecast difficulties facing the management of TGP is the transmission of sediment into the reservoir from almost 1 million km^2 of catchment area, much of which has endured a long history of degradation. Sediment concentrations at Yichang averaging 1.2 kg m^{-3} (specific sediment yield of 510 t km^{-2} y^{-1}), rank highly compared to some of the world's other large rivers. Consequently, a number of studies have attempted to analyse the spatial and temporal distribution of sediment yields within the basin and identify the controlling factors (Gu et al., 1987; Gu & Douglas, 1989; Qian et al., 1993; Higgitt & Lu, 1996). The Upper Yangtze, referring to

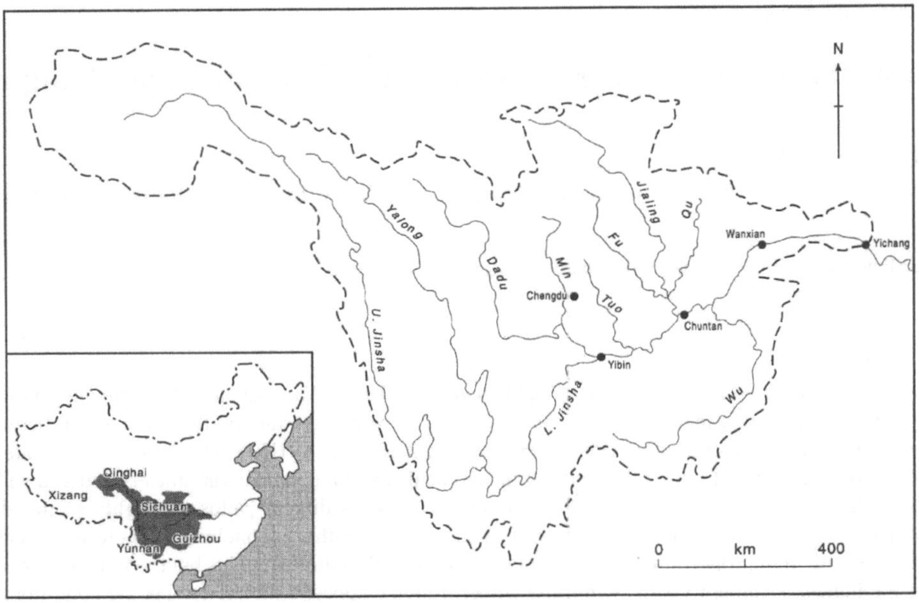

Figure 1. Principal tributaries of the Upper Yangtze.

the catchment upstream of Yichang, Hubei Province, contains diverse environments (Figure 1). Rising on the arid Qinghai-Tibet (Xizang) Plateau, the Yangtze contains a large proportion of land over 4000 m, before descending into the fertile Sichuan Basin. The Upper Jinsha, Yalong, Dadu and Min drain the mountainous west. The Tuo drains the Chengdu Plain, while the Jialing rises on the southern flanks of the Loess Plateau before traversing the agricultural lands of the Sichuan Basin where it is joined by the Fu and Qu tributaries. From the south, the Wu drains the uplands of Guizhou Province, a predominantly agricultural area underlain by karst geology.

During the last three–four decades a sequence of profound changes in agricultural practices has been superimposed on the contrasting geography of the river basin. There has been widespread deforestation and extension of agricultural land. Inventories of erosion status undertaken in the 1950s and 1980s consequently indicate a rise from 16 to 67% of land area affected by erosion in Sichuan and from 11 to 31% in Guizhou (Chen & Gao, 1988). The impetus of agricultural reform has been accompanied by a massive increase in water conservancy projects, including the construction of ponds, small check dams, ditches and headwater reservoirs, which may trap a considerable proportion of eroded sediment and prevent its transfer to the river either temporarily or permanently (Luk & Whitney, 1993). Together with larger HEP schemes, it is estimated that the by the mid-1980s the total reservoir capacity in the Upper Yangtze exceeded 16 billion m^3 (Gu et al., 1987) with some 0.3 billion m^3 capacity lost to sedimentation each year (Chen & Gao, 1988).

Despite the overwhelming evidence for the increased extent of soil erosion, particularly throughout Sichuan, a curious feature of the time series of sediment export through Yichang is the lack of a trending running mean. A more detailed examination of the temporal patterns of sediment yields and runoff (Lu & Higgitt, 1998) has demonstrated that 7.9% of incremental catchment area within the Upper Yangtze has experienced statistically significant increase in sediment yields while 2.8% has experienced decrease, much of which is downstream of major HEP schemes. Interpretation of spatial variability is complicated by the interaction of several controlling factors, year to year variation imposed by hydrological conditions and the relationship between sediment load and catchment area. Employing a specific sediment yield-catchment area regression relationship, Higgitt & Lu (1996) plotted the distributions of standardised residuals. The highest proportional contribution of sediment occurred

Table 1. Definition and source of catchment variables employed in multivariate analysis

Variable	Definition	Source
Mean elevation EV	mean of elevation of all cells within delineated catchment boundary, m	Asian 30 arcsecond DEM
Catchment area DA	catchment above gauging station, km^2	China Hydrological Yearbooks – digitized from various map sources
Basin length BL	maximum straight line distance between watershed and gauging station	Arc/Info utility
Basin relief BR	difference between maximum and minimum cell elevation within catchment boundary, m	From statistical file after clipping DEM using delineated catchment boundary
Relative Relief RR	Basin Relief/Basin Length	Derived index
Mean slope MS	mean of slope of all cells within delineated boundary, degrees	Arc/Info slope grid generation (maximum elevation changes within 9 neighbouring cells)
Agricultural Slope Index SI	see text for calculation	Arc/Info grid utility to match land cover and slope data
Population density PD	number of people km^{-2}	Asian Population database, derived from 1992 census
Precipitation PP	mean annual precipitation from 0.5 by 0.5 degree resolution, mm	Global Ecosystems Database (Version 1.0, 1992)
Runoff RO	mean annual runoff, mm	China Hydrological Year Books
Land Cover Class V1-8	regrouped land cover classifications (see Table 2)	China Land Cover Database, EROS Data Center (based on AVHRR data)

in the northern and eastern part of the Jialing tributary, on the margins of the Sichuan Basin and in tributaries of the Upper Jinsha in Yunnan Province. The generally high levels of sediment from the Jialing have suggested the importance of agricultural land as a major sediment source. Furthermore, a high proportion of the variability in sediment yields in the western tributaries can be explained in terms of topographic, climatological and hydrological factors.

By comparison, these 'natural' catchment characteristics perform less well in explaining variation in the eastern half of the basin where agricultural land prevails. It has generally been assumed (Gu & Douglas, 1989; Higgitt & Lu, 1996) that it is agricultural activity that accounts for the marked spatial pattern of sediment loads, but there have been no explicit attempts to incorporate land use information into multivariate analyses. Until recently, no high resolution information was available in the public domain. The main aim of the paper is to examine the utility of land cover information in explaining sediment yield characteristics in the Upper Yangtze and, in so doing, assess the viability of GIS approaches for modelling controls on sediment yields in large river basins.

Data acquisition and methodology

The sediment yield and runoff data are derived from 250 hydrographic stations located throughout the Upper Yangtze during the period 1956–1987. Of these, 187 stations have recorded for five or more years and 56 stations have at least 25 years of data. Stations records are derived from Chinese Hydrological Yearbooks, but information after 1987 is not in the public domain. Because of the length of time required to delineate catchment boundaries and extract catchment characteristics, multivariate analysis of factors controlling sediment yields has been limited to the 56 stations with long records plus an additional six stations from the Wu tributary with >20 years of data. Data sources and extraction procedures are described elsewhere and are summarized in Table 1.

Land cover information is derived from the China land cover database which is part of the global land cover project at EROS (Earth Resources Observation System) Data Center. The database is derived from 1 km Advanced Very High Resolution Radiometer (AVHRR) data spanning the period April 1992–March 1993, which has been combined with digital elevation data, ecosystem interpretation and county level vegetation and land use maps. There are 173 land cover types labelled in China of which 157 occur within

Table 2. Regrouping of land cover types

No	Land cover class	Categories in China land cover classification	% of ground cover for the Upper Yangtze
1	Unvegetated (water, desert, ice and snow)	1–17	3
2	Alpine meadow or steppe	18–40, 42–51, 61	25
3	Grass	52–60, 62–64, 76	9
4	Shrub/scrub woodland	65–67, 118–121, 131–135, 142	13
5	Woodland	136–137, 140–141, 143–144, 147–152, 154–156, 158–161, 163, 165–172	11
6	Cropland (mosaic <50%)	33, 68, 70, 73–74, 82–83, 124, 126–130	8
7	Cropland (mosaic >50%)	69, 71, 78–81, 85–88, 90–93, 95–96, 99–102, 106–107, 109–110, 112–113, 123	23
8	Paddy field	97–98, 103–104, 108, 111, 114–117, 122, 125	8

the Upper Yangtze catchment area. These have been simplified into eight categories (Table 2). Agricultural land is represented in three categories – paddy field, cropland mosaic >50% and cropland mosaic <50%. Two notable features about the land cover characteristics of the whole catchment are the prevalence of alpine meadow, due to the large area at high elevation and the dominance of cropland over forest cover, consistent with the recent history of deforestation.

Results

Having quantified land cover classifications for each catchment these can be incorporated into analysis of sediment yield patterns. As the influence of agriculture is of major concern, the analysis considers:

1. the relationship between the proportion of arable land, runoff and sediment yield;
2. the derivation of an arable slope index;
3. a multivariate examination of sediment yields based on factor analysis; and
4. an assessment of the influence of land use on downstream sediment delivery.

Relationship between arable land distribution, runoff and sediment yields

Any suggestion that the distribution of agricultural land explains sediment yield variability quickly evaporates when percentage agricultural land is plotted against specific sediment yield (Figure 2), regardless of whether the high density cropland cover is used (category 7) or all arable and paddy field cover combined (Category 6, 7 & 8). Many studies have resorted to grouping strategies to reduce the effects of scattering (e.g. Milliman & Syvitski, 1992; Lu & Higgitt, 1998). The 62 catchments have been grouped into four categories according to their sediment yields ($\leq$250, 251–500; 501–1000; $\geq$1001 t km^{-2} y^{-1}). The distribution of sediment yields of a particular magnitude on a plot of percentage cropland against runoff, remains confused (Figure 3a). There is a generalised grouping of data, with the lowest specific sediment yields occurring in areas of low runoff and limited arable land. The proportion of agricultural land increases in the middle sediment yield classes (Figure 3b). The highest sediment yield categories, however, are highly scattered and most do not have large coverage of arable land. Of the nine catchments in this category, the majority are relatively small catchments (<3000 km^2). Three small catchments in the Jinsha are known sites of landsliding. Three larger catchments in this category are loess-dominated areas in the upper Jialing.

An arable slope index

Percentage cropland is a crude indicator of erosion potential, not least because it is the extension of arable land onto steep slopes which is considered to be influential in the apparent rise of land area affected by serious soil erosion. Using Arc/Info functionality at a pixel level it is straightforward to produce a combined factor for the coincidence of land cover types with

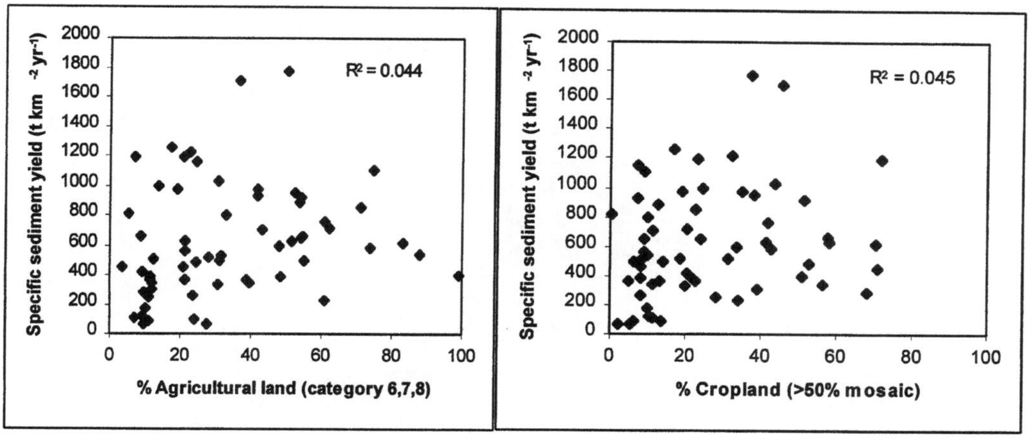

Figure 2. Scatter plots of agricultural land cover against specific sediment yield. Agricultural land can be specified as high density cropland (category 7) or by combining all arable areas (categories 6 &7) with paddy field areas (category 8).

topographic variables where

$$\mathrm{SI} = \frac{\sum_{i=1}^{n} w_i f(\theta_i)}{n}, \qquad (1)$$

where w_i=a weight for land cover type for pixel i; $f(\theta_i)$=function of mean slope (derived from DEM) for pixel i; and n=number of pixels within clipped catchment boundaries.

A simple arable sloping index (SI) has been produced where w=1 for arable mosaic land cover types (groups 6 and 7) and 0 otherwise. The function used is tan θ, which emphasises the erodibility of arable land on steep slopes. A scatter plot (Figure 4) of the arable sloping index (SI) with specific sediment yields indicates a statistically significant relationship although the degree of explanation is still low (R^2=0.14). Many of the positive residuals from the regression relationship are located in the Jialing tributary, while many of the negative residuals are from the Dadu-Min and Wu tributaries, suggesting that some refinement of weights may improve the relationship. For example the incorporation of low density arable mosaic (land cover group 6) may distort the index towards some of the marginal agricultural environments such as the Dadu-Min tributary. It should be emphasized that the derivation of slope from the 30 arc-second DEM is an extreme simplification of the topography. Improved resolution of topographic data would assist further modelling of erosion hazard.

Multivariate approaches

The results so far indicate a subtle influence of land use on sediment yields, but the incorporation of land cover types and the arable sloping index into multiple regression procedures does not improve the degree of explanation of sediment yield. An alternative strategy to multiple regression is to use factor analysis techniques which offer a means of reducing the dimensionality of the problem through examination of the correlation between variables. A simultaneous R-mode and Q-mode factor analysis was carried out following the procedure outlined by Walden & Smith (1995). R-mode methods have some similarities to principal components analysis in which the original variables are transformed into a smaller number of relatively uncorrelated factors (Zhou et al., 1983). The position of the original variables can then be plotted against any pair of the calculated factors to examine which plot closely together. The clustering of variables in factor space suggests high degrees of correlation. Q-mode methods have similarities to principal coordinates analysis, in which the aim is to identify which groups of samples plot closely together.

The distribution of variables in factor space (Figure 5) indicates several opposing characteristics which can be explained by the geography of the basin. Population density and elevation are at either end of factor 1 loadings, while drainage basin size variables and slope variables are at either end of factor 2 loadings. The distribution of the land cover classes forms a transect across the factor 1 axis, consistent with the transition

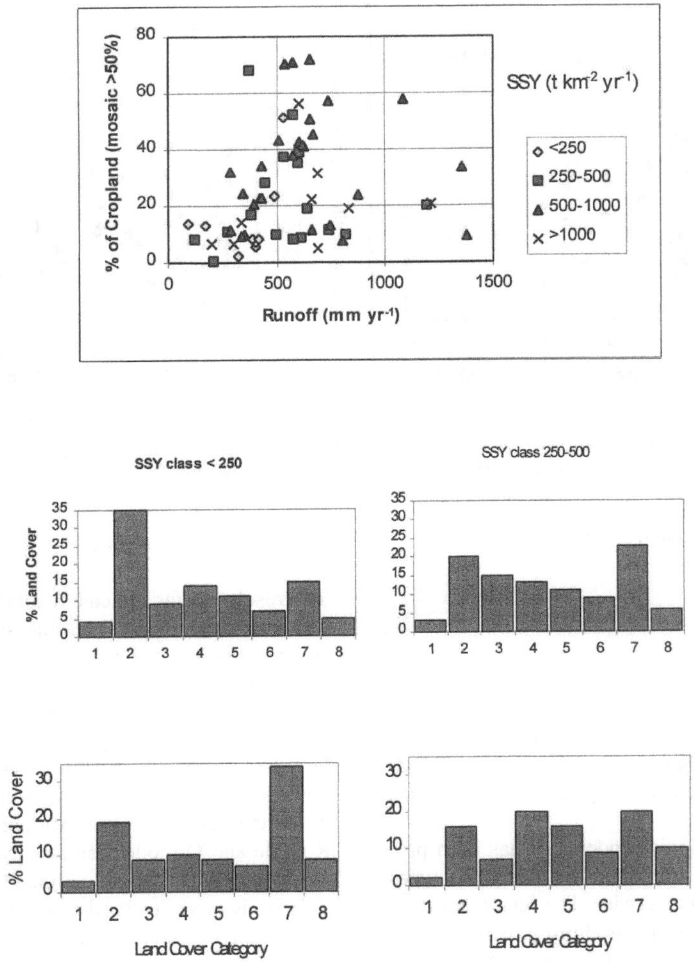

Figure 3. (a) Distribution of specific sediment yield classes in relation to cropland (>50% mosaic) and runoff . (b) Proportion of land cover types in each sediment yield class (key to land cover types in Table 2).

Table 3. The six largest Eigen-values from the 18 variables × 62 catchments matrix

Eigen-value	Total variance	Cumulative variance
4.969	27.6	27.6
3.882	21.6	49.2
2.567	14.3	63.5
1.573	8.7	72.2
1.358	7.5	79.7
0.976	5.4	85.1
others:	14.9	100.0

of alpine meadow to woodland to agricultural land with decreasing elevation and increasing population density. Eigen-values (Table 3) indicate that about half of the variability in the original data set is explained by the first two factors. Following Q-mode analysis, the position of individual catchments is plotted against these two factors. In grouping catchments by specific sediment yield classes, no clear cut link between sediment yield magnitude and controlling factors is apparent. There is some clustering of low sediment yield values towards the factor space controlled by large catchment size and high elevation variables. Sediment

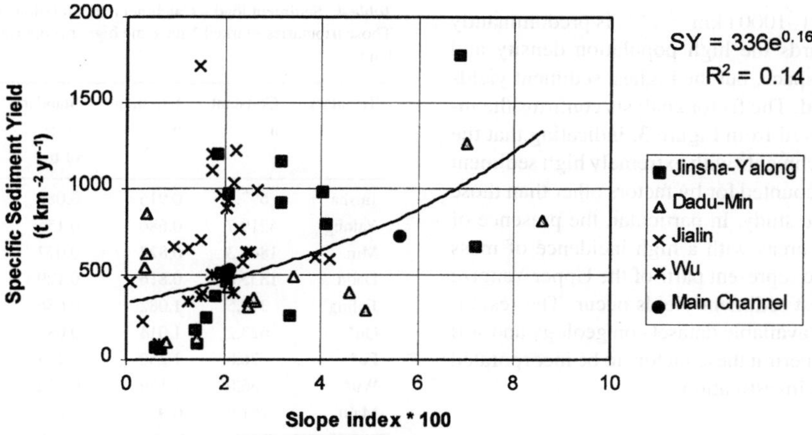

Figure 4. Relationship between arable sloping index and specific sediment yield.

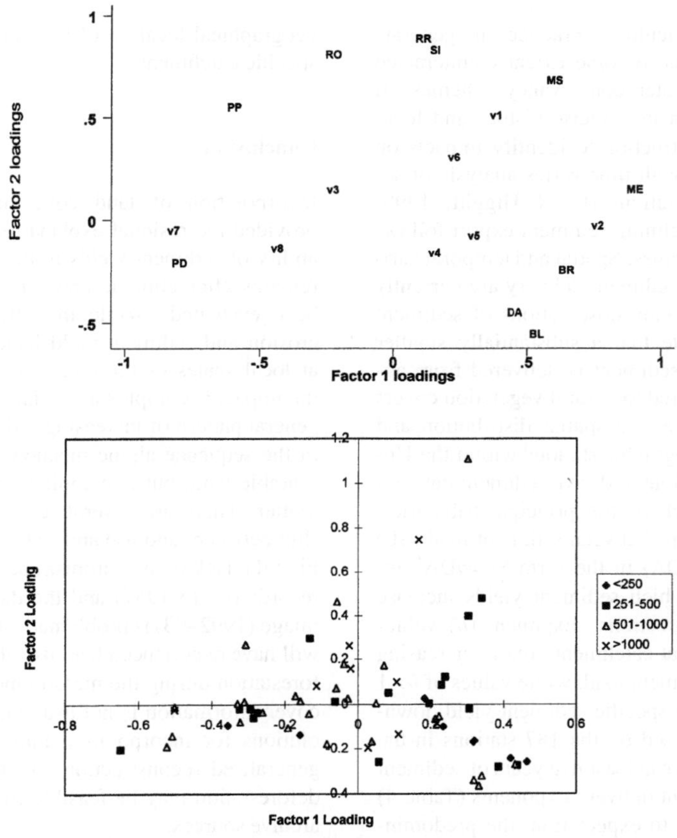

Figure 5. (a) Distribution of variables in R-mode factor space. Key for variable names in Table 2. (b) Distribution of catchments in Q-mode factor space, grouped by sediment yield class.

yield class 3 (501–1000 t km^{-2} y^{-1}) is predominantly distributed towards the high population density and cropland factor space, but the highest sediment yields are well scattered. The factor analysis confirms the interpretation derived from Figure 3, indicating that the incidence of catchments with extremely high sediment yields can be accounted for by factors other than those considered in the study. In particular, the presence of loess cover and areas with a high incidence of mass movement would represent parts of the Upper Yangtze where the highest sediment yields occur. The resolution of currently available datasets on geology and soil cover do not yet permit these factors to be incorporated into a modelling investigation.

Table 4. Sediment load – Catchment area relationships by tributary. Those tributaries marked * have the higher proportions of cultivated land

Tributary	Constant a	Exponent b	Standard error of b	n	R^2
Jinsha	697.9	0.916	0.088	39	0.744
Yalong	5219.7	0.686	0.156	12	0.659
Min	1843.3	0.827	0.077	16	0.892
Dadu	1835.0	0.816	0.129	21	0.676
Jialing*	378.5	1.082	0.086	30	0.849
Qu*	623.2	1.014	0.083	13	0.931
Fu*	76.8	1.250	0.138	12	0.891
Wu*	66.0	1.179	0.122	23	0.817
Main	751.9	0.973	0.033	18	0.982

Sediment delivery

The tendency for agricultural practices to generate sediment by erosion is to some extent counteracted by the increase in water conservancy schemes. It is difficult to establish the precise history and location of reservoir construction to identify impacts on sediment delivery, though time series analysis of selected hydrographic stations (Lu & Higgitt, 1998) have demonstrated declining sediment export following large reservoir schemes. Spatial and temporal variations in downstream sediment delivery are currently under investigation. Many observations of sediment delivery ratios indicate that a substantially smaller proportion of eroded sediment is delivered from agricultural areas compared to natural vegetation covers (Walling, 1983). Given the spatial distribution and ranging sizes of hydrographic stations within the Upper Yangtze, it is possible to derive sediment delivery characteristics for each of the principal tributaries. Regression relationships between sediment load (SL) and catchment area (DA) in the form $SL=aDA^b$ indicate the extent to which sediment yields increase in the downstream directions. Exponent (b) values less than 1 indicate that catchment area is increasing more rapidly than sediment load, while values of $b>1$ indicate an increase in specific sediment yield downstream. Analysis is based on the 187 stations in the Upper Yangtze that have at least five years of sediment load data. The sediment delivery exponents (Table 4) indicate that, contrary to expectation, the predominantly agricultural tributaries have exponent values >1, whereas the western tributaries experience a decline in specific sediment yield in larger catchments. The sediment delivery characteristics may in part reflect the geographical location of the agricultural land within specific catchments.

Conclusion

Incorporation of land cover information has not provided the residual explanation of the spatial variability of sediment yields in the Upper Yangtze, that remains after climatic and topographic factors have been evaluated. While the effects of land use on erosion and sediment yield have been demonstrated at local scales or through paired catchment studies, the impact is complex at the large catchment scale. A general pattern of increasing sediment yields emerges in the sequence alpine meadow – forest – grassland – arable land, but is complicated by high degrees of scatter. There are several reasons why the relationship between land use and sediment yields is unclear. First the lack of synchronism between sediment yield records (up to 1987) and the date of the land cover image (1992–93) is problematic as several catchments will have experienced land use change, especially deforestation during the measurement period. Past land cover information is not available at appropriate resolutions for incorporation into GIS modelling, but generalised reconstructions of the spatial pattern of deforestation may be feasible from a combination of archive sources.

Similarly, changes in the organisation of agricultural activity during a turbulent period of Chinese history will have had some impact on the land use, soil erosion and sediment delivery. Second, the classification of land cover types has involved the simplification of more than 150 types into eight groups. Chinese landscapes are frequently mosaics and the distinction between arable land, forest and paddy fields at a resolution of about 1 km causes difficulties. The planned acquisition strategy of Landsat 7 should provide a much more reliable stream of data and the operational monitoring of land cover for the whole land surface of the Earth should be feasible within the next five years. Third the grouping of hierarchical catchment areas presents some difficulties for statistical mapping analysis. Characterising the main tributaries as predominantly agricultural or non-agricultural disguises the reality that the sub-catchments are themselves heterogeneous.

Land use is one of several factors which influences the production and delivery of sediment within the Upper Yangtze. Attempts to segregate the impact of land use from other controlling factors are confounded by the interaction of variables and by scale effects. The analysis demonstrates that there is a tendency for sediment yields to increase with the proportion of agricultural land, but that the incidence of the highest rates of sediment yield can be explained by specific and varied circumstances. While the availability of global datasets in the public domain has enabled analysis of the multivariate controls within a large basin to be tackled in this study, the precise influence of land use remains somewhat inconclusive. Further improvements in both the resolution of existing datasets and the availability of information on others (e.g. geology and soils) would enable controls on sediment yields and prediction of sediment conveyance to be advanced.

References

Chen G. J. & F. H. Gao, 1988. The ecology and environment of the Yangtze and the Three Gorges and the Three Gorges Project. In Leading Group of the Research Project of Chinese Academy of Sciences, (ed.) The effect of the Yangtze Three Gorges Project on ecology and environment and countermeasures. Science Press, Beijing: 1–15 (in Chinese).

Gu, H. Y., N. S. Ai & H. L. Ma, 1987. Sediment sources and trend of sedimentation in the Three Gorges reservoir area. In (ed.) Leading Group of the Three Gorges Project Ecology and Environment Research Project, Chinese Academy of Sciences (ed.) Collected papers on ecological and environmental impact of the Three Gorges Project and countermeasures. Science Press, Beijing: 522–541 (in Chinese).

Gu, H. Y. & I. Douglas, 1989. Spatial and temporal dynamics of land degradation and fluvial erosion in the middle and upper Yangtze River basin, China. Land Degrad. Rehabil. 1: 217–235.

Higgitt, D. L. & X. X. Lu, 1996. Patterns of sediment yield in the Upper Yangtze Basin, China. In D. E. Walling & B. W. Webb (eds), Erosion and Sediment Yield: Global and Regional Perspectives (Proceedings of the Exeter Symposium), IAHS Publ. No.236. International Association of Hydrological Science, Wallingford: 205–214.

Jansen, J. M. L. & R. B. Painter, 1974. Predicting sediment yield from climate and topography. J. Hydrol. 21: 371–380.

Jansson, M. B., 1988. A global survey of sediment yield. Geog. Annaler 70A: 81–98.

Langbein, W. B. & S. A. Schumm, 1958. Yield of sediment in relation to mean annual precipitation. Trans. am. geophys. Union 39: 1076–1084.

Ludwig, W. & J. L. Probst, 1996. A global modelling of the climatic, morphological, and lithological control of river sediment discharges to the oceans. In D. E. Walling & B. W. Webb (eds), Erosion and Sediment Yield: Global and Regional Perspectives (Proceedings of the Exeter Symposium), IAHS Publ. No.236. International Association of Hydrological Science, Wallingford: 21–28.

Lu, X. X. & D. L. Higgitt, 1998. Recent changes of sediment yield in the Upper Yangtze, China. Envir. Mmt. 22: 697–709.

Luk, S. H. & J. Whitney, 1993. Unresolved issues: perspectives from China. In M. Barber & G. Ryder (eds), Damming the Three Gorges. Earthscan, London: 89–99.

Milliman, J. D. & J. P. M. Syvitski, 1992. Geomorphic/tectonic control of sediment discharge to the ocean: the importance of small mountainous rivers. J. Geol. 100: 525–544.

Qian N., R. Zhang & Z. C. Chen, 1993. Some aspects of sedimentation at the Three Gorges Project. In S-H. Luk & J. B. Whitney (eds), Megaproject: A Case Study of China's Three Gorges Project. M. E. Sharpe, Armonk: 121–160.

Summerfield, M. A. & N. J. Hulton, 1994. Natural controls of fluvial denudation rates in major world drainage basins. J. geophys. Res. 99(B7): 13,871–13,883.

Walden, J. & J. P. Smith, 1995. Factor analysis: a practical approach. In D. Maddy & J. S. Brew (eds), Statistical Modelling of Quaternary Science Data, QRA Technical Guide No. 5. Quaternary Research Association, London: 39–64.

Walling, D. E., 1983. The sediment delivery problem. J. Hydrol. 65: 209–237.

Wilson, L., 1973. Variation in mean annual sediment yield as a function of mean annual precipitation. Am. J. Sci. 273: 335–349.

Zhou, D., T. Chang & J. C. Davis, 1983. Dual extraction of R-mode and Q-mode factor solutions. J. int. Ass. math. Geol. 15: 581–606.

Hydrobiologia **410:** 279–285, 1999.
J. Garnier & J.-M. Mouchel (eds), Man and River Systems.

Bank erosion in a Danish lowland stream system

A. Laubel, L. M. Svendsen, B. Kronvang & S. E. Larsen
National Environmental Research Institute, Department of Streams and Riparian Areas, Vejlsøvej 25, DK-8600 Silkeborg, Denmark

Key words: bank erosion, stream, suspended sediment, erosion pins, land use

Abstract

This paper deals with quantification of bank erosion in an entire lowland stream system and its contribution to suspended sediment export from the stream basin. Bank erosion was measured at 33 stream reaches over a one-year period and suspended sediment transport concomitantly determined near the outlet of the stream system. There was considerable spatial variation in bank erosion within the stream system, erosion being most severe on the unprotected lower parts of the banks. No relationship could be established between stream bank erosion and land use, although erosion was lower in forest streams than in streams on land used for grazing cattle. For the stream system as a whole the erosion rate was 11 mm y^{-1} for each bank, or 0.020 m^3 per m stream reach. This corresponds to a total of 2000 tonnes bank material, of which the <63 μm fraction accounted for 20–30% (no sample pre-treatment). In comparison, total annual export of suspended sediment from the stream system amounted to 675 tonnes.

Introduction

River and stream banks often act as a key sediment source, bank erosion commonly accounting for more than 50% of catchment sediment export (e.g. Coldwell, 1957; Carson et al., 1973; Walling & Woodward, 1992; Svendsen et al., 1995). Mobilization and transport of fine sediment and associated substances such as phosphorus and heavy metals can have a harmful impact on the aquatic environment, eutrophication of lakes in Denmark and many other European countries thus being controlled by phosphorus input from wastewater and agriculture (European Environment Agency, 1994). Coarser migrating sediments such as sand can also have a harmful impact on freshwater ecology. For example, spawning grounds can be ruined because their pores become blocked and faunal species composition can be altered if a gravely substratum becomes covered with sand (Madsen, 1995). Moreover, 25% of the Danish streams that fail to meet their quality objectives do so because of poor physical conditions (Skriver et al., 1997). It is therefore important to take bank erosion into account when considering measures to improve water and habitat quality.

The bank erosion process is influenced by a number of factors. The primary natural variables are peak discharge and antecedent bank moisture (Hooke, 1979). Other important factors are freeze-thaw processes producing needle ice with incorporated bank material (Lawler, 1993b), livestock activity and farming (e.g. Trimble, 1994), bank geometry and soil type (e.g. Schumm & Thorne, 1989), channel morphology (e.g. Klaassen & Masselink, 1992), boating (e.g. Bradbury et al., 1995) and bank vegetation cover (e.g. Gray & MacDonald, 1989).

The direct measurement of short-time scale stream bank erosion in entire stream systems requires considerable field work and is subject to difficulty because of spatial variation in erosion. Since the late 1950s, bank erosion has been commonly measured using erosion pins, sometimes even at a large number of sites (e.g. Cummins & Potter, 1972; Carson et al., 1973; Bradbury et al., 1995). Erosion pins are long thin sticks that are permanently inserted into stream banks thereby allowing bank retreat to be followed relative to a fixed position on the erosion pin.

As the erosion pin method is point-specific, a high degree of systematic and random spatial variability is to be expected. Coefficients of variation often exceed 100% and great care should be taken when deriving area, volumetric or gravimetric estimates of bank retreat (Lawler, 1993a). The present study examines the importance of stream bank erosion for suspended sediment export from the Gjern stream system and the relationship between erosion rate and descriptive stream reach parameters.

Study site

The study was undertaken from mid June 1994 to mid June 1995 in the 113.5 km^2 lowland Gjern stream basin located in central Jutland, Denmark (Figure 1). The stream system comprises approximately 85 km stream channels which can be divided into four stream orders using a 1: 25 000 map (*sensu* Strahler, 1957). 33 stream reaches were randomly selected to represent the various physical, chemical and biological conditions pertaining in the watershed (Table 1). Mean stream width ranged from 1.3 m (1st order) to 4.7 m (4th order).

Most streams within the stream basin have been physically modified and channelized, mainly in connection with land reclamation for agricultural purposes, although some parts of the 4th order stream still follow the natural meandering course. Of the 33 stream reaches studied, six have meanders – one 1st order reach and five 4th order reaches. Excluding 1st order reaches, the streams are seldom dredged any longer due to the introduction of more gentle stream maintenance practice over the last decade (Madsen, 1995).

The stream basin has a gentle topography ranging from 20 to 135 m.a.s.l. Seventy seven per cent of the basin is intensively cultivated agricultural land, 14% is forest, 5% towns and villages and 4% meadows and wetlands. Precipitation and runoff in the study year were 962 mm and 363 mm, respectively, the corresponding long-term averages being 770 mm (1961–1990) and 333 mm (1974–1994). Precipitation and runoff were also higher than normal in the preceding year (1993/94). Daily mean runoff ranged from 4–51 l s^{-1} km^{-2} during the study year at the outlet of the catchment. In comparison, the median annual daily minimum in the long-term period 1974–1994 was 4 l s^{-1} km^{-2} and the median annual daily maximum was 33 l s^{-1} km^{-1}.

The soils consist of clayey till and sandy glaciofluvial deposits (Weichsel) (Svendsen et al., 1995). Their textural composition ranges from sand to loam and their swelling–shrinking potential is poor. Illite dominates the clay fraction of Danish soils with the amount of smectite ranging from a few per cent to above 30%. The bulk density of Danish soils typically ranges from 1.0 to 1.6 g cm^3 (e.g. Dalsgaard, 1983).

Methods

264 erosion pins (60 cm long; 6 mm diameter) were inserted into the bank at 33 randomly selected 30 m stream reaches in June 1994 and bank retreat measured four times during the study period (August, September, March and June). The erosion pins were inserted above the water level during the summer low-flow period. The pins were inserted horizontally (right-angled) in the case of vertical bank faces and at an angle of approximately 60–80° in the case of less step banks, the angle depending on the inclination of the bank faces. They were arranged in two groups of four, each group forming a 0.5×0.5 m square. The two groups were spaced 20 m apart, two erosion pins in each group being assigned to upper bank erosion and two to lower bank erosion. During high-flow periods in the winter season, the pins were partly or completely submerged. To ensure a reference point when measuring bank retreat, a hard plastic square (45×45 mm; 3 mm thick) with a hole in the middle was slid along each erosion pin until it was flush with the stream bank. The squares were held in place by metal clips and realigned with the bank after each measurement. The latter were made as 'topside readings' before and after moving the square, i.e. readings along the upper side of the erosion pin from the free end to the plastic square.

The bank erosion rates presented here are solely estimated from the erosion pin readings. Major bank collapses were not systematically recorded and are therefore not included in the erosion estimates. Such collapses were observed at two stream reaches, however and would have increased the erosion estimates if they had been included in the analysis. Eight erosion pins disappeared during the course of the study. As it was uncertain whether their disappearance was due to bank collapse or human disturbance, all associated data were excluded from the dataset.

Channel width, bank height and bank angle were measured at each stream reach by cross-sectional surveys. Bank vegetation density was characterized at each erosion pin group using three classes: No vegetation, sparse vegetation and dense vegetation.

Discharge and suspended sediment concentration were measured at a monitoring station located near the outlet of the stream system. Water level was re-

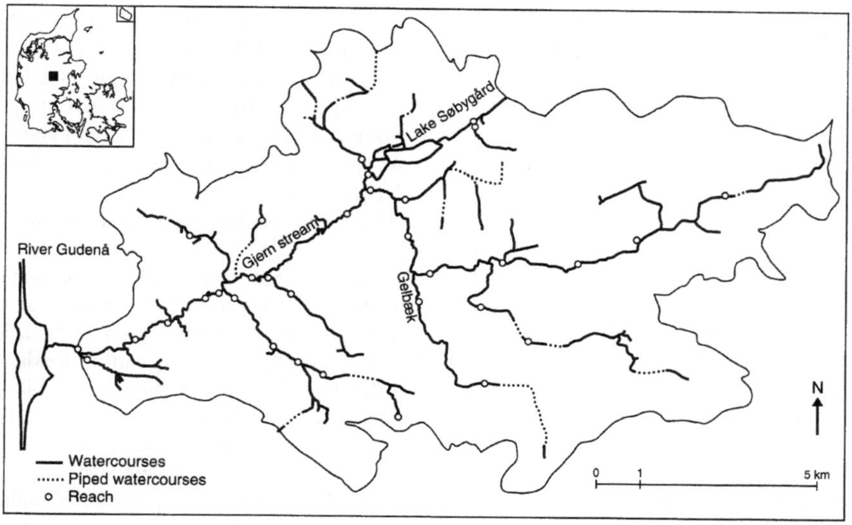

Figure 1. Diagram of the Gjern stream system and watershed showing the location of the 33 stream reaches studied.

Table 1. Descriptive characteristics of the 33 stream reaches studied in the Gjern stream basin grouped according to stream order (*sensu* Strahler, 1957). Ranges are shown in parentheses. Stream length, slope and width are from Svendsen et al. (1998), who also describe other characteristics

Stream order	Total length (km)	Number of reaches	Mean slope (%)	Mean width (m)	Mean bank height (m)	Mean bank angle (%)
1st	45.2	10	7.7 (0.5–24.9)	1.3 (0.5–2.0)	0.8 (0.6–1.3)	37 (24–61)
2nd	23.0	12	3.5 (0.4–12.1)	1.7 (0.7–2.6)	0.9 (0.5–1.4)	37 (23–51)
3rd	6.1	4	0.23 (0.1–0.8)	3.0 (1.7–3.7)	0.9 (0.6–1.1)	42 (27–48)
4th	10.1	7	1.4 (0.5–2.4)	4.7 (3.9–6.5)	1.1 (0.9–1.6)	28 (22–37)

corded continuously and calibrated against discharge by measuring the latter once fortnightly. Hourly water samples were collected using an automatic sampler and pooled into weekly or daily composite water samples. Weekly composite water samples (each covering 168 hourly samples) were analyzed except during storm-flow events, when water samples were pooled on a daily basis. Suspended sediment concentration was measured by filtering 100–500 ml of water sample through pre-combusted and pre-weighed 1.2 μm glass microfibre filters. Suspended sediment transport was calculated by summing the products of pooled weekly concentration and weekly discharge for the period in question. During storm-flow periods, suspended sediment transport was calculated for shorter, i.e. daily periods.

The bank erosion rate for the stream system as a whole (mm y^{-1} per bank side) was estimated using two different but simple methods:

1. Erosion pin method: Erosion at each pin was cumulated and the annual bank erosion rate for the stream system estimated as the geometric mean of the 256 cumulated erosion rates.

2. Stream reach method: Bank erosion at each stream reach was estimated for each inspection period as the geometric mean of the erosion pin readings and the annual bank erosion rate for each stream reach determined by cumulation. The annual erosion rate of the stream system was then calculated as the geometric mean of the annual reach bank erosion rates for the 33 reaches. This method also allows comparison between stream reaches and correla-

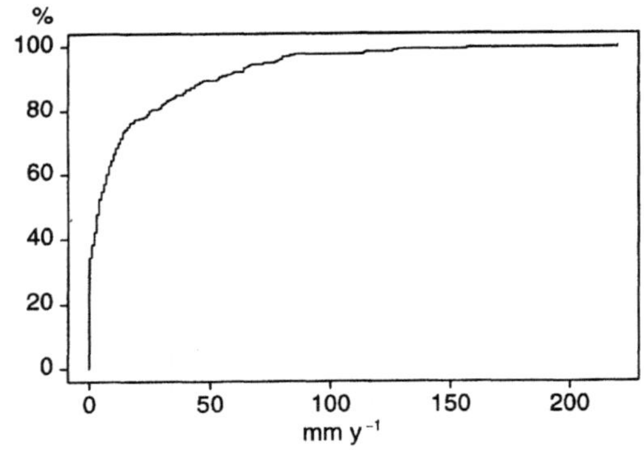

Figure 2. Cumulative curve of bank erosion measured at 256 erosion pins in the Gjern stream system during the study year 1994/95.

tion to reach-specific parameters such as land use class.

A rough comparison between bank erosion and suspended sediment export from the entire stream system was carried out based on two assumptions. Bulk density of all bank materials were conservatively estimated to be 1.2 g cm^{-3} based on national soil data (e.g. Dalsgaard, 1983). The bank material was texturally described *in situ* at all the stream reaches and bank samples representing the most common soil types were taken from three reaches for determination of grain size distribution by wet sieving (no sample pre-treatment). The <63 μm fraction of the bank material was therefore estimated to account for 20–30% (no sample pre-treatment).

Annual stream reach erosion rates were compared with reach-specific parameters such as bank height, bank angle, stream width, vegetation class, land use class and stream order by means of Spearman correlation analysis or non parametric Kruskal-Wallis test (Kruskal & Wallis, 1954).

Results and discussion

Bank erosion rates

Our results indicate that bank erosion is subject to considerable spatial variation as often reported previously (e.g. Hooke, 1979; Lawler, 1993a) (Figure 2). At 42% of the erosion pins, erosion was negligible (i.e. $\leq$2 mm) during the study period. At 90% of the pins, erosion was less than 50 mm.

The bank erosion rate for the stream system as a whole was 10 mm y^{-1} (one bank side) when estimated using the simple erosion pin method (CV=179%, n=256) and 11 mm y^{-1} when estimated using the stream reach method. At average bank height (0.9 m), 11 mm y^{-1} corresponds to 0.020 m^3 per m stream reach and 1700 m^3 for the stream system as a whole. At an estimated bulk density of the bank material of 1.2 g cm^3, this corresponds to total annual erosion of 2000 tonnes bank material from the stream system as a whole.

Mean stream reach bank erosion was significantly related (p=0.03) to stream width, but not to other reach-specific parameters such as bank height, bank angle, stream slope, vegetation density and stream order. This does not rule out the possibility that a relationship could exist at the level of individual sampling sites, however, since descriptive data were neither collected nor analyzed at that level.

Bank erosion was significantly higher from the lower part of the banks ($p<0.01$) than from the upper part of the banks (geometric mean 19 mm y^{-1} *versus* 5 mm y^{-1}; Table 2). This indicates that the banks are slowly being undercut. The lower parts of the banks are generally less protected by vegetation cover and submerged for longer periods than the upper part of the banks. The lower bank parts are therefore subject to more or less continual frictional erosion processes whereas the upper bank parts are less eroded, but can be subject to sudden bank erosion in the form of bank collapse. For the lower bank parts, the bank erosion rate was greater in the 3rd and 4th order streams than in the 1st and 2nd order streams (p=0.04; Table 2).

Although the banks generally have a low clay content and a poor swelling–shrinking potential, the possibility that seasonally expansion and contraction of the bank material may have affected some erosion pin readings can not be ruled out especially for those placed in organic soils or the more clayey soils (loam to sandy loam). Data evaluation revealed no indication of such errors, however and we consider them to be of little, if any, importance for the overall estimation of bank erosion rates and inter-reach comparisons.

Suspended sediment budget

Total suspended sediment export from the Gjern stream basin was 675 tonnes (60 kg ha^{-1}) during the study period. 80–95% of suspended sediment was below 63 μm (no sample pre-treatment) in water samples taken under different flow regimes during late summer and autumn in the Gjern stream system (Christensen, P.M., 1998). Although the grain size distribution of

Table 2. Bank erosion and lower and upper bank erosion rates (mm y^{-1}, one bank side) shown according to stream order (*sensu* Strahler, 1957). Values are the geometric means (estimation method 1) with the absolute range in parentheses. *n*=number of erosion pins. CV=coefficient of variance. Bank erosion at upper/lower bank and corresponding CV are based on half the number of erosion pins

Stream order	*n*	Bank erosion (mm y^{-1})	CV, bank erosion (%)	Upper bank erosion (mm y^{-1})	CV, upper erosion (%)	Lower bank erosion (mm y^{-1})	CV, lower erosion (%)
1	76	6 (0–114)	182	4 (0–65)	194	8 (0–114)	168
2	96	13 (0–80)	135	9 (0–30)	106	19 (0–80)	115
3	32	26 (0–128)	147	10 (0–40)	133	65 (0–128)	120
4	52	8 (0–220)	188	1 (0–115)	332	35 (0–220)	134
River system	256	10 (0–220)	179	5 (0–115)	196	19 (0–220)	147

bank material in the stream system varies from reach to reach (e.g. at three selected reaches the fraction below 63 μm was 9%, 26% and 44%, respectively) we find it likely that 20–30% of bank eroded material can be transported as suspended sediment. The majority of bank eroded material can be transported as bedload or stored within the stream system.

Given that total bank erosion was 2000 tonnes, 20–30% corresponds to 400–600 tonnes that can be transported as suspended sediment. In comparison, total export of suspended sediment amounted to 675 tonnes. A straightforward comparison thus suggests that 60–90% of the total suspended sediment export was derived from bank erosion and the remaining part from other delivery processes such as soil surface erosion and subsurface drainage. Temporary and more permanent sediment storage should also be considered, however, thus making the comparison more difficult. For example, mean suspended sediment storage rates on a riparian area in the downstream part of the stream basin during flood events in 1992/93 ranged from 0.04–3.0 kg m^2 flooded riparian area depending on the magnitude of the flooding event (Aub-Robinson et al., 1996). Moreover, based on core samples taken from the bed of the stream reaches, net annual stream bed storage rates of suspended sediment can be estimated to be 0.1–0.3 kg m^2 stream bed in 1994–95 (Svendsen et al., 1998; Svendsen, unpublished). Such storage rates suggests that bank erosion contributes less than indicated by the straightforward comparison (60–90%). This rough estimate thus suggests that bank erosion contributes more than half but less than three quarters of the total suspended sediment export and was the single major source of suspended sediment input to the stream system in 1994/95.

If part of the annual sediment load actually becomes permanently stored within a stream system, the stream channels will slowly diminish in cross-sectional area. However, as bank erosion itself widens the stream and as the water flow (especially during the major flow events, e.g. once every 20 years) erodes and resuspends deposited sediments, permanent storage is counteracted. During the study year, discharge was higher than normal during the winter season and only five storm-flow periods in the preceding 20-year period appear to have been greater than the maximum storm flow period recorded during the study year. The preceding winter, 1993/94 experienced the highest storm-flow event of the 20-year period, exhausting a major part of the retained sediment from the stream channels. Despite such storm flows, we expect net storage and slow alteration of many of the streams in the Gjern stream system to have taken place as a consequence of the more gentle stream maintenance and weed clearance introduced during the last decade, former maintenance practice having overwidend and overdeepend the stream channels (Madsen, 1995).

This relative importance of bank erosion compared to other delivery sources revealed by the present study is in concert with studies of suspended sediment delivery from artificial tile-drainage systems and soil surface erosion in the Gelbæk stream sub-catchment of the Gjern stream system during 1994/95 (Kronvang et al., 1997). Thus only 11% of the suspended sediment export from the sub-catchment was derived from artificial subsurface drainage and nothing from soil surface erosion, the remaining 89% being attributed to internal stream erosion (i.e. bank/bed erosion). Based on a detailed mass balance for phosphorus, Svendsen et al. (1995) concluded that bank and bed erosion was a major delivery source in the Gjern stream system, annual bank erosion being estimated to be up to 3 cm (both bank sides).

This is in concert with the present findings that the major part of suspended sediment export during 1994/95 derived from bank erosion, with annual erosion being approximately 11 mm from each bank side. In three other Danish lowland catchments, Hasholt (1988) also found stream bank and bed erosion to be important, accounting for 40–80% of suspended sediment export.

Anthropogenic influences on Bank erosion

The risk of bank erosion increases markedly when stream channels are widened by excavation. In the Gelbæk stream sub-catchment, 200–400 m of the upper part of the watercourse was excavated in the autumn 1994. Sediment export from this 11.6 km^2 sub-catchment increased considerably during the following months compared with periods of similar flow in other winter seasons (Kronvang et al., 1997). In another Danish lowland stream, Gelså stream, suspended sediment export increased 2.5-fold during the first year after excavation as compared with a control reach (Kronvang et al., 1998). During unstable conditions bank erosion can thus be extremely high.

In the Gjern stream system, stream channels are rarely excavated, gentle stream management having become widespread in Denmark over the last decade (Madsen, 1995). One might therefore expect other influences of man's activities to be detected, e.g. cattle grazing. No significant relationship (p=0.40) could be established between stream bank erosion and four land use classes:

1. Natural areas,
2. Crops in rotation,
3. Permanent grassland used for grazing,
4. Forest;

and there was no significant relationship between bank erosion and the width of stream border zone in areas under crop rotation. However, banks in the forest streams only eroded 2 mm y^{-1}, as compared with 14 mm y^{-1} in the grassland streams (p=0.01). It is likely that a combination of several factors is influencing bank erosion and that their separate effect therefore is difficult to prove. A larger number of erosion pins at each reach may be needed to overcome the large spatial variation in bank erosion.

Conclusions

Bank erosion in the Gjern stream system was approximately 11 mm y^{-1} per stream bank during the study year, corresponding to 0.020 m^3 per m stream reach. There was considerable spatial variation, however. Bank erosion seems to be the major source of suspended sediment input in this lowland stream system, accounting for more than half of the catchment export. Bank erosion was lower in forest streams than in streams on grassland used for grazing cattle. Slow undercutting of the banks was a general phenomenon in this stream system and vegetation cover on the banks seemed to reduce erosion.

Measurement of erosion in the Gjern stream system will be continued in the coming years in order to further evaluate the various factors affecting bank erosion, including climatic variation, as well as to investigate the influence of potential errors caused by minor seasonal expansion and contraction of the banks. During the study year reported here (1994/95), precipitation and runoff were higher than normal. Measured bank erosion may thus have been higher than usual, but its relative importance as compared with other delivery sources lower than usual.

References

Aub-Robinson, C., Ø. Falkum, C. D. Hansen, B. Kronvang & L. M. Svendsen, 1996. Retention of particulate nutrients within aquatic macrophyte patches and on the floodplain. In B. Kronvang, L. M. Svendsen & E. Sibbesen (eds), Sediment and Phosphorus. Proceedings from an international workshop in Silkeborg, Denmark, 9–12 October, 1995: 130–134.

Bradbury, J., P. Cullen, G. Dixon & M. Pemberton, 1995. Monitoring and management of stream bank erosion and natural revegetation on the lower Gordon River, Tasmanian Wilderness World Heritage Area, Australia. Envir. Man. 19: 259–272.

Carson, M. A., C. H. Taylor & B. J. Gray, 1973. Sediment production in a small Appalachian watershed during spring runoff: the Eaton Basin, 1970–1972. Can. J. Earth Sci. 10: 1707–1734.

Christensen, P. M., 1998. Partikulær sediment transport i et vandløb. Unpublished M.Sc. report. Geologisk Institut, Aarhus University, Denmark (in Danish).

Coldwell, A. E., 1957. Importance of channel erosion as a source of sediment. Trans Am. Geophys. Union. 38: 908–912.

Cummins, W. A. & H. R. Potter, 1972. Rates of erosion in the catchment area of Cropston Reservoir, Charnwood Forest, Leicestershire. Mercian Geologist 4: 149–157.

Dalsgaard, K., 1983. Eksempler på danske jordbundsprofiler. Geokompendier nr. 22 1983. Geologisk Institut, Aarhus Universitet, Denmark (in Danish).

European Environment Agency, 1994. European rivers and lakes. Assessment of their Environmental state. EEA Environmental monographs 1, European Environment Agency, Copenhagen.

Gray, D. H. & A. MacDonald, 1989. The role of vegetation in river bank erosion. In M. A. Port (ed.), Hydraulic Engineering. Proceedings of the 1989 National Conference on Hydraulic Engineering: 218–223.

Hasholt, B., 1988. On identification of sources of suspended sediment transport in small basins with special references to particulate phosphorus. In M. P. Bordas & D. E. Walling (eds), Sediment Budgets. Proceedings, Symposium, 11–15 December 1988, Porto Alegre, Brazil. IAHS, 174: 241–250.

Hooke, J. M., 1979. An analysis of the processes of river bank erosion. J. Hydrol. 42: 39–62.

Klaassen, G. J. & G. Masselink, 1992. Planform changes of a braided river with fine sand as bed and bank material. In P. Larsen & N. Eisenhauer (eds), Sediment Management. Fifth International Symposium on River Sedimentation, Karlsruhe, 1992. Proceedings 1: 459–471.

Kronvang, B., A. Laubel & R. Grant, 1997. Suspended sediment and particulate phosphorus transport and delivery pathways in an arable catchment, Gelbæk stream, Denmark. Hydrol. Proc. 11: 627–642.

Kronvang, B., L. M. Svendsen, O. Ottosen, M. B. Nielsen & L. Johansen, 1998. Remeandering of rivers: short-term implications for sediment and nutrient transport. Verh. Int. Ver. Limnol. 26: 929–934.

Kruskal, W. H. & W. A. Wallis, 1954. J. Amer. Statist. Ass. 47: 583-621, errata, ibid. 48:907–911.

Lawler, D. M., 1993a. The measurement of river bank erosion and lateral channel change: a review. Earth. Surf. Process. Landforms 18: 777–821.

Lawler, D. M., 1993b. Needle ice processes and sediment mobilization on river banks: the river Ilston, West Glamorgan, UK. J. Hydrol. 150: 81–114.

Madsen, B. L., 1995. Danish watercourses – ten years with the new watercourse act. Ministry of Environment and Energy, Denmark. Danish Environmental Protection Agency, Copenhagen.

Schumm, S. A. & C. R. Thorne, 1989. Geologic and geomorphic controls on bank erosion. In M. A. Port (ed), Hydraulic Engineering. Proceedings of the 1989 National Conference on Hydraulic Engineering: 106–111.

Skriver, J., A. Baattrup-Pedersen & S. E. Larsen, 1997. Vandløbenes miljøtilstand. In J. Windolf (ed.), Ferske Vandområder-Vandløb og Kilder. Vandmiljøplanens overvågningsprogram 1996. Report no. 214 from the National Environmental Research Institute, Denmark (in Danish): 29–45.

Strahler, A. N., 1957. Quantitative analysis of watershed geomorphology. Trans Am. Geophys. Union. 38: 913–920.

Svendsen, L. M. & B. Kronvang, 1993. Retention of nitrogen and phosphorus in a Danish lowland river system: implications for the export from the watershed. Hydrobiologia 251: 123–135.

Svendsen, L. M., B. Kronvang, P. Kristensen & P. Græsbøll, 1995. Dynamics of phosphorus compounds in a lowland river system: importance of retention and non-point sources. Hyd. Proc. 9: 119–142.

Svendsen, L. M., B. Kronvang, A. R. Laubel, S. E. Larsen & B. Andersen, 1998. Phosphorus retention in a Danish lowland river system. XXVI Congress of International Association of theoretical and applied Limnology, July 23–29 1995, Sao Paulo, Brazil. Verh. Int. Ver. Limnol. 26: 956–962.

Trimble, S. W., 1994. Erosional effects of cattle on streambanks in Tennessee, U.S.A. Earth. Surf. Process. Landforms 19: 451–464.

Walling, D. E. & J. C. Woodward, 1992. Use of radiometric fingerprints to derive information on suspended sediment sources. In J. Bogen, D.E. Walling & T.J. Day (eds), Erosion and Sediment Transport Monitoring Programmes in River Basins. Proceedings Oslo symposium. 24–28 August 1992, IAHS publ. No. 210, IAHS, Walingford: 153–164.

Hydrobiologia **410:** 287–294, 1999.
J. Garnier & J.-M. Mouchel (eds), Man and River Systems.

Flood transport of dissolved and suspended matter in the El Abid river basin (Morocco)

O. Cherifi & M. Loudiki
Department of Biology, Faculty of Sciences Semlalia, P.O. Box 2390, 40000 Marrakech, Morocco

Key words: floods, drainage basin, solute transport, particulate transport, fluxes

Abstract

This study presents the results of an analysis of the transport of suspended matters (S.M.) and nutrients (phosphorus, nitrogen and silica) by flood waters in the El Abid river basin, Morocco. The transport dynamics of these elements are considered within the context of temporal variations in discharge, and an evaluation of the loads of these elements supplied by the El Abid river basin to the Bin El Ouidane reservoir is also carried out. An analysis of the S.M. and nutrient dynamics, and their relationship to discharge has led to the identification of three types of flood flow. The first flow type defines storm flows that mainly occur from June – October, inclusive. Despite their low discharges, these are characterized by a high S.M. (up to 250 g l^{-1}). The second flow type occurs during the wet season, generally between December – May inclusive. In contrast to the first flow type, these flood flows are characterized by a relatively higher discharge (up to 204 m^3 s^{-1}) and an S.M. concentration not exceeding 50 g l^{-1}. In contrast, the total phosphorus, nitrogen and silica levels are significantly high and indicate the importance of chemical erosion during flood flows of the wet period. Finally, there are also flood flows of an intermediate type, which are characterized by a moderate discharge (67.5 m^3 s^{-1} – 94 m^3 s^{-1}) and a shorter flood duration. This third type exhibits S.M. and nutrient load characteristics common to both other flow types.

Introduction

In order to meet its ever-increasing water requirements, Morocco now has an important infrastructure of dams with a storage capacity of 14 billion m^3. However, their semi-arid location accentuates the dam management problems of siltation and eutrophication due to characteristically high suspended and solute loads of the region (Landner & Wahlgreen, 1986; Snoussi, 1988; Loudiki, 1990; Loudiki et al., 1994; Haida et al., 1996). Any attempt to estimate the fluvial load of semi-arid regions is complicated by the extreme hydrological fluctuations (Talib, 1988; Haida et al., 1996). This study will attempt to estimate the suspended and solute loads received by the Bin El Ouidane reservoir from the El Abid basin based on the summation of flood and low-water fluxes. Particular attention is given to the behaviour of suspended and dissolved matters during key hydrological periods identified during two contrasting hydrological years: 1995–1996 and 1996–1997.

Description of the El Abid river basin

The El Abid is the most important tributary in the Oum-Er-Rbia basin and drains a large proportion of the waters from the North side of the central Atlas mountains between the Saharan watershed to the South and the crest line of the Middle Atlas range to the North (Figure 1).

The El Abid basin can be sub-divided into two basins: the more important sub-basin (2635 km^2) whose main tributary is the El Abid and the sub-basin drained by the Assif Melloul (2030 km^2).

The hydrological regime is of the snow-rainfall type and can be divided into two periods. The wet winter–spring season lasts from December to May, when average monthly precipitation can reach 80 mm, with average monthly temperatures varying between 10 °C and 17 °C. The dry season from June to October with an average monthly precipitation not exceeding 5 mm and average monthly temperatures between 17 °C and 28 °C.

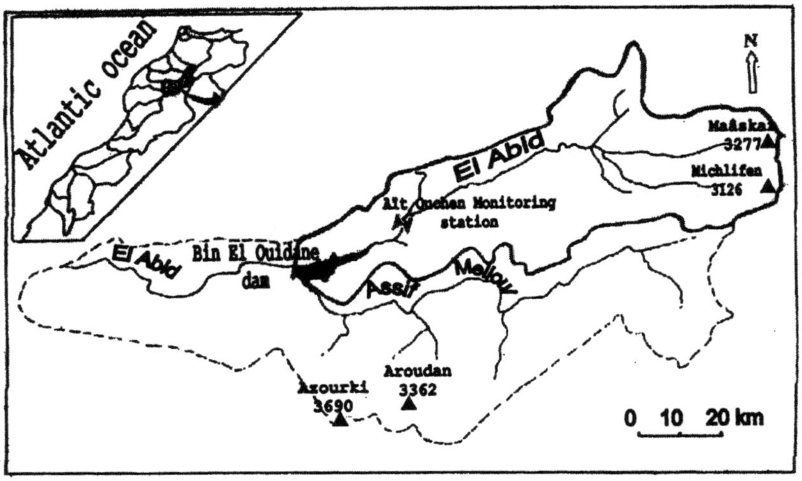

Figure 1. Location of El Abid river basin and Bin El Ouidane reservoir.

Long term hydrological records for the El Abid (1970–1988) demonstrate the annual and seasonal variability in discharge with winter and spring periods typically representing 87% of the flow.

Material and methods

An evaluation of the loads to the reservoir by the flood flows of the El Abid has been carried out for two consecutive years with contrasting climatological and hydrological characteristics. The region experienced an exceptional drought with below annual average precipitation in 1995, while 1996 was a very wet year producing high discharges in El Abid (Figure 2).

During these two hydrological cycles, it has been possible to study a total of 28 flood flows of the El Abid river between July 1995 and April 1997. Of these floods, 18 occurred in the winter–spring period when mean discharge varied between 18 $m^3 s^{-1}$ and 187 $m^3 s^{-1}$, and 10 occurred in the summer–autumn period when the mean discharge varied between 2 $m^3 s^{-1}$ and 56 $m^3 s^{-1}$. Whole water samples were taken at three hourly intervals from each flood flow at the main hydrological station (Aït Ouchen) situated at an altitude of 1200 m and 60 km up stream from the reservoir (Figure 1). During the study period, a total of 284 water samples were collected.

The suspended matters were first removed from the water by centrifuge at 6000 r.p.m. and then dried in a desiccating oven at a temperature of 105 °C. Total phosphorus (T.P.) and total dissolved phosphorus (T.D.P.) were determined after mineralization of, respectively, no-filtered and filtered samples in the autoclave in presence of sodium persulfate. The orthophosphates obtained were measured according to the A.F.N.O.R. method (T90-023) (A.F.N.O.R., 1994). Total particulate phosphorus (T.P.P.) was deducted from the concentration of T.P. and T.D.P. The total nitrogenous was measured after mineralization of the samples and transformation in nitrates (Rodier, 1984). The nitrate and ammonium were determined according to, respectively, A.F.N.O.R. method T90-013 and T90-015. Finally the silica content was determined following the procedure described by Rodier (1984).

There are many methods of determining the fluxes of matters from drainage basins: statistical, deterministic (Etchanch,1988; Kattan, 1989), interpolation and extrapolation methods (Webb et al., 1997). But in practice there are two main types which are most commonly used: the statistical probability methods based on the use of averages, and the deterministic methods based on the relationship between discharge and concentration. In both cases, the accuracy of the load estimates increases with the number of samples analysed. Also, in semi-arid areas, the frequency of sampling needs to be particularly high due to the dynamic and irregular nature of hydrological event.

In this study, annual flux (Fa) is calculated as:

$$F_a = \sum_{i=1}^{n} F^i{}_c + \sum_{j=1}^{n} F^i{}_e \qquad (1)$$

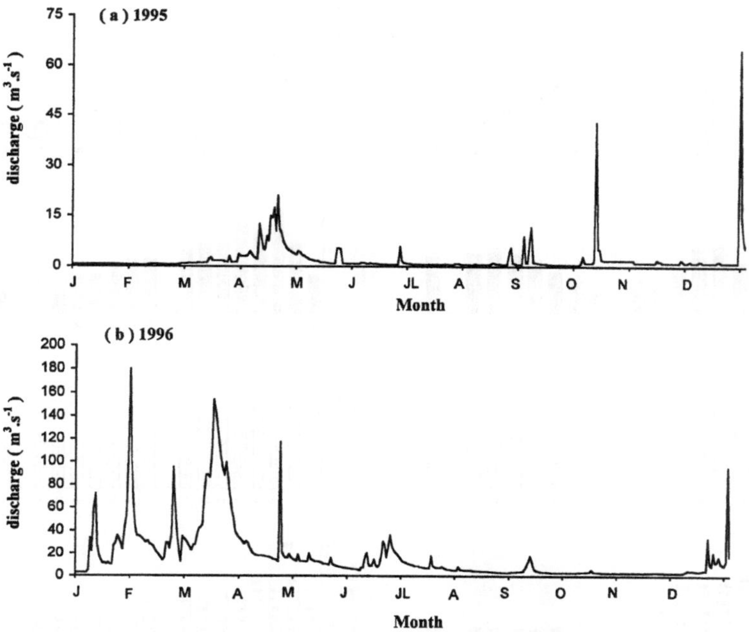

Figure 2. The mean daily discharges at Aït Ouchen hydrologic station during 1995 and 1996.

Fa: annual flux (t y^{-1}); $F^i{}_c$: flux of floods (t y^{-1}); $F^i{}_e$: low-water flux (t y^{-1}).

The evaluation of the amounts of matter at the outlet of the reservoir outside the flood periods showed that the suspended and dissolved matters are mainly carried out during flood flows (more than 90% of the measured parameters). Indeed, under semi-arid climate conditions, the major part of flow is in the form of flood events and the contributions from low-water periods can be regarded as negligible. Consequently, the annual loads correspond in practice to the flux of flood flows.

$$F_a = \sum_{i=1}^{n} F^i{}_c = \sum_{i=1}^{n} (C_{\mathrm{x}} \sum_{t=0}^{t} Qli) \quad (2)$$

$$C = \sum_{t=0}^{t} Ci \times Qli / \sum_{t=0}^{t} Qli \quad (3)$$

Qli: instantaneous discharge of flood event (m^3 s^{-1}); Ci: instantaneous concentration; C: weighted average concentrations according to the discharge (g l^{-1}); n: number of flood events; t: duration of a flood event.

Results

Dynamics of suspended and dissolved matter as a function of discharge

The analysis of discharge dynamics, precipitation type, and also the dynamics of suspended and dissolved matter as a function of discharge has allowed the identification of different types of flood flows and associated solute and suspended matter behaviour.

Summer–autumn flood flows (period 1 flood flows)
It has been possible to study 8 floods which took place in the period from June to October in 1995, all essentially due to storms. This period was characterized by a scattered precipitation with a short duration and a little volume, a large interval floods, a low peak discharge (not exceeding 28 m^3 s^{-1}), a short flood duration and hydrograph rise time (Table 1).

Figure 3 illustrates the fluctuations of the S.M. and the nutrients as a function of discharge during the two storm events occurring in 1995. An analysis of these events showed that despite their low discharge volumes (between 2 m^3 s^{-1} and 35 m^3 s^{-1}), the concentrations of S.M. reached extremely high levels (up to 250 g l^{-1}) while the flood duration are fairly short

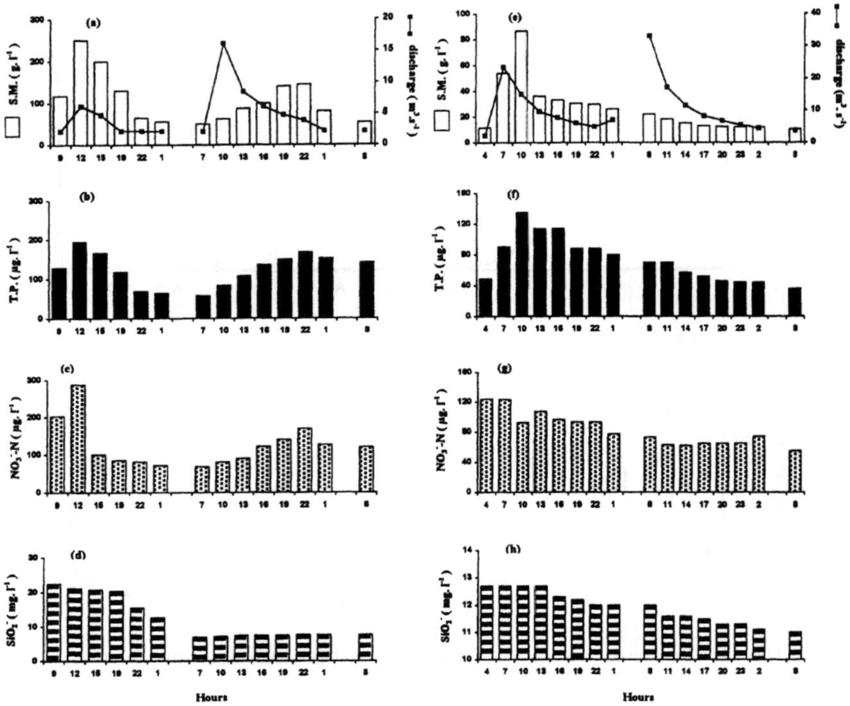

Figure 3. The variation of suspended matter (S.M.) concentration, total phosphorus and dissolved element concentration with discharge during storm events of 24th–26th August (a–d) and 4th–6th September 1995 (e–h).

Table 1. Discharge dynamics during the three periods identified

Parameters Period	Mean food interval (in day)	Mean flood duration (in hour)	Mean hydrograph rise time (in hour)	Mean peak discharge ($m^3\ s^{-1}$)	precipitation ($mm\ d^{-1}$)
Summer–autumn floods 1995	28	14	5	28	0.3 – 22.4
Winter–spring floods 1996	11	18	8	106	0.5 – 46
Intermediate period	29	12	3	55	1 – 22

(14 hours on average). Floods occurring during August and September 1995, characteristically after a long period of drought, show the highest S.M. content (80 g l^{-1} to 250 g l^{-1}). However, exhaustion effects are demonstrated by closely spaced subsequent storms as there is a net reduction in the S.M. content of the second event irrespective of its power. The total nitrogen and phosphorus concentrations follow the same trend as the S.M. content although the nitrate-nitrogen and silica concentrations are diluted during the storm flows.

Winter–spring flood flows (period 2 flood flows)

The 18 floods studied from December to May, are a response to both rainfall and meltwater from snow. This period was characterized by an important and frequent precipitation, a short interval floods, a high flood

Table 2. Fluxes (in tonnes) of the main measured parameters during the flood season at the Aït Ouchen hydrologic station

Period	S.M.	T.P.P.	T.P.	NO_3^--N	T.N.	NH_4^+-N	SiO_2^-
Summer–autumn floods 1995	2.4×10^5	2	3	1	98	4	97
Winter–spring floods 1996	9.9×10^5	167	173	310	1008	7	891
Intermediate period	4.3×10^5	103	105	16	104	1	184
Year: 1995	2.4×10^5	2	3	1	98	4	97
Year: 1996	14.2×10^5	270	277	326	1112	8	1075

Table 3. Fluxes (in kg km^{-2} y^{-1}) of the main measured parameters during the flood season at the Aït Ouchen hydrologic station

Period	S.M.	T.P.P.	T.P.	NO_3^--N	T.N.	NH_4^+-N	SiO_2^-
Summer–autumn floods 1995	91 000	0.8	1.1	0.4	37	1.5	37
Winter–spring floods 1996	376 000	63	66	118	382	2.7	338
Intermediate period	163 000	39	40	6	39	0.4	70
Year: 1995	91 000	0.8	1.1	0.4	37	1.5	37
Year: 1996	540 000	102	106	124	421	3.1	408

duration, hydrograph rise time and peak discharge (Table 1).

Figure 4 illustrates the fluctuations in the S.M. and nutrients as a function of the discharge during the period of 9 – 16 January 1996. In contrast to summer–autumn events, period 2 flood flows exhibited much higher discharge values (>90 m^3 s^{-1}), a longer base flow time and a lower S.M. content. Moreover, the fluctuations in the quantity of elements in suspension appeared to be more closely related to the temporal variation in discharge.

Intermediate period

The two floods studied in summer–autumn 1996 (24th June and 19th December) exhibits dynamics discharge, S.M. and nutrients dynamics common to both other flow types (Table 1).

Flow of suspended and dissolved matter transported to the Bin El Ouidane reservoir

Period 1 flood flows

One of the driest years of the century in Morocco was 1995 and because of this, the quantities of suspended and dissolved matter transported by the El Abid river during 1995 winter–spring period were very small.

However, the period 1 flood flows of 1995, despite their very low discharges, transported 2.4×10^5 tonnes of suspended matter, giving a specific flux of 91 t km^{-2} y^{-1}; 2.6 tonnes of total phosphorus; 97.6 tonnes of total nitrogen content; 1.1 tonnes of nitrates and 96.7 tonnes of silica during a flow time of 218 h, or 5% of the summer–autumn period (Tables 2 & 3).

Intermediate period

During 1996 (June & December), storm events, with a relatively high peak discharge, brought almost double the quantity of suspended matter, silica loads and the same quantity of total nitrogen as had been carried by the storm events in 1995. However, they transported 40 times more T.P. and 15 times more nitrates during just 101 h of flow-time, which represent 2.3% of the summer–autumn period (Table 3).

Period 2 flood flows

The period 2 flood flows of 1996, characterized by an important peak discharge, brought 1.5 times as much suspended matter, 1.6 times more total phosphorus, 18.5 times more nitrates, 7.6 times more total nitrogen and three times more silica than during the 1995+1996 Summer–Autumn combined (Table 2).

Discussion

In general, the period 1 flood flows were found to contain considerably more particles of suspended mat-

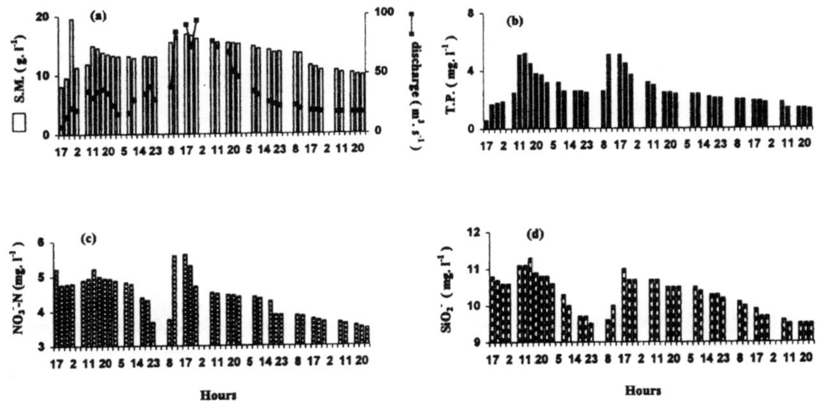

Figure 4. Variation of suspended matter (S.M.) concentration, total phosphorus and dissolved nitrate-nitrogen + silica with discharge during the flood period: 9th–16th January 1996.

Table 4. Correlation (r) between S.M. concentration with discharge and S.M. contents with total phosphorus concentration during the period of study

Date	Correlation between S.M. and discharge	Correlation between S.M. and total phosphorus
Summer–autumn floods 1995	0.1 – 0.6	0.7 – 0.9
Winter–spring floods 1996	0.3 – 0.9	0.3 – 0.9
Intermediate period	0.7 – 0.99	0.96 – 0.99

ter (up to 250 g l^{-1}) despite their weak discharge. Apart from sediment supplied by erosion and leaching, the high S.M. content observed in this period can be explained by the return to suspension of particles deposited in the river during the low-water period. Alternatively, this finding could also be explained by the severe localised erosion not only of friable and highly degraded soils but also of the banks and the bed of the river whose silty, marno-calcarious nature make it soft and easily movable during storms and rainfall. This fact has often been observed by other workers (Probst & Bazerbachi, 1986; Dorioz et al., 1989; Loudiki et al., 1994; Campy & Meybeck, 1995). In addition, all these phenomena do not necessarily coincide with the peak discharge and probably explain that 60% of storm floods, which commonly had a low discharge (below 20 m^3 s^{-1}) presented no significant correlation between discharge and concentration. This result has also been observed by certain authors (Landner & Wahlgreen, 1986; Probst & Bazerbachi, 1986) who have stated that weak discharges (<10 m^3 s^{-1}) do not allow any significant correlation between discharge and concentration to be recorded. But all the floods that occurred during this period present a positive correlation between S.M. and total phosphorus concentration. However, some floods taking place at an intermediate point in time between the true dry and wet seasons showed a positive correlation between the discharge and the S.M. content (r >0.7) (Table 4).

While the concentration of S.M. in the period 2 flood flows was very much lower than those reported for storm events, in the summer–autumn period, the total phosphorus, nitrogen, nitrate and silica concentrations were considerably higher. This is due to the importance of leaching and erosion that occurred during high-water periods. 78% of period 2 flood flows showed a positive correlation between discharge and the S.M. concentration and 67% of this floods showed also a positive correlation between S.M. and total phosphorus contents (Table 4).

The transport of solid matter by the El Abid river (expressed in t km^{-2} y^{-1}) during the period studied

Table 5. Sediment discharges calculated in this study for the El Abid river basin and those obtained by the different authors for other rivers in Morocco (t km^{-2} y^{-1})

Parameters Basin of:	S.M.	T.P.	T.N.	NO_3^--N	NH_4^+-N	Reference
El Abid river (1996)	540	0.1	0.42	0.12	0.003	This study
Lakhdar river (1987–88)	957	0.18	4.2	3.04	0.09	*a*
Bou Regreg (1978–79)	425	0.56	——	0.53	——	*b*
Sebou	929	——	——	——	——	*c,d,e*
Souss	262	——	——	——	——	*c,d,e*
Moulouya	240	——	——	——	——	*f*
Tensift	200	——	——	——	——	*g*

[a]Loudiki (1990); [b]Landner & Wahlgreen (1986); [c]Snoussi (1988), [d]Snoussi et al. (1989); [e]Snoussi et al. (1990); [f]Talib (1988); [g]Haida et al. (1996).

is fairly high in comparison with values recorded in basins of other semi-arid rivers: For example, the Bou Regreg (Landner & Wahlgreen, 1986), Tensift and N'Fis (Haida et al., 1996) and the upper reaches of La Garonne (Probst & Bazerbachi, 1986). Consequently, we notice important losts in S.M., which accentuates the Bin El Ouidane reservoir's silting rate (1.2×10^6 m^3 for 1996). This result is lower than that found by Lahlou (1986) (1.5×10^6 m^3 for the the two sub-basins El Abid and Assif Melloul). Conversely, the quantity of matter transported is smaller when compared to that recorded for the Lakhdar and Bernat basins (Loudiki et al., 1994) (Table 5).

In summer–autumn period, the chemical erosion (total phosphorus, total nitrogenous and silica) represent only 0.1% of mechanical erosion. In winter–spring period, the chemical erosion is, still, very low; it represents 0.2% of mechanical one's, but, in comparison with the chemical erosion of period 1 flood flows, it represents more than 90%. The fluxes of orthophosphorus represent only 0.6% of total phosphorus indicating that dissolved phosphorus fraction is negligible. This results agrees with those obtained from basins in arid and semi-arid regions (Landner & Wahlgreen, 1986; Loudiki, 1990). In temperate zone basins, the fluxes of dissolved material are clearly higher than those of solid matter, according to the classification set up by Meybeck (1976).

However, the low lost in nutritive elements has an effect on the water's quality of Bin El Ouidane's reservoir. Indeed; the use of O.C.D.E. diagram (1982), based on water's mean transparency (3.26 m) and mean concentration of total phosphorus (29.8 μg/l), shows that Bin El Ouidane is a mesotrophic lake. Yet, the measure of mean phytoplankton biomass (1.7 μg/l of total Chlorophyll) shows that Bin El Ouidane reservoir is an oligotrophic one. The discard observed between the trophic category is explained by the importance of particular phosphorus form (non bio-available) which principally linked to the solid load.

Conclusion

The loads measured at the outlet of the Bin El Ouidane reservoir are made up almost entirely of matter brought in by flood flows (more than 90% of the total matter brought in) and the dynamic of hydrochemical and biological components in the lake showed a high influence of loads from the basin on the system's trophic evolution. The trophic category of the lake differs, depending on the parameter and the model considered. If we approach the floods separately, we find that more than 60% of them, occurring during the study's period, exhibit a positive correlation between the discharge and S.M. and between total phosphorus and S.M. concentration.

The concentrations of S.M. in the winter–spring flood flows were very much lower than those reported for the summer–autumn period, but the total phosphorus, nitrogen, nitrate and silica concentrations were considerably higher. This latter finding is probably due to the increased importance of leaching and erosion that occurred during high-water periods.

Acknowledgements

This study was made possible by the provision of logistical support from the Provincial Office of Public Works in Azilal, the Regional Office of the National

Potable Water Board (O.N.E.P.) in Beni Mellal and from the O.N.E.P. central laboratory in Rabat.

References

A.F.N.O.R., 1994. Recueil des Normes Françaises: Qualité de l'eau, Environnement. Paris.

Campy, M. & Meybeck, M., 1995. Les sédiments lacustres. In R. Pourriot & M. Meybeck 1995 (eds), limnologie générale, Edition Masson, Ecologie générale 25: 185–226.

Dorioz, J. M., E. Pilleboue & A. Fehri, 1989. Dynamique du phosphore dans les bassins versants: importance des phénomènes de rétention dans les sédiments. Wat. Res. 23: 147–158.

Etchanchu, D., 1988. Géochimie et hydrologie des eaux fluviatiles des bassins de la Moselle et de la Mossing. Transports dissous et particulaires. Cycle biogéochimique des éléments. Thèse 3e cycle. Strasbourg: 220 pp.

Haida, S., M. Snoussi, C. Latouche & J. L. Probst, 1996. Géodynamique actuelle du bassin versant de l'oued Tensift, MAROC): érosion et bilan des transports solides fluviaux. Sci. Géol. Bull. 49: 1–4.

Kattan, Z., 1989. Géochimie des eaux du bassin de la Garonne. Transfert de matières dissoutes et particulaires vers l'océan Atlantique. Thèse 3e cycle, Toulouse. 156pp.

Lahlou, A., 1986. Envasement des barrages au Maroc. Societé Marocaine d'édition Wallada. 286 pp.

Landner, L. & U. Wahlgreen, 1986. L'eutrophisation des lacs réservoirs en climat chaud. O.M.S., P.N.UD. 170 pp.

Loudiki, M., 1990. Etude limnologique d'un hydrosystème récemment aménagé dans la région de Marrakech (Maroc). Doct. d'etat, Univ. Aix-Marseille III. 353 pp.

Loudiki, M., R. Bouterfas & K. Mouhri, 1994. Dynamique et évaluation des apports du bassin versant et de l'état trophique du lac réservoir Hassan I (Maroc). Rev. Fac. Sci. Marrakech. 8: 45–57.

Meybeck, M., 1976. Total minéral dissolved transport by world major rivers. Bull. Sci. Hydrol. XXI: 265–284.

O.C.D.E., 1982. Eutrophisation des eaux: méthodes de surveillance, d'évaluation et de lutte. O.C.D.E., Paris. 165 pp.

Probst, J. L. & J. L. Bazerbachi, 1986. Transport en solution et en suspension par la Garonne supérieure. Sci. Céol. 39: 79–98, Strasbourg.

Rodier, J., 1984. L'analyse de l'eau. Edit. Dunod. 1365 pp.

Snoussi, M., 1988. Nature, estimation et comparaison des flux de matières issus des bassins versants de l'Adour (France) du Sebou, de l'Oum-Er-Rbia et du Souss (Maroc). Impact du climat sur les apports fluviatiles à l'océan. Mém. Inst. Géol. Bassin d'Aquitaine, Bordeaux. 22: 459.

Snoussi, M., J. M. Jouanneau & C. Latouche, 1989. Impact du climat sur les apports fluviatiles. Etude Comparatives des flux de l'Adour (France), du Sebou et du Souss (Maroc). Bull. Inst. Géol. Bassin d'Aquitaine, Bordeaux. 46: 119–126.

Talib, E. H., 1988. Envasement du barrage Mohamed V. Dipl. Ing. d'Etat, option hydraulique. 80p.+ annexe.

Webb, B. W., J. M. Phillips, D. E. Walling, I. G. Littlewood, C. D. Watts & G. J. L. Leeks, 1997. Load estimation methodologies for British rivers and their relevance to the LOIS RAS (R) programme. Sci. Total Envir. 194: 379–389.

Hydrobiologia **410:** 295–306, 1999.
J. Garnier & J.-M. Mouchel (eds), Man and River Systems.

Spatial and temporal variability of Total Suspended Solids in the Seine basin

M. Meybeck[1], Z. Idlafkih[1], N. Fauchon[2] & V. Andreassian[3]
[1]*UMR Sisyphe CNRS/UPMC, place Jussieu, 75252 Paris, France*
[2]*CGE–Vivendi, 32 Place Ronde, quartier Valmy, 92982 Paris La Défense cedex 82, France*
[3]*Cemagref, Parc de Tourvoie, 92164 Antony France*

Key words: Total Suspended Solids, space variability, statistical distribution, Seine river

Abstract

We analyze the TSS distribution over the whole Seine basin (67 500 km^2, stream order 8) from: 1. a set of 236 stations sampled quaterly to bimonthly from the french national water quality monitoring network over the 1971–1997 period, 2. four stations sampled daily over 3 and 4 years located on stream orders 1, 5, 6 and 8, 3. a weekly survey of the exceptional 1994/95 high water stage at 4 stations upstream and downstream of Greater Paris (10 M. people). Due to very low relief and even rainfall distribution over the year, the Seine and its river network from order 3 to 8, are characterized by very low TSS: 79% of medians (C50%) are between 8 and 32 mg l^{-1} and maximum TSS barely reach 300 mg l^{-1}. Due to similar relief distributions, runoff patterns and geology in all sub-basins, major tributaries have near-identical long-term TSS regimes and seasonal variations during the 1994/95 flood stage. Second order TSS variations are linked to lithology: streams draining argillaceous and marl terrains are up to 3 and 4 times more turbid than those draining limestones and chalks. Basin size was also tested: the TSS range (quantiles C1% to C99%) decreases from order 1 to 8, and quantiles levels C10% to C75% double from order 3 to 8. Human impacts on TSS levels are quite limited: Greater Paris influence on longitudinal profiles is not observed; in periurban streams, where population density reach 1000 p km^{-2}, TSS levels are twice those observed in rural conditions (40 p km^{-2}): C75% are 32 ± 12 and 17.5 ± 9 mg l^{-1}, respectively. In orders 6 to 8, the lower TSS quantiles (C10% and C25%) are higher than in orders 3 – 5, this can be attributed to eutrophication and/or to an important fluvial traffic. No significant trend was observed on the TSS distributions at the river mouth from 1971 to 1997. Comparison with a previous daily survey in 1863–1866 showed present marked decrease of average TSS and TSS yearly range attributed mostly to locks.

Introduction

Total suspended solids (TSS), one of the most commonly surveyed variables in river waters (Meybeck, 1986), are used in many fields as freshwater ecology (Billen et al., 1994; Petts & Calow, 1996), geomorphology and coastal ocean science (Sundborg & Rapp, 1986; Milliman & Syvitski, 1992), environmental sciences (Chapman, 1992). For aquatic ecologists, TSS regulates the water turbidity which limits the light penetration and the photosynthesis, it is also the prime food supply of filter feeders. For physical geographers, they are regarded as a means to calculate erosion budgets over river basins while for oceanographers TSS are a major source of material to the coastal zone including some major pollutants as metals and some toxic organics. TSS are more and more considered as a key variable by water quality specialists who promote the combination of TSS budgets and particulates analyses for organic carbon, nutrients, metals, PCBs, PAHs etc (Ongley, 1992; Thomas & Meybeck, 1992; Horowitz, 1995). In some countries such as France, TSS are also a regulatory water quality criteria, used in pollution taxes establisment, and are monitored at each water quality station. Yet studies of TSS distribution in time and space are uncommon at the whole basin scale: most scientists and engineers focus on more 'appealing' variables as nutrients, metals or toxics, and the use of TSS data is only minor, generally to set up the general water quality background at the station.

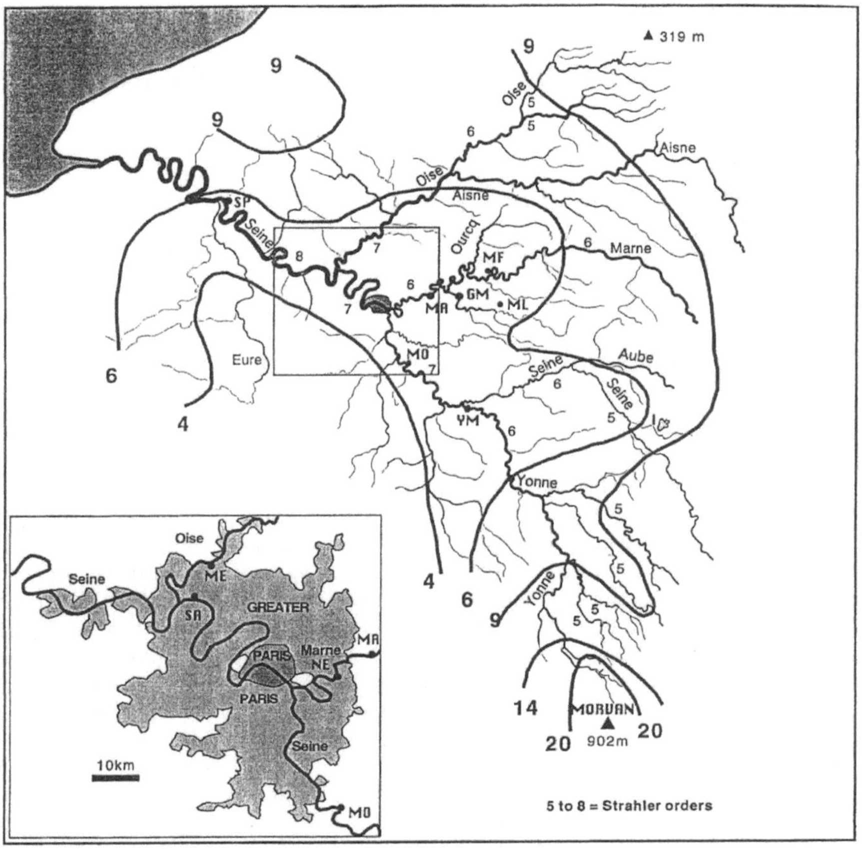

Figure 1. Hydrological network with isorunoff lines (l s^{-1} km^{-2}), upper Strahler stream orders and key stations. ML = Mélarchez, MA = Marne at Annet, NE = Marne at Neuilly, MO = Seine at Morvang, ME = Oise at Méry, SP = Seine at Poses, GM = Grand Morin. I to III: Seine, Aube and Marne reservoirs. SA = Seine Aval sewage treatment plant at Achères. YM = Yonne at Montereau, MF = Marne at La Ferté.

Table 1. Hydrological regime of the Seine and its tributaries (runoff in l s^{-1} km^{-2}) (1992–1995)

Months	Jan	Feb	Mar	Apr	May	Jun	Jul	Aug	Sep	Oct	Nov	Dec	Year
1.Marne	17.68	15.14	12.38	11.65	6.55	4.79	4.14	3.94	5.07	6.61	7.70	11.19	8.90
2. Seine	13.59	13.00	9.97	9.37	6.24	4.14	3.09	2.68	3.86	5.74	6.31	8.61	7.22
3. Oise	15.67	15.46	11.77	10.61	7.07	5.32	3.45	2.80	3.18	3.80	5.10	11.19	7.95
4. Seine	13.94	13.39	10.55	9.54	6.21	4.7	3.44	2.89	3.95	5.25	5.61	9.03	7.38

(1): at Noisiel, 12 640 km^2, (2): at Paris 32 450 km^2, (3): at Creil 17 330 km^2, (4): at Poses 67 500 km^2.

In this paper we consider the whole Seine basin with various data sources: 4 years of daily measurement at 4 stations, weakly measurements during an exceptionnal high flow (1994/95) at 4 stations, and monthly records over 27 years at 236 stations. Our objectives are:

1. to set up the general TSS levels and their spatial distribution with lithology and relief,
2. to determine the TSS seasonal regime at few key stations and its relationship with water discharge,
3. to check the possible influence of hydrographic stream order,

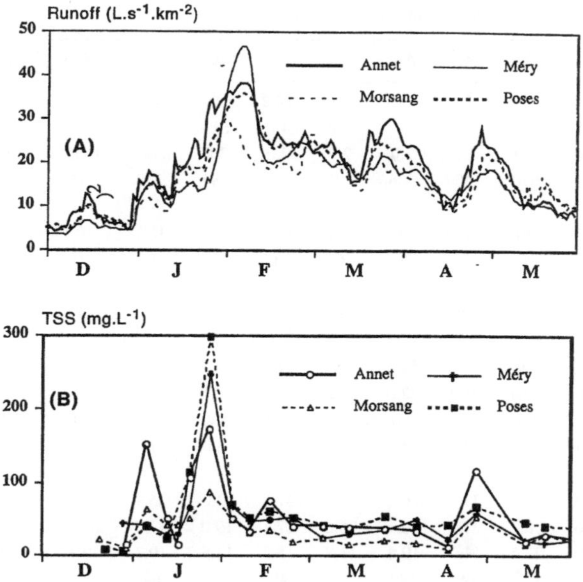

Figure 2. Evolution of daily runoff (A) and of sampled TSS (B) at 4 stations in the Seine basin during the 1994/95 high water stage. (Morsang on Middle Seine, Annet on Marne, Méry on Oise and Poses on Lower Seine). (location see Figure 1).

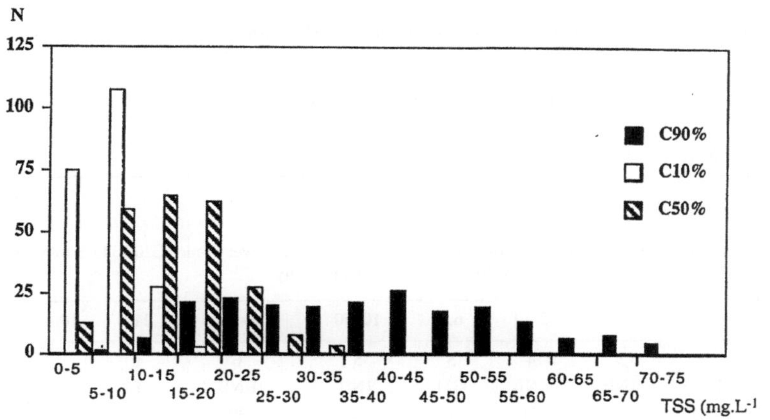

Figure 3. Distribution of TSS quantiles at 236 stations in the Seine basin. Lower (C10%) and upper deciles (C90%) and medians (C50%) for the period 1971–1997 (RNB network). *N* = number of stations in each TSS class.

4. to assess the human impacts particularly from the Paris megacity and from fluvial traffic.

Presentation of the basin

We are considering the basin upstream the last lock before the estuarine section at the Poses station, 67 500 km^2 (Figure 1). The Eure river (7000 km^2), the last major tributary, which reaches directly the estuary, is therefore not included. The basin extends from the upper Oise, at the France-Belgium border (300 m altitude) to the upper Yonne (total drainage area 11 250 km^2) in Burgundy Morvan hills where the maximum Seine basin altitude of 902 m is reached. The Aisne (7890 km^2), the Marne (13 160 km^2), the Aube (4750 km^2) are also characterized by low altitude plateaus (less than 400 m). The median altitude of the whole

Table 2. TSS regimes at four stations of increasing basin size (mg l^{-1})

		Jan	Feb	Mar	Apr	May	Jun	Jul	Aug	Sep	Oct	Nov	Dec	Mean
Melarchez	C90	6.8	250	14	28	37.5	14.2	28	44	18.6	12.8	26.4	21	41.8
(1975–77)	C50	2.4	15	7	7	5.9	4.0	10.4	5.75	2.8	3.9	3.0	3.2	5.86
At Melarchez	C10	0.4	3.0	2.2	2.0	2.0	2.0	3.0	0.00*	0.00*	1.0	0.8	0.4	1.40
A = 7km^2	cmax	130	670	45	386	196	70	184	219	44	31	184	231	199
Grand Morin	C90	192	80	40	76	22.8	13.1	8	10.7	10.1	20	13	194	56.6
(1992–95)	C50	10	6	7	7	6	6	5	6	6	7	5	6	6.42
At Villiers	C10	4	3	3	2	2	2	3	3	3	4	2	2	2.75
A =1200km^2	cmax	1000	300	149	184	36	123	34	31	30	120	600	1250	321
Marne	C90	197	69	63.8	55.3	29	21.1	22.2	24.2	25	48.5	45.7	131	60.9
at Neuilly	C50	40	19.4	28.3	22.0	14.0	14.5	14.5	16.0	12.0	18.0	16.8	16.6	19.3
(1992–95)	C10	8	6.2	8	7.0	7.0	9.0	5.4	9.0	8.0	7.7	8.9	6	7.5
A =12 500km^2	cmax	460	120	150	150	60	80	34	54	84	166	85	362	150
Seine	C90	97.4	87	37.9	76.1	67.8	41.8	28	25.9	21.5	27.3	32	42.4	48.7
at Poses	C50	34.6	38.2	21.9	36.1	35.6	27.1	19.0	17.9	16.9	15.3	15.2	9.8	23.9
(1982–85)	C10	11.0	23.5	11.9	19.4	11.8	15.2	11.8	10.3	9.4	10.5	8.0	6.1	12.4
A =67 500km^2	cmax	165	122	49	142	87	49	41	62	36	58	163	95	89

C50 = median of TSS values for each month (based on 90 – 120 measurements), C10 and C90 = lower and upper deciles, cmax = maximum TSS observed during the period of records, 0.00* = dryness.

Table 3. Range of the quantiles values of TSS (mg l^{-1}) distributions for the 1971–97 period of record at the 19 best documented stations (orders 5 – 8).

Time quantile	C1%	C5%	C10%	C25%	C50%	C75%	C90%	C95%	C99%
Min. observed	2	3.5	5	8	13	22	35	52	100
Max. observed	6	10	12	16	22	36	62	100	200

Table 4. Influence of basin size on quantiles of TSS time distribution (mg l^{-1}) from lower decile (C10%) to extreme values (C99.5%) at 4 stations sampled daily compared to ranges from RNB stations sampled monthly

Q (m^3 s^{-1})	0.055	0.5–2	2-10	6.3	10–50	50-100	112	100–300	>300
Stream Order	1	3	4	5	5	6	6	7	8
Station type	ML	RNB	RNB	GM	RNB	RNB	MA	RNB	RNB
Station number	1	54	74	1	52	30	1	18	10
C10	0.6	4.5±2	4±2.5	3	6±4	8.5±2	7	8±2	11±3
C25	2	7.5±3	6±3	4	10±5	12±2.5	10	10±2	15±4
C50	4.8	11±5	10±7	6	15±9	17±3	17	16±4	22±4
C75	12	18±7	18±9	10	20±12	24±6	30	24±5	30±10
C90	30	36±16*	35±20*	40	40±20	45±15	55	45±10	45±12
C95	53*			100	70±25	(80±30)	95		
C99	231*			396	140±60	(150±50)	220		
C99.5	/			600			280		

( ): limited number of stations, *: values probably too low. Daily data over 3 to 4 years: ML: Mélarchez stream, GM: Grand Morin, MA: Marne at Neuilly. Monthly data from 1971 to 1997: RNB, median and range including 80% of documented RNB stations.

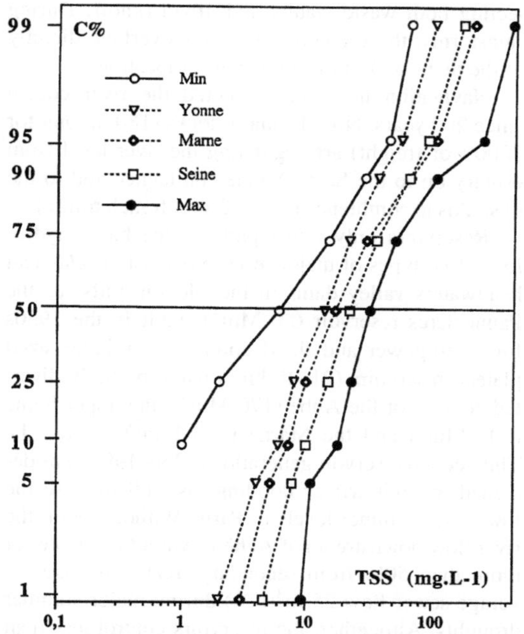

Figure 4. Typical TSS distributions in a log-probability scale at 3 stations and general range over 20 stations (RNB network 1971–1997). Seine at Poses: Marne at La Ferté; Yonne at Montereau). Min and Max = range of quantiles for 95% of stations (n = 230).

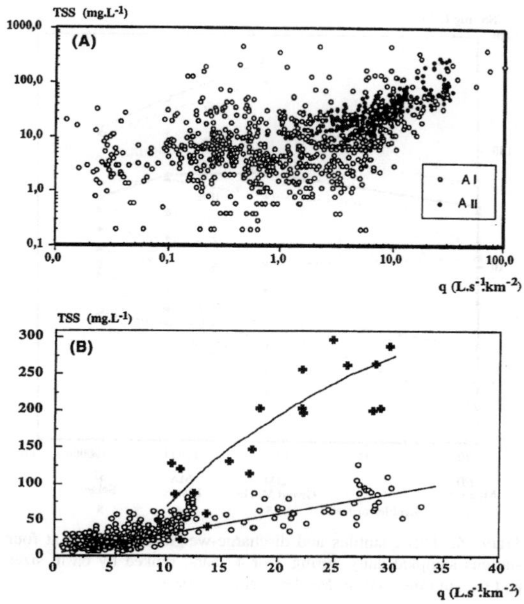

Figure 5. Daily TSS *vs* specific runoff. A1 = Mélarchez stream (7 km^2). A2 = Marne à Neuilly (12 500 km^2). (B) = Seine at Poses (67 500 km^2): + = rising stage, O = falling stage and low flows.

basin is 150 m with 99% less than 550 m (Figure 1). The Marne (stream order 6) and the Middle Seine (order 6, 32 220 km^2) meet at Paris. The Oise (order 7, 16 900 km^2) and Seine (order 7) meet at Conflans where the river gets the order 8. The average river bed slope of the upper stream orders varies from 0.3 ‰ for order 6 to 0.1 ‰ for order 8, and from 2.5 to 10 ‰ for the lower orders 1 to 3 (Meybeck et al., 1998B). Apart from the hercynian cristalline rocks of the Morvan hills, only 2.5% of the basin, the Seine basin is sedimentary and mostly underlained by carbonate rocks (22.1% chalk, 30.1% pure limestone, 27.7% mixed carbonates). Clayed carbonates (10.0%), sands and sandstone (7.6%) are of secondary importance. Mixed carbonated alluvial deposits are also found in floodplains, about 2.5% of the whole basin (Thibert, 1994).

The rainfall pattern is relatively homogeneous ranging from 550 mm y^{-1}, south of Paris, to 900 mm y^{-1} in the north and east part of the basin. The maximum reaches 1200 mm y^{-1} in the Morvan hights. The basin average is 700 mm y^{-1}. The seasonal variation of rainfall is very limited, the extreme months are ± 20% the yearly average. Due to the relatively high evapotranspiration (ETP around 500 mm y^{-1}), the surface runoff is moderate and ranges from 4 to 15 l s^{-1} km^{-2}. The Morvan is also the only part of the basin where some snowmelt influence is noted. The spatial runoff distribution presents a ring pattern (Figure 1). It is important to note that each major tributary crosses first the highest runoff ring then flows towards the center of the Seine basin where local runoff is minimum. Such ring structure is also noted for many other biophysical attributes of the Seine river basin. They are primarily due to the climate zonation and also to the sedimentary ring structure of the Bassin Parisien: the Jura limestones as the most external ring, then clays and marls, Albian sands, Cretaceous chalk, and finally tertiary limestones and sandstone of the central Bassin Parisien.

The water regime of the Seine is *'pluvial oceanic'*, the minimum flow is from July to September due to higher ETP during summer, while high flows are observed from January to March (Table 1). This regime is observed throughout the basin and the Middle Seine, the Marne and the Oise have very similar runoff patterns for long-term records (Table 1). During the

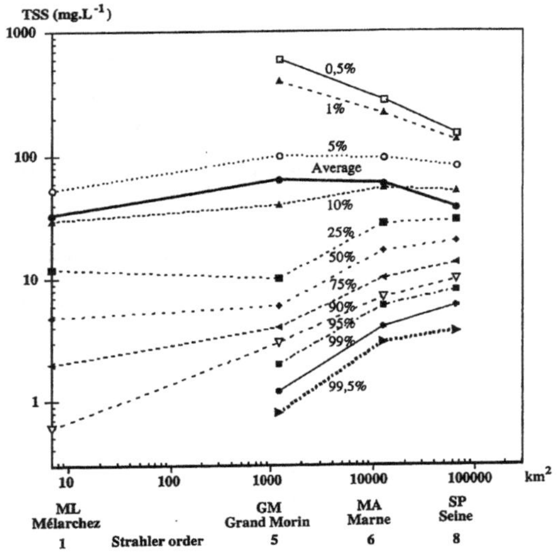

Figure 6. TSS quantiles and discharge-weighted averages at four stations sampled daily during 3 or 4 years, ranked by basin sizes and stream orders (from Meybeck et al., 1998A).

exceptional 1994/95 high flow period during which the peak runoff exceeded 35 l s^{-1} km^{-2} on the Marne and Oise the water discharge pattern at the three key stations, on the Marne, Oise and Middle Seine are very similar (Figure 2A).

Human impacts on the basin

They are multiple and date back to the Gallo-Roman period some 2000 years ago (Meybeck, 1998; Meybeck et al., 1998B). The agricultural land, represents 69% of the basin and forested area only 25%. Greater Paris corresponds to about 2500 km^2 and 10 M people, or 3.7% of the basin area and 60% of present population. Apart from the industrial corridor between Paris and the estuary, with chemical and automobile industries, food and textiles industries are also found in small and middle sized cities as Reims (200 000 agglomerated people), Troyes (125 000 p. agglomerated), Melun (77 000 p. agglomerated). Eighty per cent of waste waters of Greater Paris (population of about 10 m p.) are collected to the gigantic Seine Aval plant located at Achères (Figure 1) some 75 km downstream of Paris center and released after treatment 4 km upstream of the Seine-Oise confluence. Another station is located on the Seine, upstream of the Seine-Marne confluence. All sewers are collecting both waste waters and urban runoff. During rainstorms, these combined sewers overflow directly to the Seine and Marne at dozens of locations.

Navigation has much affected the river course since 200 years. Now 17 major locks (180 m long for 3000 t of freight) are regulating the river level from estuary up to the Seine-Yonne confluence and to the Oise-Aisne confluence (50% of total french traffic).

Reservoirs only affect parts of the basin hydrology. Two types of major reservoirs exist: 1. classical headwaters valley dams in the Morvan hills, as the Pannecières reservoir (80 Mm^3), built in the 1950s for hydropower and flood control, 2. middle sized plateau reservoirs (21–48 km^2) using partially diverted courses of the Aube (170 Mm^3), the upper Seine (205 Mm^3) and the Marne (350 Mm^3) (Figure 1). This second reservoir generation (1966–1991) was designed to limit winter flooding, as well to raise the low-water summer levels in Paris. Without them, the river flow downstream of Achères would originate for more than 50% from secondary treated sewage (average sewer flow 25 m^3 s^{-1}) during major summer droughts. Altogether, the reservoirs control less than 10% of the whole Seine basin. A detailed descriptions of the basin is found in Meybeck et al. (1998B).

Data sources and treatment

The first data set results from the regular National Basin Network (RNB), shared by the State Ministry of Environment and the Seine Basin Water Authority (Agence de l'Eau Seine-Normandie or AESN). It started in 1971 and now concerns 236 stations on the Seine network usptream of Poses, plus 100 stations on local Normandy streams, including the Eure. Typically TSS samples are taken at the water surface from bridges, at the centroid of flows, as for the other water quality variables. Their frequency varied from 4 to 12 samples per year. It is now 12 per year at most stations and up to 24 per year at very few stations, as Poses. This data bank, is freely accessible at the AESN: 12% of stations have from 30 to 50 data points, 52% from 50 to 100, 23% from 100 to 150, 6% from 150 to 300 and 6% from 300 to 697 data (at Poses). Most stations started in 1971, some were added later while only few have been discontinued: as a whole the network is continuously expending. Smallest streams (orders 1 – 3) are much less documented in the RNB, as in all other major french river basins: this national water quality network was more designed for actual or potential water users, mainly located on stream orders 4 – 8, than to determine the patrimonial value of France's aquatic habitat. Therefore the proportion of documented TSS stations is only 10% for stream order 1 and 2, 10% for order 3 (mostly in Normandy), and 80% for orders 4 – 8 (Seine basin proper) while the distribution of river network length in the Seine is respectively 65%, 15% and 20%. Our knowledge on stream water quality starts only at order 3, which is typically 12.1 km long for 145 km^2 and 900 l s^{-1} (Garnier et al., 1998).

RNB statistical data for 1971–1977 were systematically processed for all sufficient data sets (n >50 data points): medians (C50%) and quartiles (C25% and C75%) were computed for 50< n <90, deciles (C10% and C90%) for n >90, and higher quantiles (C95%, C99%) for n >250. These quantiles were in turn considered as statistical variables for which the space or time distribution was considered (e.g. the space distributions of the C75% determined at each stations; trend of the C90% at key stations, etc.

The second TSS data set is a four year daily survey at four stations of increasing stream order: 1. the Melarchez stream at Melarchez (7 km^2, order 1), located on the Grand Morin basin, a tributary of the Marne, 2. the Grand Morin at Villiers (1200 km^2, order 5), 3. the Marne at Neuilly sur Marne (12 640 km^2, order 6) and 4. the Seine at Poses (67 500 km^2, order 8) (Figure 1). The Melarchez daily TSS mean is a composite of 24 individual samples mixed together by Cemagref technicians at an automatic sampling station between 1975 and 1977, thus including the most severe 1976 spring and summer drought. The Grand Morin and Marne were sampled by the CGE-Vivendi technicians from 1992 to 1995. The Seine data at Poses have been provided by A. Ficht (Service de la Navigation de la Seine, or SNS-Rouen) for 1982 – 1985.

The last TSS set consists of four stations on orders 7 and 8, just upstream and downstream Greater Paris: the Marne at Annet, the Middle Seine at Morsang, the Oise at Méry and the Seine at Poses, during the 1994/95 year (Figure 1) where, during the flood period from December to April, weekly samples were taken from bridges at the middle of the river. All TSS samples are determined on pre-weighted 0.5 μm filters, sometimes 0.7 μm, then dried at 106 °C.

General TSS levels

The TSS variability during the 1994–95 high flow period is very much synchronous for the major tributaries, Oise, Marne and Middle Seine (Figure 2): TSS peaks were observed at each stations for the same flood events. The maximum TSS of 300 mg l^{-1} measured at Poses at the end of January 1995 had never been observed before at this station within the RNB data set, nor within the SNS three years daily survey. This maximum occurred during the rising stage of a major flood, for a 30 l s^{-1} km^{-2} runoff.

The distributions for the whole basin of the median (C50%) and of the lower deciles (C10% and C90%) determined at each of the RNB stations are presented on Figure 3: 79% of the median values are between 8 and 32 mg l^{-1}; 71% of the lower deciles are below 8 mg l^{-1}; only 11% of the upper deciles exceed 64 mg l^{-1}. Extreme TSS values determined at the 20 most documented stations (n >250 data points for the period of record) are also very much limited: the C99% quantiles are between 80 and 210 mg l^{-1}. The Seine basin is therefore characterized by a generally low to very low TSS level.

For the Yonne sub-basin located in the highest part of the basin, these figures are not much different. A longitudinal profile of 14 stations shows medians ranging from 8 to 16 mg l^{-1} and upper quartiles (C75%) from 13 to 30 mg l^{-1}. The possible effect of the relatively steeper relief is obliterated by the TSS retention in the few Morvan reservoirs, at Pannecières.

TSS seasonal regime

The TSS monthly regime is established at the four stations (Table 2) where the TSS distributions from 3 or 4 years of daily records (C10%, C50% and C90%) are computed based on 90 – 120 data points for each month. The Seine at Poses and the Marne at Neuilly have very similar TSS variability with lowest levels from July to September, and higher values from January to March, thus mimicking the runoff regime (see Table 1). Yet these monthly variations are still somewhat irregular, since only based on a limited number of years: the April values at Poses are higher than those in March. For these documented years the Marne is more turbid than the Seine at Poses with a maximum recorded of 460 mg l^{-1} compared to only 160 mg l^{-1}.

For the smaller catchment of Grand Morin, the TSS regime is more variable: summer median TSS are lower, but summer TSS peaks are higher (123 mg l^{-1} maximum recorded in June compared to only 80 mg l^{-1} for the Marne and 49 mg l^{-1} for the Seine at

Poses). For the Melarchez stream this trend is even more apparent with summer maximum up to 184 mg l^{-1} in July and 219 mg l^{-1} in August. Such peaks are associated with local rainstorms that may rise the stream runoff over 100 l s^{-1} km^{-2}. Such summer floods are not observed at higher stream orders since rainstorms are distributed locally.

Temporal distribution of TSS

Over the full period of record (1971–1997) at the 19 best documented stations, generally located on stream orders 5 – 8 (between 250 and 697 data points) the full set of TSS percentiles, from C1% to C99%, has been determined. Most of these distributions follow a log-normal distribution, represented as a straight line in a log-probability paper (Figure 4), particularly for the Seine at Poses or the Yonne at Monterau. At few stations TSS distributions present a break around C75% with relatively higher TSS levels for the upper quantiles, as for the Middle Marne at La Ferté, which could imply an additional source of TSS at the highest runoff values. Yet the overall envelope of 230 distribution curves on Figure 4 is relatively narrow, particularly for the medians, thus reflecting the great homogeneity of this basin (see also the minimum and maximum range for each quantiles for the 19 best documented stations on Table 3).

TSS and water discharge

The increase of TSS with water discharge Q, or with specific runoff q when different stations are compared, has been pointed out since long by river scientists (Forel, 1892). The TSS *vs* Q relation may be simple at some stations: TSS = a Q^b, with b>1 (Muller & Forstner, 1968), but is generally much more complex according to the sample position during the flood, the flood rank in the hydrological year, the hydrological origin of flood, the catchment heterogeneity etc. (Walling, 1977; Walling & Webbs, 1983; Williams, 1989).

In typical lowland basins as the Seine, most individual floods, when they are studied at the finest time scale, present typical TSS *vs* Q clockwise hysteresis. Such cycles are also observed at Poses (Figure 5B). It is generally assumed that the first and/or fast increase of river velocity results in a resuspension of river bed sediment deposited between two flood events. Such process is maximum for the first flood event of the hydrological year, then gradually decreases (Walling, 1977). In such basins, the TSS *vs* Q scatter at one station may be important.

At Melarchez (7 km^2) one can barely decipher any TSS *vs* Q pattern (Figure 5A): each individual flood event, lasting 1 – 3 days is characterized by different relationships which mostly depend on the agricultural practises, the soil structure at that time, etc... (Penven et al., 1998). Also, at this very local scale, the flood hydrograph should be considered at a finer time scale (few hours maximum) and daily averages are probably not appropriate to study TSS *vs* q relations. The scatter is much less for the Marne at Neuilly (12 640 km^2).

Influence of stream order

It has been checked at the four stations with daily monitoring, at stream order 1, 5, 6 and 8 where a complete distribution from C1% to C99% can be determined (Figure 6 & Table 4). There is a marked decrease of the TSS range from order 1 to 8 at both ends of the distribution: highest quantiles decrease and lowest quantiles increase. The C99% Melarchez quantile is probably well underestimated since the three year period of record is biased by the exceptionnal 1976 drought. For the Marne and Seine, the lower quantiles are probably also influenced by fluvial traffic and/or eutrophication as discussed latter. The median values C50% increase by a factor 5 from Melarchez (order 1) to the Seine (order 8), but the discharge-weighted averages are much less variable.

The RNB stations (orders 3 – 8) have been considered: for each order the quantiles of TSS distributions for the 1971–1997 period at each RNB station have been determined. For orders 3 and 4, the data base – typically 60 – 120 TSS values – is probably not appropriate to determine the highest quantiles: the short turbid events are not likely to be sampled at the monthly frequency and even the C90% are probably underestimated. Above C90%, quantiles C95% and C99% are only presented for the 19 most frequently sampled stations located on order 5 (n = 6), order 6 (n = 9) and orders 7 and 8 (n = 4). The full statistical informations for Melarchez (ML), Grand Morin (GM), Marne (MA) and Seine at Poses (SP) is also reported in Table 4. At the four key stations with daily TSS surveys (GM, MA and SP), quantiles from C10% to C90% are well within the range of the RNB values. But above C90% the RNB quantiles could be under-

Table 5. Time distribution of TSS (mg l^{-1}) for the Seine at Poses (67 500 km^2, order 8) from three different data sets

	C10	C25	C50	C75	C90	C95	C99	C99.5
A: 1983–85	9.8	14	21	33.8	50	76	121	140
B: 1971–97	10	15	22	38	65	91	170	/
C: 1959–93	14	18.5	20	30	52	85	215	240

(A) daily measurements from SNS Rouen, (B) bimonthly INP/RNB samples, (C) daily TSS reconstituted from TSS *vs* Q relationships.

estimated. At the four key stations with daily TSS surveys (GM, MA and SP), quantiles from C10% to C90% are well within the range of those determinated from regular surveys over the whole basin. They can be regarded as representative of their own stream order, which validate the observations on the general shrinking of the TSS range with increasing stream order, particularly during floods (C95% and C99%).

The validity of the regular monitoring (RNB stations) for TSS surveys can be checked here at the Poses station for which three TSS distributions based on different sets are available (Table 5):

1. the 1982–1985 daily surveys,
2. the 1971–1997 bimonthly RNB survey with 697 TSS values, and
3. a reconstituted daily record for 34 years, from 1959 to 1993 (Idlafkih, 1998).

The latter one is based on the two TSS *vs* q relationships established at Poses (see Figure 5B), valid for the whole period and taking into account the rising and falling stages of hydrographs and the flood rank within hydrological years. From C10% to C95% the three distributions are very similar. Above C95% the quantiles are lower for the first set (1983–1985) characterized by few floods compared to the long term hydrological distribution. The RNB record for 1971 – 1997 can be used to assess the distribution of TSS below C95%.

Lithology influence

It has been checked from the RNB network on orders 3 and 4, i.e. on non-navigated medium-sized basins with average water discharge between 0.5 and 10 m^3 s^{-1}. Although these basins cannot strickly be regarded as monolithologic, i.e. draining a unique rock-type, they are dominated by one rock type (more than 80%). Bigger catchment (orders 5 and up) mixing several kinds of sedimentary rocks are not appropriate. All the 35 selected basins are characterized by low population density (20–60 p km^{-2}) typical of rural catchments without major cities and are not affected by reservoirs. Stations spread all over the Seine basin, plus 5 stations located in the Eure basin, have been grouped into five categories of sedimentary rock types (Table 6). The cristalline Morvan hills, only 2.5% of the whole basin, are not considered here since much influenced by reservoirs. The differences between limestones and chalk on one side, and marls and clays on the other, is significant and attributed to differences in rock and soil erodibility although differences in slopes and in agricultural pratices, also linked to soil cover, cannot be ruled out.

Sparnacian clays found in the NW Oise basin are the most sensitive sedimentary rock in the basin. Turonian marls (N.E. Oise) and bartonian and lutetian marls. (W. Aisne and Middle Marne) drained by the Ourcq, a Marne tributary, have also higher TSS levels. For the smallest orders (1 – 4) the local lithology should be considered.

Cities influence

Some RNB stations are located on small streams (orders 3 & 4) in Greater Paris as the Orge, Yvette, Yerres, Mauldre. In their lower part, the population density well exceeds 1000 p km^{-2}. Although the related waste waters are collected and treated at Achères, Valenton or Versailles treatment plants, there is still much influence of urbanisation on these streams as observed from NH_4^+, NO_2^- and PO_4^{-3} data which may be 10 – 20 higher than in rural conditions (Meybeck, 1998). We are comparing 9 stations, with average population densities from 400 to 1000 p km^{-2} to a set of 37 streams (orders 3 & 4) in rural area over the whole Seine basin (Table 7). The urban influence is limited and more marked for the lower quantiles than for the upper ones: low flows are more turbid in periurban conditions. The lower decile C10% is 11.2 mg l^{-1}, a figure twice higher than those found for any kind of monolithological basin. Due to the mixed nature of the rock types around Paris including chalk, tertiary limestones and sands, few argilaceous layers, these differences are attributed to urbanization. Organic carbon and particulate phosphorus contents are also much higher in these river particulates (Meybeck et al., in prep).

Across Paris megacity, there is no marked variations of TSS levels from the RNB records which

Table 6. Influence of basin lithology on TSS quantiles at 35 RNB stream stations (1971–1977 period)

Dominant lithology	C10%	C50%	C75%	C90%
Upper Jurassic limestone (n = 7)	1.5±0.5	4±2	9±4	18±5
Tertiary limestones (Sannoisian) (n = 5)	3.5±0.5	7±2	12±3	20±5
Upper Cretaceous chalk (n = 8)	3±0.5	6±4	11±5	18±8
Marl and marl-limestone (Lias, Bartonian, Turonian) (n = 10)	5.5±1.5	14±6	28±8	65±20
Sparnacian clays (n = 5)	6.5±1	16±5	31±6	90±10

Ranges include more than 80% of distribution.

could be attributed to urban impacts, the slight increase is attributed to the Marne and Oise inputs at their confluences with the Seine. The treated waste waters at the Seine Aval plant are released near Achères with limited TSS (around 65 mg l^{-1}) and their impact is generally not visible on profiles. At low flows most of these waters are believed to creep on the river bed and are not much detected by conventional sampling (Meybeck et al., 1998A).

River traffic and eutrophication

Although the Lower and Middle Seine and Oise, and the Lower Marne are navigated by barges up to 3000 freight, channelized in some reaches, and equiped by a series of locks the present day impact of the river traffic is not obvious. Observed higher TSS are not linked with the river traffic: in the Marne they are observed in the upper third of the river (between Vitry le François and St Dizier), in non-navigated reaches as for the Oise.

For navigated reaches, on stream orders 6 – 8, the lower deciles (C10%) corresponding to TSS measured at summer low flows are slightly higher than those observed at stream order 4 and 5: 8 to 11 mg l^{-1} compared to 4 to 6 mg l^{-1} (Table 4, Figure 6). This could be attributed to the continuous stirring of the water column by navigation but eutrophication should also be considered.

River phytoplankton development is very high in the Seine basin (Billen et al., 1994, Garnier et al., 1998) where chlorophyll levels commonly exceed 100 $\mu g\ l^{-1}$ in the Seine, Marne and Oise at stream orders 6, 7 and 8. Maximum total pigments of 200 $\mu g\ l^{-1}$ have been measured, they correspond to 6 mg l^{-1} of autochtonous particulate organic carbon (POC), i.e. to 12 – 15 mg l^{-1} of dry particulates (Dessery et al., 1984). According to the RIVERSTRAHLER model which simulates stream phytoplankton production and bacterial respiration at the basin scale for each stream order (Billen et al., 1994; Garnier et al., 1998), the phytoplankton development starts at order 5 and reaches its maximum at orders 6 – 8. River profiles of the lower deciles on the Seine, Marne and Oise, based on 15 to 25 RNB stations, show a downstream increase in each reach. For instance, on the Seine C10% increases from 6 to 10 mg l^{-1} between Monterau, at the Seine-Yonne confluence, and Poses. On the Marne C10% is 8 ± 1 mg l^{-1} from the middle reach at Chalons en Champagne to Paris. On the Oise C10% is also 9 ± 1 mg l^{-1} throughout the orders 5, 6 and 7. At such low TSS levels the POC content, mostly autochtonous is around 13%, i.e. the autochtonous organic matter around 3 mg l^{-1} (Idlafkih, 1998).

Recent trends (1971–1997)

Trends have been checked on the RNB data set at the Poses stations over the 1971–1997 period splitted into three subsets: 1971 – 1979, 1980 – 1989 and 1990 – 1997. For each of these periods, the statistical distributions have been determined on more than 200 data points (Table 8). The three distributions are very much similar from C1% to C90% indicating a very steady TSS regime. The upper quantile C95% is somewhat higher for the last period: this can be biased by the occurrence of four humid years characterized by more turbid rivers (1990/91, 1993/94, 1994/95 and 1995/96). Also the bimonthly sampling may not fully represent the highest TSS quantiles.

Comparison with historical data (1863–1866)

Rare historical TSS records were found for the period 1863-1866 (Mangon, 1887) during which near-daily surveys of the Marne and Seine upstream their confluence have been realized during 15 and 24 months. At this period, these rivers were not yet dammed and not much regulated by sluices. The discharge weighted averages were significantly higher: 74 mg l^{-1} for the Marne and 39.7 mg l^{-1} for the Seine and these TSS levels were much more variable than for those present surveys. Recorded maximum in 1863/66 were much higher than now: on the Seine values exceeding 500 mg l^{-1} have been measured several times, including a summer event at 626 mg l^{-1} on August 17th, 1866 and an 'excessively high' (*dixit* Mangon) maximum of

Table 7. Influence of urbanization on TSS quantiles at selected RNB stream stations (orders 3 and 4) (1971–1997 record)

	C10%	C50%	C75%	C90%
Periurban streams (*d*:400 to 1,000) (n = 10)	11±2	22±12	32±12	48±15
Rural streams (*d*: 20 to 60) (*n* = 37)	5±3	12±6	17.5±9	35±25

d: population density (people.km^{-2}). Range includes 80% of distribution.

2738 mg l^{-1} on September 24th, 1866. In the Marne, the monthly weighted TSS average was up to 198 mg l^{-1} (January 1865), a factor two higher than the current situation (98 mg l^{-1} for 1992–95 in January). Minimum values were much lower: 1.35 mg l^{-1} measured during July 1864 for the Seine, and 4.6 mg l^{-1} for the weighted monthly average for the Marne in October 1864, compared to 16 mg l^{-1} for the 1992–95 period. Other human influences from secular land use and cover changes (see for examples the Rhone river evolution, Arnaud-Fassetta and Provensal, 1999) were not checked.

As the procedure followed by Mangon, a chief engineer of the Ecole des Ponts et Chaussées was very much calibrated, his measurements are probably as trustfull than those of today, both sampling being realized from bridges without depth-integration. If his results are not questioned, the Seine basin was probably exposed to higher sediment transport rates although still moderate at the global scale (about 12 t km^{-2} y^{-1} compared to about 200 t km^{-2} y^{-1} for the world mean). During the winter high water stage the TSS maximum were much higher as for few summer TSS peaks, probably resulting from severe local rain storms. These peaks are now very much buffered as a result of the succession of navigation locks on the Seine and Marne. In summer, the river traffic was very limited and eutrophication was absent; the TSS values have now been considerably raised by eutrophication, river traffic and, on some reaches, local pollution.

Conclusions

General TSS levels in the Seine are very low due to a very gentle relief and to an even distribution of rainfall through out the year. Other low-TSS basin can be found in western or eastern Europe but most of them are affected by trapping of particulates in lakes as in Finland or in the Middle Rhin basin, or in reservoirs as for the Volga, Dniepr and Don (Kimstach et al., 1998). The Seine basin is highly homogeneous in terms of geology, relief, hydrology. As a result the spatial TSS variability is very limited.

Table 8. Trend in TSS distribution at the Poses station from the RNB survey

	C1%	C5%	C10%	C25%	C50%	C75%	C90%	C95%
1971–79	5.0	10.9	13.0	17.5	26.0	39.0	61.8	79.9
1980–89	6.0	8.0	10.0	16.0	22.6	34.0	58.0	73.3
1990–97	4.4	7.6	9.2	14.0	21.0	37.1	78.6	116.3

The range of TSS variations (e.g. C1% – C99%) is one order of magnitude less for the higher hydrologic stream orders (6 – 8) than for the Melarchez representative basin (7 km^2, order 1).

Median TSS (C50%) increase from order 5 to 8. For the small orders median TSS are also dependent on basin lithology (e.g. limestones and chalk *vs* more erodible marl and argilaceous material) and on eutrophication.

Present-day anthropogenic impacts are moderate despite inputs from, industries, intensive agriculture and urbanisation. Impact of Greater Paris (10 M. people) is difficult to be set up on monthly regular sampling. Compared to other water quality variables as nutrients and some major ions (Na^+, K^+, Cl^-) the Seine Basin, TSS is not a very sensitive water quality indicator of human impacts.

At the river mouth (order 8; 67 500 km^2) the data obtained from bimonthly sampling over a long period have a statistical distribution very similar to those obtained from measured or simulated daily TSS except for very high and rare TSS levels: such regular water quality monitoring can be valorised. No trend has been remarked at this stations between 1971 and 1997. But for mid-orders streams (4 – 6) which present more variable TSS and which are only surveyed monthly the upper quantiles (C90% and above) of TSS may be underestimated.

When compared to daily TSS surveys realized in 1863–66, the present Marne and Seine TSS levels are twice lower and the natural range (1 to more than 500 mg l^{-1}) has been greatly reduced (4 – 200 mg l^{-1}) as a result of multiple human impacts, probably as the result of dams and sluices construction.

Aknowledgements

This work was realized within the PIREN-Seine programme sponsored by a pool of regional and national water institutions. RNB data base has been provided by A. Serré (AESN) and Poses daily samples by A. Ficht (SNS Rouen), both are greatly aknowledged. K. Koskina and A. Prodromos participation in data treatment and J. M. Dorioz comments (INRA, Thonon) were much appreciated.

References

A. E. S. N., 1973–78. La Seine et les cours d'eau normands. Agence de l'Eau Seine-Normandie, Nanterre, France,12 vols+ atlas.

Arnaud-Fassetta, G. & M. Provansal, 1999. High frequency variations of water flux and sediment discharge during the Little Ice Age (1586–1725 AD) in the Rhône Delta (Mediterranean France). Relationship to the catchment basin. Hydrobiologia 410: 241–250.

Billen, G., J. Garnier & Ph. Hanset, 1994. Modelling phytoplankton development in whole drainage networks: the RIVER-STRAHLER model applied to the Seine river system. Hydrobiologia 289, 119–137.

Chapman, D. (ed), 1992. Assessment of the Quality of the Aquatic Environment throught Water, Biota and Sediment, Chapman & Hall, London: 585 pp.

Dessery, S., C. Dulac, J. M. Laurenceau & M. Meybeck, 1994. Evolution du carbone organique particulaire 'algal' et 'détritique' dans trois rivières du Bassin Parisien. Arch. Hydrobiol. 100: 235–260.

Forel, F. A., 1892–1904. Le Léman. Monographie limnologique, Vol. I, II and III, F. Rouge, Lausanne.

Garnier, J., G. Billen, P. Hanset, P. Testard, M. Coste, 1998. Développement algal et eutrophisation dans le réseau hydrographique de la Seine. In M. Meybeck, G. de Marsily & E. Fustec (eds), La Seine en Son Basin. Elsevier, Paris: 593–626.

Horowitz, A. Z., 1995. The use of suspended sediment and associated trace elements in water quality studies. Int. Ass. Hydrol. Sci. Special Publ. 4: 58 pp.

Idlafkih, Z., 1998. Transport des ions majeurs, nutriments, carbone organique et métaux particulaires dans un fleuve anthropisé: la Seine. Thèse de Doctorat, Université de Paris VI.

Kimstach V., M. Meybeck, E. Baroudy (eds), 1998. A Water Quality Assessment in the Former Soviet Union. E & F.N. Spon, London: 611 pp.

Mangon, H., 1887. Expériences sur l'Emploi des Eaux dans les Irrigations sous Différents Climats et sur la Proportion des Limons Charriés par les Cours d'Eaux. Paris, Dunod.

Meybeck, M., 1986. The quality of world rivers through the GEMS programme. In E. T. Degens & A. Spitzy (eds), Transport of Carbon and Minerals in World Major Rivers, part. IV. Mitt. Geol. Palaont. Inst. Hamburg 64: 349–358.

Meybeck, M., 1998. Man and river interface: multiple impacts on water and particulate chemistry illustrated in the Seine river basin. Hydrobiologia 373/374, 1–20.

Meybeck, M., J. M. Mouchel, Z. Idlafkih, V. Andreassian & S. Thibert, 1998A. Transferts d'eau, de matières dissoute et particulaire dans le réseau fluvial. In M. Meybeck, G. de Marsily & E. Fustec (eds), La Seine en Son Basin. Elsevier, Paris: 345–389.

Meybeck, M., G. de Marsily & E. Fustec (eds), 1998B. La Seine en Son Basin: Elsevier, Paris: 749 pp.

Milliman, J. D. & P. M. Syvitski, 1992. Geomorphic/tectonic control of sediment discharges to the ocean: the importance of small mountainous rivers. J. Geol. 100: 525–544.

Müller, G. & U. Förstner, 1968. General relationship between suspended sediment concentration and water discharge in the Alpenrhein and some other rivers. Nature 217, 5125: 244–245.

Ongley, E. D., 1992. Environmental quality: changing times for sediment programmes. Int. Ass. Hydrol. Sci. Publ. 210: 379–389.

Penven, M. J., T. Muxart, F. Bartoli, P. Bonté, D. Brunstein, C. Cosandey, V. Gouy, V. Irace, T. Leviandier & S. Sogon, 1998. Petits bassins ruraux et pollutions diffuses. In M. Meybeck, G. de Marsily & E. Fustec (eds), La Seine en Son Basin. Elsevier, Paris: 159–210.

Petts, G. E. & P. Calow (eds), 1996. River Flows and Channel Forms. Blackwell Science, London: 257 pp.

Sundborg, A. & A. Rapp, 1986. Erosion and sedimentation by water: problems and prospects. Ambio 15, 215–225.

Thibert, S., 1994. Exportations naturelles et anthropiques des ions majeurs et des éléments nutritifs dans le bassin de la Seine. Ph D dissertation, univ. of Paris VI: 204 pp.

Thomas, R. & M. Meybeck, 1992. The use of particulate material. In D. Chapman (ed.), Water Quality Assessments. Chapman & Hall, London: 121–170.

Walling, D. E. & B. W. Webb, 1983. Patterns of sediment yield. In K. J. Gregory (ed.), Blackground to Palaeohydrology. John Wiley & Sons: 69–100.

Walling, D. E., 1977. Suspended sediments and solute response characteristics of the River Exe, Devon, England. In R. Davidson-Arnott & W. Nickling (eds), Research in Fluvial Geomorphology. GeoAbstracts, Norwich: 169–197.

Williams, G. P., 1989. Sediment concentrations versus mater discharge during hydrologic events in rivers. J. Hydrol. 111: 89–106.

Hydrobiologia **410:** 307–314, 1999.
J. Garnier & J.-M. Mouchel (eds), Man and River Systems.

Point and diffuse source analysis of heavy metals in the Elbe drainage area: Comparing heavy metal emissions with transported river loads

R.J. Vink[1], H. Behrendt[2] & W. Salomons[3]
[1]*Institute for Environmental Studies, Vrije Universiteit Amsterdam, De Boelelaan 1115, 1081 HV Amsterdam, The Netherlands*
[2]*Institute of Freshwater Ecology and Fisheries, Seestrasse 82, 12563 Berlin, Germany*
[3]*GKSS-Umweltforschungszentrum, Max Planckstrasse 1, 21502 Geesthacht, Germany*

Key words: Elbe, heavy metals, point and diffuse sources, river loads, retention

Abstract

Two independent methods are presented to estimate point and diffuse source contributions in emissions and riverine loads. The first method uses emission factors and areal statistics of the Elbe catchment to estimate the emissions of the diffuse and point sources. The second method relies on relationships between load–discharge and concentration–discharge, using monitoring data from several water quality stations. These statistical relationships are then used to separate point and diffuse contributions to the total riverine load, after incorporating the point waste water flow discharges. Both methods are applied to the Elbe river for the period 1992–1995 to estimate the contribution of point and diffuse sources for six heavy metals. The results of the source apportionment estimates for the Elbe river show that there are regional differences in source strength and source origin. The main inputs of metals occur in the Czech Republic nowadays, whereas for Cd the Mulde catchment contributes large amounts. The comparison of both methods shows some differences, which are caused by retention in the river system and uncertainties in emissions.

Introduction

Both point and diffuse sources contribute to the total metal load of rivers. A distinction between them is necessary for future restoration actions and determining the effect of past control measures at industrial sources. Point and diffuse sources can be distinguished by using inventories and by using basic hydrological relationships between discharge and transported load. Several studies focused on estimating the emissions and pathways of contaminants for large river basins or catchments (Behrendt, 1993 & 1997; de Wit, in prep., Feenstra et al., 1986; Werner et al., 1994). But many researchers (Hahn et al., 1994; Olendrzynski et al., 1995) only investigated one particular emission source at a certain specific location (e.g. canalization in urban areas). Whereas other studies focused on estimating riverine transported loads (Grimvall et al., 1996; Probst, 1985; Tonderski et al., 1994) by looking at discharge–concentration relationships. These discharge–concentration relationships have been used for comparisons between both estimated emissions and transported riverine loads of nutrients (Behrendt, 1993, van Dijk et al., 1997). Most studies estimating point and diffuse emissions as well as source apportionment focus on nutrients. Similar studies on heavy metals are rather scarce until now (Behrendt, 1993).

The aim of this paper is to quantify the point and diffuse emissions of several heavy metals, their distribution in the Elbe drainage area and compare these results to transported riverine heavy metal loads for the period 1992–1995.

Study area: Elbe

The Elbe drainage area covers approximately an area of 148 268 km^2, divided over four countries: Germany, Czech Republic, Poland and Austria (IKSE, 1995). This is illustrated in Figure 1. There are 25.2 million people are living in the Elbe drainage area, of which 75.7% are in Germany and 24.1% are in the Czech

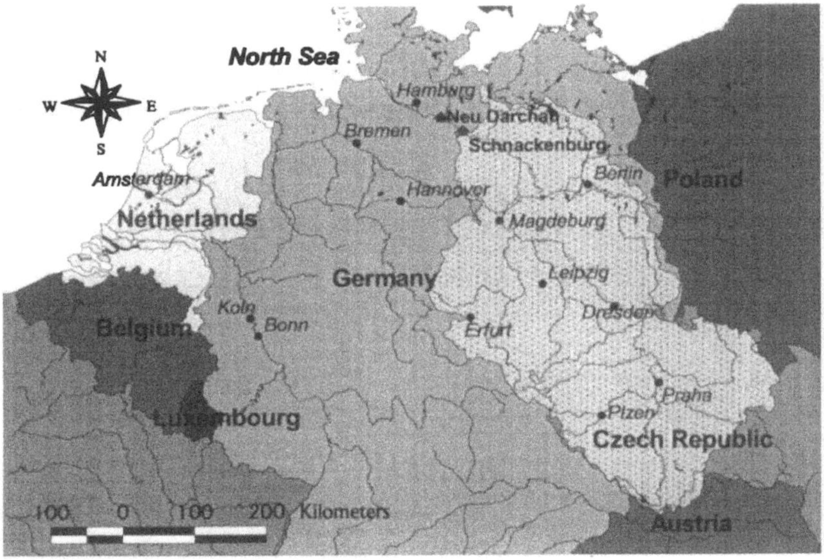

Figure 1. The Elbe drainage area.

Republic. About 56% of the total area in the Elbe catchment is in agricultural use and 27% is covered with forest. The Elbe river can be characterized as a 'rain–snow' type. The mean discharge of the Elbe river at gauging station Neu Darchau is 720 m^3/s. The history of pollution of the Elbe by heavy metals starts about 800 years ago as mining activities in the Erzgebirge developed (Guhr, 1995). Since the beginning of the industrialization the inputs by industries and urban areas have risen steadily. After the reunion of East and West Germany, the emissions have been reduced.

Emission estimates (Method 1)

The sources of chemical pollution to river systems are divided as follows (Novotny & Chesters, 1981):

- **Point sources** are *identifiable points* and are (fairly) *steady in flow and quality* (within the temporal scale of years). The magnitude of pollution is not influenced by the magnitude of meteorological factors. Major point sources under this definition included: municipal waste water effluents; industrial waste water effluents.
- **Diffuse sources** are *highly dynamic spread pollution sources* and their magnitude is *closely related to meteorological factors* such as precipitation. Major diffuse sources under this definition include: surface runoff (load from atmospheric deposition), groundwater, erosion (load from eroded material), diffuse loads of paved urban areas (atmospheric deposition, traffic, corrosion), including combined sewer overflows since these events occur discontinuous in time and are closely related to precipitation.

Point source emissions

The point sources are inventoried (IKSE, 1995a,b, 1996; Nesmerak, 1995) and divided into:

(a) sewage treatment plants with an effluent amount bigger than 20,000 inhabitant equivalent (EGW);
(b) industrial plants, divided into classes on the basis of their production (e.g. chemical industry, paper industry, metallurgical industry).

The emissions from sewage treatment plants were estimated on the basis of their treatment capacity (amount of inhabitants connected), treatment type (mechanical, biological, biological with N removal and biological with P removal) and related average efficiency of the different treatment types to remove the different heavy metals from the effluent given by Hahn & Xanthopoulos (1994), Firk (1993) and Ried (1992) based on studies performed in Karlsruhe. The product of each of these factors delivers the emissions by waste water treatment plants:

Emission per wastewater treatment plant (kg/a) = Mean input of heavy metals (kg/EGW) · Waste water treatment plant capacity (EGW) · Metal efficiency per treatment type (%)

The industrial point source emissions are based on data reported by the International Commission for the Protection of the Elbe (IKSE, 1995). These reports give estimates for the years 1989, 1991, 1992, 1993 and 1994. From these individual years estimates were derived for the period 1992–1995. The spatial distribution of the main point sources (PS) were evaluated on a sub-catchment level as shown in Table 1. The emissions from point sources of heavy metals in the Elbe drainage area were mainly emitted by the metal-processing plants and chemical industries. Most of the industry was located in the South of formerly East Germany and the Northern part of the Czech Republic. Following the unification of East and West Germany, many of the large-scale heavy industrial plants in Eastern Germany were closed and as a result the point emissions of heavy metals have been significantly reduced by the year 1992.

Diffuse source emissions

In a pristine catchment the only inputs of heavy metals will occur along the pathways of natural weathering of metal-bearing rocks, erosion and natural atmospheric deposition. Since the industrial revolution exploitation and associated processing of heavy metals have increased dramatically (Foster & Charlesworth, 1996). In addition, product use causes diffuse emissions since heavy metals dissipate into the environment. A first method presented by Feenstra & van der Most (1986) and later adapted from Behrendt (1993), Werner & Nolte (1994) and Raad et al. (1993), was used to estimate the diffuse metal emissions from various sources. The diffuse emissions are quantified on the basis of atmospheric deposition, land use, emission factors and activity data, which is defined as the amount of cars, inhabitants, fertilizers used and livestock, etc.

Direct atmospheric deposition

The direct atmospheric deposition was estimated by multiplying the total area of water surface (German statistical yearbook & Czech statistical yearbook, 1995) and total direct atmospheric deposition in a particular sub-catchment (Behrendt, 1993; Olendrzynski et al., 1995; Petersen et al., 1993; Skoda et al., 1994; van Jaarsveld, 1994).

Inputs from direct atmospheric deposition (kg/a) = Direct atmospheric deposition (kg/ha∗a)* Total water surface area (ha)

The direct atmospheric deposition (AD) is shown in Table 1. For all elements the contribution of direct atmospheric deposition to the total emissions is relatively small.

Erosion and surface runoff

The input of dissolved and particulate heavy metals was estimated using the method according to Werner et al. (1994) and Damaska et al. (1994). The method uses information on agricultural (arable) land, susceptibility to erosion of agricultural land, average metal content of fine soils (Barth, 1996) and enrichment ratio for sediments. The application of the USLE equation delivers the average long-term amount of eroded material. The emission derived from erosion and surface runoff per sub-catchment is calculated according to:

Emission by erosion and surface runoff (kg/a) = Mean soil loss of arable land (t/ha∗a)∗ Area of arable land ha* Metal content of fine soils (mg/kg)* Enrichment ratio (−) + Yearly mean direct surface runoff (m^3/a)* Mean metal concentration (μg/l)

The heavy metal load derived from erosion and surface runoff (ES) is illustrated in Table 1.

Manure and fertilizer

Agricultural activities influence the quality of surface water through the application of manure and fertilizer, since they both contain trace amounts of heavy metals. On average 0.4–3% of the applied manure and fertilizer flows directly into the surface water and drains (Raad et al., 1993). For the Elbe drainage area a run-off factor of 2% was used. On the basis of statistics on the application and usage of fertilizer and livestock, the heavy metal emissions were estimated.

Emission by manure and fertilizer (kg/a) = Run-off factor (%) * Total amount of fertilizer/manure (kg)* Metal content of fertilizer /manure (mg/kg)

Since the contribution of this diffuse source is relatively small (Vink et al., 1997), it will be neglected here.

Groundwater

When water infiltrates and reaches the groundwater, dissolved substances are also transported. The quantity of groundwater reaching the surface water was derived from Glugla (1985), who made estimates on groundwater recharge and the yearly water balance. The metal concentrations in the groundwater were derived from actual measured concentrations in the upper aquifer of the regional water authorities in Germany. The method of Dannowski et al. (1993) was

Table 1. Overview of estimated emissions (ton/a) in the Elbe drainage area.

Sub-catchment	Arsenic						Cadmium						Copper					
	PS	AD	ES	GW	UA	TE	PS	AD	ES	GW	UA	TE	PS	AD	ES	GW	UA	TE
Czech Republic[1]	9.2	1.1	5.2	23.5	1.6	40.6	1.4	0.1	0.4	0.9	0.2	3.0	43.0	1.6	9.1	62.5	3.3	120
Schwarze Elster	0.1	0.7	0.8	0.4	0.5	2.5	0	0.1	0.0	0.1	0.1	0.3	1.3	1.0	1.3	4.6	1.4	9.6
Mulde	0.7	1.0	4.3	0.6	0.9	7.5	0.1	0.1	0.2	0.2	0.1	0.7	5.1	1.3	5.4	1.6	2.3	15.7
Saale	0.9	3.2	6.6	1.9	2.3	14.9	0.4	0.3	0.3	0.8	0.4	2.2	11.2	4.4	10.4	4.9	6.0	36.9
Havel	0.7	3.2	2.3	3.0	2.6	11.8	0.1	0.3	0.2	0.8	0.4	1.8	6.8	4.4	3.9	5.0	6.9	27.0
Elbe (rest)	0.7	2.5	4.1	1.8	1.5	10.6	0.1	0.3	0.3	0.4	0.2	1.3	7.7	3.5	6.4	2.2	4.0	23.8
Total	12.3	11.7	23.3	31.1	9.4	87.8	2.0	1.2	41.5	3.2	1.4	9.3	75.1	16.2	36.5	80.8	23.9	233
Sub-catchment	**Mercury**						**Lead**						**Zinc**					
	PS	AD	ES	GW	UA	TE	PS	AD	ES	GW	UA	TE	PS	AD	ES	GW	UA	TE
Czech Republic	1.3	0.1	0.3	0.7	0.1	2.5	27.2	4.0	20.5	12.2	7.8	71.7	236	16.3	64.9	123	55.9	497
Schwarze Elster	0	0.1	0.0	0.0	0.0	0.1	0.5	2.6	3.1	0.6	2.2	9.0	6.9	10.5	7.5	37.6	80.3	143
Mulde	0.4	0.1	0.1	0.0	0.1	0.7	2.3	3.4	13.1	1.2	3.7	23.7	27.3	14.1	38.0	47.1	18.6	145
Saale	0.2	0.2	0.2	0.1	0.1	0.8	4.7	11.2	23.6	3.5	9.7	52.7	218	45.8	71.4	102	31.0	468
Havel	0.1	0.2	0.2	0.1	0.2	0.8	3.0	11.3	11.3	4.8	11.1	41.5	48.5	46.1	30.0	61.7	92.5	279
Elbe (rest)	0.1	0.2	0.2	0.1	0.1	0.7	3.3	9.0	15.7	3.0	6.4	37.4	44.7	36.6	48.2	78.9	53.6	262
Total	2	0.8	1.0	1.1	0.6	5.5	41.1	41.4	87.3	25.3	40.9	236	582	169	260	450	332	1793

PS, point source emissions;
AD, direct atmospheric deposition;
ES, erosion and surface runoff;
GW, groundwater;
UA, urban areas;
TE, total emissions;
[1] Data from Elbe Project, T.G.M. Masaryk Institute Prague, 1995.

Table 2. Estimated point and diffuse heavy metal loads (ton/a) at the catchment outlet Schnackenburg, Czech border Schmilka, Mulde catchment and Saale catchment for the period 1992–1995

Element	Schnackenburg			Czech Republic			Mulde			Saale		
	Point load	Diffuse load	Total load	Point load	Diffuse load	Total load	Point load	Diffuse load	Total load	Point load	Diffuse load	Total load
As	18	74	92	18	29	47	2.5	11	13.5	0.4	4.3	4.7
Cd	3.6	2	5.6	0	1.3	1.3	0.1	2.1	2.2	0.4	0.7	1.3
Cu	57	70	127	11	71	82	1.8	10.6	12.4	7.2	27	34.2
Hg	3	0.3	3.3	0.4	0.9	1.3	0.1	0.2	0.3			
Pb	24	65	89	3	55	58	0.3	6.5	6.8	10	15.2	25.2
Zn	654	963	1617	103	259	362	20	230	250	221	216	437

used to estimate the maximal heavy metal load which will drain to the surface water along the groundwater pathway.

Emission transported by groundwater (kg/a) = Yearly mean groundwater discharge (m3/a)* Mean groundwater concentration (μg/l)

The heavy metal load derived from groundwater (GW) is shown in Table 1.

Urban areas

Rivers running through or adjacent to urban developments are subjected to the influence of urban runoff and combined sewer overflows. Due to the different hydrological conditions (high run-off factors) prevalent in urban areas, high amounts of trace metals originate from these areas (Ellis et al., 1984, Hahn et al., 1994). Several statistical yearbooks were used to derive estimates on urban area, percentage of pavement, traffic, corrosion and the various treatment types of waste water treatment plants (Vink et al., 1997).

Emission from urban areas (kg/a) = {Metal emissions by cars (kg/a*car)* Amount of cars + Total atmospheric deposition (kg/a*ha) * Urban area (ha) + Emissions by corrosion (kg/ha*a)* Surface area of metals subjected to corrosion (ha)}*Metal efficiency per treatment type (%) + Sewer Overflow frequency (%) * {Metal emissions by cars (kg/a*car)* Amount of cars + Total atmospheric deposition (kg/a*ha) * Urban area (ha) + Emissions by corrosion (kg/ha*a)* Surface area of metals subjected to corrosion (ha)}

The heavy metal load derived from urban areas (UA) is depicted in Table 1.

Estimates from transported point and diffuse loads (Method 2)

Source apportionment of transported riverine loads was first used by Davis et al. (1978) and was elaborated and applied by Behrendt (1993), Grimvall & Stålnacke (1996) and van Dijk (1997) on various substances and catchments. In all these approaches point sources are considered to be flow-independent, whereas diffuse sources show a high correlation with flow and meteorological conditions. The first step is to establish relationships between concentration–discharge and load–discharge for the various metals and water quality stations.

The load of a pollutant can be described by the sum of all mass carried by the different runoff components (gross loads) minus losses occurring through, e.g. sedimentation:

$$L_{i,t} = L^g_{B,i,t} + L^g_{I,i,t} + L^g_{R,i,t} + L^g_{P,i,t} - \text{Losses}_{tot,t'},$$

where $L_{i,t}$ is the net pollutant load of sub-catchment i at a given time t, $L^g_{B,i,t}$ is the gross pollutant load in sub-catchment i caused by baseflow at time t, $L^g_{I,i,t}$ is the gross pollutant load in sub-catchment i caused by interflow, $L^g_{R,i,t}$ is the gross pollutant load in sub-catchment i caused by surface runoff, $L^g_{P,i,t}$ is the gross pollutant load in sub-catchment i caused by point sources at time t and $\text{Losses}_{tot,t}$ are the total pollutant losses in the surface waters of sub-catchment i at time t.

The net observed load at a monitoring station can be described according to:

$$L_t = L_{B,t} + L_{I,t} + L_{R,t} + L_{P,t'},$$

where the net individual loads caused by baseflow, interflow, surface runoff and point sources are already corrected for their losses. With this equation it is assumed that the losses and their magnitude are only dependent on the processes in the surface water and not on the source of the pollutant. In practice, it is impossible to estimate the individual runoff components and concentrations, but it is possible to separate the observed load into a part caused by diffuse sources and a part caused by point sources.

Assuming that the concentrations of surface runoff and interflow are about equal to a so-called diffuse concentration ($K_{D,t} = K_{I,t} \approx K_{R,t}$), the equation of the net observed load at a monitoring station transforms into:

$$L_i = K_{B,t} \times Q_{B,t} + K_{D,t} \times Q_{D,t} + K_{P,t} \times Q_{P,t}$$

By substitution of $Q_{D,t} = Q_t - Q_{P,t} - Q_{B,t}$ the equation reads:

$$L_t = K_{D,t} \times Q_t + L_{P,t} + L_{B,t} - K_{D,t} \times (Q_{P,t} + Q_{B,t})$$

At low discharges, where ($Q_t \approx 0$) the load at a monitoring station is dominated by inputs from baseflow and point sources and the equation can be rewritten as:

$$L_0 = L_{P,t} + +L_{B,t} - K_{D,t} \times (Q_{P,t} + Q_{B,t}).$$

This equation can be solved by establishing a relation between the measured discharge and load at a monitoring station and determining L_0 at $Q_t = 0$.

For the load discharged by point sources the following equation is valid:

$$L_P = K_P * Q_P = L_0 - L_B + K_D * (Q_P + Q_B)$$

With an inventory of point source effluents (m^3/s) it is possible to estimate the average discharge by these point sources Q_P. In order to be able to determine the point load one also has to make some assumptions about the load caused by groundwater because most of the time the necessary parameters Q_B and K_B unknown. But when the assumption is made that the weighted groundwater concentration K_B and the weighted diffuse concentration K_D are about equal, then the equation changes into a simple linear relationship

$$L_P = K_P * Q_P = L_0 + K_D * Q_P$$

From the available statistics on waste water effluents of point sources (Q_p) in sub–catchments and the established relationships of concentration–discharge and load–discharge the mean point load of the pollutant can be determined. The mean diffuse load over the considered period can be estimated from the difference between the net transported load and the estimated mean point load. The success of this method strongly depends on the quality of the used data set from the water quality stations. The data set from the water quality station Schnackenburg is homogeneous over time regarding the analytical methods. When measurements were below the detection limits the concentration was set to be half the detection limit. The waste water flows of point sources within a particular catchment area were derived from the German Statistical Yearbook (1994) and also given by Behrendt (personal comments). Ranges of point waste water flow (±20%) were used because of the uncertainty in these estimated values. The average estimates for diffuse and point sources for the station Schnackenburg, which represents the total (freshwater part of the) Elbe drainage area, is given Table 2.

The arsenic load is mainly caused by diffuse sources and is primarily of natural origin. The cadmium and the mercury load still have a relatively high point source contribution, although point emissions have declined. The water quality data for mercury and cadmium has relatively poor quality for most stations, and therefore the derived relationships and estimates should be questioned. For the elements copper, lead and zinc the highest contribution is delivered by diffuse sources.

Discussion and conclusions

Total Elbe basin

The comparison of total heavy metal loads and emissions for 1992–1995 show the following results (compare column TE in Table 1 with total load in Table 2). Firstly, the As emissions are generally under-estimated for all catchments. This could well be because of the higher background concentrations and (unknown) inputs from mining areas. Secondly, the diffuse Cd emissions are little over-estimated, which is probably caused by the high estimated inputs over the groundwater pathway. Thirdly, the diffuse Pb emissions are clearly over-estimated for all catchments. Finally, the findings for Hg should be considered with great care as the water quality data have poor quality. The emission estimates should be considered as gross loads into the river and the estimated riverine loads are net transported loads in the river system. The observed differences, e.g. higher values for the emission method to load-discharge method can be caused by temporary and permanent storage of particulate heavy metals in the river system, but also because of shortcomings of both

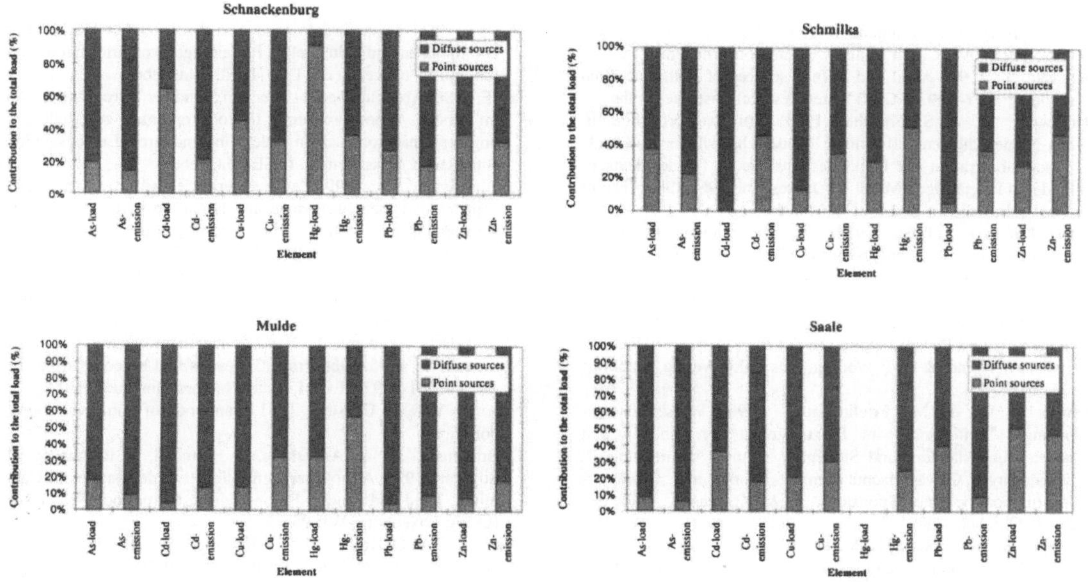

Figure 2. Comparison of heavy metal emissions with transported loads for several sub-catchments in the Elbe river basin.

estimation techniques. The explanation of material storage has first been put forward to explain differences observed for phosphorous (Behrendt, 1997). In addition, investigations in Mulde and Saale (Mueller et al., 1998) showed that mainly through sedimentation, material can be retained. Although, this retention of metals in sediments can explain a large percentage of the differences found between total emissions and total river transport, the uncertainties in specific regional data and water quality data should not be forgotten.

Source apportionment

By comparing the relative contributions of point and diffuse sources to the total load, an evaluation can be made on the capability of both methods for deriving source apportionment characteristics (see Figure 2). In this figure the relative contributions of point and diffuse sources for the different catchments are visualized. For the Czech Republic the point emissions estimated over the period 1991–1993 are clearly too high and not representative for the period 1992–1995. For the Mulde and Saale catchment we observe a clear dominance of the diffuse sources for both methods. The Saale catchment still has a high point source input for Zn, whereas the Mulde has a high input for Hg. The differences in the absolute amounts for these elements clearly differ for both methods, because of relatively poor estimates on point source emissions over the considered period and also because of the radical changes still occurring during this period.

With regard to the diffuse sources, the emissions by groundwater represent the dominant pathway for all of the heavy metals with exception of lead. Contrary to the results of Rhine (Behrendt, 1993) the emissions by urban areas are not so dominant. This study shows that there is no clear preference for either method. Both methods have their own value but should preferably be used in conjunction. The observed differences in point and diffuse load estimates show the urgent need for more reliable data on inputs as well as of water quality parameters. This will also make it possible to better quantify the permanent or temporary storage of metals in the catchment.

References

Barth, N., 1996. Bodenatlas des Freistaates Sachsen, Teil 1: Hintergrundwerte für Schwermetalle und Arsen in landwirtschaftlich genutzten Böden. Radebeul: Sächisches Landesamt für Umwelt und Geologie.

Behrendt, H., 1993. Point and diffuse loads of selected pollutants in the River Rhine and its main tributaries, Laxenburg: IIASA. RR-93-1.

Behrendt, H., 1997. Inventories of point and diffuse sources and estimated nutrient loads - a comparison for different river basins in Central Europe. Wat. Sci. Techn. 33: 99–107.

Czech Ministry of Environment, 1996. Czech Environmental Yearbook 1993/1994. Czech Ministry of Environment, Praha.

Damaska, J., 1994, Areal and diffuse sources of pollution -final report of 1991–1993 T.G.M. Water Research Institute, Praha.

Dannowski, R. & S. Fritsche, 1993. Grundwasserneubildung aus Niederschlägen, effecktieve Niederschlagshöhe, und reale Evapotranspiration im Lockergesteinsbereich Nordostdeutschlands im langjährigen Mittel der Jahresreihe 1951/1980. Institut fuer Hydrologie im ZALF e. V., Müncheberg.

Davis, J.S. & J. Zobrist, (1978). The interrelationships among parameters in rivers - analysing the effect of natural and anthropogenic sources. Prog. Water Technol. 10: 65–78.

Deumlich, D. and M. Frielinghaus, 1994. Eintragspfade Bodenerosion und Oberflächenabfluss im Lockergesteinsbereich. In: Stickstoff- und Phosphateintrag in die Fliessgewässer Deutschlands, W. Werner & H.-P. Wodsak, (eds) DLG-Verlag, Frankfurt (Main) 48–84.

Deumlich, D. & M. Frielinghaus, 1996. Wassererosionsbedingte Stoffeinträge im Einzugsgebiet der Elbe in den neuen Bundesländern und Strategien zu ihrer Vermeidung. 7. Magdeburger Gewässerschutzseminar- Vorseminar Gewässer-Informationssysteme. Spoustova, J. & K. Janson, Budweis: 21–25 Oktober, 1996.

van Dijk, S., J. Knoop, M.J.M. de Wit & R.J. Leeuwis, 1997. Source apportionment and quantification of nitrogen transport and retention in the River Rhine. RIVM report no. 733 008 004.

Ellis, J.B., D.O. Harrop & D.M. Revitt, (1986) Hydrological control of pollutant removal from highway surfaces. Wat. Res. 20: 589–595.

Feenstra, J.F. & P.F.J. van der Most, 1986. Diffuse bronnen van waterverontreinigingen. CUVWO VI.

Firk, W. & B. Boehnke, 1983. Untersuchungen über die Konzentration und Frachten von Arsen, Blei, Cadmium, Chrom, Cobalt, Kupfer, Mangan, Nickel, Quecksilber, Selen, Silber, Zink und Zinn im Zulauf und Ablauf im anfallenden Klärschlamm der Kläranlagen Aachen-Soers, Düren und Stolberg-Steinfurt. Karlsruhe:Institut für Siedlungswasserwirtschaft.

Foster, I.D.L. & S.M. Charlesworth, 1996. Heavy metals in the hydrological cycle: trends and explanation Hydrol. Processes 10: 227–261.

Glugla, G., 1985. Anwenderrichtlinie zur Nutzung der neuen Karten des Gesamtabflusses m 1:200.000 (Deckfolien zur Abflusskarte des N-A-U-Kartenwerkes über das Gebiet der DDR). Institut für Wasserwirtschaft, Berlin.

Grimvall, A. & P. Stålnacke, 1996. Statistical methods for source apportionment of riverine loads of pollutants. Environmetrics 7: 201–213.

Guhr, H., 1995. Sources of heavy metal pollution in the drainage area of the River Elbe in the former GDR. In: R.-D. Wilken, U. Foerstner & A. Knoechel, (eds) Heavy metals in the Environment. Edinburgh: CEP Consultants Ltd., 3: 72–75.

Hahn, H. & C. Xanthopoulos, 1994. Niederschlagsbedingte Schmutzbelastung der Gewässer aus Städtischen befestigten Flächen unter besonderer Berücksichtigung der Stadtentwässerung in den neuen Bundesländern. Institut für Siedlungswasserwirtschaft, Karlsruhe.

IKSE, 1995a. Aktionsprogramm Elbe. IKSE, Magdeburg.

IKSE, 1995b. Bestandsaufnahme von bedeutenden punktuellen kommunalen und industriellen Einleitungen von prioritären Stoffen im Einzugsgebiet der Elbe. IKSE, Magdeburg.

IKSE, 1996. Abschlussbericht über den Stand der Durchführung der im 'Ersten Aktionsprogramm (Sofortprogramm) zur Reduzierung der Schadstofffrachten in der Elbe und ihrem Einzugsgebiet' enthaltenen Massnahmen. IKSE, Magdeburg.

van Jaarsveld, J.A., 1994. Atmospheric deposition of cadmium, copper, lead, benzo(a)pyrene and lindane over Europe and its surrounding marine areas. -Background report of the ESQUAD study. RIVM, Bilthoven: RIVM 22401002.

Mueller, A., A. Arnold, C. Hänsel, K. Hanisch, K. Jendryschik & L. Zerling, 1998. Der Bitterfelder Muldestausee als Schadstoffsenke -Entwicklung der Schwermetallbelastung 1992 bis 1997. Sächsische Akademie der Wissenschaften zu Leipzig.

Nesmerak, I., 1995. Elbe Project: Results and outcome from the period 1991-1993.. T.G.M. Water Research Institute, Praha.

Novotny, V. & G. Chesters, 1981. Handbook of non-point source pollution.

Olendrzynski, K., S. Anderberg, J. Bartnicki, J. Pacyna & W. Stigliani, 1995. Atmospheric emissions and depositions of Cadmium, Lead and Zinc in Europe during the period 1955–1987. IIASA, Laxenburg: WP-95-35.

Petersen, G. & O. Krueger, 1993. Untersuchung und Bewertung des Schadstoffeintrags über die Atmosphäre im Rahmen von PARCOM (Nordsee) und HELCOM (Ostsee)- Teilvorhaben: Modellierung des grossräumigen Transports von Spurenmetallen. Geesthacht: GKSS 93/E/28.

Probst, J.L. 1985. Nitrogen and phosphorous exportation in the Garonne basin (France) J. Hydrol 76: 281–305.

Raad, J.S., R.A.L. Peerboom, C. Dorland, M.C. Hanegraaf & J.F. Feenstra, 1993. Diffuse bronnen van waterverontreinigingen in het oppervlaktewater in Friesland. Inventarisatie en kwantificering. Institute for Environmental Studies, Amsterdam: E93/12.

Ried, A., 1992. Analytik von Schwermetallen und EDTA im Abwasser und deren Verhalten während der kommunalen Abwasserreinigung. Marburg, Philipps-Universität.

Skoda, J. & J. Santroch, 1994. Total atmospheric deposition by selected parameters: final report of 1991–1993. Praha: CHMU.

Statistisches Bundesamt Statistisches Jahrbuch 1990, 1991. Statistisches Bundesamt, Wiesbaden.

Statistisches Bundesamt Statistisches Jahrbuch 1994, 1995. Statistisches Bundesamt, Wiesbaden.

Tonderski, A., A. Grimvall, K. Sundblad & P. Stålnacke, 1994. An East–West perspective on riverine loads of nutrients in the Vistula and Rhine basins. Wat. Sci. Techn. 30: 121–130.

Vink, R.J., H. Behrendt & W. Salomons, 1997. Point and diffuse source analysis of heavy metals in the Elbe drainage area: comparing heavy metal emissions with transported river loads GKSS-report, Germany.

Werner, W. & H.-P. Wodsak, 1994. Stickstoff- und Phosphateintrag im Fliessgewässer Deutschlands unter besonderer Berücksichtigung des Eintragsgeschehens im Lockergesteinsbereich der ehemahligen DDR. DLG-Verlag, Frankfurt (Main) Agrarspektrum 22.

de Wit, M.J.M. 1998. Nitrogen and phosphorous emissions in the Rhine basin. PhD thesis University of Utrecht (in prep.).

Hydrobiologia **410:** 315–323, 1999.
J. Garnier & J.-M. Mouchel (eds), Man and River Systems.

Annual emissions of pollutants from mine stone applications in drainage basins of Dutch rivers

R. S. E. W. Leuven, P. H. Nienhuis, J. M. A. Kesseleer & W. A. Zwart
Department of Environmental Studies, University of Nijmegen, P.O. Box 9010, 6500 GL Nijmegen, The Netherlands

Key words: hydraulic engineering, leaching models, mine stone, river pollution

Abstract

The present paper estimates the annual emissions of eight (heavy) metals and six polycyclic aromatic hydrocarbons (PAHs) from mine stone applications in hydraulic engineering structures in Dutch river basins over the periods 1980 – 2010. Two leaching models (a diffusion and a percolation model) and three application scenarios (i.e. the stop, current perception and increase scenario) have been used. It is concluded that under the current scenarios, the contribution of mine stone to pollution of Dutch river systems is rather low in comparison with emissions from industrial, communal and diffuse sources. However, continued and widespread large-scale applications of mine stone will lead to increased leaching of pollutants to surface waters, and may require that further cuts be made from other sources in order to meet regional, national and international emission reduction targets.

Introduction

Dutch environmental policy intends to scale down the utilization of primary building materials (e.g. sand and gravel) in hydraulic engineering structures, and to promote the substitution of these resources by alternative materials (Aalbers et al., 1996; De Wilde et al., 1997). The desire to use bulky wastes (such as mine stone) as secondary raw materials in hydraulic engineering structures has been motivated by the environmental hazards connected with the mining of surface minerals, the overexploitation of primary resources and the lack of landfill capacity for wastes (Rainbow, 1990; Skarzynska, 1993).

Since 1980 mine stone is a common building material in hydraulic engineering structures in drainage basins of Dutch rivers. Mine stone is a by-product of coal mining (synonyms of mine stone are coal mining waste, colliery spoil, colliery discard, grained carbon waste and quarry waste) and consists of fragments of the rock accompanying coal beds or is brought to the surface during the construction of mine shafts (Skarzynska, 1995a,b).

Mine stone can be used in a variety of ways, e.g. in constructing dams, dikes, breakwaters, quays and in protective layers which prevent damage to river beds, embankments and shores exposed to erosion (Rainbow, 1990; Skarzynska, 1993). The (potential) utilization of this building material has important and undeniable national implications with respect to economic and environmental impact (Skarzynska, 1995a,b). On the one hand, mine stone is abundant and inexpensive, and the possibilities for transport along water routes are relatively good due to the geographical location of coal mining (Hart, 1990; Sleeman, 1990; Skarzynska, 1995a). On the other hand, mine stone contains various harmful substances (e.g. metals and organic micropollutants), which leach into the groundwater and river water when it is used in hydraulic engineering structures and fills (Gonzalez, 1990; Hauss & Mayer, 1990; Twardowska et al., 1990; Skarzynska, 1995b; Aalbers et al., 1996).

In order to minimize pollution and to prevent environmental damage, the utilization of secondary building materials in Dutch surface waters is regulated by the Pollution of Surface Waters Act (PSWA) and the Soil Protection Act (SPA). Until now, the competent authorities for environmental permitting have only considered the leaching of toxic substances for each single project. The impact of a single project on water quality is generally negligible, due to low leaching rates and high water flow in rivers and estuaries. The

Table 1. Average chemical composition of sorted and unsorted mine stone (Aalbers et al., 1996)

Metal	Unsorted			Sorted		
	N	[X]	SD	*N*	[X]	SD
As	30	13.84	11.19	11	4.19	1.60
Cd	30	0.83	0.40	10	0.63	0.48
Cr	30	32.98	21.47	10	40.07	20.06
Cu	30	35.55	13.61	9	27.03	10.27
Hg	7	0.17	0.17	10	0.11	0.14
Ni	29	38.96	14.24	12	29.27	14.08
Pb	30	26.71	22.38	11	15.00	3.60
Zn	30	138.66	119.98	11	67.00	27.90

N: number of samples; [X]: average content in mg kg^{-1}; SD: standard deviation.

consequences of cumulative emissions for a river basin or on a national scale are unknown. According to the jurisprudence of the Dutch Council of State, they must be taken into account in the case of continued application of mine stone (Raad van State, 1994).

The aim of the present paper is to estimate the utilization of mine stone over the period 1980 – 2010 and to determine the annual emissions of eight metals (arsenic, cadmium, chromium, copper, lead, mercury, nickel and zinc) and six high molucular PAHs (list of Borneff: benzo(b)fluoranthene, benzo(k)fluoranthene, benzo(a)pyrene, benzo(ghi)perylene, fluoranthene, indeno(123cd)pyrene) from mine stone in hydraulic engineering structures in the Dutch parts of the drainage basins and estuaries of the rivers Rhine and Meuse. The metals and PAHs were selected on availability of consistent data for modelling the leaching process. The consequences of various utilization scenarios for the feasibility of national and international emission reduction targets will be discussed.

Materials and methods

Data

Historical data on the utilization of mine stone in hydraulic engineering structures were collected from environmental permits, archives of several non-governmental organizations, notifications in the Dutch Gazette, import files and literature (Dutch Ministry of Environment, 1984; Saft 1992; Zwart, 1996).

Data concerning the metal contents of sorted mine stone (sizes 10 – 125 mm) and unsorted mine stone (sizes 0 – 70 mm) were obtained from inventories prepared by Aalbers et al. (1996). Average values and statistical specifications were summarized in Table 1. PAHs contents were determined by Technisch Adviesbureau Unie van Waterschappen (TAUW, 1986). The average PAH6 content (i.e. sum of the average concentrations of six PAHs: abovementioned list of Borneff) used for sorted as well as unsorted mine stone was 0.7 mg kg^{-1}.

Maximum leaching percentages of (heavy) metals and PAHs were determined by TAUW (1987, 1988), using pulverized mine stone (< 3 mm) in stirring tests (4 h) with demineralized water (acidified till pH=4) and liquid/solid ratio (L/S) of 100. L/S is the ratio between water volume (in l) and mine stone weight (in kg^{-1}) in leaching tests. Data with respect to leaching in batch tests (time series 6 h – 64 d and L/S=10) were extracted from Gemeentewerken Rotterdam (1991). Avarage data with respect to leaching in column tests (L/S=10) were obtained from Aalbers et al. (1996). Batch and column tests simulate diffusion and percolation, respectively. The leaching tests were performed according to the Dutch standards 7343 and 7345 (Nederlands Normalisatie-instituut, 1995a,b). Because there is considerable similarity between the chemical properties of mine stone from different sources and countries (Skarzynska, 1995a), average values of the most frequently applied mine stones were used for the (maximum) leaching percentages. All data used for modelling were described by Zwart (1996).

Application scenarios

Three scenarios for the expected annual utilization of mine stone over the periods 1980 – 2010 were designed: the stop scenario (SS), current perception scenario (CPS) and increase scenario (IS). All scenarios distinguish between the utilization of mine stone in structures predominantly exposed to river water (e.g. dams, breakwaters, groins, quays and protective layers which prevent damage to river beds) and works predominantly exposed to groundwater (e.g. fills).

The SS assumes the utilization of mine stone in works predominantly exposed to river water and groundwater to have ceased in 1996. In the CPS, the estimated utilization of mine stone in works predominantly exposed to river water in 1996 is based on the application procedures for environmental permits, while the average of the annual utilization of mine stone over the period 1980 – 1995 (real data) is used for the years 1997 – 2010. In the IS, the utilization of mine stone in works predominantly exposed to river water in 1996 equals that of the CPS, and gradually increases to twice that of the CPS in the period 1997 – 2010. This increase can be expected if Dutch parliament accepts liberalization of quality standards for secondary building materials (best professional judgement). The CPS and IS assume that the utilization of mine stone in works predominantly exposed to groundwater is according to the outlook in an environmental impact assessment for reclamation of some harbours as building sites (Gemeentewerken Rotterdam, 1991).

Modelling and data processing

Based on the three scenarios, the annual emissions of (heavy) metals and PAHs into surface water were calculated. Average data of leaching in batch and column tests were transformed to field conditions by test-specific modelling: i.e. diffusion and percolation model (Zwart, 1996). Both models included subroutines to calculate emissions from works with sorted and unsorted mine stone, as well as from works predominantly exposed to river water and predominantly exposed to groundwater, because of differences in leaching of chemical substances under these conditions. Models simulated annual leaching under freshwater conditions, because available quantitative data mainly concerned works in rivers and groundwater with a low salt content and consistent experimental data on leaching in brackish and sea water were lacking.

The diffusion model took into account the utilization of mine stone in the past and the substance specific initial and post-initial (constant) leaching in batch tests (L/S=10), the duration of leaching processes and an extrapolation factor for transformation batch tests to field conditions. The proportions and duration of initial and constant leaching were determined on the basis of time series in batch tests and maximum leaching in stirring tests. All diffusion model formulations and calculations of proportion and duration of substance specific initial and constant leaching and extrapolation factors can be found in Zwart (1996).

The percolation model took into account the utilization of mine stone per year, the leaching in column tests (L/S=10) and an extrapolation factor for transformation of L/S ratio in column tests to field conditions. Extrapolation was required because for practical reasons column tests simulated percolation with relatively high L/S ratios and short exposure times. The extrapolation factor was determined by the initial concentration, the leaching characteristics in relation to the exposure time and the L/S ratio under field conditions. All percolation model formulations and substance specific extrapolation factors can be found in Aalbers et al. (1996) and Zwart (1996). Lowest and highest L/S ratios used for field conditions were 0.01 and 0.11, respectively. These rates were determined as a function of bulk density of mine stone, depth of the hydraulic engineering structure, percolation of water through the work and exposure time. Lowest and highest L/S ratios were simulated using the Monte Carlo technique (5000 iterations), assuming triangular distributions for the depth of works and percolation and accepting 95% certainty (Zwart, 1996). The triangular distribution describes a situation were the minimum and maximum values of parameters are fixed and the most likely value falls between these values, forming a triangular-shaped probability, which shows that the values near the minimum and maximum are less likely to occur than those near the most likely value (Decisioneering Inc., 1993). The minimum, maximum and most likely values used for the depth of a work (0.1, 0.7 and 5.1 m, respectively) and percolation (30, 50 and 110 mm y^{-1}, respectively) were obtained from environmental permits and literature (Zwart, 1996; Aalbers et al., 1996). For works predominantly exposed to groundwater (i.e. fills outside the riverbed), the percolation model assumed that 10% of the water percolated through a work contributed to river water (exfiltration) and 90% remained in groundwater (infiltration). This assumption is based on best professional judgements (cf. Gemeentewerken Rotterdam, 1991).

The model formulations were programmed in the spreadsheet program Microsoft Excel 5.0® (Microsoft, 1994). Monte Carlo simulations of percolation were performed with Crystall Ball® (Decisioneering Inc., 1993). The STIPT® version 5.2 program (Frigge, 1996) was used for geographical mapping of known locations of mine stone applications.

Results

Table 1 illustrates differences in chemical composition of sorted and unsorted mine stone. Zinc shows the highest average content in both types of mine stone. Cadmium and mercury contents are rather low. Nickel, lead, chromium and copper occupy intermediate rank

Table 2. Initial and (duration of) constant leaching in batch tests (L/S=10)

Pollutant	Unsorted			Sorted		
	Initial leaching (μg kg^{-1})	Constant leaching (μg kg^{-1})	D (y)	Constant leaching (μg kg^{-1})	Constant leaching (μg kg^{-1})	D (y)
As	118	18	81	118	18	21
Cd	12	0	0	10	0	0
Cr	330	0	0	401	0	0
Cu	617	18	48	617	18	29
Hg	6	2	1	6	0	0
Ni	494	37	15	494	40	9
Pb	370	37	26	370	40	10
Zn	1605	221	60	1605	221	26
PAH6	5	0	0	5	4	<1

D: Duration of constant leaching; PAH6: six polycyclic aromatic hydrocarbons (list of Borneff).

Table 3. Annual utilization of mine stone in works predominantly exposed to groundwater

Scenario	Period/Year	Annual utilization (10^6 kg y^{-1})
All scenarios	1980 – 1989	0
	1990 – 1992	50
	1993	5
	1994 – 1995	0
Stop scenario	1996 – 2010	0
Current perception and increase scenario	1996	3050
	1997 -2005	2600
	2006 – 2010	0

orders. Metal contents (except chromium) of unsorted mine stone are higher than of sorted minestone.

Table 2 presents the calculations of substance specific initial and constant leaching in batch tests. Cadmium, chromium, mercury and PAHs predominantly show initial leaching. For arcenic, copper, lead, nickel and zinc initial leaching, as well as constant leaching must be regarded as important parameters for modelling annual emissions. The duration of constant leaching for these pollutants from unsorted and sorted mine stone varies between 15–81 y and 9–26 y, respectively.

Figure 1 shows the known locations of hydraulic engineering structures in which mine stone has been used as a building material. Most locations are situated in the south-western part of the country (i.e. in the lower reaches and estuaries of rivers Rhine and Meuse).

Figure 2 illustrates the total annual utilization of sorted and unsorted mine stone in hydraulic engineering structures predominantly exposed to river water, according to the three scenarios (period 1980 – 2010). Although it is known that in the Netherlands mine stone was already used for this type of works before the 1980s (e.g. the Delta Works), quantitative data or estimates are not available to the authors. Over the period 1980 – 1990, mine stone was used in fairly large quantities. However, these figures are mainly based on estimates (VROM, 1984; Saft, 1992; Zwart, 1996), whereas real data are only available for the years 1982 and 1983. Until 1990, the competent authorities tolerated the utilization of mine stone in surface waters without PSWA-permit procedures, and the registration of secondary building materials applied was very limited. Over the period 1990 – 1995, the utilization of mine stone was reduced compared with the period 1980 – 1990, due to the enforcement of PSWA (ban on dumping waste materials in surface waters without a proper permit) and due to objections and appeals against PSWA-permits for the application of secondary building materials in surface waters.

Table 3 presents the annual utilization of mine stone in works predominantly exposed to groundwater. In this type of works, only unsorted mine stone is applied. Over the period 1996 – 2010, mine stone will be used mainly as filling material to reclaim some harbours in the city of Rotterdam as building sites for industrial and residential development.

Figure 3 shows results of calculations using the diffusion model developed. This model may be regarded as worst case approach, because under field conditions leaching only partly depends on diffusion. The annual emission of zinc from mine stone in hydraulic engineering structures and fills will be up to 3778 kg in the year 2010, based on the IS. The annual emissions of cadmium, mercury and PAHs are rather low (all scenarios < 10 kg y^{-1}), whereas arsenic, chromium, copper, lead and nickel show intermediate emissions.

The leaching of arsenic, copper, lead, nickel and zinc will continue for a long time, even in the case of the SS (see Figure 3). This is related to the duration of constant leaching (as shown in Table 2). Annual emissions of pollutants with predominantly initial leaching (i.e. cadmium, chromium, mercury and PAHs) show quick and large responses on changes in the annual

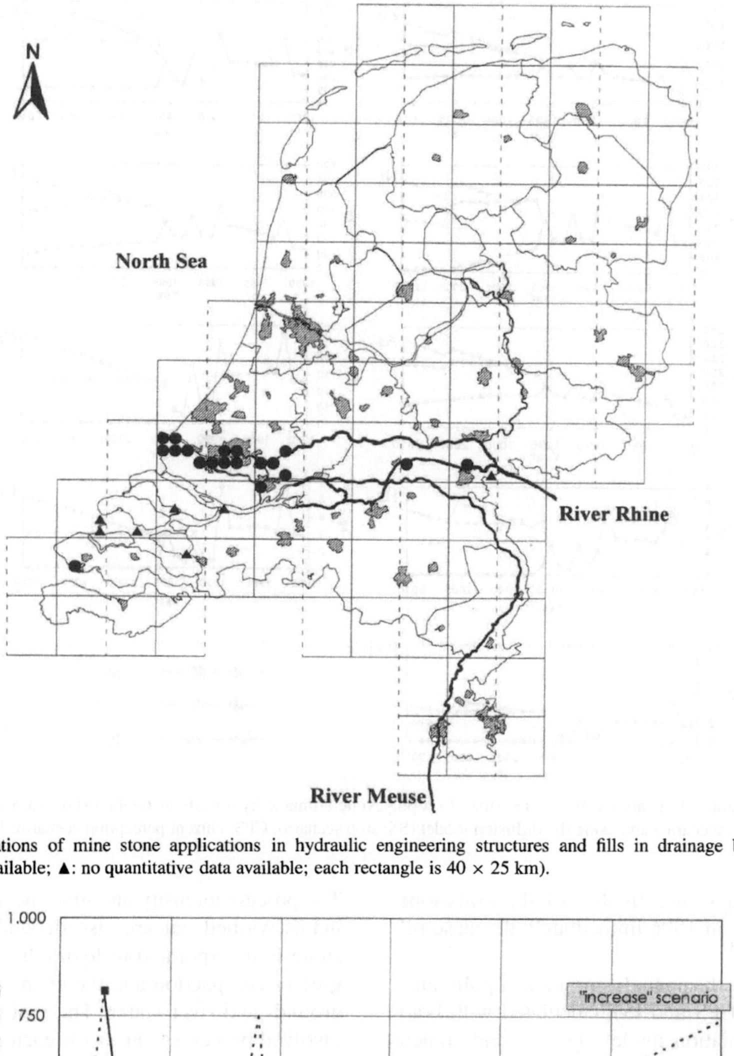

Figure 1. Known locations of mine stone applications in hydraulic engineering structures and fills in drainage basins of Dutch rivers (•: quantitative data available; ▲: no quantitative data available; each rectangle is 40 × 25 km).

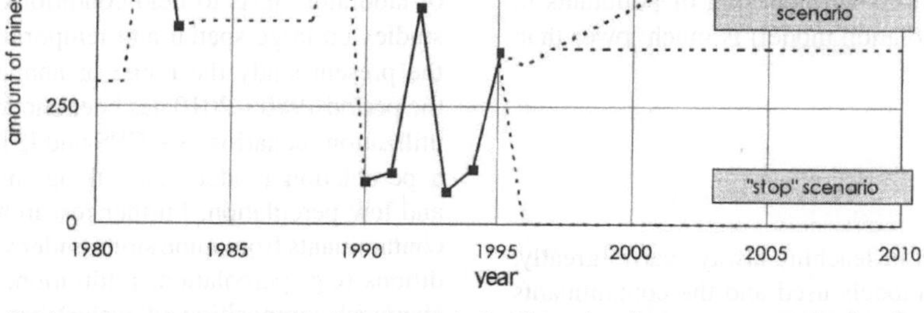

Figure 2. Annual utilization of sorted and unsorted mine stone in hydraulic engineering structures, predominantly exposed to river water (■—■: real data; - - - -: estimates).

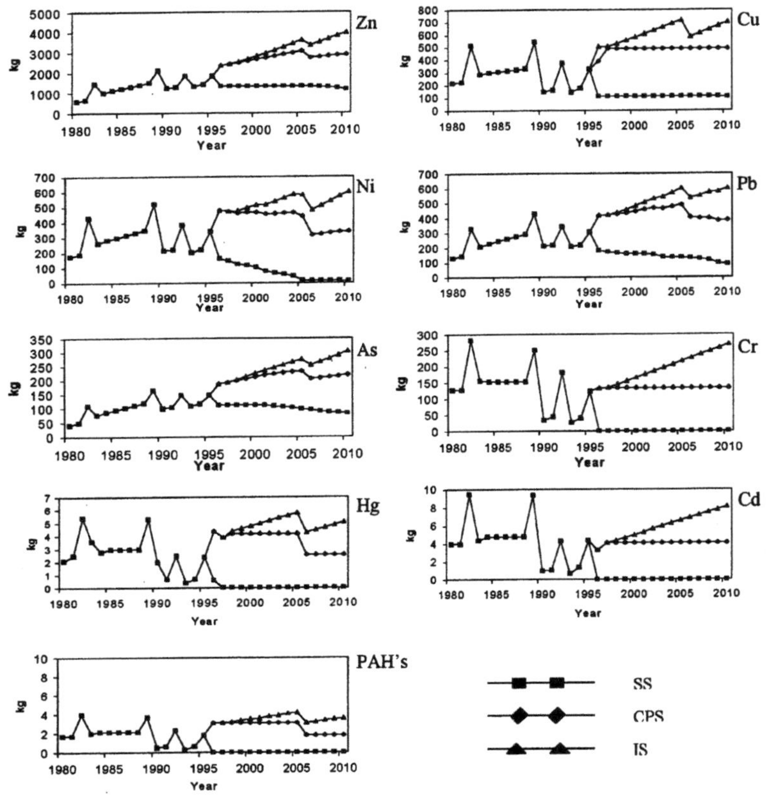

Figure 3. Annual emissions of metals and PAH6 (i.e. sum of six polycyclic aromatic hydrocarbons on the list of Borneff) in the period 1980 – 2010, based on the three scenarios and using the diffusion model (SS: stop scenario; CPS: current perception scenario; IS: increase scenario).

utilization of mine stone. In the SS the emissions of these substances in 1996 immediately decrease till zero.

Table 4 presents the annual emissions of pollutants in the year 2010 (CPS and IS), calculated with both models. The percolation model always yields much lower emissions than the diffusion model. This is due to low L/S ratios assumed in the percolation model for field conditions. Moreover, leaching of pollutants in column tests (percolation model) is much lower than in batch tests.

Discussion

The predicted level of leaching always varies greatly, depending on the models used and the contaminants considered. Under field conditions, leaching will depend on a combination of diffusion and percolation. The process intensity and ultimate amount of contaminants washed out are also dependent on how mine stone is incorporated in hydraulic structures, the degree of compaction and the chemical composition of ground- and river water. The real contribution of all involved processes varies for each application, and is actually unknown. Model assumptions and uncertainties of input parameters are intrinsic to extrapolation of laboratory tests to field conditions and to scenario studies on large spatial and temporal scale levels. In the present study the range of annual leaching over the period 1980 – 2010 has been modelled using three utilization scenarios (SS, CPS and IS), a diffusion and a percolation model, and taking into account high and low percolation. Further research on leaching of contaminants from mine stone under various field conditions (e.g. percolation, infiltration, exfiltration and chemical composition of water) can validate model

Table 4. Annual emissions of pollutants (in kg) from mine stone in the year 2010

Pollutant	Current perception scenario			Increase scenario		
	Percolation model		Diffusion model	Percolation model		Diffusion model
	Minimum	Maximum		Minimum	Maximum	
As	2	16	218	3	21	304
Cd	<1	1	4	<1	1	8
Cr	1	6	134	1	8	269
Cu	1	7	431	2	9	700
Hg	<1	<1	3	<1	<1	5
Ni	1	3	340	1	5	600
Pb	1	6	386	1	8	602
Zn	3	16	2894	4	21	3978
PAH6	n.d.	n.d.	2	n.d.	n.d.	4

n.d.: not determined; PAH6: sum of six polycyclic aromatic hydrocarbons (list of Borneff).

assumptions and will reduce uncertainties in model results.

Some difficulties were encountered during the collection of historical data. Before 1990, the competent authorities usually tolerated the application of secondary building materials without environmental permitting or formal notification procedures. In several cases (e.g. the Delta Works), the amounts, sources and properties of the mine stone used were not or incompletely documented. Due to this lack of historical data, the present study probably underestimates the utilization of mine stone in hydraulic engineering structures and annual emissions of pollutants from this works over the period 1980 – 1990. Moreover, this may also result in an underestimation of the annual emissions over the period 1990 – 2010 of substances with a long duration of leaching processes (i.e. arsenic, copper, lead, nickel and zinc).

Mine stone used in the past will remain for a long time an emission source of arsenic, copper, lead, nickel and zinc and continued utilization of this building material will add new sources of river water pollution. On the basis of the IS and the diffusion model (worst case approach), in 2010 mine stone will have a maximum share in annual Dutch emissions of arsenic and nickel to surface water of 3.7 and 1.3% respectively. This is related to the total emissions in 1995 from industrial, communal and diffuse sources, i.e. 8200 kg arsenic and 46 000 kg nickel according to estimates of the Dutch Ministry of Transport and Public Works (1994). The maximum share in the annual emission of other pollutants will be much lower (< 1%). As most of the mine stone has been applied in the south-west part of the country comparison with regional emissions would be more relevant and would reveal higher relative shares, but local or regional fluxes of pollutants are not available to the authors. According to the Dutch *National Policy Document on Water Management* and the international *Rhine Action Programme* and *North Sea Action Plan*, emissions of (heavy) metals and organic micropollutants from diffuse and point sources must be reduced with 70 – 90% over the period 1985 – 2010 (Dutch Ministry of Transport and Public Works, 1989). In the Netherlands, the emission reduction targets for the period 1985 – 1995 (50 – 70%) reduction have not been met for several pollutants (e.g. copper, lead, mercury, nickel, PAHs and zinc), and meeting the reduction targets and water quality objectives for the period 1985 – 2010 will be very difficult (Dutch Ministry of Transport and Public Works, 1994, 1996). The results of the present scenario study imply that the annual emissions of pollutants from minestone in hydraulic engineering structures, e.g. arsenic and nickel, may be a factor contributing to these problems. When the new General Administrative Order on Construction Materials Soil and Water Protection fully comes into force in 1999, it will allow the use of mine stone and several other secondary building materials (e.g. metal and phosphorous slags) after a simple notification procedure, as long as product quality standards for organic composition and for leaching of inorganic substances are met (Aalbers et al., 1996). The liberalization of quality standards may lead to a strong increase in the application of secondary building materials (cf. De Wilde, 1997). Therefore, the real utilization of mine stone over the period 1996 – 2010 is expected to be higher than that predicted by the CPS, and may be nearer to or even higher than that in the IS. This means that extra efforts will have to be made to reduce other sources of water pollution if we are to meet the regional, national and international emission reduction targets. Therefore, it is recommended that policy makers should weigh up all 'pros' (e.g. less use of primary materials and cost reduction), 'cons' (e.g. additional effort to meet emission reduction targets) and their quantitative uncertainties on a national level, especially if future developments force emission reduction targets to be set even lower than in present plans. Most of the mine stone used in the Netherlands is imported from Germany and Belgium. It should be noted that storing minestone abroad also causes serious environmental problems (Hauss & Mayer, 1990; Sleeman, 1990) and leaching from dumps and colliery spoil heaps may contribute to transboundary river pollution.

Conclusions

The annual emissions of cadmium, mercury and PAHs from mine stone applications in Dutch river basins are rather low (all models and scenarios applied < 10 kg y^{-1}). Mine stone will remain for a long time an emission source of arsenic, copper, lead, nickel and zinc. Annual emissions of arsenic and nickel in 2010 will have a share in total Dutch emissions to surface water of 3.7 and 1.3%, respectively (worst case approach). Continued and large scale utilization of mine stone in hydraulic engineering means that extra efforts will have to be made to reduce other sources of water pollution if we are to meet the regional, national and international emission reduction targets.

Acknowledgements

The authors sincerely thank Mr J. Klerkx and Mr A. M. J. Ragas for reviewing an earlier draft of the manuscript, and Mr G. W. Geerling and Mr H. J. R. Lenders for help with computer programs.

References

Aalbers, Th. G., P. G. M. De Wilde, G. A. Rood, P. H. M. Vermij, R. J. Saft, A. I. M. Van De Beek, M. H. Broekman, P. Masereeuw, Ch. Kamphuis, P. M. Dekker & E. A. Valentijn, 1996. Environmental quality of primary and secondary construction materials in relation to re-use and protection of soil and surface water. Report 771402007, National Institute of Public Health and the Environment, Bilthoven. 686 pp.

De Wilde, P. G. M., I. H. Anthonissen, A. I. M. Van De Beek & J. Keijzer, 1997. Afzet afvalstoffen als secundaire grondstoffen: milieuhygiënische kwaliteit van secundaire bouwstoffen. Rijksinstituut voor Volksgezondheid en Milieuhygiëne, Bilthoven (in Dutch).

Decisioneering Inc., 1993. Cristall Ball Version 3.0. Forecasting and risk analysis for spreadsheet users. Decisioneering Inc, Denver: 244 pp.

Dutch Ministry of Environment, 1984. Kwantitatieve inventarisatie gebruik van secundaire grondstoffen. Ministerie van Volkshuisvesting, Ruimtelijke Ordening en Milieubeheer. DHV/Broers & Partners, Den Haag (in Dutch).

Dutch Ministry of Transport and Public Works, 1989. Water in the Netherlands: a time for action: national policy document on water management. Ministry of Transport and Public Works, The Hague.

Dutch Ministry of Transport and Public Works, 1994. Evaluatienota Water. Tweede Kamer der Staten Generaal, vergaderjaar 1993 – 1994, 21 250 nrs. 27 – 28, Den Haag: 164 pp. (in Dutch).

Dutch Ministry of Transport and Public Works, 1996. Nota watersysteem-verkenningen: Toekomst voor water. Ministerie van Verkeer en Waterstaat, Directoraat-Generaal Rijkswaterstaat, Den Haag: 415 pp. (in Dutch).

Frigge, P. J., 1996. Gebruikershandleiding STIPT versie 5.2 en IN versie 1.1. Informatie en Kennis Centrum Natuurbeheer, Wageningen: 11 pp. (in Dutch).

Gemeentewerken Rotterdam, 1991. MER-mijnsteen (Achtergronddokument 2 en 4). Ingenieursbureau Havenwerken, Rotterdam: 35 and 37 pp. (in Dutch).

Gonzalez Canibano, J., J. A. Fernandez Valcarce, A. Falcon, J. L. Ibarzabel, J. M. Rodriguez Ortiz & J. A. Hinojosa, 1990. Static leaching of coal mining wastes. In A. K. M. Rainbow (ed.), Reclamation, Treatment and Utilization of Coal Mining Wastes. Balkema, Rotterdam: 165–175.

Hart, P. A., 1990. An assessment of the marine application of mine stone. In A. K. M. Rainbow (ed.), Reclamation, Treatment and Utilization of Coal Mining Wastes. Balkema, Rotterdam: 257–266.

Hauss, K. D. & B. K. Mayer, 1990. Tests concerning reclamation, treatment and utilization of coal-mining wastes in waterway engineering. In A. K. M. Rainbow (ed.), Reclamation, Treatment and Utilization of Coal Mining Wastes. Balkema, Rotterdam: 243–247.

Microsoft, 1994. Microsoft Excel Version 5.0, Microsoft Corporation.

Nederlands Normalisatie-instituut, 1995a. Leaching characteristics of solid earthy and stony building and waste materials. Leaching tests. Determination of the leaching of inorganic components from granular materials with the column test. Nederlandse Normalisatie-instituut, Delft, Nederlandse Norm 7343: 1–11.

Nederlands Normalisatie-instituut, 1995b. Leaching characteristics of solid earthy and stony building and waste materials. Leaching tests. Determination of the leaching of inorganic components from building and monolitic waste materials with the diffusion test. Nederlandse Normalisatie-instituut, Delft, Nederlandse Norm 7345: 1–19.

Raad van State, 1994. Uitspraak in het geschil tussen de Stichting Natuur en Milieu te Utrecht en de Minister van Verkeer en Waterstaat. No. G05.91.1407, Raad van State, Afdeling Bestuursrechtspraak, Den Haag: 9 pp. (in Dutch).

Rainbow, A. K. M. (ed.), 1990. Reclamation, Treatment and Utilization of Coal Mining Wastes. Balkema, Rotterdam.

Saft, R. J., 1992. Inventarisatie van het gebruik van secundaire materialen in de waterbouw. Rijkswaterstaat, Lelystad: 5 pp. (in Dutch).

Skarzynska, K. M. (ed.), 1993. Proceedings of the Fourth International Symposium on the Reclamation, Treatment and Utilization of Coal Mining Wastes, Volumes 1 and 2. University of Agriculture, Kraków: 987 pp.

Skarzynska, K. M., 1995a. Reuse of coal mining wastes in civil engineering – Part 1: properties of minestone, Waste Mgmt. 15/1: 3–42.

Skarzynska, K. M., 1995b. Reuse of coal mining wastes in civil engineering – Part 2: utilization of minestone, Waste Mgmt. 15/2: 83–126.

Sleeman, W., 1990. Environmental effects of the utilization of coal mining wastes. In A. K. M. Rainbow (ed.), Reclamation, Treatment and Utilization of Coal Mining Wastes. Balkema, Rotterdam: 65–69.

Technisch Adviesbureau Unie van Waterschappen (TAUW), 1986. Uitloging van mijnsteen, TAUW Infra Consult/Rijkswaterstaat, Deventer/Lelystad: 46 pp. (in Dutch).

Technisch Adviesbureau Unie van Waterschappen (TAUW), 1987. Milieuhygiënische kwaliteit mijnsteen Auguste Victoria 0 – 150 mm. TAUW Infra Consult, Deventer (in Dutch).

Technisch Adviesbureau Unie van Waterschappen (TAUW), 1988. Milieuhygiënische kwaliteit mijnsteen Auguste Victoria, TAUW Infra Consult, Deventer (in Dutch).

Twardowska, I., A. Szczepanski & J. Tejszerski, 1990. Prognosis of contaminants leaching from colliery spoils and its effect on the aquatic environment. In A. K. M. Rainbow (ed.), Reclamation, Treatment and Utilization of Coal Mining Wastes. Balkema, Rotterdam: 153–163.

Zwart, W. A., 1996. Milieubelasting van mijnsteen als grondstof in de waterbouw, Katholieke Universiteit Nijmegen, Verslagen Milieukunde 124: 1–94 (in Dutch).

Hydrobiologia **410:** 325–329, 1999.
J. Garnier & J.-M. Mouchel (eds), Man and River Systems.

Role of the dams on the River Lot on two types of pollution:point-sources (cadmium) and non-point sources (phosphorus)

Alain Dauta[1], Yvon Lapaquellerie[2] & Noële Maillet[2]
[1]*Centre d'Ecologie des Systèmes Aquatiques Continentaux, UMR C5576/UPS, 118 route de Narbonne, 31062 Toulouse Cedex 4, France*
[2]*Département de Géologie et Océanographie, URA CNRS 197, Université de Bordeaux I, Avenue des Facultés, 33305 Talence, France*

Key words: river sediments, cadmium, available phosphate, storage

Abstract

The concentrations of cadmium and bioavailable phosphorus were determined in sediments samples of a large section of the River Lot (a 248-km section of 57 artificial stretches and dams). Phosphorus inputs include an important agricultural watershed (diffuse sources) and main point-sources such as towns. Cadmium pollution is mainly due to an industrial activity in the upper reach: high concentrations existed some years ago, the pollution is now lower. The concentration data related to a lithological cartography of the river bottom (sand/gravels/mud) and of the thickness of soft deposits, provide a longitudinal assessment of Cd and P storage in the whole river. The results show the location in the river where the two elements are associated to the sediment muds which are mainly stored in the largest dams situated in the lower section of the river. The sedimentary amounts represent potential sources (i) of metallic pollution in case of river flood or dam emptying (e.g. 69 tons of Cd in the last lower dam), (ii) for eutrophication processes in the lower course (near 10 000 tons of bioavailable-P in the last five dams). The cadmium flux leaving the watershed of river Lot is still important and remains greater than the industrial source (strongly reduced since 1986) of the pollutant located in the upper part of the river.

Introduction

Located in the South-West of France, the River Lot (Figure 1), a tributary of the River Garonne, has a watershed of 11 840 km^2. Over a 248-km length, 57 dams determine a succession of reservoirs of different depth and length. In relation with water velocity, some reaches are areas of deposits with high fine fraction. The inputs of phosphorus are due to soils run-off (diffuse sources) and some medium-sized towns (Decazeville, Capdenac, Cahors, Villeneuve) or industries (point-sources) distributed on the whole river. Conversely, only one pollution source exists for cadmium (stretch no. 2) in the upper part of river where a small tributary drains metallic pollution from an industrial area (zinc and cadmium factories of Decazeville). Although very high before 1986 (total cadmium representing 14.6 t $year^{-1}$), the Cd flux was strongly reduced (Roux & Simonet, 1987) and is now lowered to approximately 1.8 t $year^{-1}$. However, 50 km downstream from the pollution point, 80% of the Cd was found (Latouche et al., 1992) to be associated to particulate matter, representing the main pathway for pollution. Along the river reaches, the sediment has built up and still stores both types of pollution. The present work summarises the results of investigations aimed to assess the pools of Cd and bioavailable-P. Estimation of the sedimentary pools and the gradient of concentrations of Cd and bioavailable-P led to a distribution model of the potential pollutants stocks related to the sediment deposits along the various stretches. It is very interesting to compare the behaviour of the two types of pollution because cadmium is specifically a point-source pollution while phosphorus is coming from the whole watershed.

Material and methods

Sediment cartography

The sediment mass was estimated in the 57 stretches

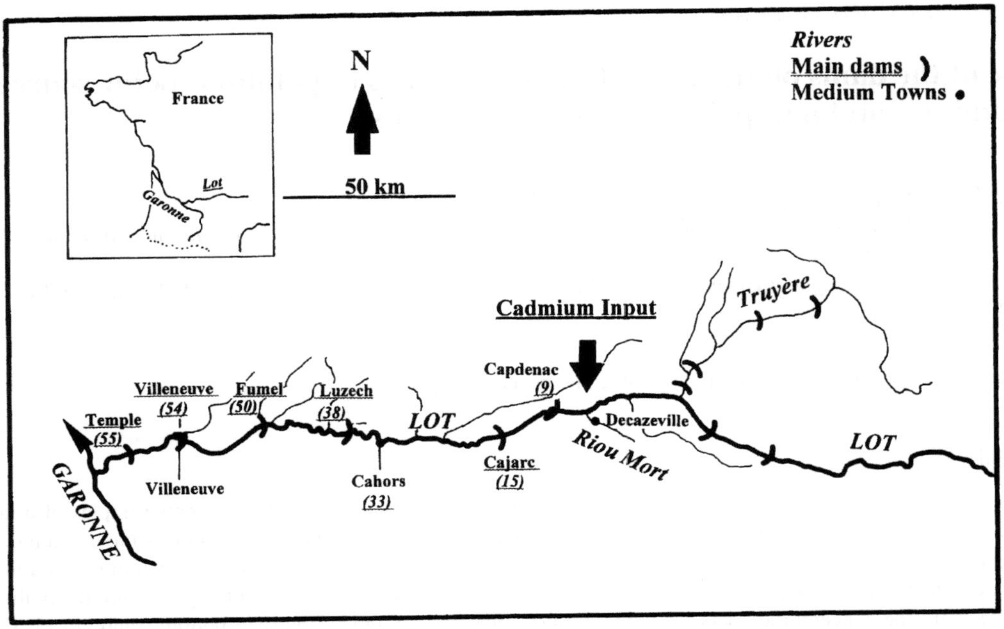

Figure 1. River Lot: map of watershed, main towns. Numbers indicate the position of the main stretches.

of the river Lot. Two methods were used (Blanc et al., 1999; Lapaquellerie et al., 1995; Latouche et al., 1992; Qotbi et al., 1996): (i) for the upstream section (191 km) where the river is generally shallow, the lithological map of the river were obtained using echosounding techniques, with three sediment types: gravels and sand, sandy mud, mud. Next, the sediment depth was estimated in the areas with soft sediment using profile from cores (dredge) and manual sounding, (ii) in the lower part (57 km long) where the depth is generally greater, the thickness of soft sediments was determined using a integration net (100-m step) of seismic profiles in addition to sediments cores.

Cadmium concentrations

Sampling was adapted to the sediment thickness. Cores were sampled in the reservoirs (20–40 cores for a reservoir) over the whole depth of soft sediments (from 0.40 to 3.5 m) with a 2.5 m space between the cores. When the thickness of sediment was low, the upper level was sampled with a box-corer integrating the 0–20 cm layer (20–300 samples depending on the reach). The sediment samples were conditioned under nitrogen atmosphere at 4°C. The sample was dried: a part (100 °C) for porosity study and another part (40 °C) for various analyses after homogenisation by grinding.

Cd concentrations in the particulate phases were measured (Danielson et al., 1986; Loring et al., 1986; Rantala & Loring, 1986) after a dissolution/extraction phase: sample (25 mg, dry weight) were treated with a tri-acid mixture (750 μl HCl 37% + 250 μl HNO_3 70% + 2 ml HF 46%) using teflon flasks (Savilex). The content of the flask was dry-evaporated for 24 h at 100°C; the residue was then treated with 150 μl HNO_3 2%. Cadmium was analyzed using atomic absorption spectrophotometry (AAS 3030, Perkin-Elmer) equipped with HGA 600 graphite tube atomizer and autosampler. Samples of 20 μl were taken for determination. The detection limit was 50 ng l^{-1}. The reproducibility of analyses was controlled using international standards (Nat. Research Council Canada).

We calculated sediment mass of the three lithological types (gravels and sand, sandy mud, mud) by determining the area and mass of each type from the sediment core network. Cd concentrations was the determined as an average per sediment type, balanced by the respective total mass of each sediment type.

The particulate organic carbon determination was made on all samples in induction furnace (Cauwet, 1975).

Bioavailable-Phosphorus

Samples were taken with Eckman dredge. About 100 samples taken in mud deposits were used in this study. The sediment samples were immediately frozen. After drying (100 °C), a fraction was used to perform the bioassay. The algal bioavailability of P contained in sediment samples was estimated according to the method of Golterman et al. (1969) and Cowen & Lee (1976): incubation of a test algae with sediment as the only source of P. The response on the bioassay is based on the algal yield (Fabre et al., 1996; Hanna & Dauta, 1983) deduced by measuring the optical density (678 nm) of the culture. The sediment samples were resuspended in a P-free medium (Dauta, 1982) and incubated with a P-starved *Scenedesmus crassus* culture in suboptimal conditions. Experimental tests were prepared for each sample by adding 200 ml P-free medium and about 0.1 g of dried sediment into a 250-ml Erlenmeyer flask. The cultures were kept until stationary growth due to P limitation, all other nutrients being in excess. The results were compared to standard curves performed with final concentrations of K_2HPO_4 ranging from 10 to 1000 $\mu g\ l^{-1}$.

Results

Sediment mass and Cd mass

For the whole river, the sediment stock was evaluated to 12.9×10^6 tons (Lapaquellerie et al., 1995) with 48% sand and gravels, 28% for sandy muds and 24% mud respectively. The amount of soft sediments stored in the stretches associated to the five largest dams (Figure 2) is very important, representing 65% of the stock. These stretches where the flow rate decreases are areas characterised by fine-grained sediments. Furthermore, the organic matter content is significantly higher. Vertical gradients (cores) of Cd concentrations are important. Generally, the concentrations in lower layers were high because they corresponded to the higher industrial activity. Considering the whole river, horizontal gradient is also important but difficult to estimate over one reach. These two data sets were used to calculate averaged Cd concentrations for each reach (Latouche et al., 1992).

Bioavailable-P and Cd concentrations in sediments

The longitudinal change of bioavailable-P (Figure 3) appears to be determined by the relative position of the larger dams and the interactions with the inputs of the main point-pollution sources as towns (Figure 1).

The longitudinal change of cadmium shows (Figure 4) decreasing values from upstream (40 mg kg^{-1}) to downstream (2.8 mg. Kg^{-1}). Granulometric studies done on the five larger dams (Cajarc, Luzech, Fumel, Villeneuve, Le Temple) showed (i) a decrease of median grain size, from 23 (upstream) to 18 (downstream), (ii) the fraction $>2\mu m$, representing clay, is not abundant, less than 0.5% in Cajarc, Luzech, Fumel, less than 1% in Villeneuve and Le Temple. High percentages of particulate organic carbon were

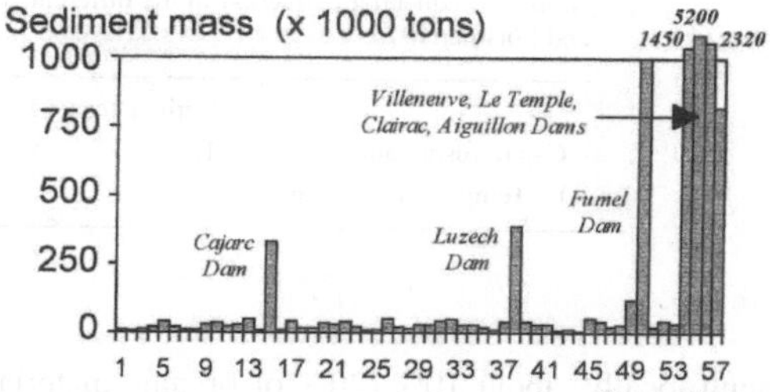

Figure 2. Deposits along the river Lot: sediment mass (10^3 tons) along the stretches.

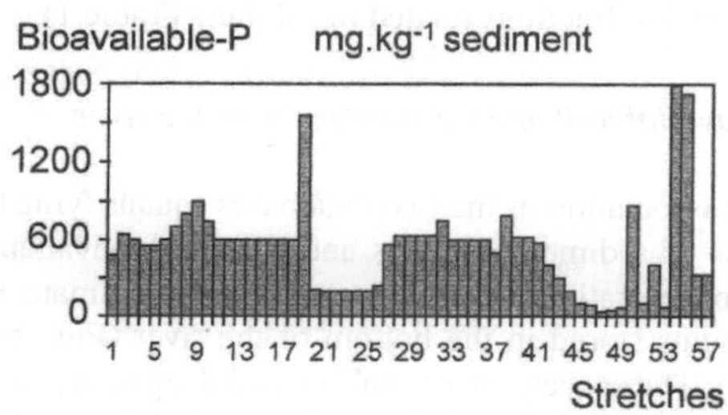

Figure 3. Bioavailable-phosphorus concentrations in the sediment: longitudinal change.

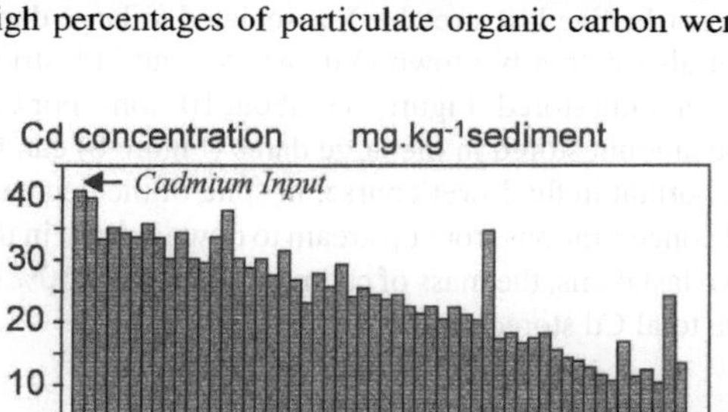

Figure 4. Cadmium concentrations in the sediment: longitudinal change.

Table 1. Sequential extraction in the different fractions of Lot sediments, according to Tessier et al. (1979) and Förstner 1976.

	Carbonates (%)	Hydroxides (%)	Org mat sulphur (%)	Silicates (%)
Cajarc (usptream)	16	34	41	9
Le Temple (downstream)	8	10	62	20

found locally, about 10% (20% of organic material) in these dams. Cd concentrations associated to this phase are important: the results from Latouche et al. (1992) showed the main role of cadmium absorption by organic fractions related to sulphurs (Table 1).

Bioavailable-P and Cd storage along the river

The association of the two data bases (quantifying the mass of sediment deposits and Cd and bioavailable-P concentrations) makes it possible to estimate the amounts stored in the bottom of the river (Figures 5 & 6). The succession of shallow and deep dams (from 2 to 20 m high), with flow rate ranging from 0.1 to 0.0025 m s^{-1} has significant effects both on the deposition rate and the storage of these compounds: this behaviour is very clear on the five largest dams. With respect to phosphorus, the amount stored (Figure 7), about 10^4 tons in the lower course of the river (the five last stretches), is associated with some large dams but also with the Villeneuve urban area. For Cd, the amount stored in the larger dams (Figure 8) is important in the final section, in spite of the decrease of concentrations from upstream to downstream; in the two last dams, the mass of cadmium represents 50% of the total of the stored Cd in the sediments of the River Lot.

Considering phosphorus, in the lower course of the river (the five last stretches), associated to large dams but also with a big town (Villeneuve) and industries, the amount stored (Figure 7) is about 10^4 tons. For Cd, the amount stored in the large dams (Figure 8) can be important in the lower course, in spite of the decrease of concentrations from upstream to downstream: in the two last dams, the mass of cadmium represents 50% of the total Cd stored on the whole river.

Discussion

The dams were built quite recently (after 1940). Using ^{210}Pb and ^{137}Cs data methods, it was established that

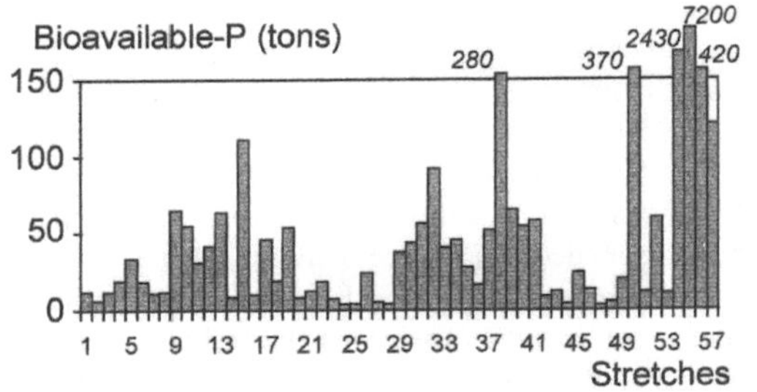

Figure 5. Bioavailable-phosphorus mass: estimation for each stretch.

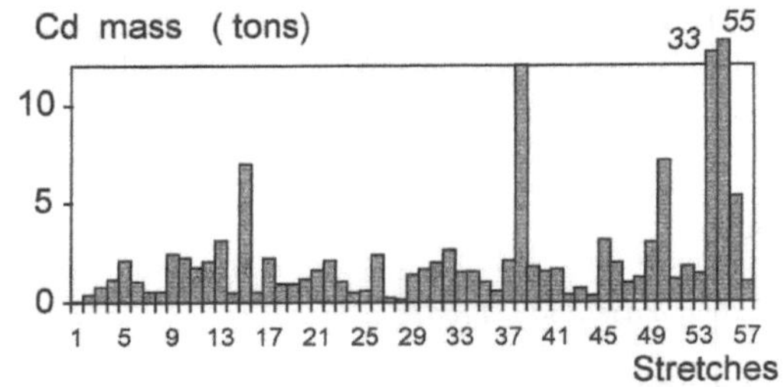

Figure 6. Cadmium mass: estimation for each stretch.

Cd concentrations in the bottom of the cores ranged from 2 to 10 mg kg^{-1}.

Upstream the pollution point, the background concentrations are less than 1 mg kg^{-1} (Latouche et al. 1992). The succession of artificial stretches has greatly

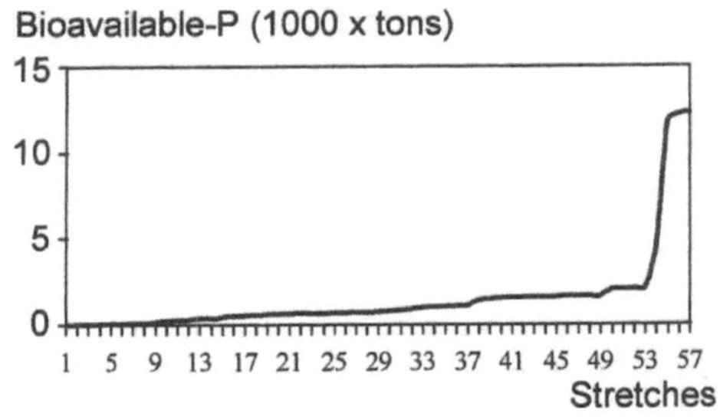

Figure 7. Bioavailable-phosphorus: cumulated mass from upstream to downstream.

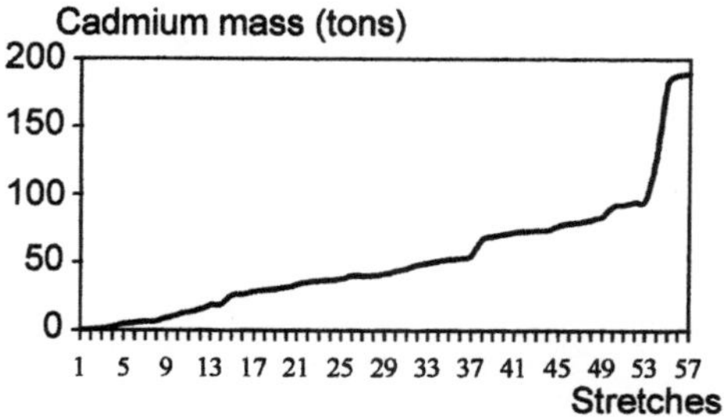

Figure 8. Cadmium: cumulated mass from upstream to downstream.

modified the river, leading to decrease the flow rate in the middle and the lower part of the river, particularly in summer. Some dams appear to have a great influence on the retention of sediment deposits. The relationship between finest grain size classes and content on available P is now well established (Fink & Santiago, 1991). Similar results were found for the River Lot where sediments present a large variation in bioavailable phosphorus (60–1900 μg kg^{-1}). This variation is explained by the relative position of the reaches along the river course and in the watershed which influence both the storage of mud and the richness of sediment. The bioavailable-phosphorus pool appeared to be very important in the lower part of the river. This pool is the main endogenous source of phosphorus representing about 100 times the calculated amount of the annual P-total input (point-sources) in the river (Qotbi, 1996) and probably a major cause of the river eutrophication.

The large dams have also led to the accumulation of sediment highly polluted by cadmium. A part of the sedimentary pools is liable to be partially reworked in case of heavy flooding or during emptying dams. They are still important potential sources of metallic pollution (about 69 tons of Cd in Le Temple dam in 1992). It has been shown that the flux of cadmium leaving the Lot and entering in the River Garonne (Lapaquellerie et al., 1996) is greater than the present input.

Acknowledgements

This research has been supported by the Agence de l'Eau Adour-Garonne, Toulouse.

References

Blanc G., Y. Lapaquellerie, N. Maillet & P. Anschutz, 1999. A cadmium budget for the Lot-Garonne fluvial system (France). Hydrobiologia 410: 331–341.

Cauwet G., 1975. Optimisation d'une technique de dosage du carbone organique particulaire des sédiments. Chem. Geol. Amsterdam 16: 59–63.

Cowen, W. F. & G. F. Lee, 1976. Phosphorus availability in particulate materials transported by urban runoff. J. Wat. Pollut. Cont. Fed. 48: 580–591.

Dauta, A. 1982. Conditions de développement du phytoplancton. Eude comparative de huit espèces en culture. Ann. Limnol. 18: 217–262.

Danielson L.G., B. Magnusson, S. Westerlund & K. Zhang, 1982. Trace metal determination in estuarine waters by electrothermal atomic absorption spectrometry after extraction of dithiocarbamate complexes into freon. Anal. Chim. Acta 144: 183–188.

Fabre, A., A. Qotbi, A. Dauta & V. Baldy, 1996. Relation between available phosphate in the sediments of the river Garonne and chemically-determined phosphate fractions. Hydrobiologia 335: 43–48.

Fink, A. & S. Santiago, 1991. Fertilité des sédiments en suspension comparée au moyen de biotests avec le phytoplancton naturel; et Scenedesmus communis. Actes 3ème CILEF, Morges (Suisse), P. Vernet (ed.).

Förstner U., 1976. Metal concentrations in freshwater sediments. In: Golterman, H.L. (ed.) Proceedings of International Symposium held at Amsterdam, The Netherlands, Sept. 1976, Interactions between sediments and freshwater, W. Junk & Pudol: 94–103.

Golterman, H., C.L. Bakels & J. Jacobs-Mogelin, 1969. Availability of mud phosphates for algal growth. Verh. int. Ver. Limnol. 17: 467–479.

Hanna M. & A. Dauta, 1983. Bioassays: a comparative study of the three parameters related to phosphorus bioavailability (yield, growth rate and intracellular concentration of phosphorus). Ann Limnol. 19: 59–66.

Lapaquellerie, Y., J.M. Jouanneau, N. Maillet & C. Latouche, 1995. Pollution en Cadmium des sédiments du Lot (France) et calcul du stock de polluant. Envir Techn. 16: 1145–1154.

Lapaquellerie, Y., N. Maillet, J.M. Jouanneau J.P. Coakley & C. Latouche, 1996. Flux de matières en suspension et de cadmium dans le Lot. Hydroécol. Appl. 8: 173–191.

Latouche, C., J.M. Jouanneau, N. Maillet & Y. Lapaquellerie, 1992. Evaluation du Cd stocké dans les sédiments fins de la zone aval du bassin du Lot: étude de son éventuelle remobilisation. Rapp. Entente Interdépartementale Aménagement Hydraulique et le Développement Touristique du Lot: 1–115.

Loring, D.H., 1986. ICES intercalibration for Trace Metals in marine sediment. Ra. Dept. Fisheries and Ocean Atlantic Oceanography, Dortsmouth, N.S., B2Ya2, Canada: 1–64.

Qotbi, A., 1996. Etude de la dynamique du phytoplancton et des nutriments dans une rivière aménagée: le Lot (France). Modélisation mathématique et simulations de scénarios. Thèse Université Toulouse III, 350 pp.

Qotbi, A., A. Dauta, V. Sarpédonti & S. Promeyrat, 1996. Estimation of the available phosphate for phytoplankton stored in the soft sediments of the river Lot. Hydrobiologia 335: 49–53.

Rantal R. & D.H. Loring, 1996. Cadmium in marine sediment: determination by graphite furnace atomic absorption spectroscopy. Techn. mar. envir. Sci. 3: 3–9.

Roux, M. & F. Simonet, 1987. Pollution accidentelle du Lot. Le flux polluant. Revue de l'Agence de l'eau Adour-Garonne 34: 7–9.

Tessier A., P.G.C. Campbell & M. Bisson, 1979. Sequential extraction procedure for the speciation of particulate trace metals. Analyt. Chem. 51: 884–851.

Hydrobiologia **410:** 331–341, 1999.
J. Garnier & J.-M. Mouchel (eds), Man and River Systems.

A cadmium budget for the Lot-Garonne fluvial system (France)

Gérard Blanc, Yvon Lapaquellerie, Noèle Maillet & Pierre Anschutz
Département de Géologie et Océanographie (D.G.O.), Université Bordeaux 1, UMR CNRS 5805 EPOC., Avenue des Facultés, 33405 Talence cedex, France
E-mail: blanc@geocean.u-bordeaux.fr

Key words: pollution, cadmium budget, Lot-Garonne rivers

Abstract

Routine measurements of river discharge and total suspended sediment concentration (TSS) are combined with regular analyses for particulate and dissolved cadmium to produce a box model that allows us to propose a cadmium mass balance for the Lot-Garonne man-influenced river system (8400 km^2). Nearly half the cadmium in the Garonne river is supplied by the tributary Lot river. Cadmium input onto the Lot river comprises wet deposition from the atmosphere, molecular diffusion at the sediment-water interface, surface-water runoff and discharge from the leaching of waste at a zinc refining plant. Approximately 85% of the cadmium in the Lot river is derived from anthropogenic origin. Cadmium in the industrial discharge is 80% dissolved and 20% in the particulate phase (4.2 and 1.1 t yr^{-1}, respectively). Total inputs are estimated at 4.81 t yr^{-1} and 1.54 t yr^{-1} for the dissolved cadmium and for the particulate phase, respectively. Budgeting estimates an output onto the Garonne river of 0.54 t yr^{-1} for the dissolved cadmium (about 8%) and 6.13 t yr^{-1} for the particulate cadmium (about 92%) indicating that downstream sediment-associated cadmium fluxes are enhanced by the 4.27 t yr^{-1} removed from solution and the 0.32 t yr^{-1} remobilized by the erosion of sediment blanketing the Lot river bed. These figures are found to be comparable with those generated by a dilution model which suggests that 97% of dissolved cadmium is taken up by the particulate phase over 0.5 km downstream from the primary anthropogenic source.

Introduction

Although the Aquitaine basin is one of the least industrialised regions in France, the Lot-Garonne-Gironde fluvial system can be considered as a natural model for pollution in non-ferrous metals such as Cd, Zn, Ba, As (Roux & Simonet, 1987). Historically, chronic pollution has resulted from industrial and mining activities that started to develop in the late nineteenth century and in particular, cadmium, which has a high toxicity for biota, is considered as the major polluting metal of this fluvial system. Cadmium pollution was first recognised in 1979 within the framework of the 'National Observation Network' (Latouche, 1988) which documented the high cadmium content in the flesh of the Gironde oysters (up to 100 μg g^{-1} of dry matter). These concentrations are 10 times higher than those typically found in oysters from French non-industrial areas (Boutier, 1981) and reflect the high kinetics accumulation in oysters. As oysters are very sensitive to this pollution, the Marennes-Oleron oyster farming located ca. 30 km to the north of the mouth of the Gironde estuary is considered at risk (Jouanneau et al., 1993).

The primary source for present day cadmium pollution is probably attributed to leaching from the waste area of a now abandoned zinc ore manufacturing facility (Latouche, 1992). This plant was founded in 1840 and is located near Decazeville, a town on the Riou-Mort that tributes to the Lot river ca. 200 km upstream from the Lot-Garonne confluence. However, several large hydroelectric dams which were built from 1945 to 1960 in the upper reaches of the Lot valley, together with the 62 smaller storage structures located downstream from the Riou-Mort confluence, decrease the magnitude of winter and spring floods (Massio, 1976). This has resulted in significant sediment and cadmium accumulation in the lower reaches of the Lot river. Lapaquellerie et al. (1995) estimated that 0.47 ± 0.9 × 10^6 m^3 of sediment and about 200 tons of cadmium accumulated in the Lot valley between the Riou-Mort and Garonne river in 1991.

This paper aims to establish a macroscopic description of cadmium behaviour in the Lot-Garonne fluvial system (8400 km^2) by using the annual (1996) flux of both the particulate and dissolved phases. By quantifying the contributions from the primary source (i.e. the Riou-Mort river) and the secondary source (i.e. the sediment-associated cadmium in the Lot river), together with internal transfert processes and the overall output, we propose a tentative cadmium budget for the Lot-Garonne study basin.

Sampling and methods

Because the Lot river is the major polluting tributary of the Garonne, and then of the Gironde, it was necessary to obtain a representative estimation of its total suspended sediment (TSS) fluxes. This was accomplished using a total of six sampling sites (Figure 1). Samples from Temple (Lot) were collected automatically using a Buhler PB-MOS sampler, whilst samples from the five remaining sites were collected manually. The automatic sampler collects a daily cumulative 1 l sample consisting of 16 elementary samples. The frequency of TSS measurements and analyses for particulate and dissolved cadmium is reported in Table 1. Transverse and depth profiling was undertaken to test the representativeness of our local sampling at Temple and La Reole sites (cf. Lapaquellerie et al., 1996). At both sites, these tests showed that variations in TSS concentrations due to depth and distance from the bank did not exceed 6% and 2.3% during flood and base flows, respectively. TSS concentrations were determined using filtration (Whatmann GF/F, 0.70 μm filters) and each sample set typically showed a standard deviation of 1.5% about the mean. The annual mean values of the TSS concentrations for each site are reported in Table 2.

The mean annual discharge (1996) for the Lot and Garonne rivers was calculated using the daily measurements undertaken by the autonomous port of Bordeaux and the hydrological office of Toulouse (see Table 2). The precision is generally about 5% for water flows higher than 40 $m^3\ s^{-1}$.

Sediment-associated cadmium concentrations were determined from bulk suspended sediment samples retrieved by centrifuging between 500 and 1000 l of water using a Wesfalia separator. The cadmium contamination from this equipment is much less than the analytical error which is about 5% (Etcheber & Jouanneau, 1980).

The water samples were collected and filtered for cadmium analyses using a clean technique. Surface-water samples were taken from a small rubber dinghy, by immersing a teflon sampler specially designed to avoid any contact with the atmosphere. This sampler holds a 100 cm^3 acid-cleaned polypropylene vial. The water samples were immediately filtered through acid-cleaned (Ø 47 mm diameter, 0.45 μm pore-size) Nuclepore polycarbonate membranes.

Triplicate analyses for the particulate and dissolved cadmium concentrations of individual samples were undertaken using a graphite furnace atomic absorption spectrometer. Particulate cadmium analyses required the complete dissolution of individual dried solid samples. The total digestion used by this study involved the dissolution of 30 mg of each sample in 750 μl HCl (12N), 2 ml HF (26N) and 250 μl HNO_3 (14N) in small Teflon bombs. The bomb were then heated at 110 °C for 2 h, and after cooling, the digested solution was evaporated to dryness and the residue dissolved in 150 μl concentrated HNO_3. Each sample was left on a hotplate until the residue was completely dissolved and then brought to 5 ml in a volumetric flask using double-deionized water. Prior to the determination of dissolved cadmium content, the samples were preconcentrated in a class 100 clean room by complexating with ammonium pyrrolidine dithiocarbamate/ diethylammonium diethyl dithiocarbamate (APDC/DDDC), extracting into freon and back extracting with HNO_3 into deionized water (Danielsson et al., 1982). In the Riou-Mort tributary, dissolved concentrations were high enough to be analyzed directly, without preconcentration. Accuracy of the analyses was determined by comparisons to river sediment (CRM320) and water (SLRS3) international standards, and was within 5% for cadmium. The precision of the analyses, as determined from duplicate samples was generally 5% for cadmium at concentrations 10 times higher than the detection limit. The content of cadmium in solid phases is given in milligrams per kg, whereas the dissolved cadmium content is given in nanograms per l. The results of the cadmium analysis for 79 suspended sediment and 99 water samples collected during year 1996 are reported in Tables 3 and 4.

Results

River discharge, total suspended sediment (TSS) and cadmium concentrations are the major parameters for

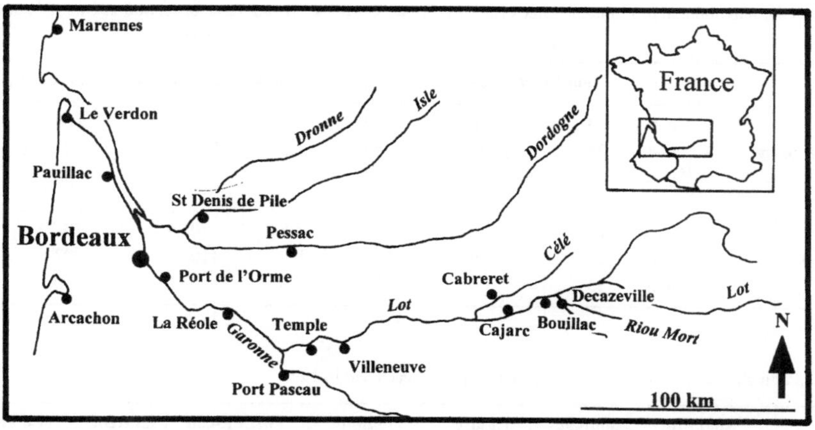

Figure 1. Sampling locations in the Garonne river, Lot river, and its tributaries (Riou-Mort, Célé), and also at the Verdon (estuary mouth).

Table 1. Location and sampling frequencies for the study sites, x and y are the coordinates in Lambert III projection, South Area

Sites	Map	Location x / y	Sampling Type	Sampling strategy TSS	dis.Cd	part.CD
Riou-Mort	IGN 2238 W	560.05 / 3241.09	manual	8 yr^{-1}	8 yr^{-1}	8 yr^{-1}
Bouillac	IGN 2338	589.00 / 3253.45	manual	8 yr^{-1}	6 yr^{-1}	5 yr^{-1}
Cajarc	IGN 2238 W	560.05 / 3241.09	manual	5 yr^{-1}	39 yr^{-1}	5 yr^{-1}
Temple sur Lot	IGN 1839 W	453.55 / 3234.00	Automatic sampler	365 yr^{-1}	18 yr^{-1}	18 yr^{-1}
Port Pascau	IGN 1739 E	439.07 / 3323.00	manual	365 yr^{-1}	18 yr^{-1}	18 yr^{-1}
La Réole	IGN XV-37 (3–4)	374.22 / 280.20	manual	365 yr^{-1}	18 yr^{-1}	18 yr^{-1}
Le Verdon	IGN XIV-33 (1–2)	334.62 / 62.18	manual	4 yr^{-1}	5 yr^{-1}	6 yr^{-1}

the estimation of dissolved and particulate matter fluxes. The river discharge can vary by factors of about 45 and 30 for the Garonne and Lot rivers, respectively. At the beginning of a flood such maximum changes in river discharge can usually be observed during a single day. In addition, the TSS concentrations change by a factor greater than one thousand between the flood and the lowest water level periods. At La Réole, the TSS concentration reached up to 1300 mg l^{-1} in December, whereas it fell down to 0.3 mg l^{-1} in September (Table 2). The annual mean value of TSS concentration varies spatially, increasing from 25 mg l^{-1} the Riou-Mort to 110 mg l^{-1} Cajarc. Conversely, the TSS concentration at Temple shows an annual mean value of 98 mg l^{-1}. This lower value probably results from the settling of the suspended sediment into the numerous reservoirs. In 1996, the TSS concentration at Port-Pascau (Garonne) was approximately twice that of the downstream Lot. The changes in concentration of the particulate and dissolved cadmium show a temporal variability less than that of TSS concentration. Thus, it was not necessary to collect a sample every day, although the floods were sampled.

In 1996, the mean particulate cadmium concentrations decreased progressively from 460 mg kg^{-1} at the Riou-Mort to 12.5 mg kg^{-1} at Temple (Table 3). They were 10 times lower at Port-Pascau, and, consequently, a mean cadmium concentration of 2.6 mg kg^{-1} was determined at La Réole, downstream of the junction of the Lot and Garonne rivers. Because of the chlorinity increases in the Gironde estuary, the particulate cadmium concentrations reach a minimum mean value of 0.43 mg kg^{-1} at Le Verdon. The mean particulate cadmium concentration of 2.59 mg kg^{-1}, determined in 1996 at La Réole, is comparable to that given by Elbaz-Poulichet (1988) for the Rhône river, and that given by Chiffoleau et al. (1997) for the Seine river (i.e. 2.8 mg kg^{-1}), but much lower the

Table 2. 1996 mean, maximum and minimum values for river discharge ($m^3.s^{-1}$) and TSS concentration ($mg.l^{-1}$) at each sampling site

		Riou-Mort	Bouillac	Cajarc	Temple	Port Pascau	La Réole	Le Verdon
	mean	3	80	89	171	637	808	1200
River discharge (m^3 s^{-1})	max	35	431	480	922	3523	3928	4748
	min	8	18	22	30	32	85	170
	mean	25.4	99.9	110.4	98.5	207.4	182.9	151.9
TSS (mg l^{-1})	max	81	196	205	390	1105	1297	313
	min	13	45	10	6	0.2	0.3	35

Table 3. Particulate cadmium concentrations determined at each site in 1996

	Riou Mort ($mg\ kg^{-1}$)	**Bouillac** ($mg\ kg^{-1}$)	**Cajarc** ($mg\ kg^{-1}$)	**Temple** ($mg\ kg^{-1}$)	**Port Pascau** ($mg\ kg^{-1}$)	**La Réole** ($mg\ kg^{-1}$)	**Le Verdon** ($mg\ kg^{-1}$)
Jan 04				11.3	0.69	2.53	
Jan 12				15.2	0.34	2.34	
Jan 24				16.1	0.33	1.91	
Jan 28							0.56
Feb 5				15.9	0.74	1.47	
Feb 28				12.3	1.32	1.58	
Mar 28				10.4	1.24	2.34	
Apr 17	484	17.3	23.1				
Apr 29				16.0	1.2	3.6	
May 16							0.94
May 23				10.8	1.31	3.52	
May 24	504	29.3	17.1				
Jun 10							0.48
Jun 28	767			7.7	1.59	2.34	
Jul 10	350	22.5	26.7				
Jul 18				8.1	1.69	2.45	
Aug 28				13.6	1.39	2.33	
Sep 02	244						
Sep 06							0.21
Sep 10	689	29.9	21.7				
Sep 26				16.9	1.67	3.02	
Oct 08				16.9	2.29	3.8	
Oct 15							0.18
Oct 28	286			10.2	2.56	2.22	
Nov 07				7.4	1.15	3.3	
Nov 22				12.1	1.86	2.01	0.21
Dec 04	386	33.1	14.5				
Dec 10				14.6	2.44	3.1	
Dec 19				9.4	2.13	2.17	
Dec 23	432						
Mean value	460	26.4	20.6	12.5	1.44	2.59	0.43

Table 4. Dissolved cadmium concentrations determined at each site in 1996

	Riou-Mort ($ng\ l^{-1}$)	Bouillac ($ng\ l^{-1}$)	Cajarc ($ng\ l^{-1}$)	Temple ($ng\ l^{-1}$)	Port Pascau ($ng\ l^{-1}$)	La Réole ($ng\ l^{-1}$)	Le Verdon ($ng\ l^{-1}$)
Jan 04			340	65	16	29	
Jan 12			134	48	48	33	
Jan 24			420	72	48	28	
Jan 28							385
Feb 5			532	97	49	64	
Feb 28			158	105	54	42	
Mar 28			245	64	58	95	
Apr 17	25 000	139	593				
Apr 29			169	43	32	88	
May 23			433	15	24	59	
May 24	25 600	975	150				
Jun 03			176				
Jun 04			134				
Jun 10							481
Jun 28	49 800	810		255	56	200	
Jul 10	34 300		346				
Jul 18			628	230	120	185	
Aug 26			245				
Aug 27			284				
Aug 28			179	110	110	315	
Sep 02			534				
Sep 03			157				
Sep 04			608				
Sep 06							
Sep 10	80 300	1860					467
Sep 26			303	20	64	52	
Oct 08			252	210	83	32	
Oct 15							320
Oct 28	16 500		128	240	45	41	
Nov 07			640	35	87	34	
Nov 14			635	94	19	38	
Nov 22							245
Dec 04	67 280	1135	325				
Dec 10			436	110	70	44	
Dec 19				87	99	32	
Dec 23	53 600	1230					
Mean value	44 048	1025	340	106	60	78	380

32 mg kg^{-1} given by Van der Weijden et al. (1989) for the Rhine. However, Idlafkih et al. (1995) gave a value of 4.9 mg kg^{-1} for the Seine river.

The dissolved cadmium concentration in 1996 averaged 44 000 ng l^{-1}in the Riou-Mort (Table 4). This high content rapidly decreased downstream and reached a value of 106 ng l^{-1} at Temple. The mean dissolved cadmium concentrations at the Garonne river sites in 1996 were slightly lower, and thereafter increased by a factor of four (i.e. 380 ng l^{-1}) at Le Verdon, where the salinity ranged from 14 to 24 ‰. An average value of 400 ng l^{-1} was previously determined for a comparable salinity range (Jouanneau et al., 1990), although Kraepiel et al. (1997) meas-

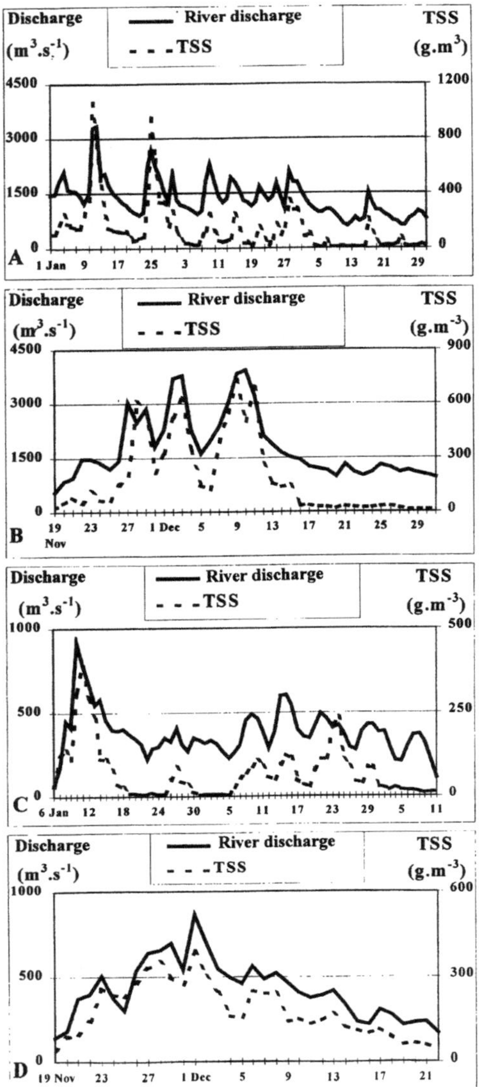

Figure 2. Temporal variations in 1996 peak discharge and TSS concentrations for the Garonne (A–B) and Lot (C–D) rivers.

ured values three times lower in February 1994 when high floods occurred. The mean concentration of the dissolved cadmium at La Réole, endpoint of the fresh water before the Gironde estuary, was 78 ng l^{-1}. This value is higher than the world average (i.e. 20 ng l^{-1}) estimated for river water (Martin & Whitfield, 1983), and is comparable to that of the Mississippi river (i.e. 90 ng l^{-1}) (Trefry & Presley,1976). The Cd concentration of 78 ng l^{-1} for the Garonne is, however, higher than values measured in other major French rivers. In the Seine, Idlafkih et al. (1995) and Chiffoleau et al. (1997) gave values of 27 ng l^{-1} and 26 ng l^{-1}, respectively. These values for the Seine are similar to those in the Rhine river (Elbaz-Poulichet, 1988), whereas a value of 48 ng l^{-1} was given by Valenta et al. (1984) for the Rhône river.

Minimum discharge values of 1500 and 500 m^3 s^{-1} were used to identify a total of 45 and 19 flood days during 1996 for the Garonne and Lot rivers, respectively. The maximum values for the river water flows and the TSS concentrations were observed in January–February and November–December (Figure 2). Note that the higher TSS concentrations correspond to the higher river water flows, as previously recorded for numerous rivers (Meybeck, 1992). The river water flow data were recorded for the 1973–1996 period at Temple (Lot) and for the 1959–1996 period at La Réole (Garonne), and gave mean respective values of 160 and 640 m^3 s^{-1}. These values are comparable to those determined in 1996: 170 m^3 s^{-1} and 810 m^3 s^{-1}, respectively. The mean discharge estimated for 1996 is, therefore, representative of the longer term hydrology of the study basin.

Discussion

As the frequency of TSS sampling is high, the quantification of the annual TSS fluxes (F_{TSS}) in the Lot-Garonne system is based on the assumption that the TSS concentration remained constant between two consecutive samples (i.e. day after day). Thus, the annual F_{TSS} are defined as the sum over 365 days of the following products (Meybeck, 1992):

$$F_{TSS} = \Sigma_1^{365} Q_{riv} \cdot C_{TSS}, \qquad (1)$$

where Q_{riv} is the daily mean river water discharge and C_{TSS} is the measured TSS concentration for each day. Considering the errors due to the sampling and analytical methods and those deriving from the Qriv determination, the precision is about 15% for the F_{TSS} at Temple (Lot), Port Pascau and La Reole (Garonne).

As the daily mean TSS fluxes are well known, the annual particulate cadmium fluxes (F_{partCd}) can be estimated from the assumption of the constant concentration given by Meybeck et al. (1994) as follows:

$$F_{\mathrm{partCd}} = \Sigma_1^{365} F_{\mathrm{TSS}} \cdot C_{\mathrm{partCd}}, \qquad (2)$$

where F_{TSS} is the daily mean TSS fluxes, C_{partCd} is the particulate Cd concentration which is considered as a constant between each sampling episode.

Table 5. 1996 river discharge and cadmium flux estimates for Temple (downstream Lot river), and percentage of cadmium delivered by the Lot to the Garonne river

	1996
River discharge (m^3 s^{-1})	171
TSS flux (10^6 t yr^{-1})	0.531
Dis. Cd flux (t yr^{-1})	0.54
Part. Cd flux (t yr^{-1})	6.13
Total Cd flux (t yr^{-1})	6.67
Lot river contribution (%)	48

Because the annual variability of the river discharge is well established, the mean annual dissolved cadmium flux (F_{disCd}) was calculated using the weighted volume concentrations (Meybeck et al., 1994) as follows:

$$F_{\text{disCd}} = Q'.(\Sigma Qi.Ci/\Sigma Qi),$$

where Q' is the annual water discharge, Ci is the dissolved cadmium concentration in a sample (i), and Qi is the instantaneous river water flow measured at the time when the sample (i) was collected.

The total cadmium fluxes given in t yr^{-1} is obtained by the summation of the F_{partCd} and F_{disCd}. These calculated fluxes are reported in Table 5. The results calculated for 1996 show that about 48% of cadmium flux through the Garonne is contributed by the Lot river. This result is comparable to others obtained since 1990 (Lapaquellerie et al., 1996). Our results also show that 90% of the cadmium output from the Lot is in the particulate phase. In order to define roughly the internal reactivity in the Lot river, it is useful to compare the cadmium output with the input onto this fluvial system.

Sources of cadmium

The annual atmospheric cadmium flux (1994) was estimated by determining the cadmium concentration of rain water collected by two rain gauges installed at Temple (downstream Lot) and at Castelnau Lassout (upstream Lot) (Figure 1). These measurements gave values averaging 37 ng l^{-1} at Castelnau Lassout and 39 ng l^{-1} at Temple for the cadmium concentrations, and 3.5 and 2 ng cm^{-2} yr^{-1} for the cadmium fluxes. Maneux et al. (1998) reported annual cadmium fluxes of 9 ng cm^{-2} yr^{-1} in 1994 from the rain deposition to the Arcachon Lagoon, which is a coastal zone in southwestern France, 60 km from Bordeaux. A maximum annual cadmium flux of 0.3 t yr^{-1} was then calculated for the entire Lot catchment (i.e. 8400. 10^{10} cm^2). Because 1994 was the most humid year since 1996, the magnitude of the atmospheric cadmium flux probably leads to a small error in the atmospheric cadmium input for 1996. However, the annual water height at Castelnau Lassout was two times higher than that at Temple in 1994, while the cadmium concentrations were comparable in both sites.

The annual cadmium flux deriving from the natural run-off was estimated from a 1991–1996 database. Concentrations of dissolved and particulate cadmium were determined at Boisse Penchot, located 1 km upstream from the junction of the Riou-Mort and Lot rivers, and at Cabreret located 5 km from the junction of the Célé tributary and the Lot river (Figure 1). At Boisse Penchot, the mean concentrations of the dissolved and particulate cadmium were 65 ng l^{-1}, and 1.2 mg kg^{-1}, respectively. These values are comparable to those measured since 1990 at Port Pascau for the Garonne basin (i.e. 51 ng l^{-1} and 1.4 mg kg^{-1}) (Lapaquellerie et al. , 1996). Taking a mean river water flow of 77 m^3 s^{-1} in 1996, the dissolved cadmium flux was 0.16 t yr^{-1}. Because the variability of the TSS concentrations is poorly documented at Boisse Penchot, we assume that the TSS flux at this site is equal to the difference between the TSS flux determined at Bouillac located 2 km downstream (i.e. 0.252 10^6 t yr^{-1}) and that of the Riou-Mort river (i.e. 0.0024 10^6 t yr^{-1}). Thus, our estimation of 0.28 t yr^{-1} for the particulate cadmium flux must be considered as an approximate average value. At Cabreret, the concentrations show mean values of 98 ng l^{-1} for dissolved cadmium, 18.5 mg kg^{-1} for particulate cadmium, and 49 mg l^{-1} for TSS, leading to the calculated fluxes of 9 kt yr^{-1} for TSS, 0.16 t $yr^{-1}$for particulate and 0.02 t yr^{-1} for dissolved cadmium. Thus, the dissolved and the particulate run-off discharges from the Lot catchment are 0.18 t yr^{-1} and 0.44 t yr^{-1}, respectively.

The next point is the estimation of the main cadmium source: the Riou-Mort. In 1996, two surveys of 24 d each were carried out using an automatic sampler. In addition to the values given in Tables 3 and 4, we calculated annual fluxes for TSS, dissolved and particulate cadmium of 2.4 kt yr^{-1}, 4.17 t yr^{-1} and 1.10 t yr^{-1}, respectively. Although the temporal variability

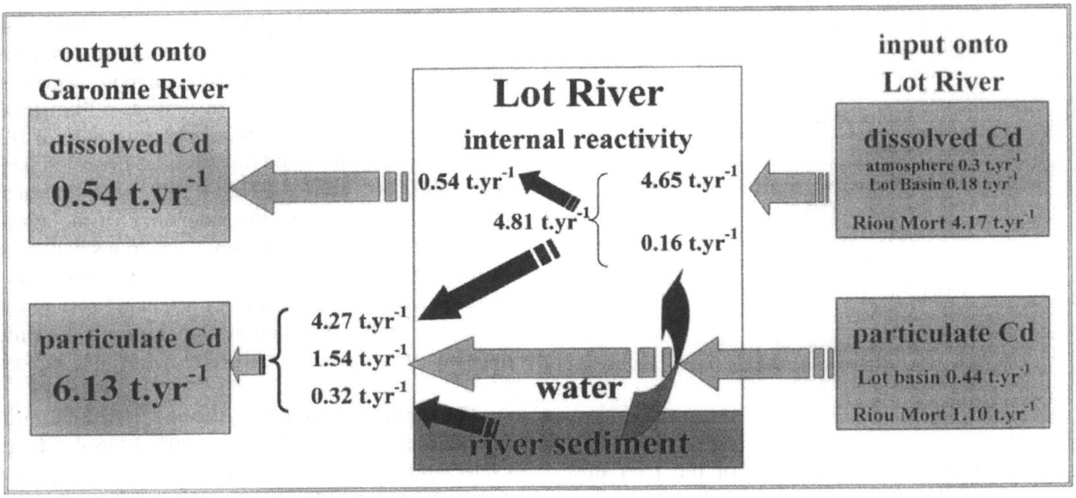

Figure 3. Box model of the cadmium budget for the Lot-Garonne study basin.

of the concentrations is underdocumented in comparison with that obtained at Temple, these results are comparable to those obtained from two short surveys in February and June 1994 made by the "Agence de l'Eau Adour-Garonne" (1995). Furthermore, we note that only about 20% of the cadmium is transported by the Riou-Mort river in particulate forms. Thus, the temporal variability of the TSS concentrations has a smaller influence on the estimation of the total cadmium input to the Lot river. From these results, we assume that the cadmium input is characterized by fluxes of 4.65 t yr^{-1} in dissolved form and 1.54 t yr^{-1} in particulate form. However, an additional source of dissolved cadmium needs to be evaluated: this is the molecular diffusion flux at the water-sediment interface. Two concentration-depth profiles for the dissolved cadmium were obtained from two sediment cores recovered in 1996 at Cajarc and Temple. If we assume that only vertical transport through interstitial waters is important, then the flux of a chemical constituent can be described by the Fick law. Using a corrected temperature diffusion coefficient of 3.3 10^{-6} cm^2 s^{-1} for cadmium (Lerman, 1979), dissolved cadmium diffusion fluxes of 25 10^{-6} and 9.2 10^{-6} ng cm^{-2} s^{-1} were computed from the concentration gradients obtained between the interstitial and the deep water at Cajarc and Temple, respectively. Two other cores recovered at the same sites in 1992 gave fluxes slightly higher, 41 10^{-6} and 16.2 10^{-6} ng cm^{-2} s^{-1}. The sediment surface within the Lot river channel between its confluences with the Riou-Mort and the Garonne was estimated at 22 km^2 (Lapaquellerie et al., 1995). Thus, we obtain a rough estimate of 0.16 t $yr^{-1}$for the annual cadmium flux associated with molecular diffusion. Consequently, the dissolved cadmium input could be 4.81 t yr^{-1}.

Since the dissolved cadmium output was 0.54 t yr^{-1}, the suspended material must act as a sink for dissolved cadmium, which is removed from the solution (Figure 3). The magnitude of this removal is 4.27 t yr^{-1}, which makes up about 70% of the particulate cadmium output. By adding 1.54 t of particulate cadmium transported by the Lot river in 1996, we obtain a deficit of 0.32 t yr^{-1} for the particulate cadmium output. Note that the cadmium input is approximately equal to the output in 1996. This small deficit can be interpreted, however, in terms of cadmium mobilization removed from the Lot river sediment by erosion.

Cadmium mass balance

This result can be tested by a simple model that shows the change of the cadmium concentrations as a function of dilution from the Riou-Mort to the Gironde estuary. The dissolved concentrations are diluted by the progressive downstream increase in the river discharge for the Lot, whereas the dilution of the particulate concentrations is related to the TSS fluxes. Thus, it is convenient to use a framework in which dissolved cadmium and particulate cadmium are con-

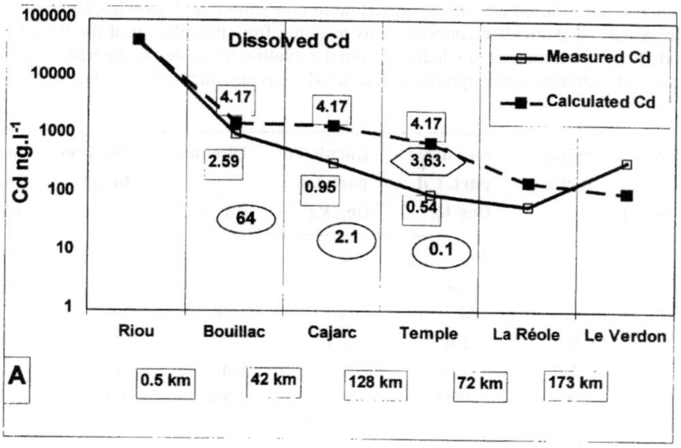

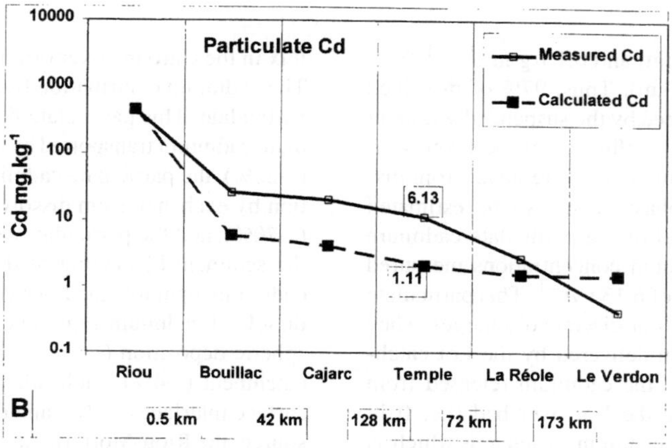

Figure 4. Dilution model for dissolved (A) and the particulate (B) cadmium concentrations. Numbers into open squares are in t yr^{-1}, numbers in open circles are in $\mu g\ m^{-1}\ s^{-1}$.

sidered separately. The parameters for calculation are presented in Table 6, and the calculated concentrations are compared to those measured in Figure 4. This comparison shows that the calculated dissolved cadmium concentrations are higher than those measured from the Riou-Mort to La Réole, whereas the measured concentrations are higher than those calculated in the estuary. The pattern is reversed for the particulate cadmium. Qualitatively, Figure 4 shows that cadmium is removed from the solution between the Riou-Mort and La Réole, then the cadmium undergoes estuarine dissolved-particulate reactions which lead to its addition to the dissolved phase. The quantification of this addition cannot be assessed with accuracy because the cadmium output at the estuary mouth is not adequately documented in this study. Nevertheless, it is useful to look at the difference between the calculated and measured concentrations along the Lot river. From these concentrations, dissolved cadmium fluxes can be estimated for each sample site (Figure 4A). At Temple, the flux difference (e.g. 4.17 t yr^{-1} minus 0.54 t yr^{-1}) yields a value of 3.63 t yr^{-1} which corresponds to the annual removal of the Riou-Mort dissolved cadmium input. By adding the cadmium input not taken into account in the dilution model (i.e. atmosphere, run-off, and diffusion inputs), we find 4.27 t yr^{-1}, equal to the removal value determined in the box model. By considering the distance between each site along the Lot river, we deduced that the cadmium removal from solution decreases rapidly in the downstream direction, with a rate of 64 $\mu g\ m^{-1}\ s^{-1}$from the Riou-Mort junction to Bouillac (0.5 km), 2.1 $\mu g\ m^{-1}\ s^{-1}$from

Table 6. A comparison between measured and modelled cadmium concentrations for each sampling sites. Parameters used for the calculation of the dissolved and particulate concentrations resulting from the dilution of the Riou-Mort cadmium input. The calculated particulate cadmium concentrations are deduced from the dilution factor due to the additional TSS fluxes along the Lot river. The calculated dissolved cadmium concentrations are deduced from the dilution factor due to the additional river discharge along the Lot river

	TSS flux (10^6 t yr^{-1})	Dilution factor	measured part. Cd (mg kg^{-1})	calculated part. Cd (mg kg^{-1})	Discharge (m^3 s^{-1})	Dilution factor	measured dis. Cd (ng l^{-1})	calculated dis. Cd ng l^{-1})
Riou-Mort	0.0024		460	460	3		44 048	44 048
Bouillac	0.252	77	26.4	6	80	27	1025	1652
Cajarc	0.31	114	20.7	4.1	89	30	340	1485
Temple	0.531	221	11.6	2.1	171	57	101	773
La Réole	4.661	1942	2.69	1.53	808	269	64	164
Le Verdon	5.75	2395	0.43	1.5	1200	400	380	110

Bouillac to Cajarc (42 km) and 0.1 μg m^{-1} s^{-1} from Cajarc to Temple (128 km). Thus, 97% of dissolved cadmium input is absorbed by the suspended sediment within 0.5 km after the confluence of the Riou-Mort and the Lot. The magnitude of the removal from dissolved to particulate phases can also be estimated using the dilution model of the particulate cadmium (Figure 4B). The cadmium concentrations measured at Temple lead to a flux of 6.13 t yr^{-1}. This particulate cadmium flux included four discrete discharges. They are the natural cadmium delivered by the Lot catchment (i.e. 0.44 t yr^{-1}), the cadmium released from the sediment blanketing the Lot river bed (i.e. 0.32 t yr^{-1}), the Riou-Mort cadmium discharge which is the value given by the dilution model (1.11 t yr^{-1}), and the cadmium which is deduced to be removed from solution (4.26 t yr^{-1}). This value is clearly comparable to that determined using the box model. Thus, the budgeting approach developed in this study is efficient to describe the macroscpic behaviour of cadmium in the Lot-Garonne study basin.

Conclusion

The annual river discharge for the Lot-Garonne fluvial system during 1996 is representative of the longer-term mean. The flux estimates presented in this study are based upon regular high frequency measurement for a large study basin and algoritihms recommended by Meybeck et al. (1994). These algorithms are similar to those used for British rivers (Webb et al., 1997). Bearing this point in mind, the tentative budget produced by this study estimates that 48% of the cadmium flux in the Garonne river originates from the Lot river. The cadmium contribution from the Lot is about 90% particulate. This particulate flux comprises the particulate cadmium transported by the entire Lot catchment (~25%), the particulate cadmium removed from solution by exchange from dissolved to particulate phases (~70%), and the particulate cadmium mobilized from the sediment blanketing of the Lot bed (~5%). The cadmium input into the Lot river is 75% dissolved. The dissolved cadmium input overall includes wet atmospheric deposition (~6%), surface run-off of the Lot catchment (~4%), molecular diffusion at the water-surface interface (~3%), and the major anthropogenic source: the Riou-Mort tributary (~87%). Within a distance of 0.5 km from this anthropogenic source, about 97% of dissolved cadmium input is taken up by the suspended solid phase.

This study indicates that two areas of the Lot-Garonne study basin should be targeted by any future investigations of change in metal speciations between the particulate, dissolved and colloidal phases. These are between the Riou-Mort and Bouillac on the Lot river, and from La Réole to Bordeaux at the Garonne-Gironde fresh/salt water interface.

Acknowledgements

This study is a scientific contribution no 1271 from Bordeaux 1 University, D.G.O, UMR CNRS 5805 EPOC and from the 'IMPACTS' program of the 'GIS ECOBAG' (Groupement d'Intérêt Scientifique: Ecologie du Bassin Adour-Garonne). We wish to thank Roger Lohou, 'Président' of this GIS. Claude

Latouche, Michel Roux and François Simonet, members of the scientific committee of the 'Agence de l'Eau Adour-Garonne' are also greatfully acknowledged. This work was supported by grants from 'l'Agence de l'eau Adour-Garonne' and 'le Ministère Français de l'Environnement'. We wish to thank R. H. Meade and anonymous reviewers for their thoughtful comments and suggestions which greatly improved the form and the content of the paper. Effective and efficient guest editors are also greatfully acknowledged.

References

Agence de l'eau Adour-Garonne, 1995. Estimation des flux de cadmium et de zinc vehiculés par le Riou-Mort. Rapport interne février: 1–13.

Boutier, B., 1981. Synthèse des résultats de la surveillance des micropolluants dans la matière vivante. Ministère de l'Environnement, Bulletin du Réseau National d'Observation. 17: 115–174.

Chiffoleau, J. F., D.Auger, E. Chartier & I. Truquet, 1997. Identification et devenir des apports intra-estuariens de metaux-traces dans l'estuaire de la Seine. In Programme scientifique Seine-aval, Rapport 1996/FIN-3: 5–23.

Danielsson, L. G., B. Magnusson, S. Westerlund & K. Zhang, 1982. Trace metal determination in estuarine waters by electrothermal AAS after extraction of dithiocarbonate complexes in freon. Anal. Chim. Acta. 144: 159–175.

Elbaz-Poulichet, F., 1988. Apports fluviatiles et estuariens de plomb, cadmium et cuivre aux océans, comparaisons avec l'apport atmosphérique. Thèse, Université Paris VI. 210pp.

Etcheber, H. & J. M. Jouanneau, 1980. Comparison of the different methods for the recovery of suspended matter from estuarine waters deposition, filtration and centrifugation; consequences for the determination of some heavy metals. Estuar. Coast. mar. sc. II: 701–707.

Idlafkih, Z., D. Cossa & M. Meybeck, 1995. Comportements des contaminants en trace dissous et particulaires (As, Cd, Cu, Hg, Pb, Zn) dans la Seine. Hydroecol. Appl. 7: 127–150.

Jouanneau, J. M., B. Boutier, J. F. Chiffoleau, C. Latouche & I. Phillips, 1990. Cadmium in the Gironde fluvioestuarine system: behaviour and flow. Sci. Total Envir. 97/98: 465–469.

Jouanneau, J. M., Y. Lapaquellerie & C. Latouche, 1993. Origin and pathways of Cd contamination in the Gironde estuary, Garonne river and tributaries. In J. P. Vernet (ed.), Environment Contamination. Elsevier: 373–389.

Kraepiel, A. M. L., J. F. Chiffoleau, J-M. Martin & F. M. M. Morel, 1997. Geochemistry of trace metals in the Gironde estuary. Geochimica et Cosmochimica Acta.61: 1421–1436.

Lapaquellerie, Y., J. M. Jouanneau, N. Maillet & C. Latouche, 1995. Pollution en cadmium dans les sédiments du Lot (France) et calcul du stock de polluant. Envir. Tech. 16: 1145–1154.

Lapaquellerie, Y., N. Maillet, J. M. Jouanneau, J. P. Coakley & C. Latouche, 1996. Flux de matières en suspension et de cadmium dans le Lot. Hydroecol. Appl. 8: 173–191.

Latouche, C., 1988. La pollution en cadmium de l'Estuaire de la Gironde, Bull. Inst. Géol. Bassin d'Aquitaine 44: 15–21.

Latouche, C., 1992. La pollution par le cadmium des huitres sauvages de l'Estuaire de la Gironde. Origine. Mécanismes responsables de la fixation du cadmium. Ichtyophysiol. Acta 15: 139–152.

Lerman, A., 1979. Geochemical processes water and sediment environments. Wiley-Interscience Publication, John Wiley & sons: 481pp.

Maneux, E., F. E. Grousset, P. Buat-Ménard, G. Lavaux, P. Rimmelin & Y. Lapaquellerie, 1998. High frequency variability of the wet deposition of anthropogenic heavy metals: the Arcachon Lagoon (SW france). Estuarine, Coastal and Shelf Science. (in press).

Martin , J. M. & M. Whitfield, 1983. The significance of the river input of chemical elements to the oceans. In C. S. Wong, E. Boyle, K. W. Bruland, J. D. Burton & E. D. Goldberg (eds) Trace metals in Sea Water. New-York: Plenum: 265–296.

Massio, J. C., 1976. Facteurs d'évolution des matières en suspension minérales et organiques dans les eaux du Lot. Thèse Université P. Sabatier, Toulouse: 135 pp.

Meybeck, M., 1992. Etablissement des flux polluants. Rapport final, Agence de l'Eau Adour-Garonne.

Meybeck, M., A. Pasco & A. Ragu, 1994. Etablissement des flux polluants dans les rivières: pourquoi, comment et à quel prix. 4ème Rencontres de l'Agence Régionale pour l'Environnement. Provence-Alpes-Côte d'Azur. Colloque scientifique sur les charges polluantes véhiculées par les fleuves et les rivières en Méditerranée.Collection 1991–1992–1993.

Roux, M. & F. Simonet, 1987. Pollutions accidentelles du Lot: Flux polluant. Revue de l'Agence de l'Eau Adour-Garonne. 34: 7–9.

Trefry, J. H. & B. J. Presley, 1976. Heavy metal transport from the Mississippi River to the Gulf of Mexico. In H. L. Windom & R. A. Duce (eds), Marine Pollutant Transport, Mass: Lexington Books: 39–76.

Valenta, P., M. L. S. Simoes-Goncalves & M. Sugawara, 1984. Voltametric studies on the speciation of Cd and Zn by amino acids in sea water. In C. J. M. Kramer & J. C. Duinker (eds), Complexation of Trace Metals in Natural Water. Martinus Nijhoff Publishers: 357–365.

Van der Weijden, C. H. & M. Middelburg, 1989. Hydrogeochemistry of the river Rhine. Wat. Res. 10: 1247–1266.

Webb, B. W., J. M. Phillips, D. E. Walling, I. G. Littlewood, C. Watts & G. J. L. Leeks, 1997. Load estimation methodologies for British rivers and their relevance to the LOIS RACS (R) programme. Sci. Total Envir. 194/195: 379–389.

Hydrobiologia **410:** 343–348, 1999.
J. Garnier & J.-M. Mouchel (eds), Man and River Systems.

Contamination by Polycyclic Aromatic Hydrocarbons (PAHs) in French and European rivers*

M. B. Fernandes[1], M.-A. Sicre[2,*], I. Broyelle[3], A. Lorre[3] & D. Pont[4]
[1]*Instituto de Química, Universidade Federal do Rio de Janeiro, Cidade Universitária, Centro de Tecnologia, Bloco A, Sala 607, 21949-900, Rio de Janeiro, RJ, Brazil. Present address: FIOCRUZ/ ENSP/CESTEH, Rua Leopoldo Bulhoes 1480, Rio de Janeiro RJ, 21041-210, Brazil*
[2]*Laboratoire des Sciences du Climat et de l'Environnement, Domaine du CNRS, Ave. de la Terrasse, 91191 Gif-sur-Yvette Cedex, France. E-mail: Marie-alexandrine.Sicre@Lsce.onrs-gif.fr*
[3]*Laboratoire de Physique et Chimie Marines, Université Pierre et Marie Curie, 4 place Jussieu, 75252 Paris Cedex 05, France*
[4]*CNRS ESA523, Université de Lyon I, 43 Bd. du 11 novembre 1918, 69622 Villerbanne Cedex, France*
*(*author for correspondence)*

Key words: PAHs, Seine, Rhône, European rivers

Abstract

PAH contamination levels were determined in the suspended matter of two French rivers, the Seine and the Rhône, and compared to those available for other major European rivers. Hydrodynamic characteristics, such as the amount of suspended matter and the water discharge, appear to strongly influence PAH levels in these systems. Sampling strategies to better monitor riverine pollution and refine budget calculations are also discussed.

Introduction

PAHs (Polycyclic Aromatic Hydrocarbons) are ubiquitous organic pollutants in the environment, which are found in fossil fuels or produced upon combustion (domestic or industrial) (Figure 1) (Neff, 1979). They are known to induce mutagenic effects on the DNA of animals and humans (Zedeck, 1980). Even though PAHs can affect the aquatic fauna and flora, the environmental legislation in France and in Europe does not quantitatively regulate their occurrence in waterways. With some exceptions, this situation is encountered worldwide. Numerous studies have been undertaken in order to define aquatic ecotoxicological criteria for PAHs, but toxic levels are still not clearly defined and data on their riverine distributions are scarce. The four major French rivers are the Rhône, Gironde, Loire and Seine. In this work, we present original data on particulate PAHs from the lower Seine riverstream and from a one year survey conducted at a fixed station in the upper delta of the river Rhône (Figure 2a,b). These estuaries differ in their sediment dynamics. The river Seine, discharging into the English Channel, has a tide-dominated estuary (macro-tidal) where a broad zone of high suspended sediment, known as the turbidity maximum zone, occurs. In contrast, the river Rhône, flowing into the Mediterranean Sea, has a micro-tidal estuary. Its effluents are delivered to the sea through surface water and partly via a benthic nepheloid layer developing in the delta (Bouloubassi & Saliot, 1994). The PAH levels measured in these two French rivers are compared to those of other European fluvial hydrosystems (Figure 2c) in an attempt to determine the relationship between PAHs and hydrodynamical features such as water discharge and suspended matter load (SM).

* This is LSCE contribution number 0294.

Materials and methods

The detailed analytical procedure employed for sampling, extraction, fractionation and analysis of Seine and Rhône suspended matter is described elsewhere (Fernandes et al., 1997). Briefly, sampling of

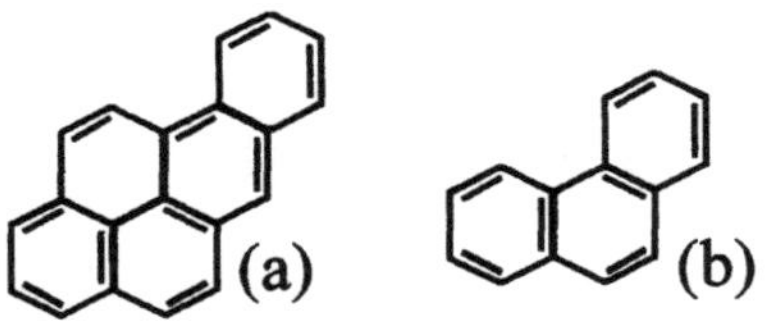

Figure 1. Example of common PAHs released in the environment: benzo(a)pyrene, M.W.=252 (a) and phenanthrene, M.W.=178 (b).

surface waters involved *in situ* pumping, followed by filtration on pre-extracted Whatman GF/F glass fiber filters (0.7 μm). Particulate lipids were extracted three times by ultra-sonication in 200 ml of a mixture of CH_3OH:CH_2Cl_2(1:3 v/v) for 15 min. PAHs were isolated from the lipid mixture by silica gel chromatography using solvents of increasing polarity. They eluted with a mixture of 25% toluene in hexane (Conde, 1989). All samples were spiked with recovery standards prior to extraction. The mean recovery of PAHs varied between 65 and 90% depending on their molecular weight (M.W.). Particulate PAHs were analyzed using a CP-Sil-8 CB Chrompack fused silica capillary column (50 m $\times$ 0.32 mm i.d. $\times$ 0.25 μm film thickness) mounted in a Girdel 3000 gas chromatograph programmed from 80 to 300 °C at 2 °C min^{-1}. Helium was used as carrier gas. Quantitation of individual PAHs was based on comparison of peak areas with those of deuterated standards. Perdeuterated phenanthrene was used to quantify phenanthrene, while all other PAHs were quantified from perdeuterated pyrene. PAH identification was performed by comparison of their retention time with those of authentic standards. Structural assignment was confirmed on selected samples by gas chromatography/mass spectrometry (GC/MS). GC/MS analyses were carried out on a Varian 3400 GC CX Series coupled to an Ion Trap Varian Saturn 3. Samples were injected on a DB5 J & W Scientific fused silica capillary column (30 m $\times$ 0.25 mm i.d. $\times$ 0.25 μm film thickness) programmed from 80 to 300 °C at 3 °C min^{-1}.

Table 1 compares data from the literature with those obtained for the Rhône and the Seine. Analytical methods used for PAH assessment as well as details on the PAHs summed up are given in this Table. When hydrodynamical factors such as water discharge and suspended matter were not provided with the literature PAH data, annual mean values were used.

Results and discussion

Seine river

PAHs were analyzed in the suspended matter (SM) collected in the river Seine surface waters from site P till the Seine Bay, and during tidal cycles in the estuary, at mooring stations A and B, in October 1993 (Figure 2a) (Fernandes et al., 1997). The average river flow at the time of sampling was 570 m^3 s^{-1}. Samples revealed a uniform PAH composition and particulate load till the river mouth ($\sim$5 μg g^{-1}). However, total particulate PAH concentrations expressed per volume unit varied by two orders of magnitude, from 2 to 687 ng l^{-1} (Figure 3a).

Spatial distribution indicated an increase of the contamination of fluvial waters between Poses (station P) and the city of Vieux Port (station 30). The observed decrease further downstream reflected estuarine mixing as well as sedimentation. At the river mouth, other particle pools from marine waters or resuspended in the turbidity maximum zone, admixed to the riverine particles. Changes in the amount of SM during tidal cycles played an important role in PAH short-term temporal variations within the bay. PAH concentrations were also strongly correlated with SM along the entire riverine/estuarine transect (Figure 3a). The mean PAH distribution pattern showed a dominance of non-alkylated PAHs with M.W. 202 and 252 (Figure 3b), characteristic of high temperature combustion. The composition of the PAH assemblages suggested coal combustion and car exhausts as major PAH sources.

Rhône river

On the river Rhône, surface waters were sampled twice a month downriver from the city of Arles (Figure 2b). Additional samples were collected during flood episodes. The water discharge over this one year survey varied from 833 to 9430 m^3 s^{-1}. Total particulate PAH levels spanned a range from 12 to 3765 ng l^{-1} (or 0.03 to 7.1 μg g^{-1}). PAH maxima (in ng l^{-1}) were recorded during flood episodes (Figure 4a). Conversely, particles appeared to be depleted in PAHs (in μg g^{-1}) at high river flow and high SM levels. PAH-enriched particles seem to be diluted by PAH-depleted SM during floods due to the high runoff. Changes in the partition coefficient between the particulate and dissolved phases during these extreme episodes may also play a role (Olsen et al., 1982).

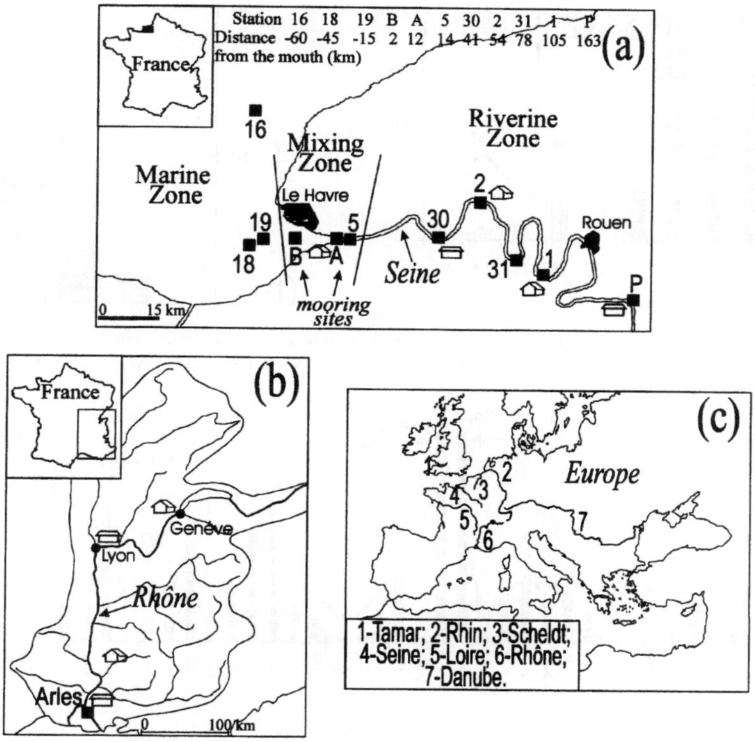

Figure 2. Sampling stations along the Seine estuary (a), collection site in the upper Rhône delta (b), and location of major European rivers for which particulate PAH data are available (c).

Table 1. Experimental procedures for some PAH reports in European rivers.

Rivers	Sampling season	Number of samples	Quantification method	Sum of PAHs
Seine (this work)	fall	25	GC	1, 4–12
Rhône (this work)	4 seasons	33	GC	1, 3–13
Loire (Troncyznski, 1985)	4 seasons	—	GC	4–12
Scheldt (van Zoest & van Eck, 1990)	4 seasons	—	HPLC	1, 2, 4–13
Rhin (van Zoest & van Eck, 1990)	—	—	—	2, 4, 10
Tamar (Readman et al., 1982)	spring	23	HPLC	1, 2, 4–8, 10
Danube (Klyuev et al., 1993)	summer–fall	15	GC/MS	1, 4, 5, 9, 10

1: phenanthrene, 2: anthracene, 3: methyl-phenanthrenes, 4: fluoranthene, 5: pyrene, 6: benz(a)anthracene, 7: chrysene, 8: benzofluoranthenes, 9: benzo(e)pyrene, 10: benzo(a)pyrene, 11: indeno(1,2,3-cd)pyrene, 12: benzo(ghi)perylene, 13: dibenz(a,h)anthracene.
GC: Gas Chromatography, HPLC: High Performance Liquid Chromatography, GC/MS: Gas Chromatography/Mass Spectrometry.
—: data not available.

The most intense flood of November accounted for 77% of the particulate PAH annual deliver. Flux calculations indicated that 96% of the PAH annual discharge occurred during flood events (for water discharge > 3000 $m^3\ s^{-1}$). The mean solid phase PAH distribution was similar to that of the river Seine (Figure 4b): non-alkylated compounds with M.W. 202 and

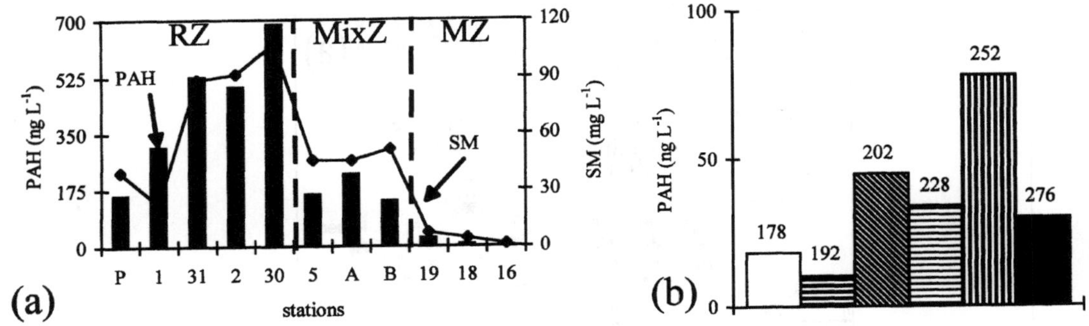

Figure 3. (a) Total particulate PAH and suspended matter (SM) levels in the riverine zone (RZ), mixing zone (MixZ) and marine zone (MZ) of the Seine estuary and (b) Seine estuary mean particulate PAH distribution as a function of molecular weight (178 – 276).

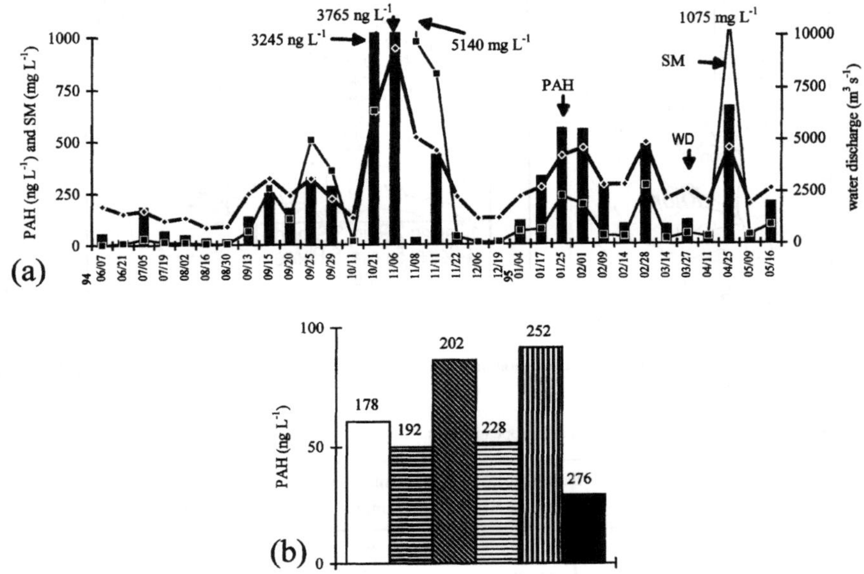

Figure 4. (a) Temporal evolution of total particulate PAH levels, SM and water discharge (WD) for the river Rhône from June 1994 to May 1995 and (b) river Rhône mean particulate PAH distribution as a function of molecular weight (178 – 276).

Table 2. Water discharge (WD), suspended matter (SM) and particulate organic carbon (POC) in some European rivers.

Rivers	WD ($m^3\ s^{-1}$)	SM ($mg\ l^{-1}$)	POC (%)
Seine (this work)	570	48	5.8
Rhône (this work)	2869	352	3.4
Loire (Figueres et al., 1985; Saliot et al., 1988)	920	276	6.1
Scheldt (van Zoest & van Eck, 1990)	157	—	2–10
Rhin (Fernandes, 1996)	2200	—	—
Tamar (Readman et al., 1982; Uncles & Stephens, 1989, 1993)	5–40	10–70	4.0
Danube (Milliman & Meade)	6530	—	—

—: data not available.

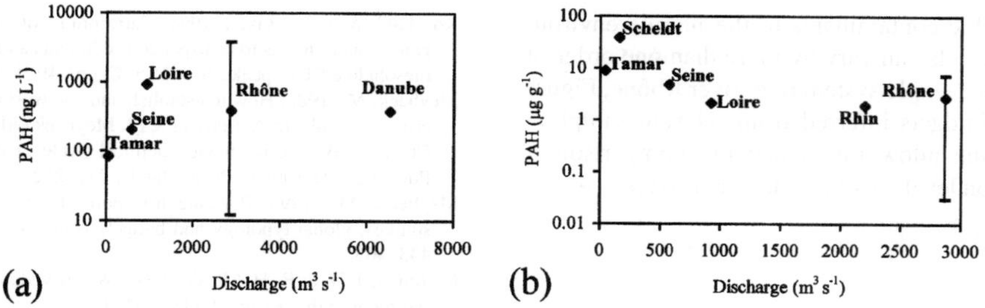

Figure 5. Mean total PAH concentrations in ng l^{-1} (a) and PAH loads in $\mu g\ g^{-1}$ (b) as a function of the water discharge for several European rivers: Seine and Rhône (this work), Loire (Tronczynski, 1985), Scheldt and Rhin (van Zoest & van Eck, 1990), Tamar (Readman et al., 1982) and Danube (Klyuev et al., 1993). Maximum and minimum PAH values for the river Rhône are also indicated.

252 prevailed, indicating the importance of pyrolytic sources.

Other French and European rivers

Some characteristics of European rivers for which particulate PAH data are available are listed in Table 2. Their location on the European continent is shown in Figure 2c.

Direct comparison of PAH datasets from distinct research groups is a difficult task due to differences in sampling and analytical procedures. Being aware of these limitations, we attempted to outline some tendencies and interpret them. The comparison of our results with those of other European rivers showed enhanced PAH concentrations per volume unit at high water discharge (Figure 5a). In contrast, PAH concentrations in $\mu g\ g^{-1}$ tended to decrease as the SM levels increased with the river flow (Figure 5b), while OC% values remained relatively constant. PAH particle loads varied by a factor of approximately 20, while OC% increased by less than a factor of 2 for the entire data set (Table 2). Rivers that drain Europe are generally small (Milliman & Meade, 1983) and cross areas of similar soil/vegetation including similar morphoclimatic types (Meybeck, 1988, 1993). Increased leaching/erosion of the land in larger catchment basins results in the entrainment of higher amounts of SM and therefore of anthropogenic compounds deposited on superficial soil layers via atmospheric fallout, such as combustion-derived PAHs.

PAH inputs depend on the range of land use and urban and industrial developments on the drainage basins (e.g. Hites et al., 1980; Readman et al., 1982; Smith et al., 1985; Bates et al., 1987; Murphy et al., 1988; Kayal & Connell, 1989; van Zoest & van Eck, 1990; Bouloubassi & Saliot, 1991; Domagalski & Kuivila, 1993; Klyuev et al., 1993; Ko & Baker, 1995; Yunker & Macdonald, 1995; Fernandes et al., 1997). With the exception of the Danube, the European rivers discussed here are situated in countries of the European Community where economical activities are similar (France, United Kingdom, Germany, Belgium, Netherlands). Therefore, inputs are expected to be comparable. This hypothesis is corroborated by similar PAH distributions as those of the rivers Seine and Rhône (Figures 3b & 4b). The observed differences in the PAH levels may therefore be explained by different SM and water discharge leading to variable dilution of fluvial inputs, as discussed above. Consequently, rivers with low water flow, such as the Scheldt (Belgium/Netherlands) and the Tamar (United Kingdom) tend to transport particles enriched in PAHs (Figure 5b), representing an immediate toxic hazard to the local fauna and flora. On the other hand, in high water discharge rivers, such as the Rhône (France) and the Danube (Eastern Europe), the amount of PAHs transported per gram of particles is lower but the total discharge of particulate PAHs to the sea (Figure 5a) may be higher.

Conclusions

Riverine PAH levels seem to be related not only to anthropogenic activities, but also water discharge and suspended matter load. Water discharge and SM latter being interdependent, higher contamination of riverine particles is expected in low water discharge riverstreams. The survey conducted in the river Rhône revealed that, on a drainage basin scale, flood events are responsible for most of the PAH annual yield to the ocean. Sampling strategies should be directed towards such episodes in order to better evaluate inputs of hydrophobic contaminants to the marine environment. PAH levels can vary by more than one order of magnitude in a single system (e.g. river Rhône, Figure 5). Annual budgets inferred from relevant sampling sets would thus allow a more accurate comparison of contamination levels between different rivers.

References

Bates, T. S., P. P. Murphy, H. C. Curl Jr & R. A. Feely, 1987. Hydrocarbon distributions and transport in an urban estuary. Environ. Sci. Technol. 21: 193–198.

Bouloubassi, I. & A. Saliot, 1991.Composition and sources of dissolved and particulate PAH in surface waters from the Rhône Delta (NW Mediterranean). Mar. Pollut. Bull. 22: 588–594.

Bouloubassi, I. & A. Saliot, 1994. Implications of a multi-layered nepheloid system for hydrocarbon distributions and transport in a microtidal Mediterranean estuary. Mar. Chem. 46: 101–117.

Conde, A., 1989. Mise au point d'un protocole analytique de séparation des lipides et application à l'étude des apports atmosphériques de matière organique à la Méditerranée Occidentale. Rapport de DEA, Université Pierre et Marie Curie, Paris: 21 pp.

Domagalski, J. L. & K. M. Kuivila, 1993. Distributions of pesticides and organic contaminants between water and suspended sediment, San Francisco Bay, California. Estuaries 16: 416–426.

Fernandes, M. B., 1996. Contribution à l'étude du comportement des hydrocarbures non aromatiques et aromatiques en milieu estuarien. Ph.D. thesis, Université Pierre et Marie Curie, Paris: 256 pp.

Fernandes, M. B., M.-A. Sicre, A. Boireau & J. Tronczynski, 1997. Polyaromatic hydrocarbon (PAH) distributions in the Seine River and its estuary. Mar. Pollut. Bull. 34: 857–867.

Figueres, G., J.-M. Martin, M. Meybeck & P. Seyler, 1985. A comparative study of mercury contamination in the Tagus estuary (Portugal) and major French estuaries (Gironde, Loire, Rhône). Estuar. Coast. Shelf Sci. 20: 183–203.

Hites, R. A., R. E. Laflamme & J. G. Windsor Jr, 1980. Polycyclic aromatic hydrocarbons in marine/aquatic sediments: Their ubiquity. In L. Petrakis & F. T. Weiss (eds), Petroleum in the Marine Environment, Advances in Chemistry Series 185. American Chemical Society, Washington: 289–311.

Kayal, S. I. & D. W. Connell, 1989. Occurrence and distribution of polycyclic aromatic hydrocarbons in surface sediments and water from the Brisbane River estuary, Australia. Estuar. Coast. Shelf Sci. 29: 473–487.

Klyuev, N. A., E. S. Brodskii, N. V. Murenets, O. G. Tarasova & V. G. Zhil'nikov, 1993. Determination of organic pollutants in Danube water by the chromatography-mass spectrometry method. Water Resourc. 20: 83–91.

Ko, F.-C. & J. E. Baker, 1995. Partitioning of hydrophobic organic contaminants to resuspended sediments and plankton in the mesohaline Chesapeake Bay. Mar. Chem. 49: 171–188.

Meybeck, M., 1988. How to establish and use world budgets of riverine materials. In A. Lerman & M. Meybeck (eds), Physical and Chemical Weathering in Geochemical Cycles. Kluwer Academic Publishers, Dordrecht, Netherlands: 247–272.

Meybeck, M., 1993. Riverine transport of atmospheric carbon: sources, global typology and budget. Water Air Soil Pollut. 70: 443–463.

Milliman, J. D. & R. H. Meade, 1983. World-wide delivery of river sediment to the oceans. J. Geol. 91: 1–21.

Murphy, P. P., T. S. Bates, H. C. Curl Jr, R. A. Feely & R. S. Burger, 1988. The transport and fate of particulate hydrocarbons in an urban fjord-like estuary. Estuar. Coast. Shelf Sci. 27: 461–482.

Neff, J. M., 1979. Polycyclic Aromatic Hydrocarbons in the Aquatic Environment: Sources, Fates and Biological Effects. Applied Science, Essex, England: 262 pp.

Olsen, C. R., N. H. Cutshall & I. L. Larsen, 1982. Pollutant-particle associations and dynamics in coastal marine environments: a review. Mar. Chem. 11: 501–533.

Readman, J. W., R. F. C. Mantoura, M. M. Rhead & L. Brown, 1982. Aquatic distribution and heterotrophic degradation of polycyclic aromatic hydrocarbons (PAH) in the Tamar estuary. Estuar. Coast. Shelf Sci. 14: 369–389.

Saliot, A., J. Tronczynski, P. Scribe & R. Letolle, 1988. The application of isotopic and biogeochemical markers to the study of the biochemistry of organic matter in a macrotidal estuary, the Loire, France. Estuar. Coast. Shelf Sci. 27: 645–669.

Smith, J. D., J. Y. Hauser & J. Bagg, 1985. Polycyclic aromatic hydrocarbons in sediments of the Great Barrier Reef Region, Australia. Mar. Pollut. Bull. 16: 110–114.

Tronczynski, J., 1985. Biogéochimie de la matière organique dans l'estuaire de la Loire: origine, transport et évolution des hydrocarbures et des acides gras, Thèse de Doctorat de 3ème cycle de l'Université Paris 6, Paris: 176 pp.

Uncles, R. J. & J. A. Stephens, 1989. Distributions of suspended sediment at high water in a macrotidal estuary. J. Geophys. Res. 94 (C10): 14395–14405.

Uncles, R. J. & J. A. Stephens, 1993. Nature of the turbidity maximum in the Tamar estuary, U.K. Estuar. Coast. Shelf Sci. 36, 413–431.

van Zoest, R. & G. T. M. van Eck, 1990. Behaviour of particulate polychlorinated biphenyls and polycyclic aromatic hydrocarbons in the Scheldt estuary. Neth. J. Sea Res. 26: 89–96.

Yunker, M. B. & R. W. Macdonald, 1995. Composition and origins of polycyclic aromatic hydrocarbons in the Mackenzie River and on the Beaufort Sea Shelf. Arctic 48: 118–129.

Zedeck, M. S., 1980. Polycyclic aromatic hydrocarbons: a review. J. Environ. Path. Toxic. 3: 537–567.

Hydrobiologia **410**: 349–355, 1999.
J. Garnier & J.-M. Mouchel (eds), Man and River Systems.

Variability of herbicides (triazines, phenylureas) and tentative mass balance as a function of stream order, in the river Marne basin (France)

Triazine and phenylurea flux and stream order

M. Chevreuil[1], M. Garmouma[1] & N. Fauchon[2]
[1]*Institut d'Hydrologie et de Climatologie/Laboratoire de Géologie Appliquée B: 122, UPMC, 4 Place Jussieu, 75252, Paris Cedex 05, France*
[2]*Compagnie Générale des Eaux, 32 Place Ronde, 92982, Paris La Défense Cedex, France*

Key words: river Marne, herbicides, stream order, season

Abstract

The magnitude of the contamination of the river Marne by various herbicides of agricultural origin (triazines and phenylureas), led the CGE (Compagnie Générale des Eaux) and the IHC/LGA (Institut d'Hydrologie et de Climatologie/Laboratoire de Géologie Appliquée) to investigate the spatial and temporal distribution of these contaminants. The field work was carried out from 1991 through 1994 on the river Marne, particularly in the sub-basin of the Grand Morin and its tributaries, direct (Orgeval) or secondary (Mélarchez). The chronic contamination of the streams by atrazine and simazine with a background concentration around 100 ng l^{-1}, results from contamination of groundwaters in the Calcaire de Champigny and Calcaire de Brie layers (CGE, 1993). The maximum atrazine and simazine concentrations (> 500 ng l^{-1}) were usually found from April to June. During this period the atrazine concentration peaks might reach 2000 ng l^{-1} in the agricultural brooks as well as in the river Marne. Isoproturon was also detected throughout the year, except during the summer low water period. The maximum concentrations of isoproturon were found between February and April: ranging from 500 ng l^{-1} to 1200 ng l^{-1}, depending on the rainfalls. The annual balances for 1991 and 1992, estimated from the exportation fluxes at the elementary catchments basin outlets, showed that the exportation levels of atrazine, simazine and isoproturon did not exceed 0.3% of the mass applied agriculturally in the Mélarchez basin and 1.5% of the mass of atrazine applied in the Grand Morin river basin. The evolution of herbicide concentrations, from the elementary brook in a rural area to the river Marne outlet in an urban area, showed that there was no longitudinal decrease of contamination. In fact, the waters of the river Marne lower course, and of its minor tributaries in the urban area, might display herbicide concentrations similar or even higher than those of rivers in the agricultural area.

Introduction

France is one of the world's largest users of pesticides (averaging 100 000 t per year; UIPP, 1991), second only to the USA. The use of pesticides in the catchment basin of the river Seine in the 'Ile-de-France' district is most important and most varied (Faasen, 1995), amongst the CEE areas. Pollution hazards for drinking water supplies and aquatic ecosystems in this area are mainly related to triazines (atrazine, simazine, terbutylazine, etc.) and phenylureas (isoproturon, diuron, etc.). About 12% of the pesticides utilized in France are represented by these two classes of herbicides. Deterioration of the water quality was primarily observed in the river Marne, a main tributary of the river Seine and an important resource of drinking water for the Paris urban center via the water filtration plants (Noisy-le Grand and Neuilly-sur-Marne). This degradation arises mainly from atrazine, the most widely used triazine in France, particularly upon corn cultivations. Although corn production decreased (around 65%) from 1980 to 1990 in the river Marne basin, atrazine and simazine contamination remains a problem. Data of the Direction Générale de l'Environnement (DIREN) Ile-de-France, show that the present concentration of atrazine in three main tributaries of the Seine, the Yonne, the rivers Marne and Oise, exceed 100 ng l^{-1} and reach 250 ng l^{-1} for a period of several months (in spring and sometimes in autumn). These concentrations exceed the maximal concentration limit (MCL) in Europe (100 ng l^{-1}) recommended for drinking water, and consequently cause problems for the water filtration plants (AESN, 1996). Because of the combination of relatively high water solubilities (> 10 mg l^{-1}) for those herbicides and low suspended sediment concentration (20 mg l^{-1}) in the study area, 99% of the contaminant transport occurs in the dissolved phase.

To better understand and help protect the drinking water resource from contamination that mainly arises from agriculture, the Compagnie Générale des Eaux (CGE) and the Institut d'Hydrologie Climatologie/Laboratoire de Géologie Appliquée (IHC/LGA) undertook an investigation program focusing on the spatial and temporal fate of these compounds.

Materials and methods

The study was performed from 1991 to 1994 in the catchment area of the river Marne, particularly in the basin of its main tributary: the Grand Morin (Figure 1). The CGE investigated the catchment area of the Grand Morin and the Aubetin sub-basin. The IHC/LGA investigated the catchments of the Orgeval, another direct tributary of the Grand Morin, and the Mélarchez, an elementary catchment basin of the Orgeval. The Grand Morin catchment area is 1170 km^2 and situated in the Brie, 60 km east of Paris. The hydromorphous cultivated soils developed on eolian silts are drained up to 90% in the Mélarchez basin (7 km^2) from which the annual average flow is only 44 l s^{-1} as compared to 7000 l s^{-1} for the entire Grand Morin. The catchment basins are classed by increasing stream order according to Strahler (1957): Mélarchez, 1; Orgeval, 4; Grand Morin, 5; Marne, 6. The distribution of land use in the elementary catchment basin of the Mélarchez is representative of that of the central part of the Grand Morin: 53% cereal cultivation (7% corn), 38% industrial cultivation (21% peas and broad beans) and 8% woods and grasslands.

In order to investigate the relationships between stream contamination to agricultural activities and hydrological conditions, two surveys were undertaken. For the investigation of seasonal variations, samples were collected with an automatic sampler that integrates measurements over 7 or 9 days. For the investigation of spatial variations related to the stream order importance (1–6), instantaneous measurements were carried out. The analytical equipment and methods for sample treatment and chromatography were previously described (Garmouma et al., 1997a).

The investigated herbicides were the triazines: atrazine, simazine, ametryne, terbutryne, terbutylazine, cyanazine and desethylatrazine, the main degradation product of atrazine (Mills & Thurman, 1994), and the phenylureas (isoproturon, diuron, chlortoluron). The triazines (in single use or as tank mixture) are primarily used in corn production. The phenylureas are primarily used on the production of wheat, barley and oats. Diuron and terbutylazine are also used on uncultivated areas.

Results and discussion

Seasonal variations of water contamination

The surface water contamination at the outlet of the Mélarchez catchment basin (Garmouma et al., 1997a) provided evidence for a continuous pollution by triazines, and nearly continuous pollution by phenylureas (except during baseflow conditions from August to October (Garmouma, 1997b). Atrazine, simazine and desthylatrazine were found in all the integrated samples (n=110). Our results are consistent with those of the CGE at the four sites of the Grand Morin catchment (CGE, 1993).

The weekly sample analyses from May 1991 to May 1992 at four stations along the Grand Morin confirmed the continuous pollution by atrazine and simazine and by isoproturon for over half the year. In spite of continuous dry weather reducing transport, the atrazine background level averaged 200 ng l^{-1} (ranging from 75 to 600 ng l^{-1}) at the upstream station Meilleray on the Grand Morin. These concentrations are consistent with those found during 1993 and 1994 (CGE, 1996). Because of the wide use (not strictly agricultural) of triazines, the sources of contamination cannot be easily determined, especially for atrazine. In 1993, in the Mélarchez water, a single high pollution period was observed for atrazine and isoproturon, during which their concentrations remained above 500 ng l^{-1}, with a maximum of nearly 2 μg l^{-1}. The maximal contamination level for atrazine occurred from mid-April to May, whereas the isoproturon peaks occurred from mid-February to May. On the contrary, during the previous period (1991–1992) the atrazine concentration showed 10 atrazine peaks above 500 ng l^{-1}, and the highest one did not correspond necessarily to the spring applications. The isoproturon

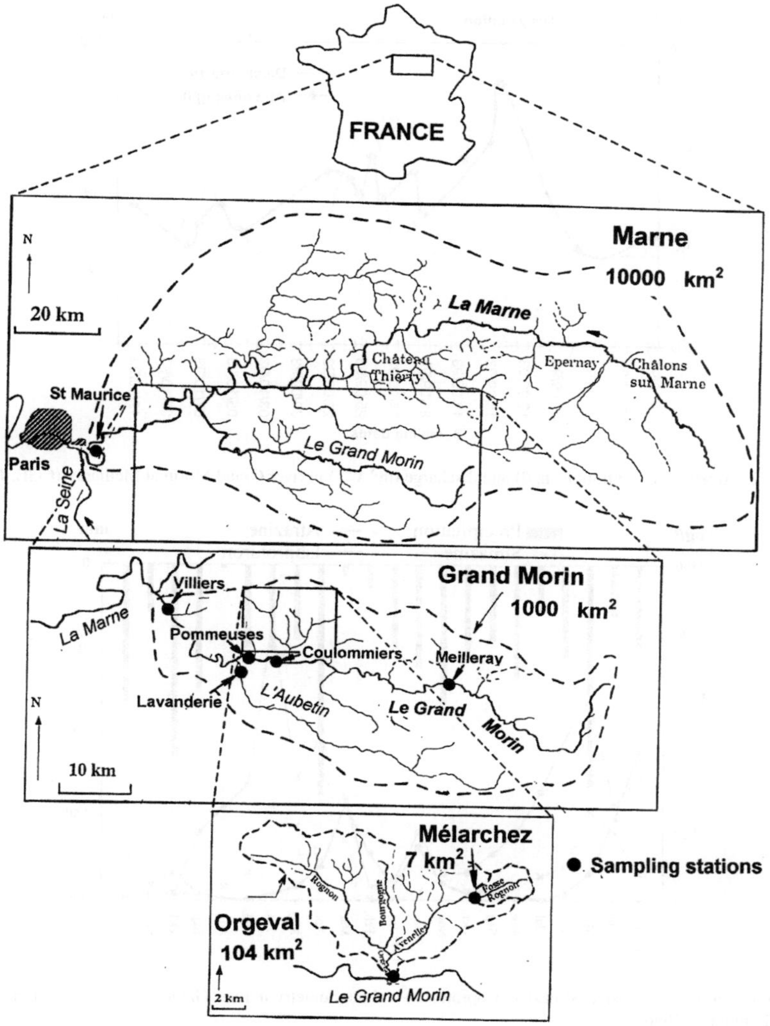

Figure 1. Basins and rivers of different stream orders of the CGE and IHC/LGA study (1991–1994)

evolution cycle was simpler, characterized by a single high contamination period of the stream water following the wheat and winter barley treatment in late winter. The magnitude of the concentration variations for each herbicide is quite different from year to year and the high concentrations did not always correspond to sequence of application, precipitation and runoff.

The two surveys led generally to the same conclusions concerning the sources of atrazine. Except for one flood in March 1992 (results from the Mélarchez and the Grand Morin), no relationship was found between the herbicide concentrations and the river flow. Most of the floods did not result in a rise of the atrazine, simazine or phenylurea concentrations. Also, there was no dilution effect following the floods mainly observed in autumn (Figure 2).

It is likely that the pollution peaks are mainly related to precipitation and slight runoff without flow modification, particularly after the application periods (April–May for the triazines and October–November for the phenylureas) when low volumes of highly contaminated runoff water are transported to the river.

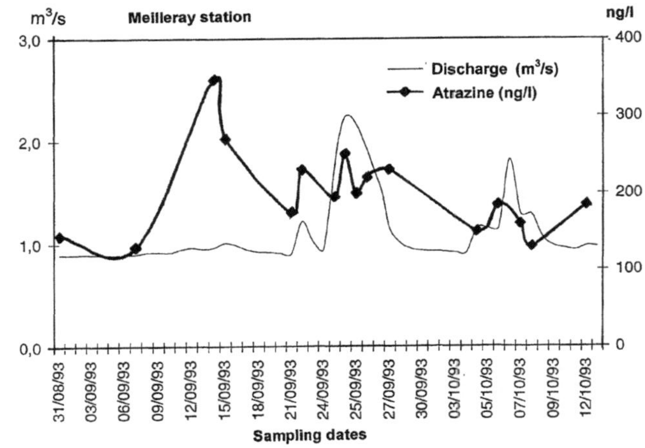

Figure 2. Atrazine concentration (ng/l) and discharge ($m^3\ s^{-1}$) in river Grand Morin at Meilleray (CGE, 1993).

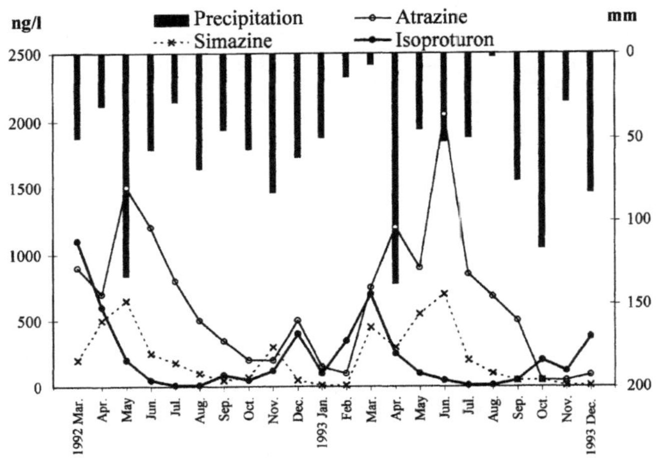

Figure 3. Herbicide concentrations (atrazine, simazine, isoproturon) and pluviometry in river Grand Morin at Coulommiers from March 1992 to December 1993 (Garmouma, 1996).

The seasonal variations of atrazine, simazine and isoproturon concentrations highlight the impact of agricultural practice upon water quality (Figure 3).

During baseflow conditions, background atrazine concentrations generally ranged from 40 to 90 ng l^{-1}, which means that all the aquifers of that catchment, both superficial and deep, were contaminated. The analyses of groundwaters support that conclusion. At the outlet of the catchment basin of the Mélarchez, the basal contamination level originated mainly from aquifer in silt (Figure 4).

Investigating hydrological processes that influence the evolution atrazine concentration of rivers, Adams and Thurman (1991) utilized the desethyl-atrazine/atrazine concentration ratio (DAR). They showed that the atrazine transport by infiltration waters through the unsaturated zone corresponded to a DAR value greater than 1, whereas the atrazine exportation by surface water runoff involved DAR values less than 1 (Thurman et al., 1992). We attempted to determine the relative importance of the different input pathways in the experimental catchment of the Mélarchez (Table 1). In spite of the complexity of

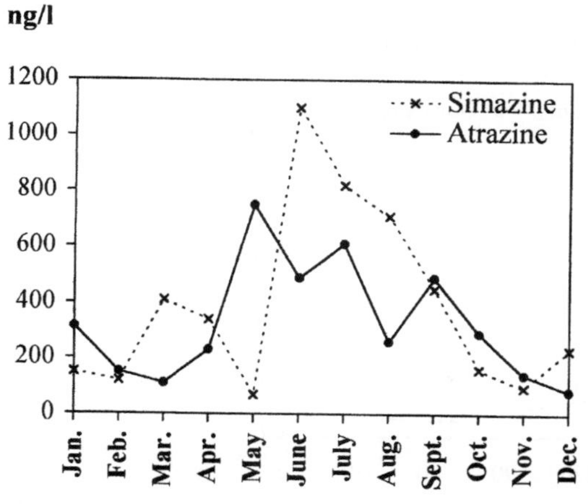

Figure 4. Concentrations of atrazine and simazine in the waters of the silt aquifer (Garmouma, 1996).

the pesticide input mechanisms in that basin, from the DAR we identified values two distinct periods, the first one corresponding to surface runoff (in spring) and the second one to groundwater (in summer).

Results from the aquifers drained by the Grand Morin catchments from September 1995 to October 1996 displayed atrazine concentrations ranging from 300 to 700 ng l^{-1} in the 'Calcaire de Brie' (upper aquifer) and from 100 to 250 ng l^{-1} in the 'Calcaire de Champigny' (lower aquifer). In contrast, phenylureas were not detected in the groundwater, which might explain their absence in the surface waters during the baseflow period (from the end of spring to the end of summer).

Exportation: dynamics and balance

The evolution of herbicide flux was different from that of their concentrations. The evolution of herbicide flux is dependent upon streamflows rather than concentrations. Therefore, the exportation of phenylureas and triazines had different evolution profiles because their application periods did not occur at the same moment of the hydrological cycle. For isoproturon, which is applied essentially in February, 50% of the exported herbicide was removed from March to May. The remaining 50% was exported from October to December. That flux evolution was mainly derived from runoff events folllowing late winter isoproturon applications on wheat and oats. Conversely, 50% of the atrazine exportation (applied in late April) was not achieved until October (high concentrations, low flows). The remaining exportation occurred during the flood in late autumn (high flows, low concentrations). Comparatively to ureas, the atrazine exportation appeared more regular throughout the year because the exportation persisted during the dry period with the draining of aquifers (contaminated by applications of atrazine from the previous year).

Table 1. Desethylatrazine/atrazine ratio in spring (April–June) and summer (July–September) in the waters of Mélarchez stream from data of Garmouma (1996).

Season (Discharge)	1991	1992	1993
Spring (> 15 l s^{-1})	0.02	0.06	0.07
Summer (< 5 l s^{-1})	1.1	5.5	3.7

However, the herbicide amounts exported by surface waters of small agricultural catchments only represent minor proportions of the quantities actually applied to the fields. At the outlet of the Mélarchez basin, the atrazine exportation ratio did not exceed 0.085% in 1992 and 0.28% in 1993 (Garmouma et al., 1997a). The exportation ratio for isoproturon was even lower, from 0.035 to 0.08%. Pereira and Rostad (1990) estimated that the atrazine exportation ratio for the Mississippi River over 1 year was 0.4% under dry weather conditions, and remained below 2% even under wet weather conditions. At the scale of the Grand Morin, the exportation ratios of atrazine were similar: 2% for the Grand Morin at Pommeuses station and 1.5%, upstream of its confluence with the Aubetin. These calculated exportation ratios were based upon the estimated agricultural use, extrapolation of the average applied dose per ha and the number of applications per year. In the Marne basin, the orders of magnitude of riverine flux as a percentage of use are consistent with those of the tributaries of the Mississippi River (Larson et al., 1995).

Spatial variations of contamination

Our investigations of the Grand Morin and river Marne basin showed heterogeneous contamination by herbicides considering the upstream/downstream river courses. The removal of herbicides by groundwater runoff is favoured by cultivated soil drainage, which is 31 and 41% in the Grand Morin (median part) and Aubetin basins.

Consequently, the increase of triazine concentrations in the Grand Morin from upstream to down-

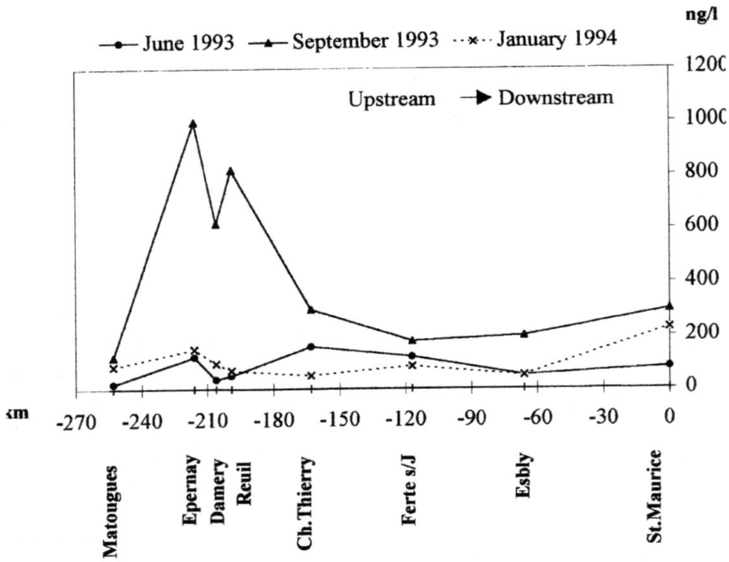

Figure 5. Atrazine concentrations along the river Marne course (Garmouma, 1996).

Table 2. Mean atrazine concentrations (ng/l) in the river Grand Morin in 1991–1992 and 1993–1994 (CGE, 1996).

Sampling period	Grand Morin			Aubetin
	Meilleray	Pommeuses	Villiers	Lavanderie
1991–1992	213	280	405	510
1993–1994	222	360	480	920

Table 3. Mean herbicide concentrations (ng/l) in affluents of different stream orders in the Marne basin.

River (stream order)	Atrazine	Simazine	Isoproturon
Mélarchez (1)	315	260	160
Orgeval (4)	270	140	145
Grand Morin (5)	600	210	175
Marne (6)	400	140	180

stream was mainly due the input from the Aubetin (CGE, 1993). Even though the atrazine concentrations in the Aubetin were 10-to 20-fold higher than in the Grand Morin above its confluence, the lower specific flows of the Aubetin minimized the overall impact on the Grand Morin (Table 2).

Comparing the levels of contamination of sub-basins with increasing size; concentrations at the outlets of the Mélarchez, the Orgeval, and the Grand Morin differed much less in their average contamination levels than they did from season to season (Table 3). The Orgeval showed lower contamination because of its lower ratio of treated surface (wooded areas).

The river Marne course, from Epernay (Champagne vineyard) to its confluence with the river Seine in urban area at Saint-Maurice showed spatial variations in herbicide concentrations. The atrazine, simazine, terbutylazine and diuron concentrations of the river Marne were at their highest levels in the section situated from 150 to 240 km upstream of the confluence with the river Seine (Garmouma, 1996), and corresponded to an area of intense vineyard production (Figure 5).

The lowest concentrations were generally found in the central part of the river Marne (Figure 5). Thus, the river Marne in the urban zone sometimes showed herbicide concentrations greater than those in the Grand Morin (January 1994), showing that pollution originated only partly from the Grand Morin inputs (low average flow, 7 m^3 s^{-1}) as compared to that of the river Marne (107 m^3 s^{-1}). Thus, urban inputs apparently arise from herbicide treatment along the roads and railways. The CGE investigations also showed the importance of riverside activities in urban areas on stream water quality. At the outlet of the river Marne catchment, the occurrence of occasional atrazine pollution peaks (above 20 $\mu g\, l^{-1}$ in July 1991, cgE, 1993) underlined the river vulnerability to pollutions other than agricultural treatments (i.e. industries, gardens, etc.).

Conclusion

Increasing scale from small streams to large rivers showed little change in average herbicide concentrations. Unexpectedly, there is no longitudinal decrease in herbicide contamination (from rural upstream to urban downstream areas) as land use changes from agricultural to urban areas. As a matter of fact, the waters of the Marne lower course and of its tributaries in periurban area, may show herbicide concentrations similar or even higher than those of agricultural areas. Lateral herbicide inputs, originating from road and railway maintenance and gardening appear to superimpose on those from intensive agricultural practices. The occurrence of high levels of diuron, a phenylurea mainly used upon uncultivated soils, supports this hypothesis. Consequently, when the impermeable surface ratio increases in periurban area, the transport of herbicide residues by runoff is more important than in rural areas. The storm overflows and the urban streams may locally enhance the magnitude of pollution. This appears even more likely as the triazine concentrations of the precipitations in urban areas may be similar or even higher than in the rivers (Chevreuil & Garmouma, 1993).

References

Adams C. & M. Thurman 1991. Formation and transport of desethylatrazine in the soil and the vadose zone. J. Environ. Qual., 20: 540–547.

AESN, 1996. Le point sur les triazines. Rapport Agence de l'Eau Seine Normandie, Direction de l'Environnement, des Etudes et de la Recherche, 42 pp.

CGE Compagnie Générale des Eaux, 1993. Etude de la qualité des eaux de la Marne et du Grand Morin. Rapport de synthèse CGE/AESN, 75 pp.

CGE Compagnie Générale des Eaux, 1996. Qualité des eaux sur le bassin du Grand Morin. Rapport de synthèse CGE/AESN, 42 pp.

Chevreuil M. & M. Garmouma 1993. Atmospheric fallout of triazines on the catchment basin of the river Marne (France), Chemosphere, 27: 1605–1608.

Faasen, R., 1995. Agricultural pesticide use: a threat to the European environment?. Eur Water Poll Contr. 5: 29–31.

Garmouma M., 1996. Transfert d'herbicides (Triazines et Phénylurées) et de produits de dégradation dans le bassin versant de la Marne, *Thèse de Doctorat de l'Université de Paris VI, Spécialité Sciences de l'Eau*, 220 pp.

Garmouma M., M. Blanchard, A. Chestérikoff, P. Ansart and M. Cheveuil, (1997a). Seasonal transport of herbicide (triazines and phenylureas) in a small stream draining an agricultural basin: Mélarchez (France), Wat. Res., 31: 1489–1503.

Garmouma, M., M. Chevreuil & Chestérikoff, (1997b). Transfert d'herbicides (Triazines et Phénylurées) et de produits de dégradation dans le bassin versant de la Marne, Rapport PIREN Seine, 1996/IV, 11 pp.

Larson, S., P. Capel, D. Goolsby, S. Zaugg, M. Sandstrom 1995. Relations between pesticides use and riverine flux in the Misissippi River basin. Chemosphere, 31: 3305–3321.

Mills M. & M. Thurman, 1994. Preferential dealkylation reactions of s-triazine herbicides in the unsaturated zone. Environ. Sci. Technol., 28: 600–605.

Pereira W. & C. Rostad, 1990. Occurrence, distribution and transport of herbicides and their degradation products in the lower Mississippi River and its tributaries. Environ. Sci. Technol, 24: 1400–1406.

Strahler A.H. 1957. Quantitative analysis of watershed geomorphology. Geophys. Union Trans. 38: 913–920.

Thurman M., D. Goolsby, M. Meyer, M. Mills, M. Pomes & D. Kolpin, 1992. A reconnaisssance study of herbicides and their metabolites in surface water of the Midwestern United State using immunoassay and gas chromatography/mass spectrometry. Environ. Sci. Technol., 26: 2440–2447.

U.I.P.P., 1991. Rapport annuel de l'Union des Industries de la Protection des Plantes, Boulogne s/Seine, 6 pp.

MIX
Papier aus verantwortungsvollen Quellen
Paper from responsible sources
FSC® C105338

If you have any concerns about our products,
you can contact us on
ProductSafety@springernature.com

In case Publisher is established outside the EU,
the EU authorized representative is:
Springer Nature Customer Service Center GmbH
Europaplatz 3, 69115 Heidelberg, Germany

Printed by Libri Plureos GmbH
in Hamburg, Germany